Cancer as a Metabolic Disease

Cancer as a Metabolic Disease

On the Origin, Management and Prevention of Cancer

Thomas N. Seyfried

A John Wiley & Sons, Inc., Publication

Published by John Wiley & Sons, Inc., Hoboken, New Jersey
Published simultaneously in Canada

Library of Congress Cataloging-in-Publication Data:

Seyfried, Thomas N., 1946–
Cancer as a metabolic disease : on the origin, management, and prevention of cancer / by Thomas N. Seyfried.
p. ; cm.
Includes bibliographical references and index.
ISBN 978-0-470-58492-7 (cloth)
I. Title.
[DNLM: 1. Neoplasms–diet therapy. 2. Neoplasms–metabolism. 3. Energy Metabolism. 4. Ketogenic Diet. 5. Mitochondria–metabolism. QZ 200]
616.99′40654–dc23

2011049798

SKY10071899_040924

This book is dedicated to the millions of people who have suffered and died from toxic cancer therapies

Contents

Foreword

Cancer persists as a major disease of mortality and is afflicting more people today than ever before. Few families remain untouched by this insidious and vicious disease. In fact, cancer is predicted to overtake heart disease as the prime cause of death in industrialized societies during this century. I have worked in the cancer metabolism field since the late 1960s and have extensively published works on the metabolic basis and properties of cancer. While I do not know Dr. Seyfried personally, I am very impressed with the excellent job he has done in highlighting abnormal energy metabolism as the central issue of the cancer problem. I recognized long ago the pivotal role of mitochondria and of aerobic glycolysis in sustaining and promoting cancer growth. The Nobel laureate, Otto Warburg, was the first to provide evidence during the early part of the last century for the involvement of disturbed respiration with compensatory fermentation (glycolysis) as a common property of cancer, thus perceived to be related to its uncontrolled growth and progression. Few subjects have been as controversial in the cancer field as Otto Warburg and his theory of cancer. It is nice to see how Seyfried shows that Warburg was largely correct in defining the nature of the disease as involving insufficient respiration with compensatory fermentation. I knew personally many of the key figures and their research mentioned in Seyfried's book, including Dean Burk, Peter Mitchell, Sidney Weinhouse, and my former Department Chair, Albert Lehninger, among others. Nevertheless, there were times in my early career when I felt almost alone in considering energy metabolism as important to the cancer problem. I even remember one of my colleagues, an expert in DNA technology, dumping Lehninger's "Warburg Flasks" in the trash as relics of a bygone era in cancer research. Fortunately for him, Lehninger was no longer the Department Chair, and fortunately for me, I salvaged many of these flasks and am now glad I did. After reading Seyfried's book, I think these flasks will become valuable as collector items.

The cancer field went seriously off course during the mid-1970s when many investigators began considering cancer as primarily a genetic disease rather than as a metabolic disease. The metabolic defects in cancer cells were thought to arise as secondary consequences of genomic instability. Seyfried provides substantial evidence documenting the inconsistencies of the gene "only" theory. He critically reevaluates the evidence linking cancer progression to a Darwinian process and raises the intriguing possibility that cancer progression is an example of Lamarckian evolution. When viewed collectively, the documented inconsistencies of the gene "only" theory make it clear why little progress has been made in the cancer war and in the development of effective nontoxic therapies. A key point made by Seyfried is that most of the genomic instability seen in cancer likely arises as a consequence rather than as the cause of the disease. When viewed more as a metabolic disease, many cost effective therapeutic strategies become recognized for cancer management. I know this first hand from our studies of 3-bromopyruvate (3BP), discovered in my laboratory by Dr. Young Ko, as a potent anticancer agent. This is a low cost drug with powerful and quick antitumor effects against multiple cancers in animal models and in cancer patients. 3BP works primarily by targeting tumor cell energy metabolism, thus depleting the energy-rich compound "ATP" essential for growth. At the effective doses used, it does this without toxicity to normal cells. Seyfried's book provides substantial evidence showing how cancer can be managed using various other drugs and diets that target energy metabolism. In addition, the restriction of glucose and glutamine, which drive cancer energy metabolism, cripples the ability of cancer cells to replicate and disseminate. The gene theory has deceived us into thinking that cancer is more than a single disease. Certainly, tumors do not all grow at the same rate. Nevertheless, cancer is a singular disease involving aberrant energy metabolism as Warburg originally showed and as I, and more recently many others, have documented in biochemical studies. Seyfried drives home this message throughout his book.

Seyfried's treatise refocuses attention on the central issue of cancer as a metabolic disease according to Warburg's original theory. The book is unique in linking nearly all aspects of the disease to respiratory insufficiency with compensatory fermentation. Cancer has remained incurable for many due largely to a general misunderstanding of its origin, biology, and metabolism. Hopefully, Seyfried's thoughtful analysis of the "cancer problem" will change our understanding of the disease and move the field in the right direction toward solutions and therapies, such as 3BP, that act much faster and more effectively than those currently available.

Dr. Peter Pedersen
Professor of Biological Chemistry
Johns Hopkins University School of Medicine
Baltimore, MD

Preface

Cancer persists as a plague in modern society. The lack of progress in either managing or preventing cancer motivated me to write this treatise. I am a biochemical geneticist and have worked on the lipid biochemistry of cancer since the early 1980s. I have developed numerous mouse models for brain tumors and for systemic metastatic cancer. Several major findings planted the seed for this treatise. First, it became clear to me that the therapeutic action of some anticancer drugs operated largely through reduced caloric intake. Second, that reduced caloric intake could target the majority of cancer hallmarks. Third, that ketone bodies can serve as an alternative fuel to glucose in most cells with normal respiratory function. Fourth, that metastatic cancer arises from cells along macrophage lineage. Fifth, that all cancer cells regardless of tissue origin express a general defect in mitochondrial energy metabolism. Finally, that cancer can be effectively managed and prevented once it becomes recognized as a metabolic disease.

In recognizing cancer as a metabolic disease, it gradually became clear to me why so many people die from the disease. Many of the current cancer treatments exacerbate tumor cell energy metabolism, thus allowing the disease to progress and eventually become unmanageable. Most cancer patients do not battle their disease but are offered toxic concoctions that can eventually undermine their physiological strength and their will to resist. Cancer treatments are often feared as much as the disease itself. The view of cancer as a genetic disease has confounded the problem and is largely responsible for the failure to develop effective therapies. The view of cancer as a genetic disease is based on the flawed notion that somatic mutations cause cancer. Substantial evidence indicates that genomic instability is linked to protracted respiratory insufficiency. Once cancer becomes recognized as a metabolic disease with metabolic solutions, more humane and effective treatment strategies will emerge. My treatise highlights cancer as a metabolic disease and identifies the inconsistencies of the gene theory of cancer. Moreover, my treatise addresses most of the so-called provocative questions raised by the National Cancer Institute regarding outstanding issues in cancer research. This treatise lays the foundation for the eventual resolution of the disease.

I would like to thank my many students and colleagues for helping me in producing the data and in developing the concepts for this treatise. I thank my former graduate students Mary Louise Roy (MS, 1987), Michelle Cottericho (MS, 1992), Mohga El-Abbadi (PhD, 1995), Hong Wei Bai (PhD, 1996), John Brigande (BS, 1989; MS, 1992; PhD, 1997), Jeffrey Ecsedy (PhD, 1998), Mark Manfredi

(PhD, 1999), Michaela Ranes (BS, 1998; MS, 2000), Dia Banerjee (MS, 2001), Michael Drage (MS, 2006), Christine Denny (BS, 2005; MS, 2006), Weihua Zhou (MS, 2006), Laura Abate (PhD, 2006), Michael Kiebish (PhD, 2008), Leanne Huysentruyt (PhD, 2008), John Mantis (PhD, 2010), and Laura Shelton (PhD, 2010). I would also like to acknowledge my current students Linh Ta and Zeynep Akgoc for their continued productivity. I would also like to thank the following undergraduate students for their input and help, including Katherine Holthause, Jeremy Marsh, Jeffery Ling, Will Markis, Tiernan Mulrooney, Todd Sanderson, Todd Theman, Lisa Mahoney, Michelle Levine, Emily Coggins, Erin Wolfe, Ivan Urits, Taryn LeRoy, and Emily Gaudiano. I would like to thank those students from my BI503 class on *Current Topics in Cancer Research* for their input.

I would like to thank faculty colleagues in the Boston College Biology Department, including Drs. Thomas Chiles, Fr. Richard McGowan SJ, and Jeffery Chuang. I would like to thank Dr. Robert K. Yu, Dr. James Fox and my son Dr. Nicholas T. Seyfried for technical assistance. I would like to thank Avtar Roopa for provocative discussion. I would like to thank the late Drs. Sanford Palay, Harry Zimmerman, and Allan Yates for their encouragement and assistance. I would also like to give special acknowledgement to Dr. Purna Mukherjee and Roberto Flores. Purna was the first to make me aware of the powerful therapeutic action of calorie restriction. She is superbly trained in the areas of angiogenesis and inflammation and her work provided seminal information on the mechanisms by which dietary energy reduction can both treat and prevent cancer. Roberto Flores is exceptional in his dedication to finding the truth underlying the metabolic origin of cancer and in questioning the metabolic origin of cancer. Finally, I would like to thank my institution, Boston College, for providing animal care support over the first 23 years of my employment there (1985–2008). The data collected supporting my treatise would not have been possible without this invaluable institutional support. This support was consistent with the Ignatian philosophy of service to others.

Chapter 1

Images of Cancer

Cancer is a devastating disease both physically and emotionally and is projected to overtake heart disease as the leading killer of people in industrialized societies. Cancer is complex. The disease involves multiple time- and space-dependent changes in the health status of cells, which ultimately lead to malignant tumors. Abnormal cell growth (neoplasia) is the biological endpoint of the disease. Tumor cell invasion of surrounding tissues and spread to distant organs is the primary cause of morbidity and mortality in most cancer patients. This phenomenon is referred to as *metastasis*. The biological process by which normal cells are transformed into malignant cancer cells has been the subject of an enormous research effort in the biomedical sciences for more than a century. Despite this effort, cures or long-term management strategies for metastatic cancers are as challenging today as they were 40 years ago when President Richard M. Nixon declared a war on cancer with the National Cancer Act (1–3). According to the American Cancer Society, 569,490 people died in the United States from cancer in 2010 (4). This comes to about 1500 people each day! Remarkably, the number of deaths in 2002 was 555,500 providing quantitative evidence of no real progress in management over a 8-year period (5). All one needs to do is read the obituary pages from any local newspaper to know that the "cancer war" is not going well.

How is it possible that we are not winning the cancer war when this disease is under constant investigation in many major pharmaceutical companies and in most leading medical centers throughout the world? One would think that effective nontoxic therapies would be readily available from all this attention. We constantly hear in the media of new breakthroughs in the fight against cancer, yet high profile celebrities and politicians continue to die from the disease. If the breakthroughs are real or meaningful, shouldnt the wealthy and powerful have access to any potential life-saving therapy? That these folks are just as vulnerable as the rest of us to the ravages of the disease clearly indicates that the war is not won. The road to the cancer front is littered with major breakthroughs that never materialized into effective solutions. A plateau in overall death rates for some cancers has been due more to better awareness and avoidance of risk factors, for example, smoking for

Cancer as a Metabolic Disease: On the Origin, Management and Prevention of Cancer, First Edition.
Thomas Seyfried.

lung cancer, than to any real advances in the management of systemic metastasis, the most deadly feature of the disease (6, 7). Clearly, we are not wining the war on cancer, as Guy Faguet has emphasized (8).

HOW CANCER IS VIEWED

The image of cancer depends on your perspective. It depends on whether you are a cancer patient, a friend or family member of a patient, an oncologist, a pathologist, a statistician, or a person who does basic research on the disease. The image of cancer can be framed from these various perspectives.

Figure 1.1a shows the number of genetic alterations detected through sequencing and copy number analyses in each of the 24 different pancreatic cancers. According to the figure, point mutations are more common in pancreatic cancer than are larger deletions or amplifications. The authors of this study, and of many similar studies, believe that the cataloguing of mutations found in various tumors will be important for disease identification and management. While cataloguing cancer genetic defects is interesting, it is important to recognize that the defects often vary from one neoplastic cell to another within the same tumor (12).

Figure 1.1b shows the percentage of genetic alterations found in brain tumors (glioblastoma multiforme). Similar kinds of alterations are found in pancreatic and ovarian cancers. Primary sequence alterations and significant copy number changes for components of the RTK/RAS/PI(3)K (A), p53 (B), and RB (C) signaling pathways are shown. The different shades of gray are indicative of different degrees of genetic alteration (13). For each altered component of a particular pathway, the nature of the alteration and the percentage of tumors affected are indicated. Boxes contain the final percentages of glioblastomas containing alterations in at least one known component gene of the designated pathway. It is also interesting to note that no alterations in any of the pathways occur in about 15% of glioblastomas despite similarity in histological presentation. It remains unclear how these genomic alterations relate to the origin or progression of the disease.

Akt (v-Akt murine thymoma viral oncogene) or PKB (protein kinase-B) is a serine/threonine kinase that is involved in mediating various biological responses, such as inhibition of programmed cell death (apoptosis), stimulation of cell proliferation, and enhancement of tumor energy metabolism (Fig. 1.2). Akt expression is generally greater in cancer cells than in normal cells. Although targeting of Akt-related pathways is part of cancer drug development, the simple restriction of calorie intake will reduce Akt expression in tumors (14). This image is synthesized from information on the molecular biology of cancer. I refer to these types of cancer images as *balloons on strings*. They convey an ordered arrangement of pathways for a disease that is biologically chaotic. SABiosciences is a QIAGEN company specializing in molecular array technologies that can help analyze gene expression changes, epigenomic patterns, microRNA expressions, and so on.

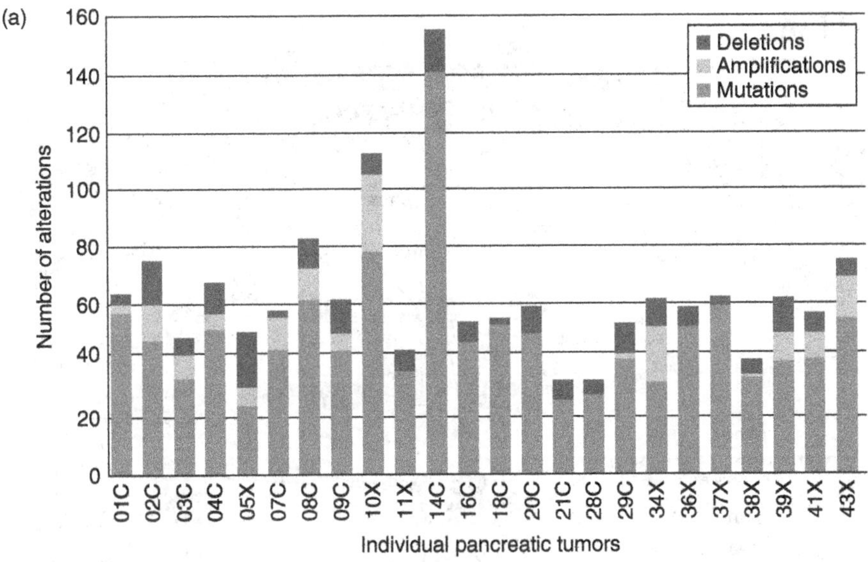

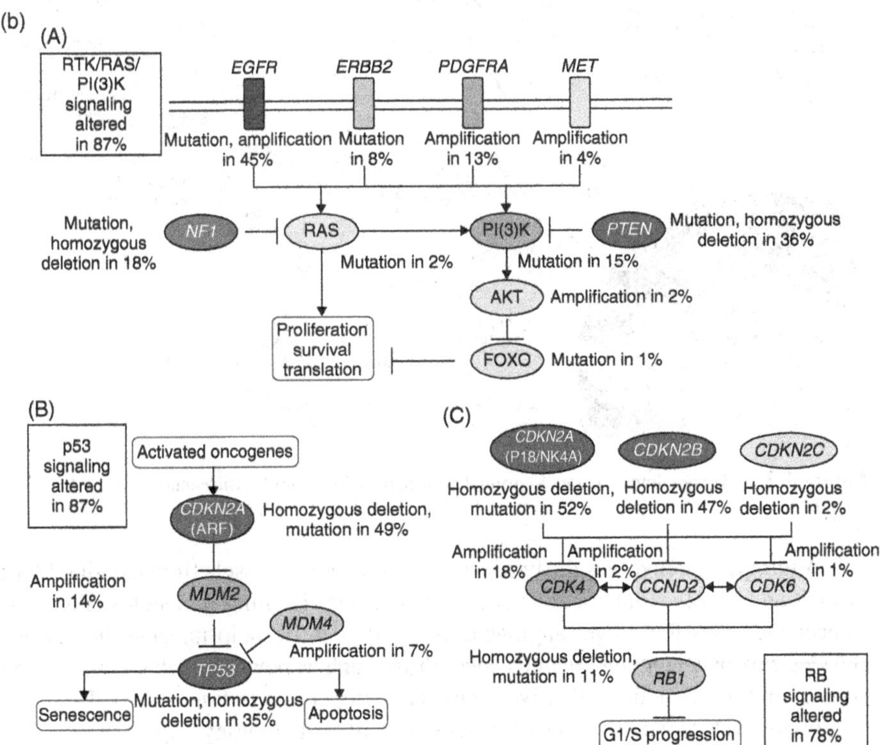

Figure 1.1 Cancer images from cancer genome projects. *Source:* (a) Modified from Jones et al. (13); (b) Reprinted from Jones et al (13). To see this figure in color please go to ftp://ftp.wiley.com/public/sci_tech_med/cancer_metabolic_disease.

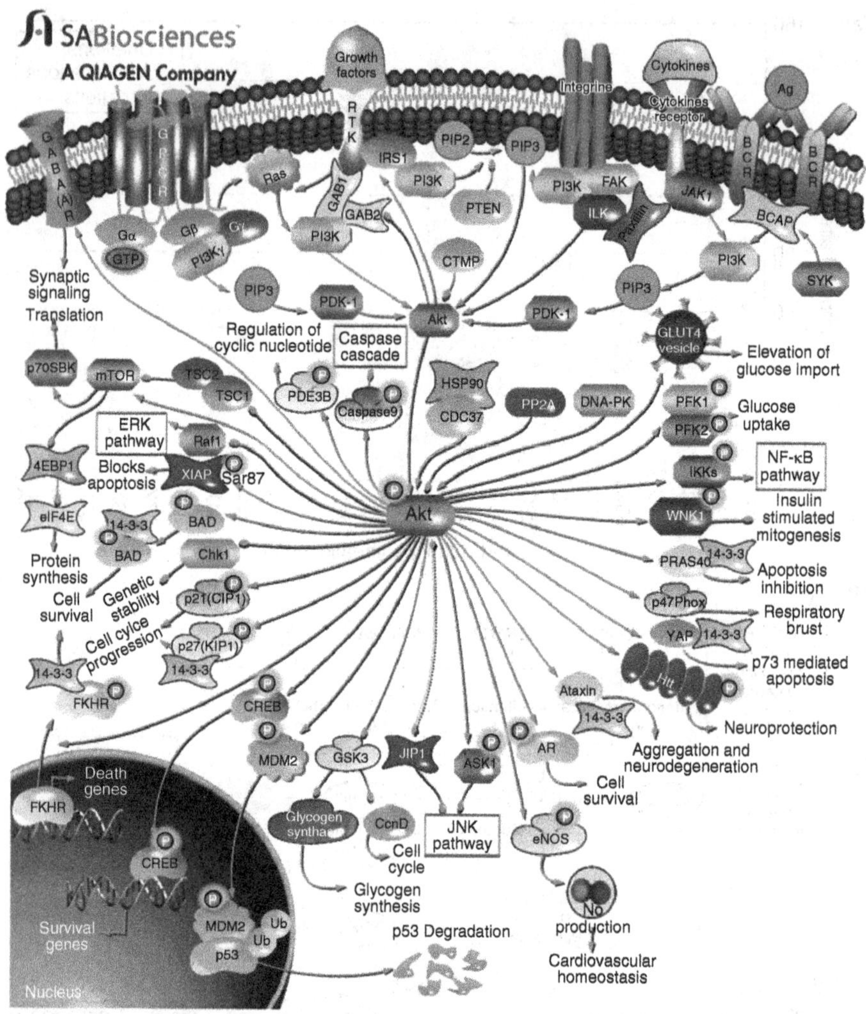

Figure 1.2 Akt signaling. *Source:* Reprinted with permission from SABiosciences. See color insert.

Angiogenesis involves the production of new blood vessels from existing blood vessels and involves interactions among numerous signaling molecules (Fig. 1.3). Cancer therapies that target angiogenesis are thought to help manage the disease. Besides expensive antiangiogenic cancer drugs such as bevacizumab (Avastin) (15), simple calorie restriction effectively targets angiogenesis in tumors (16, 17).

Figure 1.4 depicts the cancer images of cellular pathology.

The following is a list of the mortality rate of different cancers:

- Breast cancer killed about 40,170 women in 2010 (4).
- Lung and bronchus cancer killed about 159,390 persons in 2010 (4).

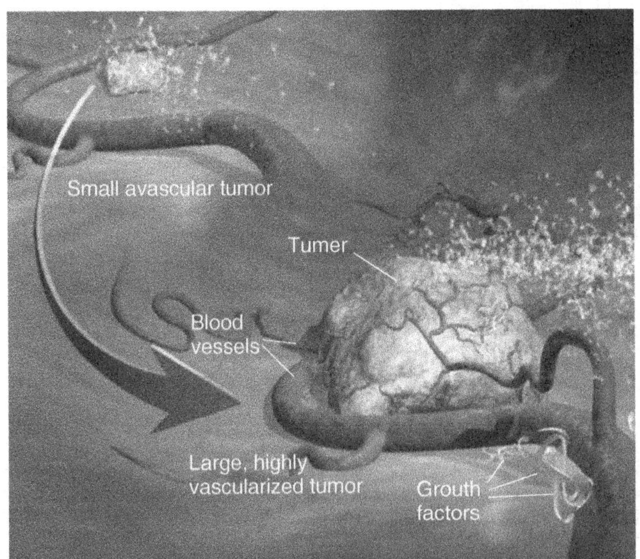

Figure 1.3 Tumor angiogenesis. *Source:* Reprinted with permission from BioOncology. To see this figure in color please go to ftp://ftp.wiley.com/public/sci_tech_med/cancer_metabolic_disease.

- Colon/rectum cancer killed about 49,920 persons in 2010 (4).
- Skin cancer killed about 11,590 persons in 2010 (4).
- Brain and nervous system cancer killed about 12,920 persons in 2010 (3).
- Liver and bile duct cancer killed about 18,910 people (4).

Cancer images of organ pathology are shown in Figure 1.5.

I think the artwork of Robert Pope, who died from the adverse effects of chemotherapy and radiation, is especially powerful in conveying the image of cancer from the perspective of the patient, the family, and the physician (19, 20). I also think the Commentary by Donald Cohodes on the experience of chemotherapy should be read as a supplement to Pope's book (21). I have included below a few of Pope's many paintings and drawings.

In the painting in Figure 1.6, Pope depicts the subtleties of communication among cancer doctors. The doctors talk among themselves about cancer differently than they do to the patient or to the patient's family so as not to alarm the sensitivity of the layperson. In the hallway, the communication is considered scientific, blunt, and factual, while in the room it is considered more nurturing and emotional. Although many patients view cancer doctors as secular priests in today's society, the toxic therapies doctors use to treat cancer are often counterproductive to the long-term well-being of cancer patients.

The image in Figure 1.7 is an acrylic on canvas depicting a man lying underneath a radiation machine. Radiation therapy is given to many cancer patients. Radiation will kill both cancer cells and normal cells. Some normal cells that are not killed outright can be metabolically transformed into tumor cells. Moreover,

(a)

(b)

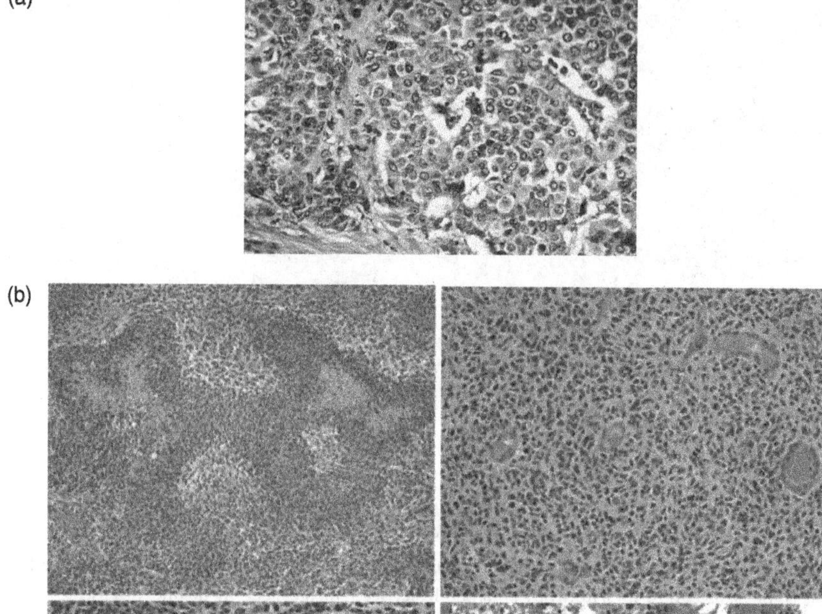

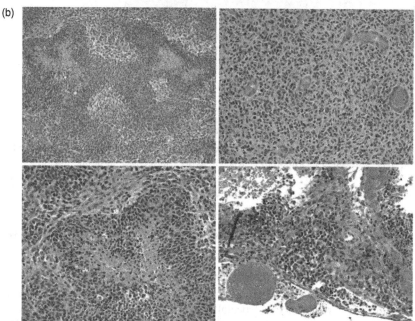

Figure 1.4 (a) Histological image of breast cancer. *Source:* Reprinted with permission from the NCI. (b) Histological images of glioblastoma multiforme. *Source:* Reprinted with permission from Reference 18. To see this figure in color please go to ftp://ftp.wiley.com/public/sci_tech_med/cancer_metabolic_disease.

those tumor cells that survive the radiation treatment will sometimes grow back as more aggressive and less manageable cancers in the future.

Figure 1.8 is also an acrylic on canvas that conveys the psychological impact of cancer drugs. The chemical in the syringe is Adriamycin (*doxorubicin*), which Pope received along with other drugs during his battle with cancer. In this painting, Pope depicts an older woman with lymphatic cancer who is getting chemotherapy. The woman is wearing a turban to hide her baldness caused from the drug treatments. Pope attempts to convey the patient's thoughts about the drug. The drug within the syringe elicits thoughts of either life or alarm. According to Pope, the painting shows the human encounter with poisonous drug therapy, an all-too-familiar scene for the cancer patient.

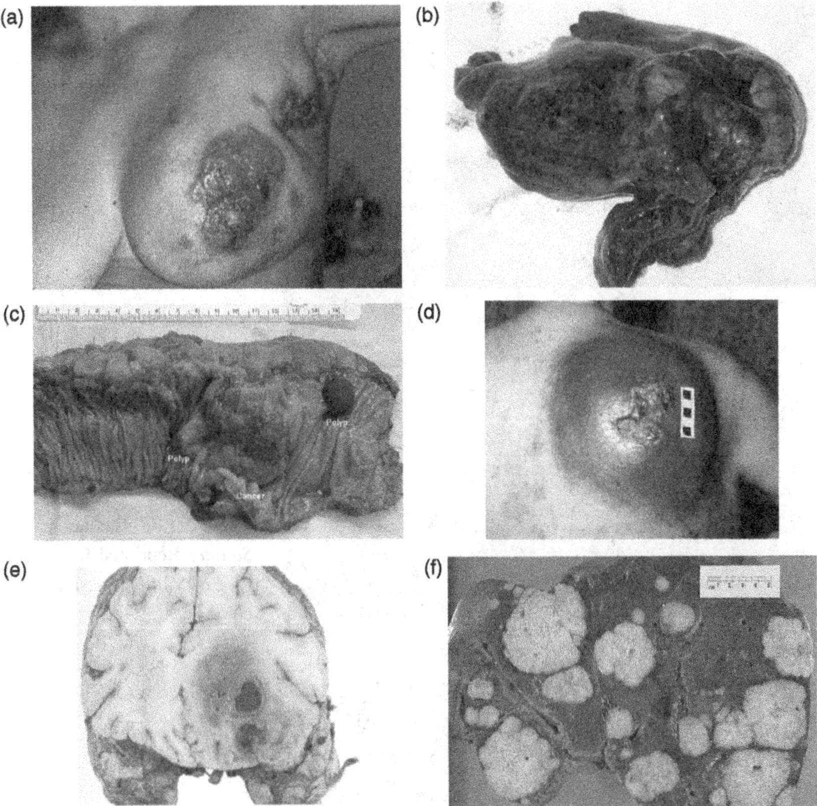

Figure 1.5 (a) Breast cancer, (b) lung cancer, (c) colon cancer, (d) melanoma, (e) glioblastoma, and (f) liver cancer. See color insert for (a, d). To see figures (b, c, e, f) in color please go to ftp://ftp.wiley.com/public/sci_tech_med/cancer_metabolic_disease.

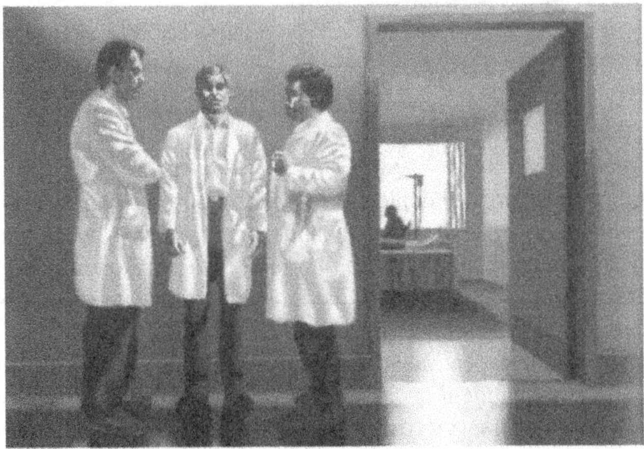

Figure 1.6 The Conference. *Source:* Reprinted from Pope (p. 113) with permission. See color insert.

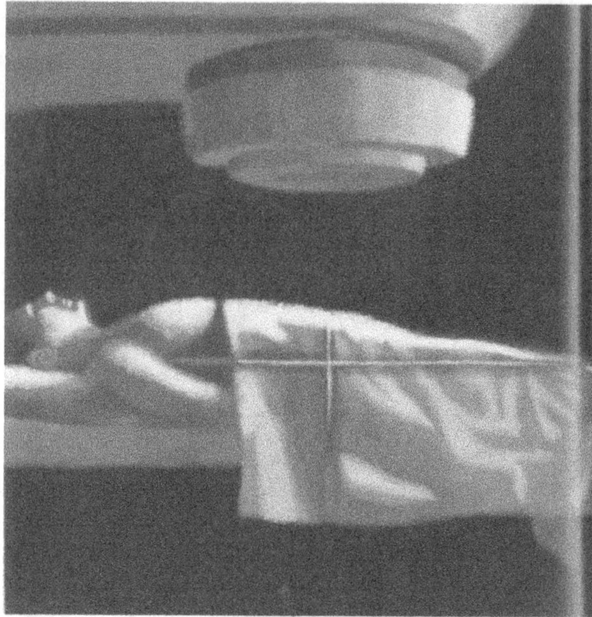

Figure 1.7 Radiation.
Source: Reprinted from Pope (p. 52) with permission See color insert.

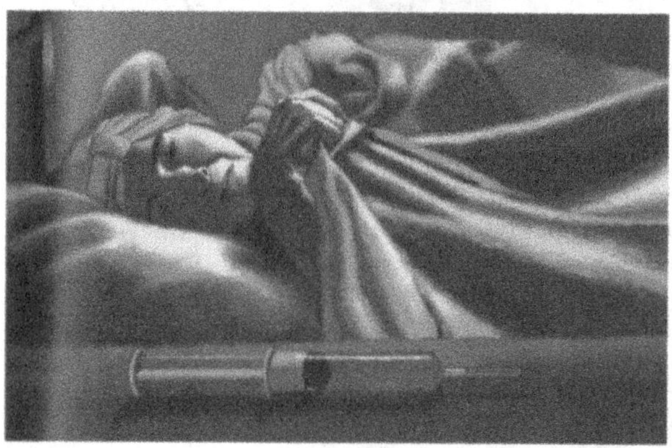

Figure 1.8 Chemotherapy.
Source: Reprinted from Pope (p. 47) with permission. See color insert.

The ink on paper image in Figure 1.9 depicts the suffering of a woman receiving her scheduled chemotherapy. Pope recalled that the injection days were the worst days of his life. The woman pictured winces in pain as the poisonous drug is administered. In contrast to the treated patient, the mask and gloves protect the nurse from the toxic effects of the chemotherapy.

Figure 1.10 is also an ink on paper image that conveys Pope's memories of his sickness from chemotherapy treatment and the responses of his father (driving) and brother (in back seat) to Pope's suffering. Many cancer patients and their family members continue to experience these emotions. Indeed, these sufferings have become even worse with some of the newer drugs available (15, 22).

Figure 1.9 Chemotherapy injection. *Source:* Reprinted from Pope (p. 62) with permission.

Figure 1.10 Three men. *Source:* Reprinted from Pope (p. 89) with permission.

Figure 1.11 Mastectomy. *Source:* Reprinted from Pope (p. 101) with permission.

Another ink on paper image in Figure 1.11 conveys a woman's emotional trauma associated with mastectomy, which involves the surgical removal of a breast to prevent the spread of cancer.

Figure 1.12 is a charcoal on paper image that conveys the suffering of a young girl from the ravages of chemotherapy. She gently touches the instrument of her suffering, while her doll in the background and the metal pan in foreground are reminders of the comfort and pain in her life.

Figure 1.13 depicts a son's artistic impression of the neurological devastation of glioblastoma in his father.

In addition to these pictorial images of cancer, we can also obtain a literary image of cancer from a paraphrase of Herman Melville's "Moby-Dick," when captain Ahab (played by the actor Gregory Peck) utters these words:

> *Look ye, Starbuck, all visible objects are but as pasteboard masks. Some inscrutable yet reasoning thing puts forth the molding of their features. The white whale tasks me; he heaps me. Yet he is but a mask. 'Tis the thing behind the mask I chiefly hate; the malignant thing that has plagued mankind since time began; the thing that maws and mutilates our race, not killing us outright but letting us live on, with half a heart and half a lung.*

More personal accounts of cancer images can be found in the 2010 HBO movie, *Wit*, starring Emma Thompson, and in the popular books by physicians

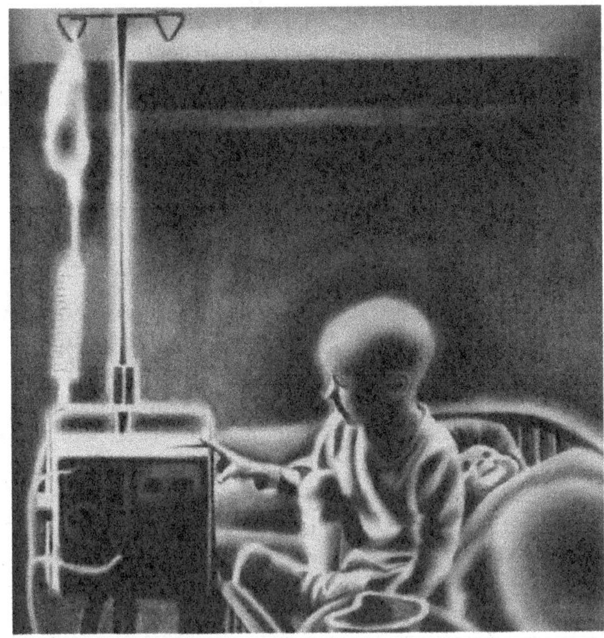

Figure 1.12 Erica.
Source: Reprinted from Pope (p. 80) with permission.

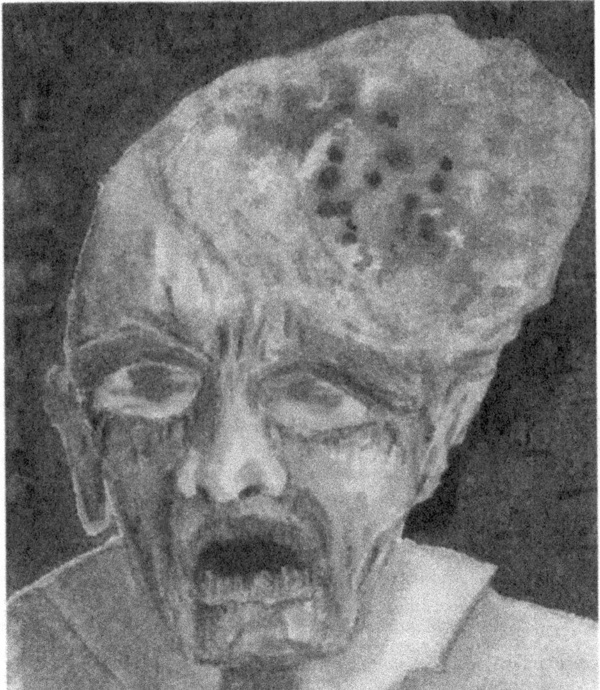

Figure 1.13 Fading away.
Source: Reprinted with permission from Gupta and Sarin (23). See color insert.

David Servan-Schreiber ("Anticancer: A New Way of Life") (24) and Siddhartha Mukherjee ("The Emperor of All Maladies: A Biography of Cancer") (25).

Synopsis

The images of cancer have changed little for more than a hundred years. If anything, they have become worse in this new century. The data in Table 1.1 show that we are not winning the war on cancer, regardless of what the pundits say (8). The promises of new drugs based on improved understanding of cancer genetics and biology have not materialized (26–28). As each new "miracle" cancer drug is discontinued due to no efficacy or unacceptable toxicity, a new "miracle" drug with similar disappointing effects quickly takes its place (15, 29). The media feeds into this process, providing false hope and misinformation (30). When will this continuum end? It will end, in my opinion, only after we come to recognize cancer as a metabolic disease that can be effectively managed with nontoxic metabolic therapies (31). My goal is to provide scientific evidence supporting this view.

Table 1.1 Cancer Statistics from 1990 to 2010

Year	Number of new cases	Number of deaths per year	Number of deaths per day
1990[a]	1,040,000	510,000	1397
1996[b]	1,359,150	554,740	1520
2002[c]	1,284,900	555,500	1522
2003[c]	1,334,100	556,500	1525
2004[c]	1,368,030	563,700	1544
2005[c]	1,372,910	570,280	1562
2006[c]	1,399,790	564,830	1547
2007[c]	1,444,920	559,650	1533
2008[c]	1,437,180	565,650	1549
2009[c]	1,479,350	562,340	1541
2010[c]	1,529,560	569,490	1560

The data show that the number of new cancer cases and deaths per year is increasing, while the number of deaths per day has remained fairly constant from 1996 until 2010. The numbers clearly indicate that the war on cancer is not going well. Indeed, the number of new cases, deaths per year, and deaths per day for cancer in 2010 was greater than the number of total casualties (1,076,245), total deaths (405,399), and deaths per day (416) suffered by all US military forces during the Second World War (1941–1945; data from http://en.wikipedia.org/wiki/United_States_military_casualties_of_war). What does this say about the leadership of those who are directing the war on cancer? The persistent high number of cancer deaths per year is especially disheartening considering that the budget for the National Cancer Institute (NCI) increased from $4.12 billion in 2002 to $5.10 billion in 2010. The 24% increase in the NCI budget is comparable to the 19% increase in new cancer cases.

[a] Data from Silverberg et al., http://caonline.amcancersoc.org/cgi/reprint/40/1/9.

[b] Data from Parker et al., http://caonline.amcancersoc.org/cgi/reprint/46/1/5.

[c] Data from Jamal et al. (4, 5, 7, 9–11).

REFERENCES

1. Kiberstis P, Marshall E. Cancer crusade at 40. Celebrating an anniversary. Introduction. Science. 2011;331:1539.
2. Anand P, Kunnumakkara AB, Sundaram C, Harikumar KB, Tharakan ST, Lai OS, et al. Cancer is a preventable disease that requires major lifestyle changes. Pharm Res. 2008;25: 2097–116.
3. Bailar JC, 3rd, Gornik HL. Cancer undefeated. N Engl J Med. 1997;336:1569–74.
4. Jemal A, Siegel R, Xu J, Ward E. Cancer statistics, 2010. CA Cancer J Clin. 2010;60:277–300.
5. Jemal A, Thomas A, Murray T, Thun M. Cancer statistics, 2002. CA Cancer J Clin. 2002;52: 23–47.
6. Gabor Miklos GL. The human cancer genome project–one more misstep in the war on cancer. Nat Biotechnol. 2005;23:535–37.
7. Jemal A, Center MM, Ward E, Thun MJ. Cancer occurrence. Methods Mol Biol. 2009;471: 3–29.
8. Faguet G. The War on Cancer: an Anatomy of a Failure, a Blueprint for the Future. Dordrecht, The Netherlands: Springer; 2008.
9. Jemal A, Murray T, Samuels A, Ghafoor A, Ward E, Thun MJ. Cancer statistics, 2003. CA Cancer J Clin. 2003;53:5–26.
10. Jemal A, Siegel R, Ward E, Hao Y, Xu J, Thun MJ. Cancer statistics, 2009. CA Cancer J Clin. 2009;59:225–49.
11. Jemal A, Siegel R, Ward E, Murray T, Xu J, Thun MJ. Cancer statistics, 2007. CA Cancer J Clin. 2007;57:43–66.
12. Salk JJ, Fox EJ, Loeb LA. Mutational heterogeneity in human cancers: origin and consequences. Annu Rev Pathol. 2010;5:51–75.
13. Jones S, Zhang X, Parsons DW, Lin JC, Leary RJ, Angenendt P, et al. Core signaling pathways in human pancreatic cancers revealed by global genomic analyses. Science. 2008;321:1801–6.
14. Marsh J, Mukherjee P, Seyfried TN. Akt-dependent proapoptotic effects of dietary restriction on late-stage management of a phosphatase and tensin homologue/tuberous sclerosis complex 2-deficient mouse astrocytoma. Clin Cancer Res. 2008;14:7751–62.
15. Fojo T, Parkinson DR. Biologically targeted cancer therapy and marginal benefits: are we making too much of too little or are we achieving too little by giving too much? Clin Cancer Res. 2010;16:5972–80.
16. Mukherjee P, Zhau JR, Sotnikov AV, Clinton SK. Dietary and Nutritional Modulation of Tumor Angiogenesis. In: Teicher BA, editor. Antiangiogenic Agents in Cancer Therapy. Totowa (NJ): Humana Press; 1999. p.237–61.
17. Mukherjee P, Abate LE, Seyfried TN. Antiangiogenic and proapoptotic effects of dietary restriction on experimental mouse and human brain tumors. Clin Cancer Res. 2004;10:5622–9.
18. Zuccoli G, Marcello N, Pisanello A, Servadei F, Vaccaro S, Mukherjee P, et al. Metabolic management of glioblastoma multiforme using standard therapy together with a restricted ketogenic diet: case report. Nutr Metab. 2010;7:33.
19. Carlson T. Turning sickness into art: Robert Pope and his battle with cancer. CMAJ. 1992;147: 229–32.
20. Pope R. Illness & Healing: Images of Cancer. Hantsport (NS): Lancelot Press; 1991.
21. Cohodes DR. Through the looking glass: decision making and chemotherapy. Health Aff (Millwood). 1995;14:203–8.
22. Uhm JH, Ballman KV, Wu W, Giannini C, Krauss JC, Buckner JC, et al. Phase II evaluation of gefitinib in patients with newly diagnosed grade 4 astrocytoma: Mayo/North central cancer treatment group study N0074. Int J Radiat Oncol Biol Phys. 2010;80:347–53.
23. Gupta T, Sarin R. Poor-prognosis high-grade gliomas: evolving an evidence-based standard of care. Lancet Oncol. 2002;3:557–64.
24. Servan-Schreiber D. Anticancer: A New Way of Life. New York: Viking; 2009.
25. Mukherjee S. The Emperor of all Maladies: A Biography of Cancer. New York: Scribner; 2010.

26. Hambley TW, Hait WN. Is anticancer drug development heading in the right direction? Cancer Res. 2009;69:1259–62.
27. Hanahan D, Weinberg RA. Hallmarks of cancer: the next generation. Cell. 2011;144:646–74.
28. Gibbs JB. Mechanism-based target identification and drug discovery in cancer research. Science. 2000;287:1969–73.
29. Couzin-Frankel J. Immune therapy steps up the attack. Science. 2010;330:440–3.
30. Fishman J, Ten Have T, Casarett D. Cancer and the media: how does the news report on treatment and outcomes?. Arch Intern Med. 2010;170:515–8.
31. Seyfried TN, Shelton LM. Cancer as a metabolic disease. Nutr Metab. 2010;7:7.

Chapter 2

Confusion Surrounds the Origin of Cancer

A major impediment in the effort to defeat cancer has been due, in large part, to the confusion surrounding the origin of the disease. "Make no mistake about it, the origin of cancer is far from settled." Contradictions and paradoxes continue to plague the field (1–5). Much of the confusion surrounding the origin of cancer arises from the absence of a unifying theory that can integrate the diverse observations on the nature of the disease. Without a clear idea on cancer origins, it becomes difficult to formulate a clear strategy for effective management and prevention. The failure to clearly define the origin of cancer is responsible in large part for the failure to significantly reduce the death rate from the disease.

Currently, most researchers consider cancer as a type of genetic disease where damage to a cell's DNA underlies the transformation of a normal cell into a potentially lethal cancer cell. The finding of hundreds and thousands of gene changes in different cancers has led to the idea that cancer is not a single disease, but is a collection of many different diseases. Consideration of cancer as a "disease complex" rather than as a single disease has contributed to the notion that management of various forms of the disease will require individual or "personalized" drug therapies (6–8). This therapeutic strategy would certainly be logical if, in fact, most cancers were of genetic origin. What if most cancers are not of genetic origin? What if most of the gene changes identified in tumor tissue arise as secondary downstream epiphenomena of tumor progression? What if cancer were a disease of respiratory insufficiency?

The somatic mutation theory, which has guided cancer research and drug development for over half a century, is now under attack. Carlos Sonnenschein and Anna Soto along with others have identified major inconsistencies in the evidence supporting the genetic origin of cancer (2–4, 9–12). Despite these concerns, the cancer field slogs forward with massive genome-based projects to identify all gene defects that occur in various tumor types (13–16). Gabor Miklos provided a compelling

---

Cancer as a Metabolic Disease: On the Origin, Management and Prevention of Cancer, First Edition. Thomas Seyfried.

argument for the unlikelihood that data generated from cancer genome projects will provide effective cures for the disease (14). A recent commentary in *Science* supports Miklos' argument in mentioning that little new information was uncovered from a comprehensive analysis of the ovarian cancer genome (Jocelyn Kaiser, 333:397, 2011). Is anyone listening to these arguments? Do people comprehend these messages? We have a financial crisis in the federal government and yet we are wasting enormous resources on genome projects that provide little useful information for cancer patients.

While the cancer genome projects are commendable for their technical achievement and have advanced the field of molecular biology, they have done little to defeat cancer (17–19). At the 2011 meeting of the American Association of Cancer Research, Dr. Linda Chin mentioned in her plenary lecture that improved genomic sequencing speed was a major beneficiary of the cancer genome projects. Another benefit has been the increased number of jobs created in the biotechnology sector as a result of the genome projects. How many dying cancer patients would be comforted by knowing this? While enhanced sequencing speed and creation of new jobs are certainly important and noteworthy, these achievements are not connected to curing cancer.

The information collected from the large cancer genome projects has done more to confuse than to clarify the nature of the cancer (13, 15, 20). To make matters worse, there are now suggestions for an international effort to identify all abnormal proteins in tumors, that is, a cancer proteome project (21). If the ratio of "information in to useful information out" was so low for the cancer genome

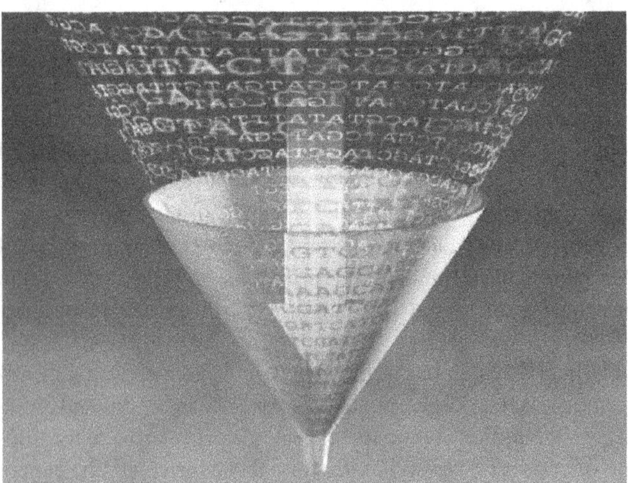

Figure 2.1 Too much in, nothing out. According to Serge Koscielny, the gene microarray bioinformatics literature is polluted with many gene expression signatures that have inadequate validation or no validation at all. Even if the expression signatures were adequately validated, the information would have little impact on the daily cancer death rate. *Source*: Reprinted with permission from Ref. 18. To see this figure in color please go to ftp://ftp.wiley.com/public/sci_tech_med/cancer_metabolic_disease.

projects (Fig. 2.1), what is the justification that the ratio would be better for a cancer proteome project? If technology improvement and new jobs creation is the justification, then this should be clearly stated, as a cure for cancer will not likely be the ultimate outcome.

In my opinion, it is wishful thinking that the vast information generated from the cancer genome atlas will someday serve as a foundation for the development of new and more effective cancer therapies despite recent arguments to the contrary (22). While gene-based targeted therapies could be effective against those few cancers that are inherited and where all cells within the tumor have a common genetic defect, most cancers are not inherited through the germ line and few cancer cells have gene defects that are expressed in all cells of the tumor (1, 8, 11, 14, 16, 17, 20). Although almost 700 targeted therapies have been developed from the cancer genome projects, no patients with solid tumor have been cured from this strategy (19). How many times must we beat the dead horse before we realize that it will not get up and walk?

Most mutations found in tumors arise sporadically, as do most cancers. The types of mutations found in one tumor cell will differ from those found in another tumor cell within the same tumor (7, 15, 23). Genetic heterogeneity and randomness is the norm rather than the exception for mutations found in most sporadic cancers. We have recently shown how the majority of cancer gene defects could arise as downstream epiphenomena of tumor progression rather than as cancer causes (24). In light of these findings, it is not likely that gene-based targeting strategies will be useful for managing most advanced cancers. Recent evidence bears this out (7, 19, 25).

It is my opinion that most genetic changes in tumors are largely irrelevant to the origin or treatment of cancer. They are but epiphenomena of biological chaos. While genomic changes might participate in disease progression, they do not cause the disease. If my prognosis is accurate, then where should one look for real solutions to the cancer problem?

Emerging evidence suggests that cancer is primarily a metabolic disease rather than a genetic disease (24). I will present evidence showing how cancer is a disease of defective cellular energy metabolism and that most of the genomic defects found in cancer cells arise as secondary downstream effects of defective energy metabolism. Most genetic defects found in tumors are "red herrings" that have diverted attention away from mitochondrial respiratory insufficiency, the central feature of the disease. I trained in classical genetics with Herman Brockman at Illinois State University and in biochemical genetics with William Daniel at the University of Illinois. I was, like many people, swept up in the hype surrounding the gene theory of cancer. Unfortunately, much of my original enthusiasm for the genetic origin of cancer has given way to skepticism and frank disbelief. This will become clear to all who read this treatise.

Regardless of cell type or tissue origin, the vast majority of cancer cells share a singular problem involving abnormal energy metabolism. While many in the cancer field consider gene defects as being responsible for the metabolic abnormalities in cancer cells, I do not share this view. In fact, I will present evidence showing how the gene defects in cancer cells can arise following damage to respiration. I predict that targeting the defective energy metabolism of tumors will eventually become the most cost-effective, nontoxic approach to cancer prevention and management. Moreover, the therapeutic efficacy of molecularly "targeted" therapies could be enhanced if combined with therapies that target energy metabolism. I will review substantial evidence supporting my views.

THE ONCOGENIC PARADOX

Although very specific processes underlie malignant transformation, a large number of unspecific influences can initiate the disease including radiation, chemicals, viruses, and inflammation. Indeed, it appears that prolonged exposure to almost any provocative agent in the environment can potentially cause cancer (26, 27). That a very specific process could be initiated in very unspecific ways was considered "*the oncogenic paradox*" by Albert Szent-Gyorgyi, a leading cancer researcher of his day (27, 28). *Oncogenesis* is the term used to describe the biological process leading to tumor formation. John Cairns also struggled with this paradox in his essay on *The Origins of Human Cancers* (29). The oncogenic paradox persists today as an unresolved issue in cancer research (26, 30). I will show how respiratory insufficiency is the origin of the oncogenic paradox.

HALLMARKS OF CANCER

In a landmark review on cancer, Drs. Hanahan and Weinberg suggested that six essential alterations in cell physiology were largely responsible for malignant cell growth (5). This review was later expanded into a book on the *Biology of Cancer* (31). These six alterations were described as the hallmarks of nearly all cancers and have guided research in the field for the last decade (32). The six hallmarks (Fig. 2.2) include the following:

1. **Self-Sufficiency in Growth Signals.** This process involves the uncontrolled proliferation of cells owing to self-induced expression of molecular growth factors. In other words, dysregulated growth would arise through abnormal expression of genes that encode growth factors. The released growth factors would then bind to receptors on the surface of the same cell (autocrine stimulation) or bind to receptors on other nearby tumor cells (paracrine stimulation), thereby locking-in signaling circuits that perpetuate continuous replication. Complicated cybernetic-type diagrams are often presented to illustrate these phenomena (Fig. 2.3). Cybernetics is generally viewed as the study of goal-directed control and communication systems (33). The abnormal circuitry in tumor cells is assumed to result in large part from the dominant expression of cancer-causing oncogenes.
2. **Insensitivity to Growth-Inhibitory (Antigrowth) Signals.** In order to carry out specific functions in mature differentiated tissues, most cells

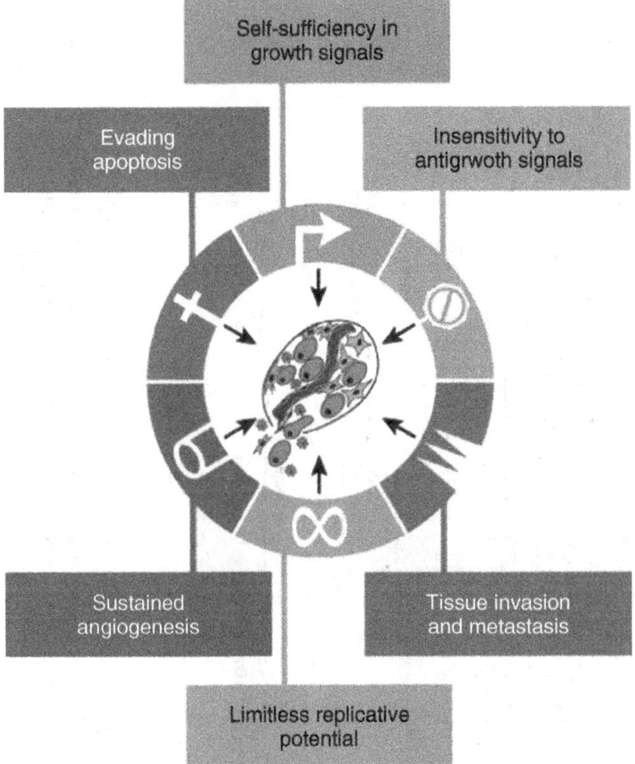

Figure 2.2 The six hallmarks of cancer from Hanahan and Weinberg. An updated version of this figure recently appeared in Ref. 32. *Source*: Reprinted with permission from Figure 1 of Hanahan and Weinberg (5). See color insert.

must remain quiescent or nonproliferative. A complex signaling circuitry involving the action of tumor-suppressor genes is necessary to maintain the quiescent state. In addition to these internal signals, interactions with other cells (cell–cell) and the external environment (cell–matrix) also act to maintain quiescence. Damage to suppressor genes or the microenvironment is assumed to dampen growth inhibition and provoke proliferation, as the cell no longer responds appropriately to the growth-inhibitory actions of these genes or molecules. Tumor cells are known to express multiple defects in tumor-suppressor genes and in cell–cell or cell–matrix interactions.

3. **Evasion of Programmed Cell Death (Apoptosis)**. Programmed cell death is an effective means of eliminating damaged or dysfunctional cells. Elimination of damaged cells is necessary in order to maintain tissue homeostasis and health. Cell damage can initiate the release of mitochondrial *cytochrome c*, a protein of the mitochondrial electron transport chain, which is a potent inducer of apoptosis in normal cells. In contrast to normal cells, however,

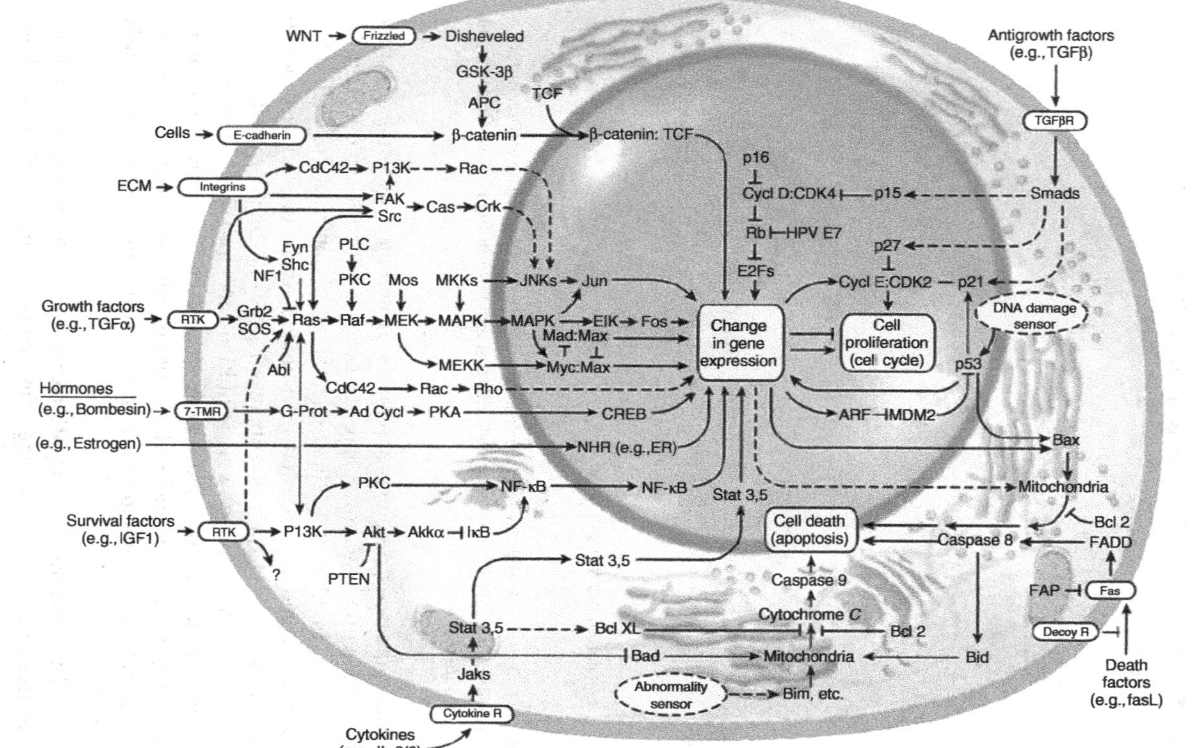

Figure 2.3 The emergent integrated circuit of the cell. Progress in dissecting signaling pathways has begun to lay out a circuitry that will likely mimic electronic integrated circuits in complexity and finesse, where transistors are replaced by proteins (e.g., kinases and phosphatases) and the electrons by phosphates and lipids, among others. In addition to the prototypical growth signaling circuit centered around Ras and coupled to a spectrum of extracellular cues, other component circuits transmit antigrowth and differentiation signals or mediate commands to live or die by apoptosis. As for the genetic reprogramming of this integrated circuit in cancer cells, some of the genes known to be functionally altered are given in gray. An updated version of this figure has appeared in Ref. 32. *Source*: Reprinted with permission from Figure 2 of Hanahan and Weinberg (5). See color insert.

tumor cells lose their sensitivity to apoptotic death signals. Consequently, tumor cells continue to live and proliferate despite damage to their nuclear DNA and respiration. Loss of tumor-suppressor genes, which sense cell damage and initiate cell death, is responsible in part for resistance of tumor cells to programmed cell death. The acquired resistance to apoptosis is a recognized hallmark of most cancers (5, 32).

4. **Limitless Replicative Potential.** All cells of a given species possess a finite number of divisions before they reach mortality. This is a cell-autonomous program that induces senescence and prevents immortality (5). Tumor cells, however, lose responsiveness to this program and continue to divide. The phenomenon of limitless replicative potential is closely connected to the first three acquired capabilities.
5. **Sustained Vascularity (Angiogenesis).** Angiogenesis involves neovascularization or the formation of new blood capillaries from existing blood vessels and is associated with the processes of tissue inflammation and wound healing. Many solid tumors have difficulty growing unless enervated with blood vessels, which can deliver nutrients while removing metabolic waste products (Fig. 1.3). The dissemination of tumor cells throughout the body is assumed to depend in part on the degree of tumor vascularization. The more blood vessels in tumors, the greater will be the potential to invade and metastasize. Tumor cells release growth factors that stimulate nearby host stromal cells (vascular endothelial cells and macrophages) to proliferate, thus providing the tumor with a vasculature and the means for more rapid growth. The endothelial cells form the vessel walls, while the local macrophages and other stromal cells degrade the microenvironment facilitating neovascularization. A switch from low vascularization to high vascularization is considered to be an essential acquired capability for tumor progression (5, 32, 34).
6. **Tissue Invasion and Metastasis.** Invasion of tumor cells into local tissue and their spread to distant organs underlies the phenomenon of metastasis. Metastasis or complications of metastasis is associated with about 90% of all cancer deaths (32, 35). The prevention of metastasis remains the single most important challenge for cancer management.

Genomic Instability

According to Hanahan and Weinberg, genome instability is considered to be the essential enabling characteristic for manifesting the six major hallmarks of cancer (5, 32). Genome instability was assumed to elicit the large numbers of mutations found in tumor cells, supporting the idea that cancer is a type of genetic disease. However, the mutation rate for most genes is low, making it unlikely that the thousands and even millions of pathogenic mutations found in cancer cells would occur sporadically within a normal human lifespan (15, 26, 36). Pathogenic mutations are those that disrupt normal cell physiology and differ from nonpathogenic mutations, which generally do not have any physiological effect on cell homeostasis. This then creates another paradox. If mutations are such rare events, then how is it possible that cancer cells can express so many different types and kinds of mutations during the development of a malignant tumor?

The loss of genomic "caretakers" or "guardians", involved in sensing and repairing DNA damage, was proposed to explain the increased mutability of tumor cells (26, 37–39). The loss of these caretaker systems would allow genomic instability, thus enabling premalignant cells to reach the six essential hallmarks of cancer (5, 32). Attempts to classify cancer mutations as either "drivers" or "passengers" have done little to clarify the situation (13, 15, 22, 40). It has been difficult to define with certainty the origin of premalignancy and the mechanisms by which the caretaker/guardian systems themselves are lost during the emergent malignant state (4, 6, 26). If the genome guardians are so essential for maintaining genomic integrity, then why are these guardians prone to such high mutability? Indeed, the p53 genome guardian is one of the most commonly mutated genes found in tumors (38). Most genes necessary for survival, for example, ubiquitin, histones etc., show little mutability across species. It is difficult for me to see how natural selection would select high mutability genes as "guardians of the genome." This would be like bank owners hiring tellers who are highly prone to corruption!

It appears that the route taken by the driver genes and their passengers to explain cancer seems more circular than straight with neither the drivers nor the passengers knowing the final destination. This is further highlighted with suggestions that some cancer genes, such as the isocitrate dehydrogenase gene 1 (*IDH1*), can act as either a tumor-provoking oncogene or as a tumor-inhibiting suppressor gene (reference *IDH1*) (41). The situation is even more confusing with suggestions that *IDH1* is both an oncogene and a tumor-suppressor gene! The view of cancer as a genetic disease reminds me of a traffic jam in Calcutta, India, where passengers direct drivers onto sidewalks and into opposite lanes of traffic in order to arrive at their destination. The attempt to link the six hallmarks of cancer to genomic instability is like a Calcutta traffic jam, but without a clear destination.

The Warburg Theory

In addition to the six recognized hallmarks of cancer, aerobic fermentation or the Warburg effect is also a robust metabolic hallmark of most tumors whether they are solid or blood born (42–47). Aerobic fermentation involves elevated glucose uptake with lactic acid production in the presence of oxygen. Elevated glucose uptake and lactic acid production is a defining characteristic of most tumors and is the basis for tumor imaging using labeled glucose analogs (48–50). Labeled glucose analogs have become an important diagnostic tool for cancer detection and management using positron emission tomography (PET). The radiolabeled glucose collects in the tumor tissue because nearly all tumors depend heavily on glucose for survival. Consequently, it is easy to detect many tumor types based on their requirement for glucose as shown in Figure 2.4.

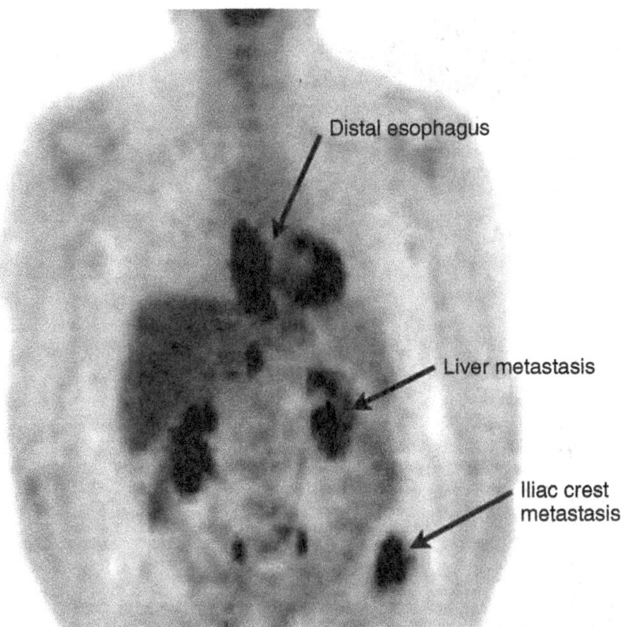

Figure 2.4 Shown here is a whole body scan of a 57-year-old man with esophageal adenocarcinoma. This FDG-PET scan shows malignancy in the distal esophagus with metastatic disease in the liver and in the superior iliac crest. *Source*: Modified from http://www.medscape.com/viewarticle/457982_4.

Although no specific gene mutation or chromosomal abnormality is common to all cancers (17, 22, 26, 51, 52), nearly all cancers express elevated fermentation, regardless of their tissue or cellular origin (24). In light of this important fact, it was good to see that Hanahan and Weinberg included information on energy metabolism in their more recent review of the subject (32). It is unfortunate, however, that the subject was not addressed in their original review or in Dr. Weinberg's textbook on the subject (5, 31).

The origin of the Warburg effect in tumor cells has been the subject of intense investigation and debate since Warburg first discovered the phenomenon during the early twentieth century (53, 54). Warburg was a pioneer in biochemistry and cell physiology and received the Nobel Prize for Physiology and Medicine in 1931 for his work on iron porphyrins in biological oxidations (Fig. 2.5).

Warburg was considered for a second Nobel Prize in 1944 for his identification of flavins and nicotinamide as hydrogen carriers, but was not chosen because of Hitler's decree forbidding German citizens from accepting Nobel Prizes (55). Prior to his work in cancer biochemistry, Warburg served with an elite Prussian cavalry regiment during the First World War. He was wounded while deployed on the Russian front and was decorated with the Iron Cross, First Class (55). Warburg felt that his service in the *Deutsches Heer* (German Army) prepared him for the rigors of a long academic career. Warburg said, "I learned to handle people; I learned to

Figure 2.5 Otto Warburg (holding pen) with Dean Burk. From Figure 10 in Krebs' book (55). Koppenol and colleagues recently provided an overview of Warburg's contribution to science and cancer research (53).

obey and to command. I was taught that one must be more than one appears to be" (55). Warburg remained in Germany during the Second World War and continued his experiments on cancer metabolism despite the fact that he was part Jewish (55). This fact, together with Warburg's arrogance in knowing how cancer arises, might have contributed in large part to the anti-Warburg sentiment in the post-war era.

Warburg initially proposed that aerobic glucose fermentation (aerobic glycolysis) was an epiphenomenon of a more fundamental problem in cancer cell physiology, that is, impaired or damaged respiration (54, 56). He used the metaphor of the plague to illustrate this connection.

> *Just as there are many remote causes of plague-heat, insects, rats-but only one common cause, the plague bacillus, there are a great many remote causes of cancer-tar, rays, arsenic, pressure, urethane- but there is only one common cause into which all other causes of cancer merge, the irreversible injuring of respiration.*

An increased dependency on energy through glucose fermentation (glycolysis) was viewed as an essential *compensatory* mechanism of energy production for cell viability following damage to respiration. If cells lose their ability to derive energy through respiration, then an alternative source of energy becomes essential for survival. Although aerobic glycolysis in cancer cells and anaerobic glycolysis in normal cells are similar in that lactate is produced under both situations, anaerobic glycolysis in normal cells arises from the absence of oxygen, whereas aerobic glycolysis in tumor cells arises as a consequence of both absence of oxygen and respiratory insufficiency (24). As oxygen reduces anaerobic glycolysis and lactate production in most normal cells because of increased respiratory activity (Pasteur effect), the continued production of lactate in the presence of oxygen in cancer cells represents an abnormal Pasteur effect. The continued production of lactate, a metabolic waste product of glucose metabolism, in the presence of oxygen is the metabolic hallmark of most tumor cells.

Warburg argued that only those body cells that are able to increase glycolysis following intermittent respiratory damage were capable of forming cancers (56). Cells unable to elevate glycolysis in response to respiratory insults, on the other hand, would perish due to energy failure. Cancer cells would therefore arise from normal body cells through a gradual and irreversible damage to their respiratory capacity. We recently expanded Warburg's concept to include energy derived through amino acid fermentation and substrate level phosphorylation in the citric acid cycle, also known as *the tricarboxylic acid* (*TCA*) *cycle* (24, 57, 58). In other words, respiratory insufficiency leads to a dependency on nonoxidative phosphorylation for energy and survival. Substrate level phosphorylation, arising from respiratory insufficiency, is the single most common phenotype found in cancer regardless of tissue origin (24). Respiratory insufficiency can arise from the cumulative effects of any number of environmental or genetic factors that alter mitochondrial function.

On the basis of metabolic data collected from numerous animal and human tumor tissue samples, Warburg proposed with insight and certainty that irreversible damage to respiration was the prime cause of cancer (54, 56, 59). Warburg investigated 35 different rat tumors, 15 different mouse tumors, and 10 different human tumors (54). His concise assessment on the origin of cancer, developed from years of rigorous experimentation, generated a firestorm of controversy in the cancer field. Warburg's theory was attacked as being too simplistic and not consistent with evidence of apparent normal respiratory function in some tumor cells (60–68). I will later show how mitochondrial fermentation can confound the appearance of normal respiration in cancer cells.

Moreover, critics argued that Warburg's hypothesis on the origin of cancer did not address the role of tumor-associated mutations, the phenomenon of metastasis, nor did it link the molecular mechanisms of uncontrolled cell growth directly to impaired respiration. Indeed, even Warburg's biographer and research associate, Hans Krebs, mentioned that Warburg's idea on the primary cause of cancer, that is, the replacement of respiration by fermentation (glycolysis), was only a symptom of cancer and not the cause (55). The primary cause was assumed to be at the level of gene expression.

A genetic origin of cancer was consistent with the early studies of Theodor Boveri, who suggested that tumors arose from the abnormal behavior of chromosomes during mitosis (1, 28, 69). A genetic origin of cancer was also consistent with evidence showing that chemical carcinogens and X rays caused mutations and that the genetic material was DNA (15, 70). It is important to mention that carcinogens and X rays also damage mitochondria and the respiratory function (56, 70–73). The view of cancer as a metabolic disease was gradually displaced with the view of cancer as a genetic disease involving damage to DNA. The origin of cancer as a genetic disease has been the rationale for the massive cancer genome projects underway currently.

REASSESSMENT

While there is now renewed interest in the energy metabolism of cancer cells, it is widely assumed that the Warburg effect and the metabolic defects expressed in cancer cells arise primarily from genomic mutability selected during tumor progression (24, 53, 74–77). In other words, the abnormal energy metabolism in cancer arises as a secondary consequence of defects in oncogenes and tumor-suppressor genes (78). Emerging evidence, however, questions the genetic origin of cancer and suggests that cancer is primarily a metabolic disease as Warburg originally described.

It is interesting in this regard that James Watson, who co-discovered DNA as the genetic material with Francis Crick in 1953, recently suggested that more attention be paid to the metabolism of cancer (79). Watson also believes that the direction of cancer research in the United States is largely offtrack and misdirected at the highest levels. The absence of major clinical breakthroughs in the cancer war over the last 40 years and the death statistics presented in Table 1.1 support Watson's contention.

My goal is to reengage the discussion of tumor cell origin and to provide evidence supporting a general hypothesis that genomic mutability and essentially all hallmarks of cancer including the Warburg effect can be linked to impaired respiration and energy metabolism. I will review evidence showing that respiratory insufficiency precedes and underlies the genome instability that accompanies tumor development. Once established, genome instability contributes to further respiratory impairment, genome mutability, and tumor progression.

I contend that most of the gene defects in natural cancers arise as downstream effects of damaged mitochondrial function. My hypothesis is based on evidence that nuclear genome integrity is largely dependent on the cell having sufficient mitochondrial respiration, and that all cells require regulated energy homeostasis to maintain their differentiated state. While Warburg recognized the centrality of impaired respiration in the origin of cancer, his research did not explain how impaired mitochondrial function was connected to what are now recognized as the hallmarks of cancer. Moreover, he did not clearly describe how cancer cells appear to respire normally, but have defective mitochondrial respiration (53). I will review evidence making these linkages and expand Warburg's ideas on how impaired energy metabolism can be exploited for tumor prevention and management. My former student, Laura Shelton, and I recently published an overview of the key issues (24). However, it was not possible in this brief review to present the detailed evidence supporting the central hypothesis of cancer as a disease of impaired respiration. The following chapters present more detailed evidence in support of the main hypothesis.

REFERENCES

1. Gibbs WW. Untangling the roots of cancer. Sci Am. 2003;289:56–65.
2. Sonnenschein C, Soto AM. Theories of carcinogenesis: an emerging perspective. Semin Cancer Biol. 2008;18:372–7.
3. Baker SG, Kramer BS. Paradoxes in carcinogenesis: new opportunities for research directions. BMC Cancer. 2007;7:151.
4. Soto AM, Sonnenschein C. The somatic mutation theory of cancer: growing problems with the paradigm?. Bioessays. 2004;26:1097–107.
5. Hanahan D, Weinberg RA. The hallmarks of cancer. Cell. 2000;100:57–70.
6. Nowell PC. The clonal evolution of tumor cell populations. Science. 1976;194:23–8.
7. Fojo T, Parkinson DR. Biologically targeted cancer therapy and marginal benefits: are we making too much of too little or are we achieving too little by giving too much?. Clin Cancer Res. 2010;16:5972–80.
8. Rosell R, Perez-Roca L, Sanchez JJ, Cobo M, Moran T, Chaib I, et al. Customized treatment in non-small-cell lung cancer based on EGFR mutations and BRCA1 mRNA expression. PloS One. 2009;4:e5133.
9. Sonnenschein C, Soto AM. The Society of Cells: Cancer and the Control of Cell Proliferation. New York: Springer; 1999.
10. Sonnenschein C, Soto AM. Somatic mutation theory of carcinogenesis: why it should be dropped and replaced. Mol Carcinog. 2000;29:205–11.
11. Tarin D. Cell and tissue interactions in carcinogenesis and metastasis and their clinical significance. Semin Cancer Biol. 2011;21:72–82.
12. Bissell MJ, Hines WC. Why don't we get more cancer? A proposed role of the microenvironment in restraining cancer progression. Nat Med. 2011;17:320–9.
13. Stratton MR, Campbell PJ, Futreal PA. The cancer genome. Nature. 2009;458:719–24.
14. Gabor Miklos GL. The human cancer genome project–one more misstep in the war on cancer. Nat Biotechnol. 2005;23:535–7.
15. Salk JJ, Fox EJ, Loeb LA. Mutational heterogeneity in human cancers: origin and consequences. Annu Rev Pathol. 2010;5:51–75.
16. Collaborative. Integrated genomic analyses of ovarian carcinoma. Nature. 2011;474:609–15.
17. Vitucci M, Hayes DN, Miller CR. Gene expression profiling of gliomas: merging genomic and histopathological classification for personalised therapy. Br J Cancer. 2010;104:545–53.
18. Koscielny S. Why most gene expression signatures of tumors have not been useful in the clinic. Sci Transl Med. 2010;2:14ps2.
19. Yin S. Experts question benefits of high-cost cancer care. Medscape Today. 2011. http://www.medscape.com/viewarticle/754808?src=iphone.
20. Greenman C, Stephens P, Smith R, Dalgliesh GL, Hunter C, Bignell G, et al. Patterns of somatic mutation in human cancer genomes. Nature. 2007;446:153–8.
21. Belda-Iniesta C, de Castro J, Perona R. Translational proteomics: what can you do for true patients? J Proteome Res. 2010;10:101–4.
22. Stratton MR. Exploring the genomes of cancer cells: progress and promise. Science. 2011;331:1553–8.

23. Shackleton M, Quintana E, Fearon ER, Morrison SJ. Heterogeneity in cancer: cancer stem cells versus clonal evolution. Cell. 2009;138:822–9.
24. Seyfried TN, Shelton LM. Cancer as a metabolic disease. Nutr Metab. 2010;7:7.
25. Kolata G. How bright promise in cancer testing fell apart. New York Times. 2011 July 7.
26. Loeb LA. A mutator phenotype in cancer. Cancer Res. 2001;61:3230–9.
27. Szent-Gyorgyi A. The living state and cancer. Proc Natl Acad Sci USA. 1977;74:2844–7.
28. Manchester K. The quest by three giants of science for an understanding of cancer. Endeavour. 1997;21:72–6.
29. Cairns J. The origin of human cancers. Nature. 1981;289:353–7.
30. Kiberstis P, Marshall E. Cancer crusade at 40. Celebrating an anniversary. Introduction. Science. 2011;331:1539.
31. Weinberg RA. The Biology of Cancer. New York: Garland Science; 2007.
32. Hanahan D, Weinberg RA. Hallmarks of cancer: the next generation. Cell. 2011;144:646–74.
33. Wiener N. Cybernetics: or the Control and Communication in the Animal and the Machine. 2nd ed. Cambridge, MA: MIT Press 1965.
34. Folkman J. Incipient angiogenesis. J Natl Cancer Inst. 2000;92:94–5.
35. Lazebnik Y. What arc thc hallmarks of cancer? Nat Rev. 2010;10:232–3.
36. Rous P. Surmise and fact on the nature of cancer. Nature. 1959;183:1357–61.
37. Lane DP. Cancer. p53, guardian of the genome. Nature. 1992;358:15–6.
38. Levine AJ. p53, the cellular gatekeeper for growth and division. Cell. 1997;88:323–31.
39. Lengauer C, Kinzler KW, Vogelstein B. Genetic instabilities in human cancers. Nature. 1998;396:643–9.
40. Parmigiani G, Boca S, Lin J, Kinzler KW, Velculescu V, Vogelstein B. Design and analysis issues in genome-wide somatic mutation studies of cancer. Genomics. 2009;93:17–21.
41. Garber K. Oncometabolite? IDH1 discoveries raise possibility of new metabolism targets in brain cancers and leukemia. J Natl Cancer Inst. 2010;102:926–8.
42. Boag JM, Beesley AH, Firth MJ, Freitas JR, Ford J, Hoffmann K, et al. Altered glucose metabolism in childhood pre-B acute lymphoblastic leukaemia. Leukemia. 2006;20:1731–7.
43. Seyfried TN, Mukherjee P. Targeting energy metabolism in brain cancer: review and hypothesis. Nutr Metab. 2005;2:30.
44. Semenza GL, Artemov D, Bedi A, Bhujwalla Z, Chiles K, Feldser D, et al. The metabolism of tumours: 70 years later. Novartis Found Symp. 2001;240:251–60. discussion 60–4.
45. Ristow M. Oxidative metabolism in cancer growth. Curr Opin Clin Nutr Metab Care. 2006;9: 339–45.
46. Gatenby RA, Gillies RJ. Why do cancers have high aerobic glycolysis?. Nat Rev. 2004;4: 891–9.
47. Gogvadze V, Orrenius S, Zhivotovsky B. Mitochondria in cancer cells: what is so special about them? Trends Cell Biol. 2008;18:165–73.
48. Frezza C, Gottlieb E. Mitochondria in cancer: not just innocent bystanders. Semin Cancer Biol. 2009;19:4–11.
49. Gatenby RA, Gillies RJ. Glycolysis in cancer: a potential target for therapy. Int J Biochem Cell Biol. 2007;39:1358–66.
50. Vander Heiden MG, Cantley LC, Thompson CB. Understanding the Warburg effect: the metabolic requirements of cell proliferation. Science. 2009;324:1029–33.
51. Wokolorczyk D, Gliniewicz B, Sikorski A, Zlowocka E, Masojc B, Debniak T, et al. A range of cancers is associated with the rs6983267 marker on chromosome 8. Cancer Res. 2008;68: 9982–6.
52. Nowell PC. Tumor progression: a brief historical perspective. Semin Cancer Biol. 2002;12:261–6.
53. Koppenol WH, Bounds PL, Dang CV. Otto Warburg's contributions to current concepts of cancer metabolism. Nat Rev. 2011;11:325–37.
54. Warburg O. The Metabolism of Tumours. New York: Richard R. Smith; 1931.
55. Krebs H. Otto Warburg: Cell Physiologist, Biochemist, and Eccentric. Oxford: Clarendon; 1981.
56. Warburg O. On the origin of cancer cells. Science. 1956;123:309–14.

57. Seyfried TN. Mitochondrial glutamine fermentation enhances ATP synthesis in murine glioblastoma cells. Proceedings of the 102nd Annual Meeting of the American Association Cancer Research; 2011; Orlando (FL); 2011.
58. Shelton LM, Strelko CL, Roberts MF, Seyfried NT. Krebs cycle substrate-level phosphorylation drives metastatic cancer cells. Proceedings of the 101st Annual Meeting of the American Association for Cancer Research; 2010; Washington (DC); 2010.
59. Warburg O. Revidsed Lindau Lectures: The prime cause of cancer and prevention-Parts 1 & 2. In: Burk D, editor. Meeting of the Nobel-Laureates Lindau, Lake Constance, Germany: K.Triltsch; 1969. p. http://www.hopeforcancer.com/OxyPlus.htm.
60. Moreno-Sanchez R, Rodriguez-Enriquez S, Saavedra E, Marin-Hernandez A, Gallardo-Perez JC. The bioenergetics of cancer: is glycolysis the main ATP supplier in all tumor cells? Biofactors. 2009;35:209–25.
61. Bonnet S, Archer SL, Allalunis-Turner J, Haromy A, Beaulieu C, Thompson R, et al. A mitochondria-K+ channel axis is suppressed in cancer and its normalization promotes apoptosis and inhibits cancer growth. Cancer Cell. 2007;11:37–51.
62. Semenza GL. HIF-1 mediates the Warburg effect in clear cell renal carcinoma. J Bioenerg Biomembr. 2007;39:231–4.
63. Moreno-Sanchez R, Rodriguez-Enriquez S, Marin-Hernandez A, Saavedra E. Energy metabolism in tumor cells. FEBS J. 2007;274:1393–418.
64. Aisenberg AC. The Glycolysis and Respiration of Tumors. New York: Academic Press; 1961.
65. Fantin VR, Leder P. Mitochondriotoxic compounds for cancer therapy. Oncogene. 2006;25: 4787–97.
66. Hervouet E, Demont J, Pecina P, Vojtiskova A, Houstek J, Simonnet H, et al. A new role for the von Hippel-Lindau tumor suppressor protein: stimulation of mitochondrial oxidative phosphorylation complex biogenesis. Carcinogenesis. 2005;26:531–9.
67. Weinhouse S. On respiratory impairment in cancer cells. Science. 1956;124:267–9.
68. Weinhouse S. The Warburg hypothesis fifty years later. Z Krebsforsch Klin Onkol Cancer Res Clin Oncol. 1976;87:115–26.
69. Wolf U. Theodor boveri and his book, on the problem of the origin of malignant tumors. In: German J, editor. Chromosomes and Cancer. New York: John Wiley & Sons, Inc; 1974. p.1–20.
70. Hadler HI, Daniel BG, Pratt RD. The induction of ATP energized mitochondrial volume changes by carcinogenic N-hydroxy-N-acetyl-aminofluorenes when combined with showdomycin. A unitary hypothesis for carcinogenesis. J Antibiot (Tokyo). 1971;24:405–17.
71. Sajan MP, Satav JG, Bhattacharya RK. Effect of aflatoxin B in vitro on rat liver mitochondrial respiratory functions. Indian J Exp Biol. 1997;35:1187–90.
72. Bhat NK, Emeh JK, Niranjan BG, Avadhani NG. Inhibition of mitochondrial protein synthesis during early stages of aflatoxin B1-induced hepatocarcinogenesis. Cancer Res. 1982;42:1876–80.
73. Smith AE, Kenyon DH. A unifying concept of carcinogenesis and its therapeutic implications. Oncology. 1973;27:459–79.
74. Kim JW, Dang CV. Cancer's molecular sweet tooth and the Warburg effect. Cancer Res. 2006;66: 8927–30.
75. Hsu PP, Sabatini DM. Cancer cell metabolism: Warburg and beyond. Cell. 2008;134:703–7.
76. Shaw RJ. Glucose metabolism and cancer. Curr Opin Cell Biol. 2006;18:598–608.
77. Jones RG, Thompson CB. Tumor suppressors and cell metabolism: a recipe for cancer growth. Genes Dev. 2009;23:537–48.
78. Kaelin WG Jr, Thompson CB. Q&A: Cancer: clues from cell metabolism. Nature. 2010;465: 562–64.
79. Watson JD. To fight cancer, know the enemy. New York Times. 2009 August 6.

Chapter 3

Cancer Models

PROBLEMS WITH SOME CANCER MODELS

Good models of cancer can provide insight into disease mechanisms and development of new therapies. Many cancer models, however, fall short of replicating the full spectrum of cancer traits especially those related to metastasis. Following the great financial crisis of 2008, economists began to question the models used to predict the dynamics of financial markets. It became clear that available models were "scrubbed clean" in the interest of theoretical elegance and were largely divorced from the way real-world economies actually function (1). In short, the available models of financial systems were inadequate in their ability to predict the impending economic crisis. The situation in the cancer field is similar in some respect to the situation in the economic field. Basically, many available cancer models do not mimic the dynamics of real-world in vivo metastasis. The situation in the fields of finance and cancer research reminds me of the joke about the guy who tried to catch a mouse in his house by baiting the mousetrap with a picture of a cheese. On checking the mousetrap the following morning, the guy was surprised to find that he caught a picture of a mouse!

Metastatic Models

While there are many good animal models for the developmental stages of cancer initiation, promotion, and progression, few good animal models exist for systemic metastasis (2). This is unfortunate, as systemic metastasis is the single-most serious aspect of the disease. Metastasis is largely responsible for the majority of cancer deaths. Yuri Lazebnik recently mentioned that all of the cancer hallmarks discussed in the Hanahan and Weinberg paper except metastasis could also be found in benign tumors (3). According to Lazebnik, the failure to recognize this fact has contributed

---

Cancer as a Metabolic Disease: On the Origin, Management and Prevention of Cancer, First Edition.
Thomas Seyfried.

in large part to the failure to win the war on cancer (3). I wholeheartedly agree with Dr. Lazebnik on this position.

Once tumor cells leave their primary site and begin to show up in distant organs or tissues, effective management and long-term patient prognosis become uncertain. Most available cancer models, however, rarely show systemic metastatic behavior. Indeed, most tumor cells will rapidly grow when implanted under the skin (subcutaneously) or in an orthotopic site (tissue of origin) but will rarely show distal invasion or spread to multiple organ systems as is often seen in the human disease (2, 3). While some tumor models might show local tumor cell invasion into the surrounding tissue or spread to a neighboring organ, they rarely show systemic metastasis encompassing multiple and diverse organ systems. If metastasis occurs, it often lacks fidelity and expediency in most animal models (2, 4–6). In other words, not every mouse inoculated with the tumor cells develops metastatic cancer. In addition, the time to systemic metastasis can vary significantly from one mouse to another. Models expressing these shortcomings are of limited value for the evaluation of new antimetastatic therapies.

To overcome these shortcomings, cancer researchers often inject tumor cells directly into the circulation (blood stream) of a host animal (7). This approach bypasses a critical step in metastasis, that is, the natural ability of metastatic tumor cells to leave their site of origin and to enter the circulation. I contend that metastasis models using vascular injection of tumor cells are not representative of the real-world situation. Truly, metastatic cancer cells should not require blood stream injection to manifest the disease. Vascular injection of tumor cells as a model of metastasis is like baiting the mousetrap with a picture of a cheese. While good models of human disease can be powerful tools for evaluating underlying mechanism and for developing effective therapies, bad models can stymie real progress and worse yet, providing misinformation on the nature of the disease, thus retarding progress.

Xenograft Models

Xenograft models involve growth of human tumor cells in nude mice or some other mice with a compromised innate and/or adaptive immune system. It is not possible to grow human tumors in mice that have normal T- and B-cell immunity because of antibody production and host tumor rejection. In addition, functional innate immunity derived from natural killer cells (NK), complement, etc., may contribute to tumor–host interactions. The normal mouse immune system will destroy implanted human cells. Most knowledgeable investigators in the cancer field know that xenograft models are unrepresentative of the real-world situation (4, 5). Nevertheless, many investigators persist with expensive studies using xenograft models of human cancer.

We have worked extensively with various mouse cancer models. Some of these are naturally invasive and metastatic, while some are neither invasive nor metastatic. The differences between the metastatic and nonmetastatic models are striking. A sharp border is seen between the tumor tissue and normal tissue in the noninvasive tumors, whereas no clear border is seen between the tumor tissue and the normal tissue in the metastatic cancers. Many xenograft models used in the cancer field are locally invasive, but rarely show systemic metastasis as seen in most human metastatic cancers (5, 6, 8).

The situation with xenograft models becomes even more bizarre, as human cells implanted into a mouse host gradually take on biochemical characteristics of mouse cells. We showed that human U87MG brain cancer cells express mouse carbohydrates on their surface when grown as a xenograft in immune-deficient mice (9). More than 65% of the sialic acid composition on the U87 tumor cells consisted of the nine-carbon sugar, *N*-glycolylneuraminic acid. Humans, however, are unable to synthesize *N*-glycolylneuraminic acid because of mutation in the gene that encodes a common mammalian hydroxylase enzyme (9, 10). The hydroxylase mutation occurred in the human genome sometime after our evolutionary split with the great apes (10). The acquisition of mouse carbohydrates and lipids will likely occur in any human tumor grown in the body of a mouse or rat. *N*-Glycolylneuraminic acid also alters the characteristics of human embryonic stem cells when grown on nonhuman feeder cells. This has been a confounding variable in the stem cell field (11).

Expression of mouse carbohydrates and lipids on human tumor cells when grown as xenografts can alter gene expression and growth behavior of the tumor cells, thus altering their response to changes in the microenvironment. The basal metabolic rate of the mouse is also sevenfold greater than that of humans (12). It is not clear how the striking difference in basal metabolic rate between mice and humans will influence tumor biology. About fifty million years of evolution separates humans from mice. Many cancer researchers are unaware of these complications. If researchers had been aware of these problems, more attention would have been given to them in the scientific literature. It might be reasonable to view the human xenograft tumor models as a type of human–mouse centaur!

We also found that food consumption is substantially greater in immunocompromised SCID mice, a common xenograft host, than that is found in the C57BL/6J mouse strain that has a normal immune system (13). Differences in food consumption are indicative of differences in energy metabolism. The NOD-SCID mice are also commonly used as a host for growing xenograft human tumors. The acronym stands for mice that are *nonobese diabetic and severely compromised immunodeficient*. These mice not only have an abnormal immune system but also express characteristics of both type-1 and type-2 diabetes (14). This is not a usual situation for most cancer patients. This experimental model might be useful for those individuals who have cancer, are genetically immunodeficient, and also suffer from both type-1 and type-2 diabetes. It is naive to assume that the growth behavior and response to therapies of human tumors grown as xenografts would be similar to the situation in the natural host.

If most of the xenograft models are flawed in representing the real-world situation, why does the cancer field persists with requirements for showing therapeutic efficacy in these animal models? The short answer is xenograft models are often required by reviewers in order to get papers published in top scientific journals or to get research grants funded. Many investigators believe that xenograft models are more representative of the human disease than natural animal models of cancer simply because the tumor cells are derived from humans. Consequently, many cancer researchers use xenografts to demonstrate therapeutic effects. Many clinical drug trials have been initiated in patients on the basis of information generated from xenograft models. Many of these drugs are later discontinued because of lack of efficacy, unacceptable toxicity, or some combination of these. Should this be surprising considering the unnatural nature of the experimental system?

Genetic Models

Besides xenograft models, a number of genetic cancer models are available, which produce tumors in various organs at specific developmental periods (2, 6). Most of these models involve mice since more is known about the mouse genome than about the genome of most other mammalian systems. Rarely, however, have targeted gene disruptions in the mouse produced cancers that show widespread invasion or metastasis outside the affected tissue (5). Occasionally, several simultaneous gene defects are required in order for some mice to develop tumors. Many of the genetic cancer models are considered valuable because cancer is thought to be a genetic disease. While these models certainly illustrate the role of genes in the origin of some cancers, only few, if any, human cancers are known, which arise from the "simultaneous" inheritance of germ line mutations in these genes. Similar to the models used to predict the dynamics of financial markets, the genetic cancer models, in my opinion, are scrubbed clean in the interest of theoretical elegance and are largely divorced from the way real-world metastasis actually occurs.

Cell Culture Models

Besides animal models, considerable information is generated on the metastatic behavior of cancer cells using cell culture model systems. Migration of cancer cells through artificial extracellular matrix materials such as Matrigel or into scratches made on the surface of culture dishes has often been used to assess the invasive behavior of various cancer cells. Are these assays reliable in predicting the invasive properties of tumor cells growing in their natural environment? The answer is unclear, as only few studies compare and contrast the invasive and metastatic behavior of tumor cells in the artificial culture environment with their invasive behavior in the natural environment. The farther the model system is from the "real-world" situation, the more caution is required in relating the observations to what actually takes place in the human body.

This is especially true for human brain cells that are grown in fetal cow serum. The blood–brain barrier evolved over millions of years to prevent molecules in the serum from entering the brain. Astrocytes protect neurons from serum molecules and become quite reactive when exposed to serum. Yet, many investigators, including me, have studied the behavior of neural tumor cells cultured in growth media containing fetal cow serum.

We found that CT-2A mouse astrocytoma cells migrate through Matrigel and into scratches on glass slides, but do not invade or show metastasis when grown in their natural genetic mouse host and in their known orthotopic site of origin (brain). This was surprising to us, as the CT-2A tumor is highly vascularized when grown in its natural host. As many invasive and metastatic human tumors are often highly vascularized, vascularization is often considered a hallmark of human metastatic cancer (15, 16). Clearly, high tumor vascularization (also referred to as *tumor angiogenesis*) does not enhance invasion or metastasis in this model. We found that growth rate was significantly faster in the more vascularized tumors than in the less vascularized tumors, but enhanced growth rate was not associated with increased metastasis. We recently showed that metastasis arises largely from the transformation of cells of myeloid origin, which already embody the capacity to invade and enhance angiogenesis (17).

Our experience with various cancer models highlights some of the problems in linking the behavior of the models to the situation in the human disease. For example, the CT-2A tumor cells are invasive (migratory) in cell culture assays of invasion but not in the natural environment. Vascularization is considered a hallmark of human metastatic tumors, but the CT-2A tumor does not invade or metastasize despite heavy vascularization and rapid growth rate. Our experience with the CT-2A brain tumor model and with other mouse models is illustrative of the inconstancies between tumor models and their human disease counterparts.

Other inconsistencies with in vitro brain tumor characteristics were recently highlighted. Brain cancer cells expressing SDH1 mutations and other well-recognized phenotypes such as *EGFR* gene amplification in the in vivo environment could not be observed in cells cultured from the tumors (18). Indeed, tumor cells with SDH1 mutations that grow rapidly in vivo do not grow or survive in vitro. How would one interpret such findings? Although cell culture models can be valuable tools for defining molecular mechanisms, it is important to remain cognizant of their limitations.

Natural Models

My experience in the cancer field was honed from decades of research using multiple in vivo and in vitro models of the disease. Most of the in vivo models we use for our research were developed in my present laboratory at Boston College or in my previous laboratory in the Neurology Department at Yale University. While any cancer model can provide information on the nature of the disease, the best cancer models in my opinion are those that *arise naturally* (spontaneously) and *are grown orthotopically* in their *syngeneic hosts*. Why use cancer animal models that do not represent the full spectrum of the disease when other models are available that display the most important features of the disease?

The spontaneous brain tumors in the inbred VM mouse strain represent a more natural model of metastatic cancer than any xenograft model (17, 19, 20).

The tumor cells arising in the VM mouse strain manifest the full spectrum of growth characteristics seen in most human metastatic cancers. The VM model can be classified as a natural spontaneous model according to the criteria of Kerbel and coworkers (2). The metastatic VM tumors also share several features in common with the fusion hybrid metastatic mouse cancers described by Kerbel (21, 22). These types of models can provide insight into the mechanisms of metastasis and are best suited for the development of effective therapies. Considering the importance of metastasis in cancer, it is unclear why the cancer field has not adapted these excellent metastasis models for screening novel antimetastatic therapies.

The National Cancer Institute web site can provide information on mouse models of various cancers (http://emice.nci.nih.gov/mouse_models). A good cancer model is one where the metastatic and invasive behavior of tumor cells is similar to that seen in the human disease (2). A good model of metastasis should be one where cancer cells will invade locally from any implanted tissue site and readily spread to multiple organ systems within a short period of time (2–4 weeks). This is seen in the VM model (Figs. 3.1 and 3.2). The invasive properties of the VM cells in

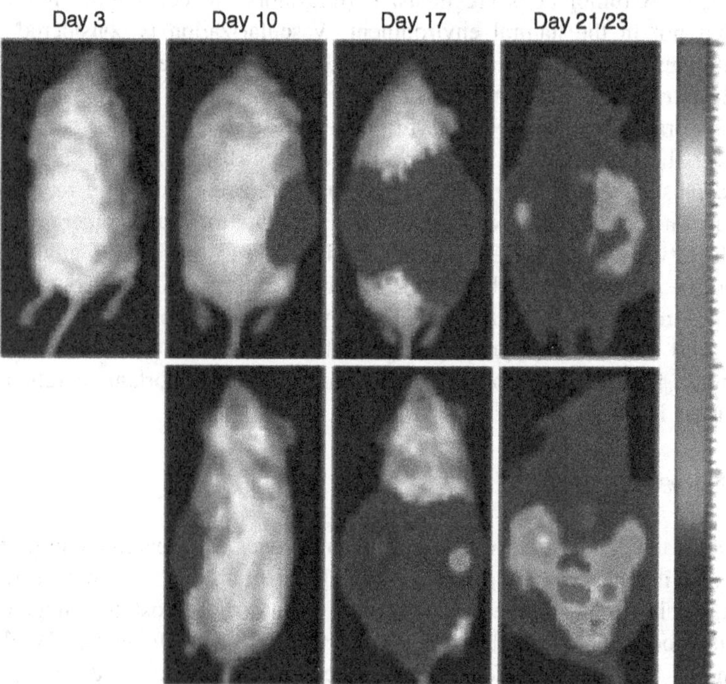

Figure 3.1 Whole body view of bioluminescence from metastatic VM-M3 tumor cells. VM-M3 tumor cells, containing the firefly luciferase gene, were implanted subcutaneously on the flank of a syngeneic VM mouse on day 0 as we described (23). Bioluminescent signal from the metastatic cells was measured in live mice using IVIS Lumina system (Caliper LS). Bioluminescence appeared throughout the mouse after 23 days indicative of widespread systemic dissemination of metastatic cancer cells. *Source*: Reprinted with permission from Ref. 24. See color insert.

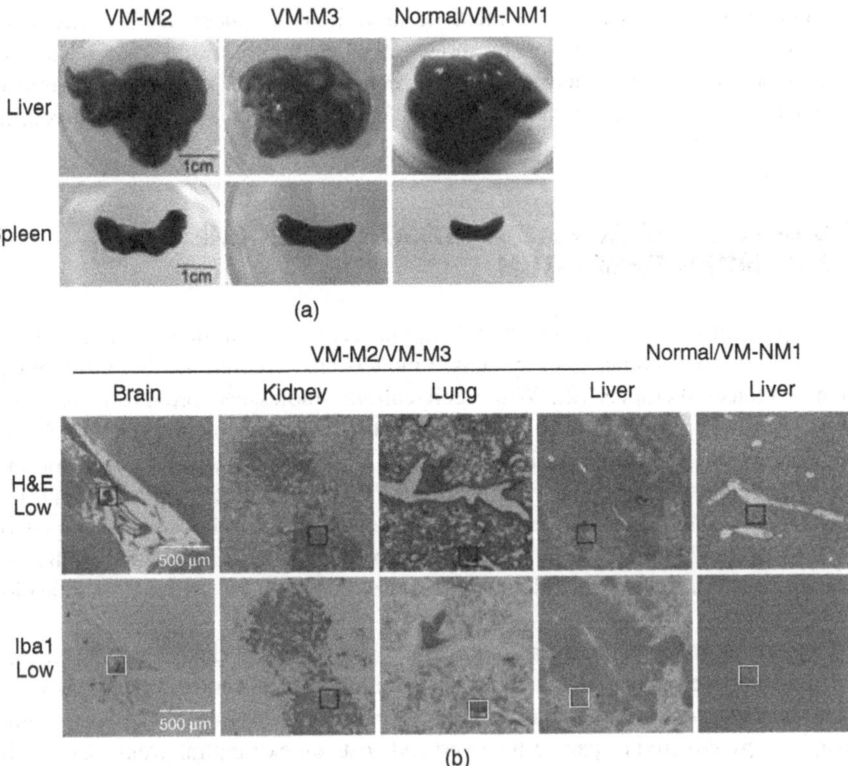

Figure 3.2 Appearance of gross (a) metastatic lesions and (b) micrometastatic lesions from mice bearing the VM-M2 and VM-M3 tumors. Mice of the inbred VM strain develop spontaneous brain tumors naturally. Shown here are the results when VM-M2 and VM-M3 tumor cells are implanted subcutaneously on the flank of syngeneic VM mice. While primary brain tumors do not often metastasize from the brain to extraneural tissues, glioblastoma can be highly metastatic if cells gain access to extraneural sites (17, 19, 25). The metastatic VM tumors (VM-M2 and VM-M3) express several characteristics of microglia/macrophages. No gross or micrometastatic lesions were found in tissues from mice implanted subcutaneously with the VM-NM1 tumor. The VM-NM1 tumor expresses stem cell markers but does not express macrophage biomarkers (19). H&E and Iba-1 staining revealed numerous micrometastatic lesions in the kidney, lung, brain meninges, and liver in mice bearing the VM-M2 tumor (b). Iba-1 is a recognized marker for cells of microglia/macrophages (19). The micrometastatic lesions are shown at × 100. The black boxes in the × 100 images (low) were shown previously at higher power (× 400) (19). The distribution, morphology, and staining of the micrometastatic lesions in mice bearing the VM-M3 appeared the same as that in the mice bearing the VM-M2 tumor (not shown). In addition to these organs, cells from the metastatic VM tumors are also found in bone marrow. *Source*: Reprinted with permission from Ref. 19. See color insert.

brain are also presented in Chapter 17. The response of the metastatic VM tumors to the antimetastatic drugs methotrexate and cisplatin is similar to the response seen in many human metastatic cancers (23). Many metastatic VM tumor cells can remain dormant following these treatments only to grow again after the therapy is terminated. If metastasis is the cause of most human cancer deaths, why study the disease in models that do not show extensive invasion and metastasis? The VM model of systemic metastatic cancer can help answer or address several of the NCI provocative questions including #15, #16, #17, and #24 (provocativequestions.nci.nih.gov).

ANIMAL CHARGES AS A MAJOR IMPEDIMENT TO CANCER RESEARCH

Although many cancer animal models might have shortcomings in reflecting the true nature of the human disease, animal models are essential for the development of new cancer therapies (2). While cell culture studies can provide insight into molecular mechanisms of action, cell culture studies are unable to provide accurate information on systems physiology associated with new therapies. Cancer not only involves defects in subcellular molecular mechanisms but also involves multiple changes to animal health and physiology. The influence of anticancer therapies on physiology can be best studied in animals harboring the disease. Animal studies are essential for translating potentially new cancer therapies into practical application in the clinic.

However, the high cost of animal-maintenance charges (cage charges) is having a major negative impact on animal cancer research. It is becoming too expensive for many investigators to include animals in their research designs. In the past, animal charges were covered as part of the overhead costs on extramural grants. Currently, the animal charges are added to research grants as a "direct cost" line item. As institutions can charge "overhead" costs on direct cost items, animal charges have now become a convenient means of enhancing institutional revenue. In other words, animal charges have become a "cash cow" for university administrators. Even though it is legal for universities to double dip on the animal charges, I consider the practice as unethical and not in the best interests of medical research.

Moreover, the activities of the animal rights movement have led to excessive federal regulations that now impede animal research. Some of these regulations can border on the absurd, for example, that only five mice weighing less than 25 g can be housed in a standard mouse cage. How might these regulations relate the natural housing of mice in the wild? The excessive rules and regulations imposed by Institutional Animal Care and Use Committees (IACUC) have become an impediment to the conquest of cancer.

It is also interesting that some cancer drugs used to treat humans in the clinic are considered too toxic to use on animals. In some institutions, veterinarians are on call (24 h/day and 7 days/week) to attend the needs of sick rodents. It appears that the animal rights organizations have achieved their mission. The quality of life of rodents housed at US universities is now better than the quality of life of most people living on the planet. Considering the high maintenance costs and excessive federal and institutional regulations, many investigators are opting out of using animals for their cancer research projects. It is simply easier and less costly to study cancer in the culture dish than in the living animal no matter how good the animal model might be. This is unfortunate, as new natural and genetic animal models of cancer will go under-utilized because of the excessive cost and the regulations involved in using live animals. The consequence of excessive animal cage charges and government regulations will mean greater numbers of human cancer deaths and suffering. Do cancer patients and their advocacy groups know about this?

PROBLEMS WITH TUMOR HISTOLOGICAL CLASSIFICATION

Too often investigators will focus more on tumor cell classification than on the biological behavior of the tumor cells. The success or failure in adapting a new cancer model can sometimes depend on how the tumor is classified histologically. This is especially the case in the brain cancer field, where neuropathology has a dominant influence on the direction of research. I came to seriously question the accuracy and importance of brain tumor classification, however, while I was on the faculty at Yale University in the early 1980s.

At that time, I initiated studies on the abnormal expression of gangliosides (complex glycosphingolipids) in various brain tumors. To conduct these studies, I obtained two in vivo brain tumor models from Dr. Harry Zimmerman. Dr. Zimmerman was head of Neuropathology at the Montiforie Hospital, which is part of the Albert Einstein College of Medicine in the Bronx, New York. Dr. Zimmerman was a distinguished neuropathologist who also developed the first department of neuropathology in the United States at Yale University during the 1930s. He also developed numerous experimental brain tumor models in mice using the chemical carcinogen, 20-methylcholanthrene. Many of these mouse tumors had histological characteristics similar to those seen in common human brain tumors (26).

Dr. Zimmerman and his associate Dr. Carl Sutton sent me several live mice with brain tumor that they previously classified as ependymoblastoma (EPEN) (26, 27). This tumor was produced originally from methylcholanthrene implantation into the cerebral ventricle of a mouse of the C57BL/6 inbred strain. Ependymal cells line the cerebral ventricles in the brain and are thought to be the origin of EPEN, a type of brain tumor. I also used Zimmerman's procedure to produce a group of brain tumors in the same mouse strain (28). The growth characteristics of one of my tumors, CT-2A (described above), were similar to that of Zimmerman's astrocytoma in expressing florid vascularization (angiogenesis) and rapid growth. This tumor arose from methylcholanthrene implantation into cerebral cortex.

The appearance and growth characteristics of the CT-2A differ markedly from that of EPEN (Fig. 3.3). In contrast to the CT-2A tumor, the EPEN tumor has fewer vessels and grows much slower than the CT-2A. In cell culture, the EPEN cells grow as cohesive islands, whereas the CT-2A cells grow as a noncohesive monolayer (Fig. 3.4). In addition to the striking differences in growth and morphology, the EPEN and CT-2A tumors also differed markedly in the composition of their

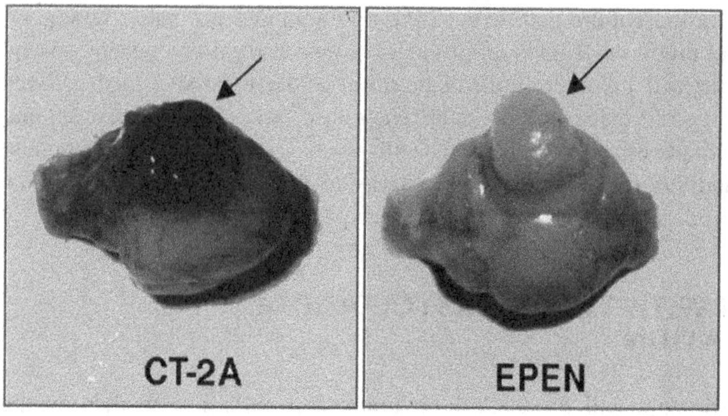

Figure 3.3 Gross morphology of the EPEN and CT-2A brain tumors growing in the cerebrum of syngeneic C57BL/6J mice. The EPEN tumor grows as a solid, cohesive, and nonhemorrhagic tissue. The CT-2A tumor grows as a soft, noncohesive, and highly hemorrhagic tissue (28). The CT-2A tumor grows significantly faster than the EPEN tumor (30). Despite these and other morphological biochemical differences, the tumors are histologically similar and are classified as poorly differentiated astrocytomas (28). See color insert.

gangliosides (28, 29). Gangliosides are a family of cell-surface glycolipids. Ganglioside GM3-NeuAc was the major ganglioside synthesized by the EPEN cells, whereas the CT-2A cells synthesized several complex gangliosides, but synthesized very little GM3 (Fig. 3.5 (30)). Considered together, these findings clearly show that the EPEN and CT-2A tumors differ strikingly in gross appearance and ganglioside biochemistry.

To further determine if the striking differences in appearance, growth rate, and ganglioside biochemistry were associated with differences in histological appearance, I had histology slides made from tissue sections of each tumor. The tumors were grown both in the brain and subcutaneously on the flank of the host C57BL mice. The histology slides were made in the Neuropathology Department, Yale University. Considering the many morphological and biochemical differences between the EPEN and CT-2A tumors, I was surprised to hear from Dr. Jung H. Kim, Yale's chief neuropathologist, at that time that the histological appearance of the two tumors was very similar (Fig. 3.6). Moreover, Dr. Kim suggested that the two tumors could be classified as soft tissue sarcomas, a type of muscle or connective tissue tumor. I was somewhat puzzled by this classification since both tumors were initiated in the central nervous system and should be of neural cell origin.

I contacted Dr. Zimmerman following Dr. Kim's classification of the two tumors. I wanted to know from Dr. Zimmerman himself whether he was certain about the classification of his EPEN. He told me he was absolutely certain about the classification and suggested that I send him the same histology slides that Dr. Kim evaluated. Consequently, I mailed the same histology slides of the two tumors to

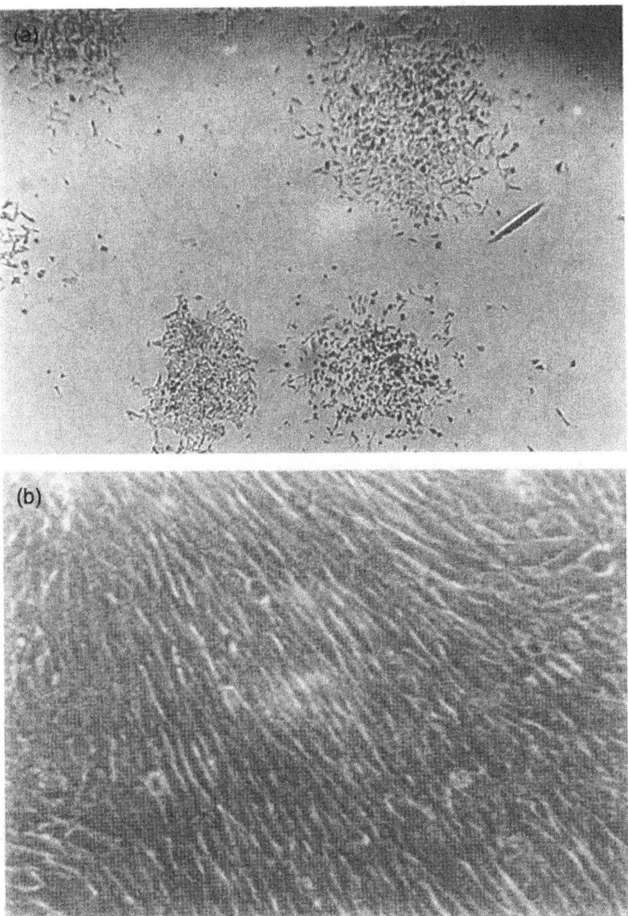

Figure 3.4 In vitro growth characteristics of the EPEN and CT-2A brain tumors. The (a) EPEN grows as clumps or islands, whereas the (b) CT-2A grows as a diffuse monolayer (30).

Dr. Zimmerman at Montiforie Hospital. After careful evaluation of these slides, Dr. Zimmerman was confident that the histological characteristics of the EPEN tumor remained the same and that this tumor was an EPEN. He also classified the CT-2A tumor as an astrocytoma, which was similar to the rapidly growing angiogenic astrocytomas that he had seen previously in his studies (26). I was therefore miffed that these two distinguished neuropathologists would have such different views of the same tumors.

The following year, I relayed my story to my friend the late Dr. Alan Yates, who at that time was the chief of neuropathology at The Ohio State University. Alan mentioned that disagreements in brain tumor classification were common among neuropathologists. He asked me if he could take a look at the slides of these two tumors. I sent him the same slides that Drs. Zimmerman and Kim had evaluated.

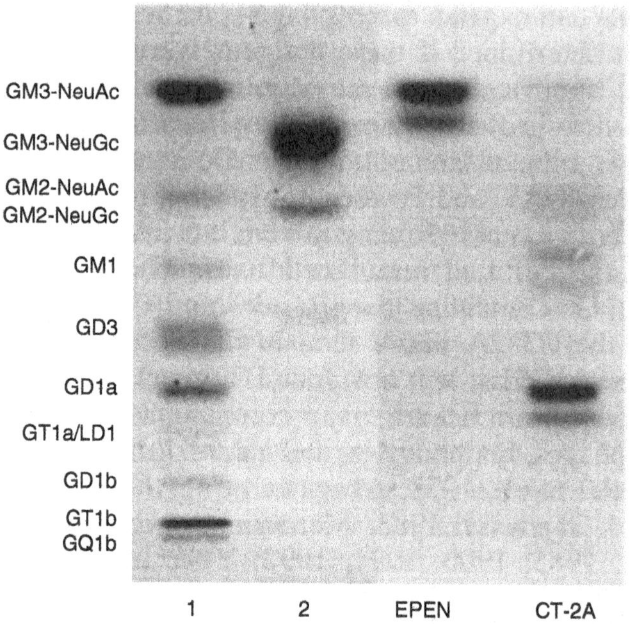

Figure 3.5 Autoradiogram of a high-performance thin-layer plate showing synthesized gangliosides in the cultured EPEN and CT-2A tumor cells.

As Dr. Kim, Alan found no real differences in the histology of the two tumors, but classified both tumors as poorly differentiated anaplastic astrocytomas (28). He was unable to confirm Zimmerman's classification of ependymoblastoma for the EPEN tumor but was fairly certain that the tumors were not sarcomas. Alan's classification of the tumors left me even more confused. How was it possible that three distinguished neuropathologists could have such different opinions on the cellular classification of the same mouse brain tumors?

I also discussed my dilemma regarding the ambiguous classification of these mouse brain tumors with Dr. Albee Messing. Albee is a neuropathologist working at the University of Wisconsin Medical School, Madison. Albee mentioned to me that he was considered the best in his group at classifying brain tumors of ambiguous cellular origin. So, I sent Albee the same tumor histology slides previously evaluated by Drs. Kim, Zimmerman, and Yates. After careful evaluation, Albee also considered both tumors as the same, but classified them as PNETs, that is, primitive neuroectodermal tumors.

Albee Messing's classification of these tumors confused me even more. How was it possible that all of these distinguished neuropathologists could come to such different opinions on the cellular origin of these mouse brain tumors? There was no question in my mind that the two tumors were strikingly different from each other in growth characteristics and ganglioside biochemistry. Anyone can see that these two tumors are different from each other (Figs. 3.3–3.5). How was it possible

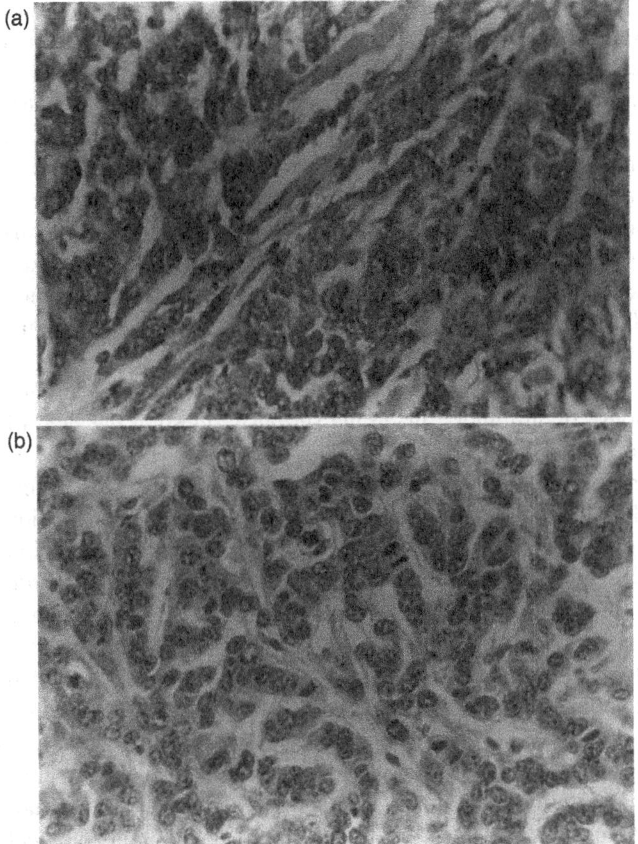

Figure 3.6 Histological appearance of 20-MC-induced brain tumors growing on the flank of the C57BL/6J mice. All of the experimental tumors were similar in histological appearance (28). The vascularity, growth rate, and ganglioside composition of the tumor in shown (a) (CBT-1) was similar to that of the EPEN tumor. The vascularity, growth rate, and ganglioside composition of the tumor in shown (b) (CBT-4) was similar to that of the CT-2A tumor. The histological appearance of the tumors was also similar whether grown subcutaneously on the flank or orthotopically in the brain (31). The findings indicate that the histological appearance of these tumors is not indicative of the tumor growth characteristics or ganglioside biochemistry.

that the histological features of these tumors could appear so similar, but other biological and biochemical characteristics appear so different?

Several years later, I discussed my experience with the classification of these mouse brain tumors with the late Dr. Sanford Palay. Sandy had joined our Boston College Biology Department as a distinguished professor in residence in 1994 after he retired as Chair of the Neuroanatomy Department at the Harvard Medical School. Sandy was a member of the National Academy of Sciences and served for many years as Editor-in-Chief of the Journal of Comparative Neurology. As

Sandy was widely recognized as one of the nation's leading experts in the field of neurocytology, I felt confident that Sandy might provide insight into this dilemma of brain tumor classification.

Sandy told me that he once tried to help some neuropathologists in the classification of brain tumors, but was unsuccessful. He told me that it was almost impossible to be certain of the cell origin of most brain tumors because the growing tumors caused major abnormalities in cytoarchitecture of the microenvironment. According to Sandy, the abnormalities in cytoarchitecture made tumor cell identification ambiguous at best. "If your assessment of this situation is correct", I asked Sandy, "how is it possible that so many neuropathologists can make such quick decisions on the classification of brain tumors?" Sandy's reply to me was "I don't know".

Sandy's reply, together with my dilemma in trying to get a histological classification of the two mouse brain tumors, made me seriously question the field of brain tumor classification. I should not have been surprised, however, as diagnosis of most cancers rests primarily on the subjective impressions of pathologists (referenced in Ref. (32)). While information on brain tumor classification might provide some insight into the tumor origin, it is not clear how tumor cell classification will influence therapy. Support for my contention comes from findings that little progress has been made in brain cancer management in more than 50 years despite extensive studies on brain tumor classification.

I consider the biological behavior of the tumor cells as more important than what they are called. That tumor cell growth behavior is more important than tumor cell classification was also the view held by the German neuropathologist, H. J. Scherer, during the early part of the last century (33). He clearly defined a number of growth behaviors seen in malignant tumors as "secondary structures." These structures were predictive of patient prognosis and were independent of histological classification. These behaviors have since been recognized as "Scherer's structures," which can provide targets for assessing effective therapies (34, 35). Hence, good cancer models should be evaluated more for their in vivo growth behaviors than for their histological classification, which can be ambiguous at best.

PERSONAL PERSPECTIVE ON CANCER

Several major events changed my perspective on the nature of cancer. The first involved our extensive studies on the role of energy restriction in tumor growth and vascularization. The second involved our analysis of spontaneous brain tumors in the VM mice. The third involved our extensive studies on mitochondrial lipids in tumors grown both in vivo in their natural hosts and as cultured cells in vitro. It gradually became clear to me that most cancer is a singular disease of energy metabolism regardless of cellular or tissue origin. Regardless of histological appearance, all tumor cells can be killed if their energy is targeted.

My view is counter to the general view that cancer therapies should be individualized based on gene signatures (36). My perspective is similar to that of Otto

Warburg, who originally proposed that all cancer is a disease of respiration. Moreover, we found that many metastatic cancers share multiple properties with cells of myeloid origin (17). These are cells of the immune system such as macrophages and leukocytes. Macrophages and leukocytes are already mesenchymal cells genetically programmed to enter and exit tissues and to survive in hypoxic environments. These are hallmarks of most metastatic tumor cells. It is not necessary to view cancer as a complicated cybernetic system. I will expand my views of cancer and show how the defective energy metabolism in tumors can be exploited for cancer prevention and management.

REFERENCES

1. Bennett D. Paradigm Lost. Boston Globe; 2008.
2. Francia G, Cruz-Munoz W, Man S, Xu P, Kerbel RS. Mouse models of advanced spontaneous metastasis for experimental therapeutics. Nat Rev. 2011;11:135–41.
3. Lazebnik Y. What are the hallmarks of cancer?. Nat Rev. 2010;10:232–3.
4. Peterson JK, Houghton PJ. Integrating pharmacology and in vivo cancer models in preclinical and clinical drug development. Eur J Cancer. 2004;40:837–44.
5. Kim IS, Baek SH. Mouse models for breast cancer metastasis. Biochem Biophys Res Commun. 2010;394:443–7.
6. Khanna C, Hunter K. Modeling metastasis in vivo. Carcinogenesis. 2005;26:513–23.
7. Fidler IJ. The pathogenesis of cancer metastasis: the 'seed and soil' hypothesis revisited. Nat Rev. 2003;3:453–8.
8. Sontheimer H. A role for glutamate in growth and invasion of primary brain tumors. J Neurochem. 2008;105:287–95.
9. Ecsedy JA, Holthaus KA, Yohe HC, Seyfried TN. Expression of mouse sialic acid on gangliosides of a human glioma grown as a xenograft in SCID mice. J Neurochem. 1999;73:254–9.
10. Chou HH, Takematsu H, Diaz S, Iber J, Nickerson E, Wright KL, et al. A mutation in human CMP-sialic acid hydroxylase occurred after the Homo-Pan divergence. Proc Natl Acad Sci USA. 1998;95:11751–6.
11. Martin MJ, Muotri A, Gage F, Varki A. Human embryonic stem cells express an immunogenic nonhuman sialic acid. Nat Med. 2005;11:228–32.
12. Mahoney LB, Denny CA, Seyfried TN. Caloric restriction in C57BL/6J mice mimics therapeutic fasting in humans. Lipids Health Dis. 2006;5:13.
13. Mukherjee P, Abate LE, Seyfried TN. Antiangiogenic and proapoptotic effects of dietary restriction on experimental mouse and human brain tumors. Clin Cancer Res. 2004;10:5622–9.
14. Chaparro RJ, Konigshofer Y, Beilhack GF, Shizuru JA, McDevitt HO, Chien YH. Nonobese diabetic mice express aspects of both type 1 and type 2 diabetes. Proc Natl Acad Sci USA. 2006;103:12475–80.
15. Bacac M, Stamenkovic I. Metastatic cancer cell. Annu Rev Pathol. 2008;3:221–47.
16. Hanahan D, Weinberg RA. The hallmarks of cancer. Cell. 2000;100:57–70.
17. Huysentruyt LC, Seyfried TN. Perspectives on the mesenchymal origin of metastatic cancer. Cancer Metastasis Rev. 2010;29:695–707.
18. Piaskowski S, Bienkowski M, Stoczynska-Fidelus E, Stawski R, Sieruta M, Szybka M, et al. Glioma cells showing IDH1 mutation cannot be propagated in standard cell culture conditions. Br J Cancer. 2011;104:968–70.
19. Huysentruyt LC, Mukherjee P, Banerjee D, Shelton LM, Seyfried TN. Metastatic cancer cells with macrophage properties: evidence from a new murine tumor model. Int J Cancer. 2008;123:73–84.

20. Shelton LM, Mukherjee P, Huysentruyt LC, Urits I, Rosenberg JA, Seyfried TN. A novel pre-clinical in vivo mouse model for malignant brain tumor growth and invasion. J Neurooncol. 2010;99:165–76.
21. Kerbel RS, Lagarde AE, Dennis JW, Donaghue TP. Spontaneous fusion in vivo between normal host and tumor cells: possible contribution to tumor progression and metastasis studied with a lectin-resistant mutant tumor. Mol Cell Biol. 1983;3:523–38.
22. Kerbel RS, Twiddy RR, Robertson DM. Induction of a tumor with greatly increased metastatic growth potential by injection of cells from a low-metastatic H-2 heterozygous tumor cell line into an H-2 incompatible parental strain. Int J Cancer. 1978;22:583–94.
23. Huysentruyt LC, Shelton LM, Seyfried TN. Influence of methotrexate and cisplatin on tumor progression and survival in the VM mouse model of systemic metastatic cancer. Int J Cancer. 2010;126:65–72.
24. Shelton LM, Huysentruyt LC, Seyfried TN. Glutamine targeting inhibits systemic metastasis in the VM-M3 murine tumor model. Int J Cancer. 2010;127:2478–85.
25. Huysentruyt LC, Akgoc Z, Seyfried TN. Hypothesis: are neoplastic macrophages/microglia present in glioblastoma multiforme?. ASN Neuro. 2011. Forthcoming.
26. Zimmerman HM, Arnold H. Experimental brain tumors: I. Tumors produced with methylcholanthrene. Cancer Res. 1941;1:919–38.
27. Rubin R, Ames RP, Sutton CH, Zimmerman HM. Virus-like particles in murine ependymoblastoma. J Neuropathol Exp Neurol. 1969;28:371–87.
28. Seyfried TN, el-Abbadi M, Roy ML. Ganglioside distribution in murine neural tumors. Mol Chem Neuropathol. 1992;17:147–67.
29. el-Abbadi M, Seyfried TN. Influence of growth environment on the ganglioside composition of an experimental mouse brain tumor. Mol Chem Neuropathol. 1994;21:273–85.
30. Bai H, Seyfried TN. Influence of ganglioside GM3 and high density lipoprotein on the cohesion of mouse brain tumor cells. J Lipid Res. 1997;38:160–72.
31. Seyfried TN, Yu RK, Saito M, Albert M. Ganglioside composition of an experimental mouse brain tumor. Cancer Res. 1987;47:3538–42.
32. Sonnenschein C, Soto AM. Somatic mutation theory of carcinogenesis: why it should be dropped and replaced. Mol Carcinog. 2000;29:205–11.
33. Scherer HJ. A critical review: The pathology of cerebral gliomas. J Neurol Neuropsychiat. 1940;3:147–77.
34. Rubinstein LJ. Tumors of the central nervous system. Washington (DC): Armed Forces Institute of Pathology; 1972.
35. Zagzag D, Esencay M, Mendez O, Yee H, Smirnova I, Huang Y, et al. Hypoxia- and vascular endothelial growth factor-induced stromal cell-derived factor-1alpha/CXCR4 expression in glioblastomas: one plausible explanation of Scherer's structures. Am J Pathol. 2008;173:545–60.
36. Kolata G. Add patience to a leap of faith to discover cancer signatures. New York Times. 2011 July 18.

Chapter 4

Energetics of Normal Cells and Cancer Cells

In order for cells to remain viable and to perform their genetically programmed functions they must produce energy. Most of this energy is commonly stored in the terminal γ and β phosphates of adenosine triphosphate (ATP), and is released during the hydrolysis of their phosphoanhydride bonds (Fig. 4.1). This energy is generally referred to as *the free energy of activation* or *ATP hydrolysis* (1–4). The standard energy of ATP hydrolysis under physiological conditions is known as $\Delta G'_{ATP}$ and is tightly regulated in all cells between −53 and −60 kJ/mol (5). G is the Gibbs free energy, Δ is the difference between two energy states, and prime represents the activated state (1, 6).

J. Willard Gibbs was a nineteenth-century mathematical physicist who first defined the principles of statistical mechanics upon which the laws of thermodynamics are based (7). $\Delta G'_{ATP}$ differs from the $\Delta G'^{\circ}_{ATP}$, which is usually described in textbooks. The $\Delta G'^{\circ}_{ATP}$ represents the free energy of activation under closed conditions where temperature, gases, and solutes are all standardized. The $\Delta G'_{ATP}$ relates more to the situation in open systems, that is, the situation in cells and tissues (2, 7, 8). Negative values for ΔG indicate that energy is released in the conversion of reactants to products. Although the free energy of ATP hydrolysis is used to power nearly all cellular activities, the majority of energy in any given cell is used to power ionic membrane pumps (1, 2, 9–11). It is the mundane membrane pumps that require constant energy to maintain viability.

METABOLIC HOMEOSTASIS

Homeostasis is the tendency of biological systems to maintain relatively stable conditions in their internal environments. Each cell and each organ contributes to the overall homeostasis of the organism. This is especially important for humans

Cancer as a Metabolic Disease: On the Origin, Management and Prevention of Cancer, First Edition.
Thomas Seyfried.

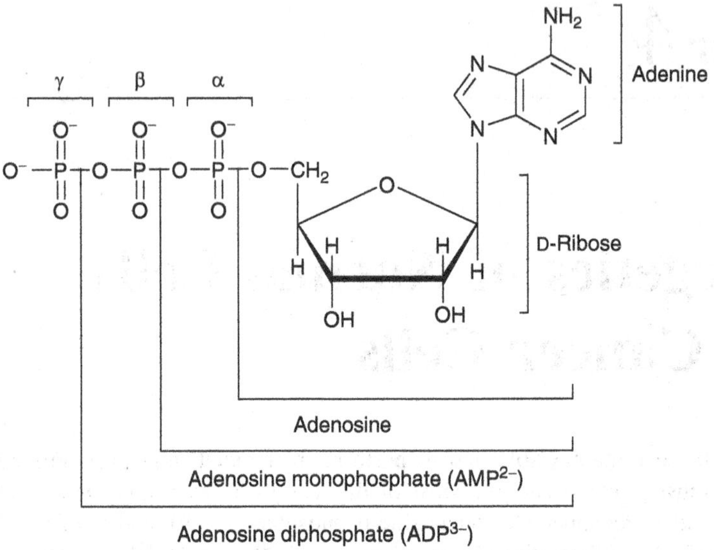

Figure 4.1 Structural formula of adenosine triphosphate (ATP) at pH 7.0. The three phosphate groups are identified by Greek letters α, β, and γ. The γ- and β-phosphate groups are linked through phosphoanhydride bonds and their hydrolysis yields a large negative $\Delta G^{\circ\prime}$, whereas the α-phosphate linked by a phosphate ester bond has a much lower negative $\Delta G^{\circ\prime}$. *In vivo* most ATP is chelated to magnesium ions (Mg· ATP^{2-}). *Source*: Reprinted with permission from Reference 6.

that follow a feast/fast schedule of nutrient supply (6). Metabolic homeostasis within cells is dependent to a large extent on the energy supply to the membrane pumps. Hormones such as insulin and glucagon can regulate global system energy homeostasis in order to maintain steady energy balance within the cells of each organ. If the energy to the cellular pumps is interrupted, the cell begins to swell. Swelling results from increased Na^+ and Ca^{2+} concentrations and decreased K^+ concentration. Because the inside of the cell is more negative than the outside, Na^+ and Ca^{2+} will naturally move down their concentration gradient from outside to inside. On the other hand, K^+, which is more concentrated inside than outside, will flow down its concentration gradient. Most cell functions are linked either directly or indirectly to the plasma membrane potential and to the $Na^+/K^+/Ca^{2+}$ gradients. Ready availability of ATP to the pumps maintains these ionic gradients. Global cellular dysfunction and ultimately organ and systems failure will arise if energy flow to the pumps is disrupted.

There are several sources of ATP synthesis that can be used to maintain membrane potentials. The mitochondria produce most of the energy in normal mammalian cells. The general structure of a mitochondrion with associated functions is shown in Figure 4.2. Other images of mitochondrial are presented in Chapter 5. In cells with functional mitochondria, ATP is derived mostly from oxidative phosphorylation (OxPhos) where approximately 89% of the total cellular energy is produced (about 32/36 total ATP molecules during the complete oxidation

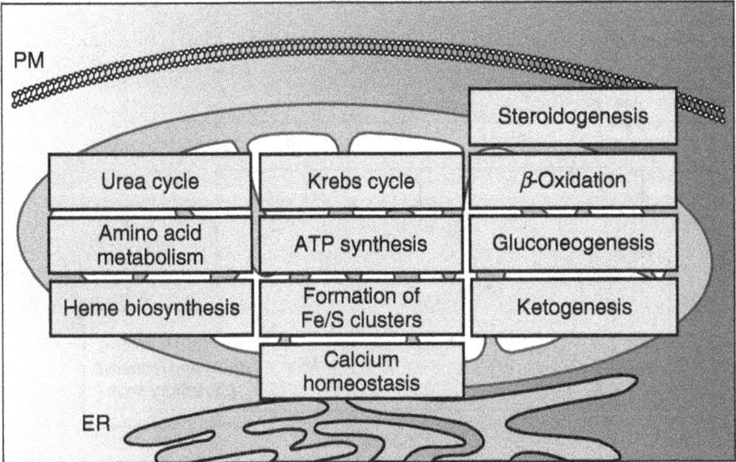

Figure 4.2 Mitochondria in cell life. Through OxPhos, mitochondria produce the bulk of intracellular ATP, and hence are considered the cell's "power plants." In addition, mitochondria regulate Ca^{2+} homeostasis and modulate several other metabolic circuitries including the Krebs cycle, the urea cycle, gluconeogenesis, ketogenesis, heme biosynthesis, fatty acid β-oxidation, steroidogenesis, metabolism of certain amino acids, and the formation of iron/sulfur clusters. ER, endoplasmic reticulum; PM, plasma membrane. *Source*: Reprinted with permission (21). To see this figure in color please go to ftp://ftp.wiley.com/public/sci_tech_med/cancer_metabolic_disease.

of glucose) (Fig. 4.3). This value can differ among different cells depending on which shuttle systems are used in the transport of cytoplasmic reducing equivalents (nicotinamide adenine dinucleotide (reduced form), NADH) from the cytoplasm to the mitochondria (6) (Table 4.1). These shuttles include the malate–aspartate shuttle, the glycerol–phosphate shuttle, and the malate–citrate shuttle. These shuttles are operational in tumor cells, but their activity can differ among the different types of tumor cells (12–19). Under OxPhos, ATP synthesis in normal cells is coupled to electron flow across the inner mitochondrial membrane through a chemiosmotic molecular mechanism (Fig. 4.4) (20)

The F_0F_1-ATPase, sometimes referred to as *complex V*, generates ATP through condensation of ADP and inorganic phosphate Pi (Fig. 4.4). Oxygen becomes the final acceptor of electrons with water as the end product. The efficiency of the process is strongly dependent on the lipid composition of the inner mitochondrial membrane where cardiolipin is a major component (25) (Chapter 5). The proton motive gradient or force of the inner mitochondrial membrane, symbolized as $\Delta\Psi m$, is required not only for ATP synthesis but also for transport functions including those for nucleotides, amino acids, Ca^{2+}, and other metabolites needed for normal mitochondrial function (6). The maintenance of this gradient is essential for normal mitochondrial function and ultimately cell function and life (7, 9). Galluzzi, Kroemer, and colleagues provide a more complete coverage of the multiple functions of mitochondria and discuss how these functions can be the gateway to tumorigenesis (21).

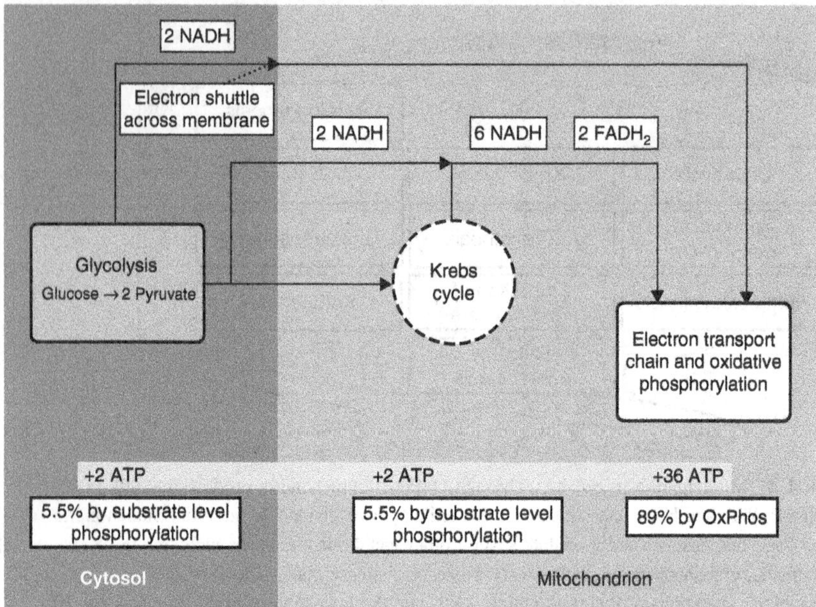

Figure 4.3 Cellular energy production through glycolysis, TCA cycle, substrate-level phosphorylation, and OxPhos. The majority of the cellular energy produced in normal cells is through OxPhos (about 89%). Glycolysis and TCA cycle, substrate-level phosphorylation contribute only about 11% of cellular energy. OxPhos energy production is less in tumor cells than in normal cells. Enhanced glycolysis and TCA cycle, substrate-level phosphorylation can compensate for insufficient OxPhos. Shuttle systems can deliver additional reducing equivalents (electrons) to the mitochondrial for OxPhos (Fig. 4.12). See color insert.

Table 4.1 Energy-Yielding Reactions in the Complete Oxidation of Glucose

Reaction	Net moles of ATP produced per mole of glucose
Glycolysis (phosphoglycerate kinase, pyruvate kinase; two ATPs are expended)	2
NADH shuttle	
Malate–aspartate shuttle	4(6)
Pyruvate dehydrogenase (NADH)	6
Succinyl CoA synthetase (ATP or GTP)	2
Succinate dehydrogenase (Succinate → fumarate + $FADH_2$)	4
Other TCA cycle reactions (isocitrate → α-ketoglutarate, α-ketoglutarate → succinyl-CoA, malate → oxaloacetate; total of 3 NADH produced)	18
Total	36(38)

Source: Modified from Reference 6.

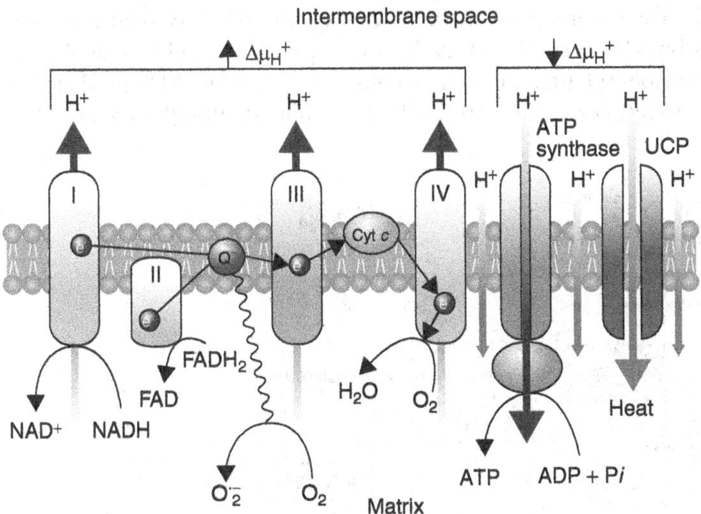

Figure 4.4 Mitochondrial electron transport chain (ETC) and the origin of chemiosmosis. Electron donors from the TCA cycle [NADH and flavin adenine dinucleotide (reduced form) ($FADH_2$)] generate a high mitochondrial membrane potential ($\Delta\mu H^+$) by pumping protons across the mitochondrial inner membrane at complexes I, III, and IV (22). This pumping generates a proton motive gradient that provides the driving force for proton influx through the F_1F_0-ATP synthase (ATP synthase). Proton influx is coupled to ATP-synthase-catalyzed phosphorylation of ADP to form ATP. At a standard metabolic rate, a fraction of the protons pumped out across the ETC can leak back into the mitochondrial matrix without synthesizing ATP (9). Proton leak effectively uncouples respiration from phosphorylation. Proton leak or back-decay (smaller thin arrow) is greater in mitochondria of tumor cells than in mitochondria of normal cells (23). Under hypoxic conditions, the ATP synthase works in reverse as an ATPase. This action couples ATP hydrolysis to proton pumping from the matrix into the intermembrane space (9). Matrix proton accumulation in hypoxia could arise from reversal of complex I or from back leak. Reverse operation of the ATP synthase is done to protect the mitochondria and to maintain the $\Delta\mu H^+$. Roberto Flores and I believe that the ATP synthase also works in reverse in highly glycolytic tumor cells under normoxia like it would in hypoxia. Succinate accumulation under hypoxia supports our contention that electrons are transferred from complex I to complex II (9). The gradient energy can also be dissipated as heat if the protons pass through an uncoupling protein (UCP) or from excessive back leak. Uncoupling proteins can be overexpressed in some cancer cells (24). Excessive heat is produced in some cancers (Chapter 5). The figure also shows the origin of free radical formation at the coenzyme *Q* couple. *Source*: Modified from Reference 22. See color insert.

Besides OxPhos, approximately 11% (4/36 total ATP molecules) of the total cellular energy is produced through substrate-level phosphorylation (Fig. 4.3). Substrate-level phosphorylation involves the transfer of a free phosphate to ADP from a metabolic substrate to form ATP. Two major metabolic pathways can produce ATP through substrate-level phosphorylation in mammalian cells and tissues. The first involves the "pay off" part of the Embden–Myerhoff glycolytic pathway in the cytosol where phosphate groups are transferred from the organic molecules, 1,3-bisphosphoglycerate and phosphoenolpyruvate (PEP), to ADP with formation of ATP (Fig. 4.5). The second pathway involves the succinyl-CoA synthetase reaction of the tricarboxylic acid (TCA) cycle (Fig. 4.6). The synthesis of ATP by substrate-level phosphorylation in normal cells can augment ATP produced by oxidative phosphorylation by about 10% (21). Most importantly, the succinyl-CoA

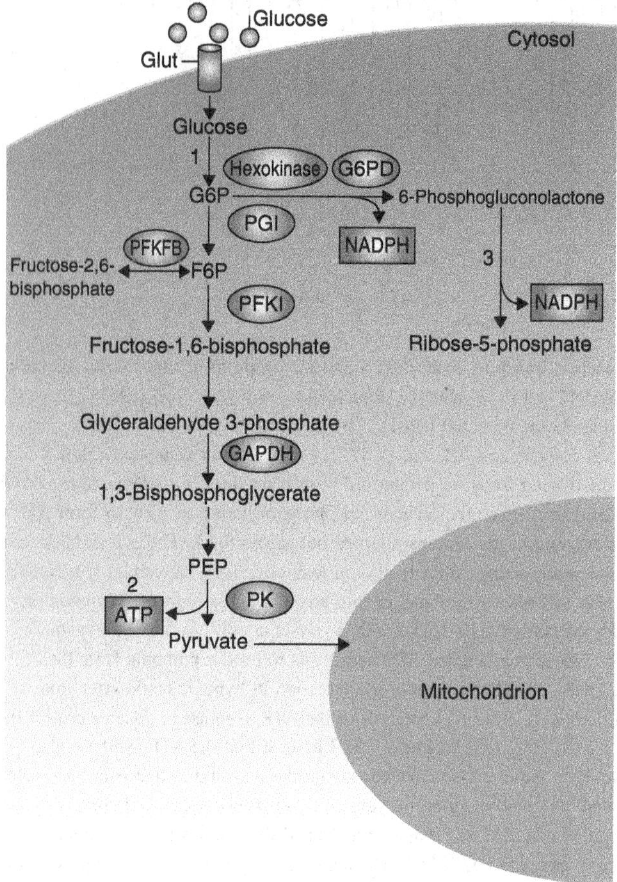

Figure 4.5 Embden–Myerhoff glycolytic pathway converts glucose to pyruvate. This pathway can provide energy in the presence or absence of oxygen, but becomes dominant under hypoxia when proton transfer no longer occurs at the various ETC complexes (Fig. 4.3). The continued production of lactate in the presence of oxygen is referred to as the Warburg effect or aerobic glycolysis. Ribose-5-phosphate used for nucleotide synthesis is synthesized through the pentose–phosphate pathway (PPP). The PPP is a source of NADPH for synthesis of glutathione and lipids. G6P, glucose-6-phosphate; G6PD, glucose-6-phosphate dehydrogenase; PGI, phosphoglucoisomerase; F6P, fructose-6-phosphate; PFKFB, 6-phosphofructo-2-kinase/fructose-2,6-biphosphatase; PFK1, phosphofructokinase 1; GAPDH, glyceraldehyde-3-phosphate dehydrogenase; PEP, phosphoenolpyruvate; PK, pyruvate kinase; NADPH, nicotinamide adenine dinucleotide phosphate; ATP, adenosine triphosphate. *Source*: Modified from Reference 31. To see this figure in color please go to ftp://ftp.wiley.com/public/sci_tech_med/cancer_metabolic_disease.

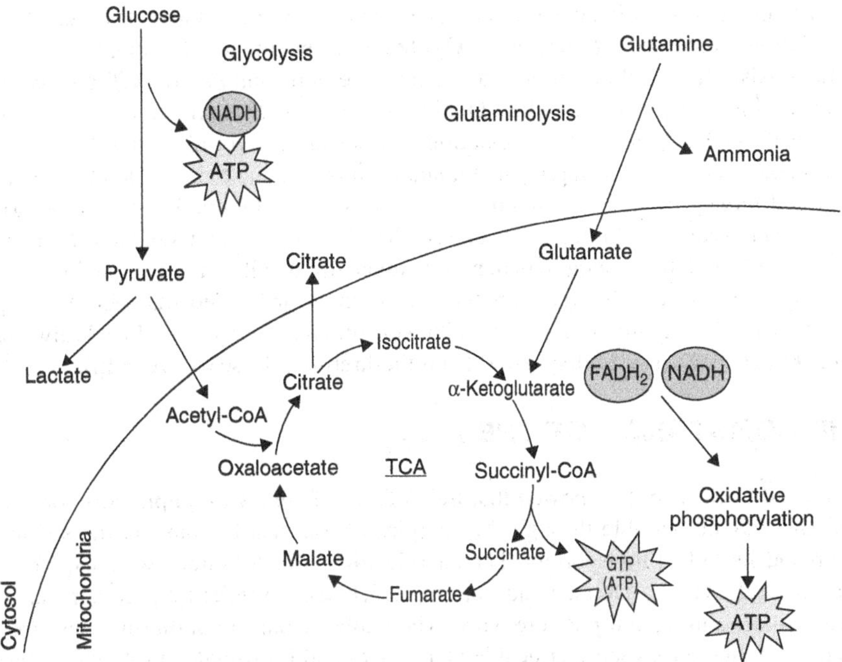

Figure 4.6 Metabolic pathways. Illustrated are the tricarboxylic acid (TCA) cycle, and the glucose and glutamine utilizing metabolic pathways of glycolysis and glutaminolysis, respectively. Reactions of the TCA cycle take place in the mitochondrial matrix, while reactions of the ETC take place in the inner mitochondrial membrane (Fig.4.4). ETC, electron transport chain, $FADH_2$, flavin adenine dinucleotide (reduced form); GTP, guanosine triphosphate; NADH, nicotinamide adenine dinucleotide (reduced form). *Source*: Modified from Reference 65. To see this figure in color please go to ftp://ftp.wiley.com/public/sci_tech_med/cancer_metabolic_disease.

synthetase reaction can provide more energy under anaerobic conditions than under aerobic conditions (26–30). Few investigators in the cancer field have discussed the role of the succinyl-CoA synthetase reaction for the nonoxidative energy production in tumor cells. We suggest that this pathway could also be a major source of energy in metastatic mouse cells (Chapter 8).

Under normal physiological conditions, two ATP molecules are produced from glycolysis in the cytoplasm and two from the succinyl-CoA synthetase reaction in the mitochondrial matrix (Fig. 4.3). In contrast to OxPhos, which involves oxygen and a membrane-regulated proton gradient, oxygen is not a requirement for ATP synthesis through substrate-level phosphorylations. A proton motive gradient could still operate, however, through a reverse action of the F1F0-ATPase. Stepien and colleagues have showed how the mitochondrial attached hexokinase II isoform provides glycolytic ATP to the mitochondria in order to maintain the proton motive gradient (32). This is important because it explains, in part, how tumor cells can produce energy and remain viable in hypoxia despite damage to mitochondrial structure and function.

The number of ATP molecules produced from TCA cycle, substrate-level phosphorylation would need to increase if OxPhos were insufficient to maintain energy homeostasis. This would be similar to the increase in the number of ATP molecules produced through glycolysis when OxPhos is reduced. Nonoxidative energy production through amino acid fermentation and substrate-level phosphorylation has been documented in developing mammalian embryos, in diving animals, and in heart and kidney tissue under hypoxia (27–29, 33–35). Figure 4.3 shows the origin of ATP synthesis through glycolysis, the TCA cycle, and OxPhos. Readers are also referred to general biochemistry texts for details of these biochemical pathways (36). I can also recommend a YouTube "rap" video that puts to song the key aspects of cellular energy metabolism (http://www.npr.org/blogs/krulwich/2011/09/14/140428189/lord-save-me-from-the-krebs-cycle?sc=fb&cc=fp).

THE CONSTANCY OF THE $\Delta G'_{\mathrm{ATP}}$

Veech and coworkers (4) showed that the $\Delta' G_{\mathrm{ATP}}$ of cells was empirically formalized and was measurable through the energies of ion distributions via the sodium pump and its linked transporters. The energies of ion distributions were explained in terms of the Gibbs–Donnan equilibrium, which was essential for producing electrical, concentration, and pressure work. The Gibbs–Donnan equilibrium describes the flow of ions across semipermeable membranes and is estimated using the Nernst equation. The Nernst equation can link the Gibbs free energy to the electric charge across a membrane.

A remarkable finding was the similarity of the $\Delta G'_{\mathrm{ATP}}$ among cells with widely differing resting membrane potentials and mechanisms of energy production. For example, the $\Delta G'_{\mathrm{ATP}}$ in heart, liver, and erythrocytes was approximately −56 kJ/mol despite having very different electrical potentials of −86, −56, and −6 mV, respectively (4). Moreover, energy production in the heart and liver, which contain many mitochondria, is largely through OxPhos, whereas energy production in the erythrocyte, which contains no nucleus or mitochondria, is entirely through glycolysis. Despite the profound differences in resting membrane potentials and in mechanisms of energy production among these disparate cell types, they all express a similar free energy of ATP hydrolysis. These observations suggest that the balance of energy consumption and production is independent of the energy source and the amount of the total ATP produced.

The constancy of the $\Delta G'_{\mathrm{ATP}}$ of approximately−56 kJ/mol is fundamental to cellular energy homeostasis and its relationship to cancer cell energy metabolism is critical. Using a phrase from T. S. Eliot's poem, *Buirnt Norton*, Veech refers to this energy value as the "the still point of the turning world" (personal communication). Why is this particular free energy of ATP hydrolysis so important for cell physiology remains unclear (9). The maintenance of the $\Delta G'_{\mathrm{ATP}}$ is the end point of both genetic and metabolic processes and any disturbance in this energy balance will compromise cell function and viability (2).

It is important to mention, however, that precise measurement of the $\Delta G'_{\mathrm{ATP}}$ within any given tumor would be challenging, as differences occur in the pH of the microenvironment and in the viability of cells within the tumor (9, 37, 38). All of these dynamic changes would reduce the accuracy of $\Delta G'_{ATP}$ measurements taken from actively growing solid tumors where entropy is accelerated. Nevertheless, it is apparent from carefully conducted proton nuclear magnetic resonance studies that normal cells have the capacity to balance energy use with energy production through both substrate-level phosphorylation and respiration to achieve a stable free energy of ATP hydrolysis (35, 38). Compared to the regulated energy homeostasis in normal cells, energy dysregulation is the hallmark of tumor cells.

Cells can die from either too little or too much energy. Too little energy leads to cell death by either necrotic or apoptotic mechanisms. Overproduction of ATP, a polyanionic Donnan active material, disrupts the Gibbs–Donnan equilibrium, alters the function of membrane pumps, and inhibits respiration and viability (4). To maintain cellular energy balance, the mitochondrial F_0F_1-ATPase can sometimes run in reverse (hydrolyzing ATP) (9, 32, 39) (Fig. 4.4). Additionally, some tumor cells release ATP into the extracellular milieu through the action of the *p*-glycoprotein, which is linked to glycolysis and is often overexpressed in tumors (40–42).

If OxPhos becomes compromised, energy production through substrate-level phosphorylation must be increased in order to maintain a stable free energy of ATP hydrolysis and cell viability (10, 38, 43). Alternatively, energy expenditure can be reduced to offset reduced energy production (9). Acute damage to respiratory function usually causes apoptotic or necrotic cell death due to membrane pump energy depletion. However, energy through substrate-level phosphorylation can gradually compensate for minor damage to OxPhos-derived energy over protracted periods. As tumors rarely occur following acute injury to respiration, considerable time is required for nonoxidative energy metabolism to displace OxPhos as the dominant energy generator in the cell.

It is important to recognize that prolonged reliance on substrate-level phosphorylation for energy production in previously normally respiring cells produces genome instability, disorder, and increased proliferation, that is, the hallmarks of cancer (33, 44–46). Entropy refers to the degree of disorder in systems and is the foundation of the second law of thermodynamics (1, 7). Szent-Gyorgyi described cancer as a state of increased entropy, where randomness and disorder predominate (46). Protracted OxPhos insufficiency coupled with persistent compensatory fermentation increases entropy. Cells that do not increase fermentation energy to compensate for insufficient OxPhos simply die off and never become neoplastic. Adaptation to fermentation allows a cell to bypass mitochondrial-induced senescence (21, 47). Cancer arises in those cells that bypass mitochondrial-induced senescence.

ATP PRODUCTION IN NORMAL CELLS AND TUMOR CELLS

Warburg showed that the total energy production in quiescent kidney cells and liver cells was similar to that produced in proliferating ascites tumor cells (Table 4.2).

Table 4.2 Comparison of the Metabolic Quotients of Some Normal Body Cells with the Metabolic Quotients of Ascites Tumor Cells

Cells	Q_{O_2}	$Q_M^{N_1}$	$Q_{ATP}^{O_2}$	$Q_{ATP}^{N_2}$	$Q_{ATP}^{N_1} + Q_{ATP}^{N_2}$
Liver	−15	1	105	1	106
Kidney	−15	1	105	1	106
Embryo	−15	25	105	25	130
Ascites tumor cells	−7	60	49	60	109

Ascites tumor cells grow in the peritoneal cavity of mice. Warburg considered the ascites cancer cells a better preparation than tumor tissue slices, as the ascites cells are not contaminated with nonneoplastic stromal cells that are present to various degrees in the tumor tissue slices. Stromal cells are expected to have normal metabolism and might therefore dilute the magnitude of metabolic deficiency in the neoplastic cells of the tumor. This can be problematic as we found that stromal cells in the form of tumor-associated macrophages (TAMs) can contribute significantly to the total cell population of some tumors (48).

In contrast to many current studies of energy metabolism in cultured tumor cells, Warburg evaluated the ascites cells maintained in a medium, which was supplemented only with glucose and bicarbonate (33). Later studies showed, however, that normal cell respiration dramatically increased in pure serum, whereas cancer cell respiration increased only slightly. The slight respiratory increase by the cancer cells most likely reflects the upper limits of their respiratory capacity. Under physiological conditions of pH and temperature, Warburg expressed his data on energy metabolism in the ascites cells as metabolic quotients (Q) (33, 49).

The Q^{O_2} quotient reflected the amount of oxygen in cubic millimeters that 1 ml of tissue (dry weight) consumes per hour at 38°C with oxygen saturation. The $Q_M^{O_2}$ quotient reflected the amount of lactic acid produced under similar conditions in the "presence" of oxygen, whereas the $Q_M^{N_2}$ quotient reflected the amount of lactic acid produced under similar conditions in the "absence" of oxygen. On the basis of Warburg's calculations, 1 mol of O_2 consumed yields approximately 7 mol of ATP, whereas 1 mol of lactic acid produced approximately 1 mol of ATP. Although these ATP values might not be completely accurate, they indicate that more energy is produced from the complete oxidation of glucose through OxPhos than from the partial oxidation of glucose through fermentation.

It is clear from an examination of Warburg's data that the metabolic quotient for total ATP production ($Q_{ATP}^{O_2} + Q_{ATP}^{N_2}$) is similar in respiring kidney and liver cells (a value of 106) and in the ascites tumor cells (a value of 109) (Table 4.2). However, more ATP is produced (over 50%) in association with lactic acid production than with oxygen consumption in the ascites cells than in the kidney or liver cells. The energy situation in the ascites cells is more similar to that in young embryos than to that in differentiated normal cells, in which significant energy is produced through fermentation. These findings indicate that the tumor cells differ from the normal cells in the origin of the energy produced rather than in the amount of energy produced. Updated numbers have 1 mol of O_2 consumed yielding approximately 3–4 ATP (~3.5) (36). This is significant as Warburg's ascites cells were producing even less energy through OxPhos than he had originally assumed. Instead of the respiratory capacity yielding about 50% of the ATP, respiration was yielding only around 25%. I will later discuss how some of the oxygen consumed in tumor cells might not be used completely for OxPhos due to uncoupling.

The findings of Warburg are also consistent with the later studies of Donnelly and Scheffler who showed that the total ATP production is similar in respiration-deficient and in respiration-normal Chinese hamster fibroblasts (50). The respiration deficiency in these cells involved a defect in the NADH-coenzyme Q reductase. This defect significantly reduces the TCA cycle function and oxygen consumption. Although energy production through glycolysis was high in both cell lines, more energy was obtained from glutamine metabolism in the wild-type cells than in the respiration-deficient cells. It appears that the mutant cells were unable to obtain much energy from glutamine. The high glycolysis in both cell types can be due, in part, to the effects of the culture environment, which alters the composition of lipids in the inner mitochondrial membrane (51). Altered mitochondrial lipids reduce the efficiency of OxPhos, thus requiring increased energy production through substrate-level phosphorylation.

ENERGY PRODUCTION THROUGH GLUCOSE FERMENTATION

Warburg was the first to describe in detail the dependence of cancer cells on glucose and glycolysis in order to maintain viability following irreversible respiratory damage (33, 49, 52). He considered respiration and fermentation as the sole producers of energy within cells, and energy alone as the central issue of tumorigenesis. "*We need to know no more of respiration and fermentation here than that they are energy-producing reactions and that they synthesize the energy-rich adenosine triphosphate, through which the energy of respiration and fermentation is then made available for life*" (33).

Warburg considered fermentation as the formation of lactate from glucose in the absence of oxygen. This type of energy is also produced in mammalian embryos and in our muscles during strenuous exercise. Instead of entering the TCA cycle for complete oxidation, pyruvate is reduced to lactate when oxygen levels are low. Lactate fermentation generates NAD^+ as an oxidizing agent for glycolysis (Fig. 4.7). The NAD^+ can be used as an electron acceptor during the oxidation of dihydroxyacetone phosphate to 1,3-diphosphoglycerate, the reaction preceding the first substrate-level phosphorylation in glycolysis (53). Failure to regenerate cytoplasmic NAD^+ reduces energy through glycolysis, which could compromise cell viability in the absence of energy through OxPhos or TCA cycle, substrate-level phosphorylation.

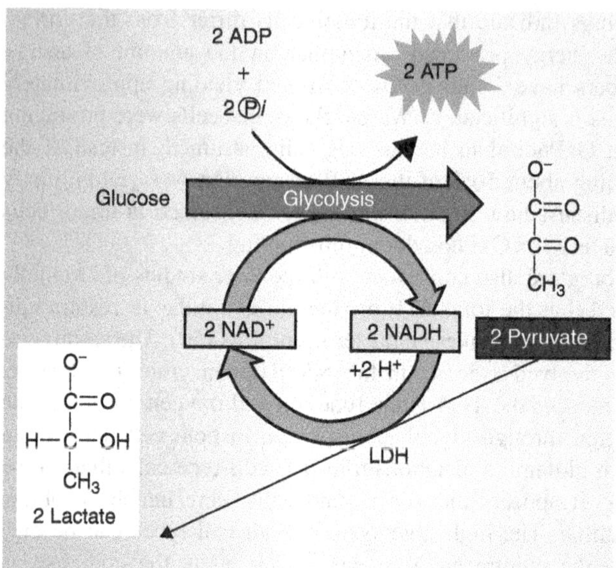

Figure 4.7 Lactic acid fermentation. Pyruvate is the end product of glycolysis and serves as an electron acceptor for oxidizing NADH back to NAD^+. The enzyme lactic acid dehydrogenase (LDH) reduces pyruvate to lactate. The NAD^+ formed can then be reused to oxidize glucose during glycolysis, which yields two net molecules of ATP by substrate-level phosphorylation during fermentation. Lactate is the common waste product formed from fermentation in mammalian cells. *Source*: Modified from Campbell p. 91 (54). To see this figure in color please go to ftp://ftp.wiley.com/public/sci_tech_med/cancer_metabolic_disease.

Lactate is basically metabolic waste from the incomplete oxidation of glucose and must be removed from the microenvironment as quickly as possible. Waste management is not only a problem for societies and organisms but also for individual cells. Most lactate enters the blood stream where it is used to synthesize glucose in the liver through what is known as the *Cori cycle*, named after its discoverers Carl and Gerty Cori. Lactate is simply excreted into the medium in cultured cells that are grown in glucose. This usually changes the color of the pH indicator dye (phenol red) from red to yellow (55). Once oxygen becomes available, glucose utilization and lactate production decreases due to the Pasteur effect, named after Louis Pasteur who first described the phenomenon.

The Pasteur effect is a common phenotype in facultative anaerobes such as yeasts and many bacteria. Facultative anaerobes ferment in the absence of O_2, but can respire in its presence. A dependence on glucose with lactate production in the presence of oxygen later became known as the Warburg effect, which is essentially aerobic glucose fermentation or the continued production of lactic acid in the presence of O_2.

Why would cancer cells continue to ferment glucose in the presence of O_2? Warburg attributed the aerobic fermentation in tumor cells to respiratory damage or respiratory insufficiency. Tumor cells grown in the presence of O_2 behave as if they were abnormal facultative anaerobes in continuing to ferment in the presence of O_2. Some tumor cells might be considered to be obligate anaerobes if they die in the presence of oxygen. My view on this phenomenon is essentially the same as Warburg's view in that respiratory damage or insufficiency underlies the behavior of the tumor cells with respect to their energy metabolism in oxygen. Cancer cells continue to ferment glucose in the presence of O_2 (aerobic glycolysis or the Warburg effect) because they cannot produce sufficient ATP through OxPhos for cellular homeostasis. The origin of the Warburg effect thus arises from damaged or insufficient respiration. I will present more evidence supporting this fact in Chapters 7 and 8 and describe in Chapters 9 and 10 how oncogene expression is needed to facilitate fermentation in response to OxPhos insufficiency.

If cancer cells could respire effectively, there would be no need for increased energy production through nonoxidative means. There are a number of concerns with the view that oncogenes upregulate fermentation in cancer cells with normal respiration. I will address these concerns in later chapters. It also appears that Warburg was unaware of possible mitochondrial amino acid fermentation in tumor cells. While O_2 exposure of tumor cells can decrease lactate production to some extent, lactate production from glucose is generally higher in tumor cells than in their normal cellular counterparts. Mitochondrial amino acid fermentation provides a possible missing metabolic link in Warburg's theory. Mitochondrial amino acid fermentation obscures the boundaries between normal respiration and fermentation and can explain much of the controversy surrounding the Warburg theory. I will discuss this concept more in Chapter 8.

Lactate accumulates as an end product of glucose fermentation. If OxPhos were normal in cancer cells, then lactate production would decrease in the presence of O_2, as pyruvate would be effectively oxidized through the TCA cycle and would no longer be available for the lactic acid dehydrogenase (LDH) reaction (Fig. 4.7). In contrast to normal cells, tumor cells continue to ferment glucose in the presence of oxygen. Cancer cells that rely more on glutamine than on glucose for energy production can produce ATP through nonoxidative processes in the mitochondria (Chapter 8). It is also important to recognize that glutamine metabolism increases ammonia in the extracellular environment. Ammonia can neutralize extracellular acidity from simultaneous glycolytic lactate production (19, 56). Caution is therefore necessary in using pH as an indicator of lactate production especially in cancer cells that use glutamine as a major fuel. We prefer to measure lactate production directly rather than use indirect methods such as changes in pH.

Normal cells are exquisitely adaptable to balancing the energy demand with energy supply (9). Tumor cells lose this ability due to mitochondrial damage or respiratory insufficiency. Aerobic fermentation (glycolysis) is considered to be the metabolic signature of cancer cells (57, 58). This phenotype is the result of insufficient respiration. There are no known highly malignant cancers to my knowledge that can produce adequate levels of ATP for cellular homeostasis through normal aerobic respiration. While respiration is not completely gone in some low malignancy cancer cells, they nevertheless produce some lactate in oxygen implying an insufficient respiratory capacity (52).

Although many tumor cells have active TCA cycles and might appear to respire, in that they consume oxygen and produce CO_2 and ATP in the mitochondria, I will present data showing that this is pseudo respiration in some cases. In other words, pseudo respiration has all the characteristics of respiration, but does not involve ATP synthesis through OxPhos. I propose that this apparent respiratory energy is derived from amino acid fermentation. Just as tumor cells ferment glucose in the presence of O_2, some tumor cells also ferment glutamine and possibly other amino acids in the presence of elevated glucose and O_2. Glucose and glutamine interact synergistically to drive tumor cell fermentation (59). Fermentation is the bioenergetic signature of tumor cells. I will address this subject more in Chapter 8.

An increase in glucose utilization with corresponding lactate production becomes necessary for tumor cell viability following respiratory injury. Warburg clearly showed this in his experiments, as did Donnelly and Scheffler in their experiments (50). Unlike most normal mammalian cells, which balance energy production to energy output, energy balance is dysregulated in cancer cells because they do not suppress ATP turnover under anoxia (9). Rather, tumor cells seem to enhance ATP turnover under hypoxia (60).

There are few topics more hotly debated or more controversial in the cancer field than the role of respiration and the Warburg effect in tumor cell energy metabolism. Some investigators contend that respiration is normal in tumor cells despite having upregulated glycolysis (61–63). I consider this possibility unlikely and will address this subject more thoroughly in Chapters 5–8. The respiratory capacity of tumor cells could also depend on the levels of glucose and glutamine available to the cells. Respiration could be greater under lower than higher glucose conditions in some tumor cells especially if glutamine is also available.

Some investigators suggest that lactic acid can be directly used as a fuel for tumor cells or normal astrocytes in the brain (55). This is somewhat controversial, however, as Allen and Attwell showed that lactate was unable to replace glucose as a metabolic fuel for brain cells under normoxia or hypoxia (64). Lactate can, however, be metabolized to glucose through the Cori cycle, which can then be used to fuel tumor cell growth. The metabolism of lactate would require reversal of the lactate dehydrogenase complex to oxidize lactate to pyruvate. Although pyruvate might then enter the mitochondria, it is unlikely to be oxidized completely, especially if OxPhos is insufficient. Pyruvate could, however, be converted in the mitochondria to PEP through oxaloacetate (OAA), but it is not clear if ATP could be produced through this pathway. Moreover, this reaction would consume NAD^+ needed for glycolysis. Reduction of pyruvate to lactate is needed to generate NAD^+ for the 3-phospho-glyceraldehyde reaction that helps drive glycolysis (Fig. 4.7).

It is not clear to me how lactate could be used as a major energy substrate for tumor cells, which manifest diminished respiration. We found that lactate could not maintain viability of our highly metastatic VM-M3 mouse tumor cells when grown for 24 h in lactate alone without serum, glucose, or glutamine. However, either glucose or glutamine alone could maintain viability (65). We do not exclude the possibility that lactate might be used as a fuel in combination with glutamine for some tumor cells, but further studies will be needed to confirm this in our metastatic cancer cells.

Mitochondrial membrane lipids are altered and energy production through OxPhos is compromised from simply growing dividing cells in culture (51) (Chapter 5). Unfortunately, many investigators fail to account for mitochondrial ATP production through amino acid fermentation and substrate-level phosphorylation in the mitochondria especially when high glucose levels are present. Consequently, monitoring extracellular pH as a marker for lactic acid production could be misleading especially if the cells are metabolizing glutamine and producing ammonia. Not all investigators consider this possibility in their evaluation of pH changes and energy metabolism in tumor cells. Hence, some confusion over the role of respiration in maintaining tumor cell viability might arise from measurements of oxygen consumption that is uncoupled to OxPhos and the failure to correlate extracellular pH with direct lactate content. It is also helpful to include nontransformed control cells and experimental designs where all energy substrates and metabolites are carefully monitored or accounted for.

GLUTAMINOLYSIS WITH OR WITHOUT LACTATE PRODUCTION

The neutral amino acid glutamine is readily taken up into cells through simple uniport mechanisms (16, 19). Glutamine can serve as a major source of metabolic fuel for generating ATP through TCA cycle, substrate-level phosphorylation when OxPhos is deficient (43, 45). Glutamine is also anapleurotic in replenishing metabolites for the TCA cycle (60, 66). We recently described how cancer cells could generate energy through mitochondrial fermentation and substrate-level phosphorylation in the TCA cycle using glutamine as a substrate (45, 59, 67) (Fig. 4.8). Glutamine is also a major energy fuel for cells of the immune system (68). As myeloid cells can be the origin of many metastatic cancers following fusion hybridizations, glutamine becomes an important fuel for driving metastasis (67, 69) (Chapter 13). Indeed, targeting glutamine can significantly inhibit systemic metastasis as we have shown (70) (Chapter 17).

McKeehan has first described glutaminolysis as the process by which glutamine metabolism produces carbon dioxide, pyruvate, and lactate through oxidative pathways (71). Under this scheme, malate would leave the mitochondria where it would be metabolized to pyruvate and then to lactate (71, 72). McKeehan has not explained how malate would leave the mitochondria. Malate usually enters the mitochondria through the malate–aspartate shuttle, which is active in cancer cells (see below). Moreadith and Lehninger (16) were unable to support McKeehan's metabolic scheme in their analysis of five different tumor types. Their data indicate that malate does not leave the mitochondria, but is instead metabolized to OAA, which then serves as a substrate for aspartate synthesis through transamination (Fig. 4.9). Under certain metabolic conditions, malate can enter the mitochondria and serve as a substrate for mitochondrial malic enzyme (ME) for the synthesis of

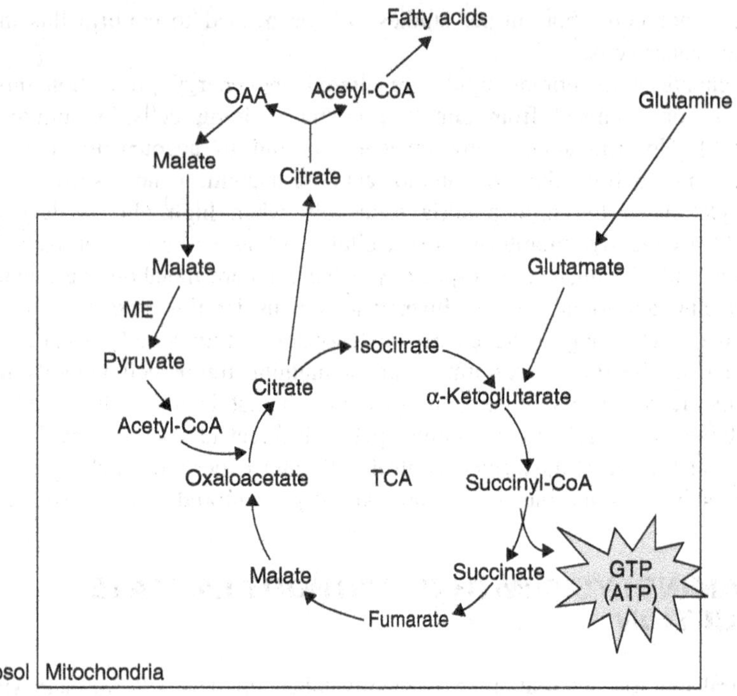

Figure 4.8 Proposed mechanism by which glutamine maintains viability in the VM-M3 cell line. Glutamine enters the TCA cycle as α-ketoglutarate, generating energy from substrate-level phosphorylation from the conversion of succinyl-CoA to succinate. Citrate from the TCA cycle is extruded from the mitochondria to the cytosol whereby it is converted to oxaloacetate (OAA) and acetyl-CoA. Acetyl-CoA is further used in fatty acid synthesis. OAA is converted to malate, which reenters the mitochondria. Once in the mitochondria, mitochondrial malic enzyme (ME) converts malate to pyruvate, which is further converted to acetyl-CoA. Acetyl-CoA can now reenter the TCA to allow for continued TCA cycling. *Source*: Reprinted with permission from Reference 65. To see this figure in color please go to ftp://ftp.wiley.com/public/sci_tech_med/cancer_metabolic_disease.

citrate (Fig. 4.8). In neither case does malate leave the mitochondria to serve as a substrate for pyruvate synthesis.

However, malate could leave the mitochondria through the little-known pyruvate malate shuttle, which is active in some cells (73). In this scheme, pyruvate would enter the mitochondria for conversion to OAA through the pyruvate carboxylase reaction. The OAA in then converted to malate, which exits the mitochondria where it is converted back to pyruvate through the action of the cytoplasmic ME (73). Significant nicotinamide adenine dinucleotide phosphate (NADPH) would be formed through this reaction for use in synthetic reactions in those cells with diminished activity of the pentose–phosphate pathway. The pyruvate formed through the cytoplasmic ME reaction would then reenter the mitochondria for another turn of the shuttle or for decarboxylation to acetyl-CoA and oxidation (73). It is not clear if malate would exit mitochondria in tumor cells using this shuttle system since the pentose phosphate pathway is usually quite robust in most cancer cells (74–77).

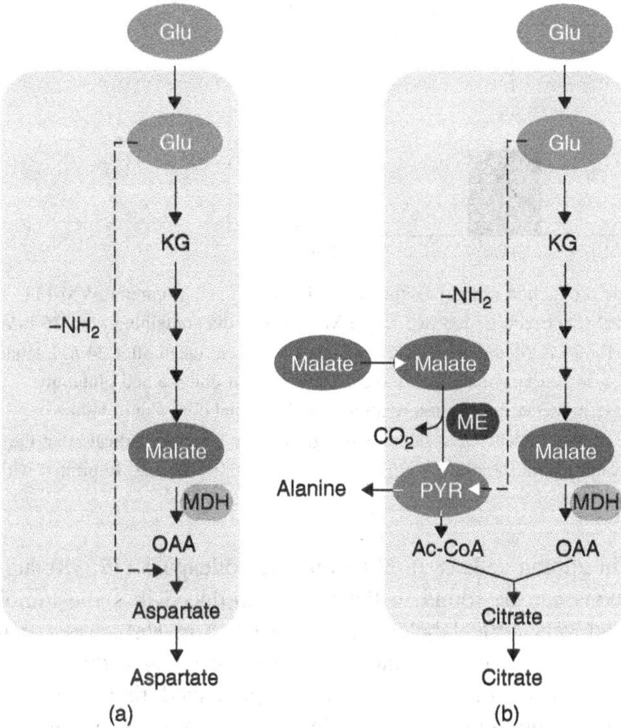

Figure 4.9 Proposed pathways of malate and glutamate oxidation in Ehrlich tumor mitochondria. (a) The pathway of glutamate oxidation in the absence of added malate in the medium. Glutamate (Glu) undergoes transamination with oxaloacetate (OAA) to yield aspartate (dashed line). In this case, malate oxidation occurs exclusively via malate dehydrogenase (MDH). (b) The pathways of glutamate and malate utilization when malate is also available in the medium. In this case, the malate derived from the medium is oxidized to pyruvate via malic enzyme (ME), whereas malate derived from glutamate is oxidized via malate dehydrogenase, in such a way that acetyl-coA and oxaloacetate are formed at equal rates en route to citrate (CIT). In this scheme, the glutamate amino group is transaminated to the pyruvate to form alanine. It is likely that both metabolic pathways are used to various extents in tumor cells. In neither pathway does malate leave the mitochondria. KG, α-ketoglutarate. *Source*: Reprinted with modification from Reference 16. See color insert.

Several investigators find little lactate production from glutamine in tumor cells (16, 60, 78, 79). We also found very little lactate production in our metastatic VM-M3 mouse tumor cells that were grown in glutamine alone (Fig. 4.10). However, lactate production was significantly greater in glucose and glutamine than in glucose alone. This illustrates the synergistic interaction between glucose and glutamine in driving fermentation energy metabolism in the VM-M3 tumor cells (59).

DeBerardinis and coworkers also found very little labeled lactate when glutamine alone was used as a metabolic substrate. However, these investigators found significant labeled lactate when labeled glucose and glutamine were used together

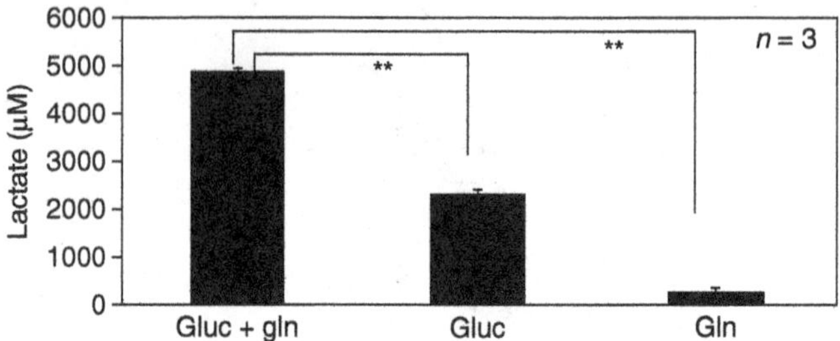

Figure 4.10 VM-M3 lactate production in the presence of both glucose and glutamine. VM-M3 cells were incubated in minimal (Dulbecco's Modified Eagle Medium) media containing both 25 mM glucose and 4 mM glutamine for 24 h. Media aliquots from each group were taken after 24 h. Lactate accumulation was determined using an enzymatic assay (65). Incubation in glucose and glutamine resulted in a significant increase in lactate production relative to either metabolite alone. Values represent the mean ±95% CI of three independent samples per group. The asterisks indicate that the gluc+gln values differ significantly from the gluc or gln values at a $p < 0.01$. *Source*: Reprinted with permission from Reference 65.

as metabolic substrates in glioma cells (80). Mazurek and colleagues (17, 18) suggest that glutamine carbons can be found in the lactate produced in some tumor cells. However, we found very little labeled lactate (<10%) in HeLa cells that were grown in unlabeled glucose and ^{14}C-labeled glutamine using a direct biochemical measurement of lactate (Ta and Seyfried, unpublished findings). Our findings indicate that lactate is not a major end product of glutamine metabolism in HeLa cells.

These and other studies make it clear that the issue of lactate production from glutamine remains unsettled in tumor cells. It is therefore surprising to me that several recent reviews have accepted the McKeehan hypothesis, showing that malate leaves the mitochondria for eventual lactate production, without mentioning the published data not supporting this hypothesis, especially the reports of Lanks and Moreadith and Lehninger (72, 79, 81–83).

Why is it important to know whether glutaminolysis is involved in cancer energy metabolism and whether glutamine carbons are found in lactate? This information can provide insight on the fate of tumor energy metabolites that can be used for energy and growth. Moreover, it will be important to determine if the metabolic changes in tumor cells are the cause or the consequence of the genetic changes that occur in the tumor cells. It is therefore important that we know the metabolic origin of lactate production in tumor cells.

TRANSAMINATION REACTIONS

It is well documented that glutamine enters the mitochondria where it is rapidly metabolized to glutamate by mitochondrial glutaminase (19, 84). Glutamate is then metabolized to α-ketoglutarate through either a transamination reactions with aspartate or alanine as products or through the action of glutamate dehydrogenase

(16, 19, 66, 84). In the glutamate dehydrogenase reaction, NH_3 becomes a toxic by-product that must be eliminated. In the transamination reactions, on the other hand, OAA accepts the NH_3 group to form either aspartate or alanine (16) (Fig. 4.11). Whether aspartate or alanine becomes the primary product of the transamination reaction depends upon the presence or absence of oxygen and malate, and whether or not respiration is sufficient or insufficient (16, 35, 85).

I consider that the transamination reactions would predominate over the glutamate dehydrogenase reaction in tumor cell mitochondria since the guanosine triphosphate (GTP), and also the ATP, formed through TCA cycle, substrate-level phosphorylation could inhibit the glutamate dehydrogenase reaction, thus reducing its activity. My suggestion comes from the previous findings of Moreadith and Lehninger (16) who showed that glutamate oxidation in tumor cell mitochondria proceeds almost entirely through transamination either with OAA or pyruvate rather than through direct dehydrogenation via glutamate dehydrogenase. This observation was observed in five different types of tumor cells making the phenomenon relevant to many types of cancer cells.

The green tea polyphenol, Epigallocatechin gallate (EGCG), which inhibits the glutamate dehydrogenase, could be used to help determine if the α-ketoglutarate arises through the action of glutamate dehydrogenase or a transamination reaction (84). Some of the aspartate formed through transamination could also be used for

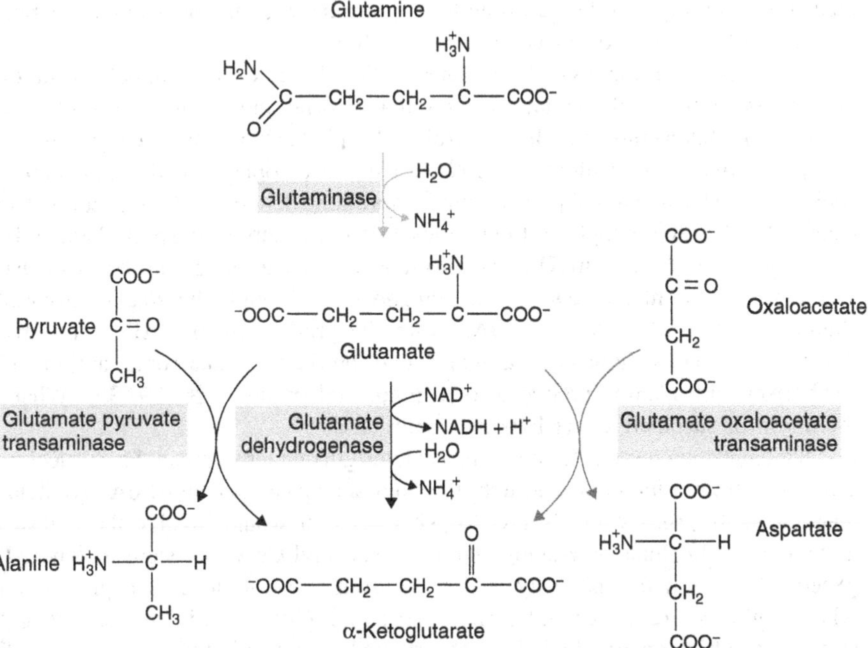

Figure 4.11 Transamination reactions. *Source*: Reprinted with permission from Cybille Mazurek (http://en.wikipedia.org/wiki/File:Glutaminolysisengl1.png). See color insert.

malate production in the cytosol. This will depend to a large extent on the activity of the malate aspartate shuttle (Fig. 4.11). The activity of the shuttle is correlated with the glycolytic activity, that is, the greater the glycolysis, the greater is the shuttle activity (12, 14, 18, 86). I use some caution in my predictions, however, as the biochemical reactions can depend on multiple variables in the growth environment and with the type of tumor cells involved (18, 85).

TCA CYCLE, SUBSTRATE-LEVEL PHOSPHORYLATION

The α-ketoglutarate formed from glutamine enters the TCA cycle where it is decarboxylated and conjugated with coenzyme A to form succinyl-CoA (Fig. 4.6). The α-ketoglutarate dehydrogenase catalyzes this reaction with the formation of NADH and CO_2. The Succinyl-CoA formed is then oxidized to succinate. A histidine residue in the enzyme becomes phosphorylated. It is during this reaction that the histidine phosphate is transferred from the enzyme itself (succinyl-CoA synthetase) to either GDP or ADP to form GTP or ATP (43, 87). The transfer of the phosphate to either ADP or GDP is dependent on the tissue and the metabolic state (87). Succinyl-CoA synthetase is capable of ATP production via substrate-level phosphorylation in the absence of oxygen (30) (Chapter 8). It is our view that this reaction can provide significant nonoxidative ATP synthesis in cells under hypoxia or hyperglycemia. We have recently suggested that mitochondrial glutamine fermentation and TCA cycle, substrate-level phosphorylation can maintain viability of a naturally metastatic cancer cell line (59, 67).

As NADH in the cytoplasm cannot directly enter the mitochondria due to impermeability and the lack of a mitochondrial membrane transporter, the malate–aspartate shuttle and the glycerol 3-phosphate shuttle are used to indirectly transport reducing equivalents from NADH in the cytoplasm to the mitochondria (Fig. 4.12). The malate–aspartate shuttle normally delivers reducing equivalents from NADH in the cytoplasm to complex I of the electron transport chain (ETC) (6). Consequently, the NADH is used to reduce OAA to malate in the cytoplasm. The cytoplasmic malate enters the mitochondria through the malate–aspartate shuttle where it is oxidized to OAA with the production of NAD^+. It is well documented that succinate is a major end product of anaerobic amino acid catabolism with alanine produced as a minor end product (26–29, 35). What is the origin of succinate under hypoxic conditions?

Succinate can have at least two possible origins under hypoxia. First, succinate can arise from glutamine through the mitochondrial succinyl-CoA synthetase reaction in the TCA cycle (Fig. 4.6). This reaction would involve the pathway: glutamine → glutamate → α-ketoglutarate → succinyl-CoA → succinate. The ATP generated through this pathway is derived from substrate-level phosphorylation. ATP synthesis from substrate-level phosphorylation contributes significantly to mitochondrial energy in kidney, heart, and tissues of diving animals under hypoxia (26, 28, 29, 34, 35, 88). As mentioned above, this pathway would end in either citrate or aspartate (Fig. 4.10). Succinate accumulation under

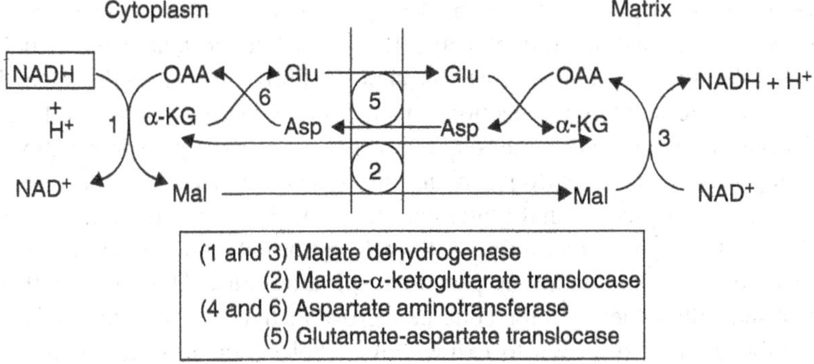

Figure 4.12 Malate–aspartate shuttle for the transport of cytoplasmic reducing equivalents across the inner membrane of mitochondria. Malate carries the reducing equivalents and is oxidized to oxaloacetate with the generation of NADH in the matrix. To complete the unidirectional cycle, oxaloacetate is transported out of the matrix as aspartate. Mal, malate; OAA, oxaloacetate; α-KG, α-ketoglutarate; Glu, glutamine; Asp, aspartate.

hypoxia could occur through the fumarate reductase reaction involving the malate → fumarate → succinate pathway. Tomitsuka and colleagues [89] provide evidence that tumor cells express an active fumarate reductase reaction. I will show in Chapter 8 how this reaction compliments the α-ketoglutarate dehydrogenase reaction to drive mitochondrial glutamine fermentation and ATP synthesis through substrate-level phosphorylation.

CHOLESTEROL SYNTHESIS AND HYPOXIA

Cholesterol is a major membrane lipid that must be synthesized for cancer cells to grow. Cholesterol synthesis requires oxygen. Oxygen is required for the squalene monooxygenase reaction of cholesterol synthesis. We found that cultured metastatic VM-M3 tumor cells grow well with glucose and glutamine as the only metabolic fuels in normoxia without serum. These cells actively synthesize cholesterol from either glucose or glutamine. However, the cells die rapidly under hypoxia without added serum. It appears that serum is required for growth under hypoxia, but is not required for growth under normoxia. Serum contains, among many factors, high levels of cholesterol. We found that the VM-M3 cells obtain cholesterol directly from the serum under hypoxia. It is not necessary for the cells to synthesize cholesterol if they can get it free from the growth environment. Hence, the VM-M3 cells can grow in hypoxia as long as they have fermentable fuels and can obtain cholesterol from an external source.

SUMMARY

All cells including tumor cells require a relatively constant level of usable ATP synthesis for maintaining viability. This appears to be a biological constant independent of cell origin or function. Energy metabolism in cancer cells differs markedly from that in normal cells. In contrast to normal cells, which generate most of their useable energy through OxPhos, cancer cells depend more heavily on fermentation reactions using nonoxidative, substrate-level phosphorylations for their ATP synthesis. These substrate-level phosphorylations occur through glycolysis in the cytoplasm and through succinyl-CoA synthetase in the mitochondria. Both glucose and glutamine can provide energy to cancer cells through substrate-level phosphorylation. Although tumor cell mitochondria might appear to respire, we refer to this as pseudo respiration since OxPhos is either reduced or absent altogether. Tumor cells can grow in hypoxia as long as they have fermentable fuels and access to extracellular cholesterol. Respiration is the bioenergetic signature of normal cells; fermentation is the bioenergetic signature of cancer cells. Fermentation drives cancer cells whether or not oxygen is present. Malignant cancer cells ferment more than they respire!

REFERENCES

1. Harold FM. The Vital Force: A Study of Bioenergetics. New York: W. H. Freeman; 1986.
2. Veech RL, Chance B, Kashiwaya Y, Lardy HA, Cahill GF, Jr. Ketone bodies, potential therapeutic uses. IUBMB Life. 2001;51:241–7.
3. Kocherginsky N. Acidic lipids, H(+)-ATPases, and mechanism of oxidative phosphorylation. Physico-chemical ideas 30 years after P. Mitchell's Nobel Prize award. Prog Biophys Mol Biol. 2009;99:20–41.
4. Veech RL, Kashiwaya Y, Gates DN, King MT, Clarke K. The energetics of ion distribution: the origin of the resting electric potential of cells. IUBMB Life. 2002;54:241–52.
5. Veech RL, Lawson JW, Cornell NW, Krebs HA. Cytosolic phosphorylation potential. J Biol Chem. 1979;254:6538–47.
6. Bhagavan NV. Medical Biochemistry. 4th ed. New York: Harcourt; 2002.
7. Schneider ED, Sagan D. Into the Cool: Energy Flow, Thermodynamics, and Life. Chicago: University of Chicago Press; 2005.
8. Banks BE, Vernon CA. Reassessment of the role of ATP *in vivo*. J Theor Biol. 1970;29:301–26.
9. Hochachka PW, Somero GN. Biochemical Adaptation: Mechanism and Process in Physiological Evolution. New York: Oxford Press; 2002.
10. Masuda R, Monahan JW, Kashiwaya Y. D-beta-hydroxybutyrate is neuroprotective against hypoxia in serum-free hippocampal primary cultures. J Neurosci Res. 2005;80:501–9.
11. Seyfried TN, Mukherjee P. Targeting energy metabolism in brain cancer: review and hypothesis. Nutr Metab. 2005;2:30.
12. Chiaretti B, Casciaro A, Minotti G, Eboli ML, Galeotti T. Quantitative evaluation of the activity of the malate-aspartate shuttle in Ehrlich ascites tumor cells. Cancer Res. 1979;39:2195–9.
13. Grivell AR, Korpelainen EI, Williams CJ, Berry MN. Substrate-dependent utilization of the glycerol 3-phosphate or malate/aspartate redox shuttles by Ehrlich ascites cells. Biochem J. 1995;310(Pt 2):665–71.
14. Greenhouse WV, Lehninger AL. Occurrence of the malate-aspartate shuttle in various tumor types. Cancer Res. 1976;36:1392–6.
15. Greenhouse WV, Lehninger AL. Magnitude of malate-aspartate reduced nicotinamide adenine dinucleotide shuttle activity in intact respiring tumor cells. Cancer Res. 1977;37:4173–81.
16. Moreadith RW, Lehninger AL. The pathways of glutamate and glutamine oxidation by tumor cell mitochondria. Role of mitochondrial NAD(P)+-dependent malic enzyme. J Biol Chem. 1984;259:6215–21.

17. Mazurek S, Boschek CB, Hugo F, Eigenbrodt E. Pyruvate kinase type M2 and its role in tumor growth and spreading. Semin Cancer Biol. 2005;15:300–8.
18. Mazurek S, Michel A, Eigenbrodt E. Effect of extracellular AMP on cell proliferation and metabolism of breast cancer cell lines with high and low glycolytic rates. J Biol Chem. 1997;272:4941–52.
19. Baggetto LG. Deviant energetic metabolism of glycolytic cancer cells. Biochimie. 1992; 74:959–74.
20. Mitchell P. A chemiosmotic molecular mechanism for proton-translocating adenosine triphosphatases. FEBS Lett. 1974;43:189–94.
21. Galluzzi L, Morselli E, Kepp O, Vitale I, Rigoni A, Vacchelli E, et al. Mitochondrial gateways to cancer. Mol Aspects Med. 2009;31:1–20.
22. Brownlee M. Biochemistry and molecular cell biology of diabetic complications. Nature. 2001;414:813–20.
23. Villalobo A, Lehninger AL. The proton stoichiometry of electron transport in Ehrlich ascites tumor mitochondria. J Biol Chem. 1979;254:4352–8.
24. Samudio I, Fiegl M, Andreeff M. Mitochondrial uncoupling and the Warburg effect: molecular basis for the reprogramming of cancer cell metabolism. Cancer Res. 2009;69:2163–6.
25. Kiebish MA, Han X, Cheng H, Chuang JH, Seyfried TN. Cardiolipin and electron transport chain abnormalities in mouse brain tumor mitochondria: lipidomic evidence supporting the Warburg theory of cancer. J Lipid Res. 2008;49:2545–56.
26. Penney DG, Cascarano J. Anaerobic rat heart. Effects of glucose and tricarboxylic acid-cycle metabolites on metabolism and physiological performance. Biochem J. 1970;118:221–7.
27. Weinberg JM, Venkatachalam MA, Roeser NF, Nissim I. Mitochondrial dysfunction during hypoxia/reoxygenation and its correction by anaerobic metabolism of citric acid cycle intermediates. Proc Natl Acad Sci USA. 2000;97:2826–31.
28. Weinberg JM, Venkatachalam MA, Roeser NF, Saikumar P, Dong Z, Senter RA, et al. Anaerobic and aerobic pathways for salvage of proximal tubules from hypoxia-induced mitochondrial injury. Am J Physiol Renal Physiol. 2000;279:F927–43.
29. Gronow GH, Cohen JJ. Substrate support for renal functions during hypoxia in the perfused rat kidney. Am J Physiol. 1984;247:F618–31.
30. Phillips D, Aponte AM, French SA, Chess DJ, Balaban RS. Succinyl-CoA synthetase is a phosphate target for the activation of mitochondrial metabolism. Biochemistry. 2009;48: 7140–9.
31. Buchakjian MR, Kornbluth S. The engine driving the ship: metabolic steering of cell proliferation and death. Nat Rev Mol Cell Biol. 2010;11:715–27.
32. Chevrollier A, Loiseau D, Chabi B, Renier G, Douay O, Malthiery Y, et al. ANT2 isoform required for cancer cell glycolysis. J Bioenerg Biomembr. 2005;37:307–16.
33. Warburg O. On the origin of cancer cells. Science. 1956;123:309–14.
34. Pisarenko OI, Solomatina ES, Ivanov VE, Studneva IM, Kapelko VI, Smirnov VN. On the mechanism of enhanced ATP formation in hypoxic myocardium caused by glutamic acid. Basic Res Cardiol. 1985;80:126–34.
35. Hochachka PW, Owen TG, Allen JF, Whittow GC. Multiple end products of anaerobiosis in diving vertebrates. Comp Biochem Physiol B. 1975;50:17–22.
36. Nelson DL, Cox MM. Lehninger: Principles of Biochemistry. 5th ed. New York: W. H. Freeman and Company; 2008.
37. Bergman C, Kashiwaya Y, Veech RL. The effect of pH and free Mg^{2+} on ATP linked enzymes and the calculation of Gibbs free energy of ATP hydrolysis. J Phys Chem B. 2010;114: 16137–46.
38. Wackerhage H, Hoffmann U, Essfeld D, Leyk D, Mueller K, Zange J. Recovery of free ADP, Pi, and free energy of ATP hydrolysis in human skeletal muscle. J Appl Physiol. 1998;85: 2140–5.
39. Chinopoulos C, Adam-Vizi V. Mitochondria as ATP consumers in cellular pathology. Biochim Biophys Acta. 2010;1802:221–7.

40. Dzhandzhugazyan KN, Kirkin AF, thor Straten P, Zeuthen J. Ecto-ATP diphosphohydrolase/CD39 is overexpressed in differentiated human melanomas. FEBS Lett. 1998;430: 227–30.
41. Wartenberg M, Richter M, Datchev A, Gunther S, Milosevic N, Bekhite MM, et al. Glycolytic pyruvate regulates P-glycoprotein expression in multicellular tumor spheroids via modulation of the intracellular redox state. J Cell Biochem. 2010;109:434–46.
42. Guidotti G. ATP transport and ABC proteins. Chem Biol. 1996;3:703–6.
43. Schwimmer C, Lefebvre-Legendre L, Rak M, Devin A, Slonimski PP, di Rago JP, et al. Increasing mitochondrial substrate-level phosphorylation can rescue respiratory growth of an ATP synthase-deficient yeast. J Biol Chem. 2005;280:30751–9.
44. Chevrollier A, Loiseau D, Gautier F, Malthiery Y, Stepien G. ANT2 expression under hypoxic conditions produces opposite cell-cycle behavior in 143B and HepG2 cancer cells. Mol Carcinog. 2005;42:1–8.
45. Seyfried TN, Shelton LM. Cancer as a metabolic disease. Nutr Metab. 2010;7:7.
46. Szent-Gyorgyi A. The living state and cancer. Proc Natl Acad Sci USA. 1977;74:2844–7.
47. Moiseeva O, Bourdeau V, Roux A, Deschenes-Simard X, Ferbeyre G. Mitochondrial dysfunction contributes to oncogene-induced senescence. Mol Cell Biol. 2009;29:4495–507.
48. Ecsedy JA, Yohe HC, Bergeron AJ, Seyfried TN. Tumor-infiltrating macrophages influence the glycosphingolipid composition of murine brain tumors. J Lipid Res. 1998;39:2218–27.
49. Warburg O. The Metabolism of Tumours. New York: Richard R. Smith; 1931.
50. Donnelly M, Scheffler IE. Energy metabolism in respiration-deficient and wild type Chinese hamster fibroblasts in culture. J Cell Physiol. 1976;89:39–51.
51. Kiebish MA, Han X, Cheng H, Seyfried TN. In vitro growth environment produces lipidomic and electron transport chain abnormalities in mitochondria from non-tumorigenic astrocytes and brain tumours. ASN Neuro. 2009;1:e00011.
52. Warburg O. Revidsed Lindau Lectures: The prime cause of cancer and prevention-Parts 1 & 2. In: Burk D, editor. Meeting of the Nobel-Laureates Lindau, Lake Constance, Germany: K.Triltsch; 1969. p. http://www.hopeforcancer.com/OxyPlus.htm.
53. Yeluri S, Madhok B, Prasad KR, Quirke P, Jayne DG. Cancer's craving for sugar: an opportunity for clinical exploitation. J Cancer Res Clin Oncol. 2009;135:867–77.
54. Campbell NA. Biology. 3rd ed. New York: Benjamin/Commings; 1993.
55. Sonveaux P, Vegran F, Schroeder T, Wergin MC, Verrax J, Rabbani ZN, et al. Targeting lactate-fueled respiration selectively kills hypoxic tumor cells in mice. J Clin Invest. 2008;118:3930–42.
56. Kelley MA, Kazemi H. Role of ammonia as a buffer in the central nervous system. Respir Physiol. 1974;22:345–59.
57. Boag JM, Beesley AH, Firth MJ, Freitas JR, Ford J, Hoffmann K, et al. Altered glucose metabolism in childhood pre-B acute lymphoblastic leukaemia. Leukemia. 2006;20:1731–7.
58. Gillies RJ, Gatenby RA. Adaptive landscapes and emergent phenotypes: why do cancers have high glycolysis? J Bioenerg Biomembr. 2007;39:251–7.
59. Seyfried TN. Mitochondrial glutamine fermentation enhances ATP synthesisin murine glioblastoma cells. Proceedings of the 102nd Annual Meeting of the American Association Cancer Research; 2011; Orlando, FL. 2011.
60. Scott DA, Richardson AD, Filipp FV, Knutzen CA, Chiang GG, Ronai ZA, et al. Comparative metabolic flux profiling of melanoma cell lines: beyond the Warburg effect. J Biol Chem. 2011;286:42626–34.
61. Weinhouse S. On respiratory impairment in cancer cells. Science. 1956;124:267–9.
62. Weinhouse S. The Warburg hypothesis fifty years later. Z Krebsforsch Klin Onkol Cancer Res Clin Oncol. 1976;87:115–26.
63. Koppenol WH, Bounds PL, Dang CV. Otto Warburg's contributions to current concepts of cancer metabolism. Nat Rev. 2011;11:325–37.
64. Allen NJ, Karadottir R, Attwell D. A preferential role for glycolysis in preventing the anoxic depolarization of rat hippocampal area CA1 pyramidal cells. J Neurosci. 2005;25:848–59.

65. Shelton LM. Targeting Energy Metabolism in Brain Cancer. Chestnut Hill: Boston College; 2010.
66. Brunengraber H, Roe CR. Anaplerotic molecules: current and future. J Inherit Metab Dis. 2006;29:327–31.
67. Shelton LM, Strelko CL, Roberts MF, Seyfried NT. Krebs cycle substrate-level phosphorylation drives metastatic cancer cells. Proceedings of the 101st Annual Meeting of the American Association for Cancer Research; 2010; Washington (DC). 2010.
68. Newsholme P. Why is L-glutamine metabolism important to cells of the immune system in health, postinjury, surgery or infection? J Nutr. 2001;131:2515S–22S. Discussion 23S–4S.
69. Huysentruyt LC, Seyfried TN. Perspectives on the mesenchymal origin of metastatic cancer. Cancer Metastasis Rev. 2010;29:695–707.
70. Shelton LM, Huysentruyt LC, Seyfried TN. Glutamine targeting inhibits systemic metastasis in the VM-M3 murine tumor model. Int J Cancer. 2010;127:2478–85.
71. McKeehan WL. Glycolysis, glutaminolysis and cell proliferation. Cell Biol Int Rep. 1982; 6:635–50.
72. Levine AJ, Puzio-Kuter AM. The control of the metabolic switch in cancers by oncogenes and tumor suppressor genes. Science. 2010;330:1340–4.
73. MacDonald MJ. Feasibility of a mitochondrial pyruvate malate shuttle in pancreatic islets. Further implication of cytosolic NADPH in insulin secretion. J Biol Chem. 1995;270:20051–8.
74. Langbein S, Frederiks WM, zur Hausen A, Popa J, Lehmann J, Weiss C, et al. Metastasis is promoted by a bioenergetic switch: new targets for progressive renal cell cancer. Int J Cancer. 2008;122:2422–8.
75. Langbein S, Zerilli M, Zur Hausen A, Staiger W, Rensch-Boschert K, Lukan N, et al. Expression of transketolase TKTL1 predicts colon and urothelial cancer patient survival: Warburg effect reinterpreted. Br J Cancer. 2006;94:578–85.
76. Otto C, Kaemmerer U, Illert B, Muehling B, Pfetzer N, Wittig R, et al. Growth of human gastric cancer cells in nude mice is delayed by a ketogenic diet supplemented with omega-3 fatty acids and medium-chain triglycerides. BMC Cancer. 2008;8:122.
77. Frohlich E, Fink I, Wahl R. Is transketolase like 1 a target for the treatment of differentiated thyroid carcinoma? A study on thyroid cancer cell lines. Invest New Drugs. 2009;27:297–303.
78. Reitzer LJ, Wice BM, Kennell D. Evidence that glutamine, not sugar, is the major energy source for cultured HeLa cells. J Biol Chem. 1979;254:2669–76.
79. Lanks KW. End products of glucose and glutamine metabolism by L929 cells. J Biol Chem. 1987;262:10093–7.
80. DeBerardinis RJ, Mancuso A, Daikhin E, Nissim I, Yudkoff M, Wehrli S, et al. Beyond aerobic glycolysis: transformed cells can engage in glutamine metabolism that exceeds the requirement for protein and nucleotide synthesis. Proc Natl Acad Sci USA. 2007;104:19345–50.
81. Dang CV. p32 (C1QBP) and cancer cell metabolism: is the Warburg effect a lot of hot air? Mol Cell Biol. 2010;30:1300–2.
82. Dang CV. Glutaminolysis: supplying carbon or nitrogen or both for cancer cells? Cell Cycle. 2010;9:3884–6.
83. Vander Heiden MG, Cantley LC, Thompson CB. Understanding the Warburg effect: the metabolic requirements of cell proliferation. Science. 2009;324:1029–33.
84. DeBerardinis RJ, Cheng T. Q's next: the diverse functions of glutamine in metabolism, cell biology and cancer. Oncogene. 2010;29:313–24.
85. Piva TJ, McEvoy-Bowe E. Oxidation of glutamine in HeLa cells: role and control of truncated TCA cycles in tumour mitochondria. J Cell Biochem. 1998;68:213–25.
86. Pedersen PL. Tumor mitochondria and the bioenergetics of cancer cells. Prog Exp Tumor Res. 1978;22:190–274.
87. Lambeth DO, Tews KN, Adkins S, Frohlich D, Milavetz BI. Expression of two succinyl-CoA synthetases with different nucleotide specificities in mammalian tissues. J Biol Chem. 2004;279:36621–4.

88. Kaufman S, Gilvarg C, Cori O, Ochoa S. Enzymatic oxidation of alpha-ketoglutarate and coupled phosphorylation. J Biol Chem. 1953;203:869–88.
89. Tomitsuka E, Kita K, Esumi H. The NADH-fumarate reductase system, a novel mitochondrial energy metabolism, is a new target for anticancer therapy in tumor microenvironments. Ann N Y Acad Sci. 2010;1201:44–9.

Chapter 5

Respiratory Dysfunction in Cancer Cells

"As we study cancer bioenergetics more in depth, we gradually realize that comprehending the true meaning of Warburg effect implicates resolving a continuously growing puzzle, which spans several fields of scientific research and occupies the mind of thousands of investigators and students."

—Leonardo M.R. Ferreira (1)

If Warburg's theory were correct, then some degree of respiratory insufficiency should occur in the neoplastic cells of all tumors. Although this treatise will present substantial evidence in support of Warburg's theory, it is not always easy to recognize mitochondrial dysfunction or respiratory insufficiency in cancer cells. Mitochondria are complex organelles responsible for cell respiration. What part of mitochondrial function is abnormal in neoplastic cells?

Warburg considered oxidative phosphorylation (OxPhos) injury or insufficiency to be the origin of cancer. OxPhos is the final stage of cellular respiration involving multiple coupled redox reactions where the energy contained in carbon–hydrogen bonds of food molecules is captured and conserved in the terminal phosphoanhydride bond of ATP. The process specifically involves the following: (i) the flow of electrons through a chain of membrane-bound carriers, (ii) the coupling of the downhill electron flow to an uphill transport of protons across a proton-impermeable membrane, thus conserving the free energy of fuel oxidation as a transmembrane electrochemical potential, and (iii) the synthesis of ATP from ADP+Pi through a membrane-bound enzymatic complex linked to the transmembrane flow of the protons down their concentration gradient (2). These processes are illustrated in Figure 4.4.

Abnormalities in any number of mitochondrial structures could potentially compromise the ability of OxPhos to provide enough energy to maintain metabolic

Cancer as a Metabolic Disease: On the Origin, Management and Prevention of Cancer, First Edition. Thomas Seyfried.

homeostasis. The difference between normal cells and cancer cells is the presence or absence of structure (3–5). Normal cells have structure, while most tumor cells are dysmorphic compared to normal cells. As the structural integrity of the mitochondria provides the energy needed to maintain cellular differentiation, it is necessary to consider the types of injuries that would reduce OxPhos in tumor cells.

NORMAL MITOCHONDRIA

Before evaluating the types of mitochondrial dysfunction in cancer cells, it would be good to first consider what constitutes a normal mitochondrion. As defined in numerous textbooks of biology and biochemistry, a mitochondrion is a threadlike or granular organelle that functions in aerobic respiration and occurs in varying numbers in all eukaryotic cells except in mature erythrocytes. The mitochondrion is bounded by two sets of membranes, a smooth outer membrane and an inner membrane that is arranged in folds, or cristae that extend into the interior matrix area of the organelle (Fig. 5.1). The complexes of the electron transport chain (ETC), which contribute to energy through OxPhos are found in the mitochondrial cristae. The cristae are swollen cisterns or sacs, with multiple narrow tubular connections to the peripheral surface of the inner membrane (called the *inner boundary membrane*), and to each other (Fig. 5.2).

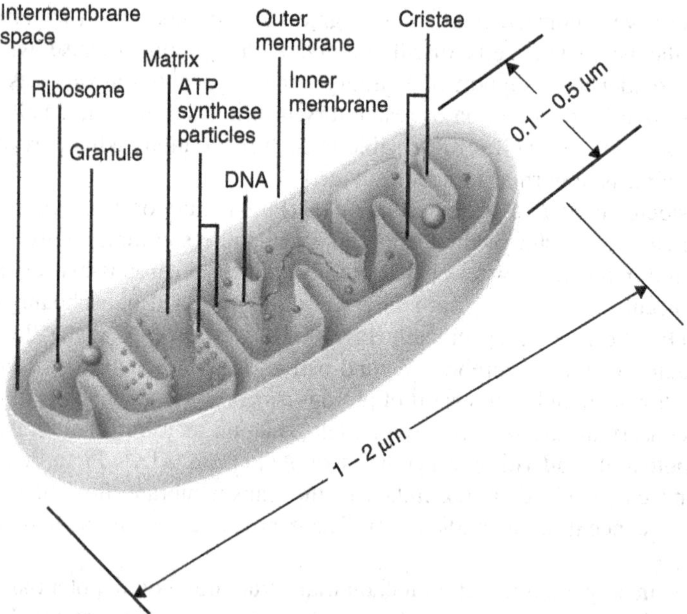

Figure 5.1 The baffle or orthodox model of mitochondria structure. This model shows the cristae with broad openings to the inter membrane space on one side of the mitochondrion and protruding across the matrix nearly to the other side (6). *Source*: Reprinted with permission from Reference 6. See color insert.

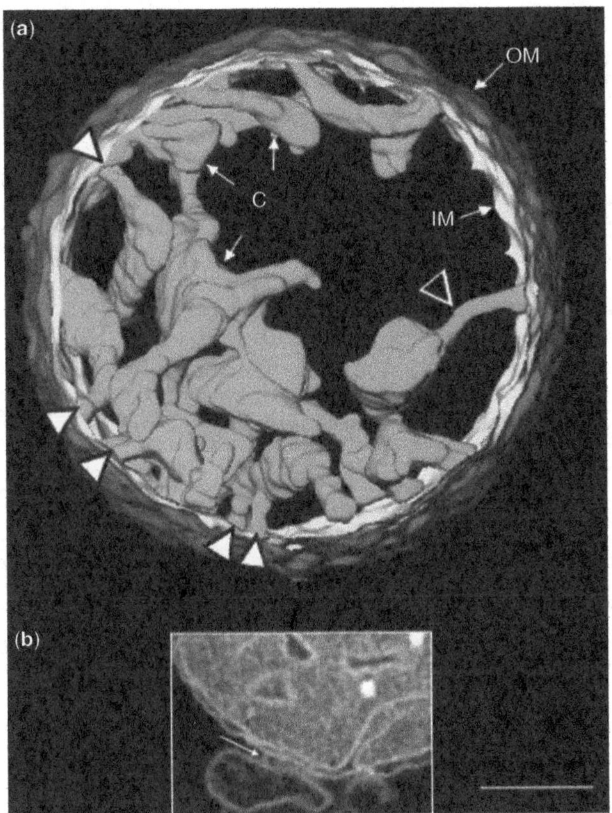

Figure 5.2 Tomogram of a rat liver mitochondrion showing morphology intermediate between that of the condensed and orthodox models. (a) Surface-rendered 3D image of an isolated rat liver mitochondrion. C, cristae; IM, inner boundary membrane; OM, outer membrane. Arrowheads point to tubular regions of cristae that connect them to IM and to each other. (b) Region of a 5-nm slice from the same tomogram showing numerous contact sites between OM and IM. Arrow points to particle bridging OM with attached vesicle of putative endoplasmic reticulum. Bar: 0.4 μm. *Source*: Reprinted with permission from References 6 and 7. See color insert.

OxPhos requires the transport of electrons through proteins imbedded in the cristae. Mitochondria contain many enzymes within the matrix involved with activities including the citric acid and fatty acid cycles as well as calcium flux (Fig. 4.2). Mitochondria also regulate intracellular calcium that can have global effects on numerous aspects of cell physiology (Fig. 4.2). Mitochondria are self-replicating and contain their own DNA, RNA polymerase, transfer RNA, and ribosomes (8). Mitochondria are also dynamic organelles that can expand and contract and undergo fission and fusions in response to the metabolic state of the cell (9–14). Figure 5.3 illustrates the fusion and fission properties of mitochondria.

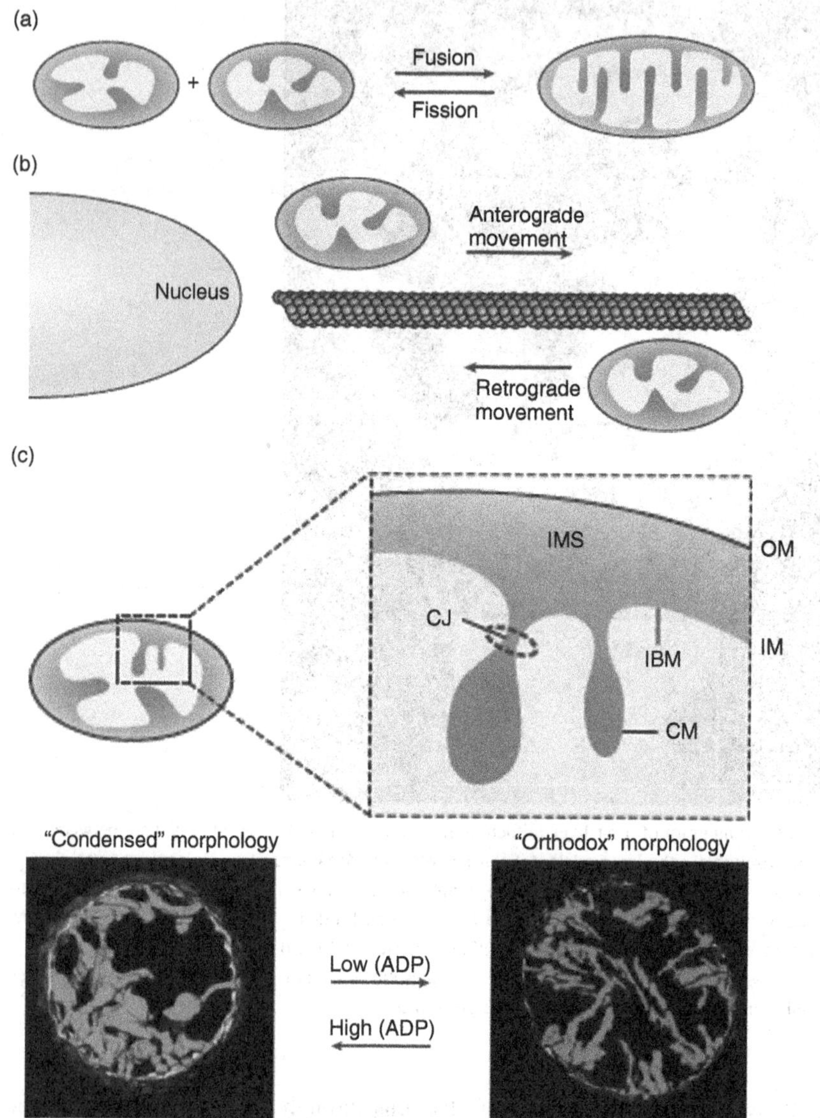

Figure 5.3 Mitochondria are dynamic organelles. (a) Mitochondrial fusion and fission control mitochondrial number and size. With fusion, two mitochondria become a single larger mitochondrion with continuous outer and inner membranes. Conversely, a single mitochondrion can divide into two distinct mitochondria by fission. (b) In mammalian systems, mitochondria are distributed throughout the cytoplasm by active transport along microtubules and actin filaments. Distinct molecular motors transport the mitochondria in anterograde or retrograde directions. (c) Inner membrane dynamics. The diagram indicates the different regions of the inner membrane. CJ, cristae junction; CM, cristae membrane; IBM, inner boundary membrane; IM, inner membrane; IMS, intermembrane space; OM, outer membrane. *Source*: Reprinted from Reference 14 with permission. See color insert.

MORPHOLOGICAL DEFECTS IN TUMOR CELL MITOCHONDRIA

Numerous studies on patient and animal cancers show that tumor mitochondria differ from normal mitochondria in number, size, and shape. Pedersen (15) summarized data from over 20 studies showing that the total number of mitochondria in tumor cells was significantly lower than the number in normal cells of origin. He also mentioned that the total respiratory capacity of tumor mitochondria was lower relative to that of normal cells. Carew and Huang (8) also reviewed evidence suggesting that abnormalities in mitochondrial DNA could compromise mitochondrial function in tumor cells.

Abnormalities in mitochondrial size and shape are correlated with mitochondrial dysfunction (10, 11, 16). In an early investigation of this subject, Potter and Ward (17) demonstrated that mitochondria from spontaneous or transplantable leukemia cells differed in number and size from mitochondria in normal lymphocytes in C58 mice. In a comparative analysis of epithelial cell lines derived from human carcinomas and nonmalignant tissues, Springer found mitochondria with longitudinal cristae arrangement in all the malignant cell lines. None of these morphological abnormalities were found in any of the lines derived from normal tissue or from tissue peripheral to tumors (18). Pedersen (15) also documented numerous morphological abnormalities of tumor mitochondria in his extensive review of the subject. Basically, no highly malignant tumors were found that contained mitochondria of normal number and morphology. On the basis of a review of the literature, Springer raised the possibility that these and other findings of mitochondrial morphological abnormalities in malignant cancers were connected to the origin of the disease.

Kim and colleagues (19) also found multiple alterations in the mitochondrial number, size, and morphology in human gastric carcinoma. The size and number of mitochondria were significantly greater in normal gastric cells (3.5 ± 0.3 μm, 23.5 ± 4 mitochondria) than in the human gastric (AGS) cancer cells (1.3 ± 0.5 μm, 16.3 ± 3 mitochondria). Moreover, the abnormalities in mitochondrial size and number were also associated with abnormalities in the mitochondrial function. In addition to the mitochondrial morphological abnormalities found in these tumor cells, mitochondrial morphological abnormalities were also reported in HeLa cells, one of the most intensely evaluated neoplastic cell types in the cancer field (20). Under examination using transmission electron microscopy, isolated HeLa cell mitochondria appeared predominantly rounded and lacked the cristae pattern of normal mitochondria (21). These mitochondrial abnormalities were similar in some respects to those reported by Pedersen (15) in isolated mitochondria from Morris hepatoma. Mitochondria of HeLa cells also responded differently from the mitochondria of noncancerous fibroblasts when grown in media containing galactose and glutamine (22). The abnormalities observed in HeLa mitochondrial morphology are suggestive of abnormalities in the function of OxPhos (16). The greater the degree of mitochondrial morphological abnormality, the greater the degree of malignancy (15).

Arismendi-Morillo and Castellano-Ramirez (13) provided further compelling electron microscopy evidence for mitochondrial morphological abnormalities in brain cancer. They evaluated mitochondrial morphology in fresh biopsy specimens from patients with various types of malignant brain tumors. The major mitochondrial changes that were detected involved swelling with disarrangement of cristae and partial or total cristolysis (the breakdown or severe reduction of the inner mitochondrial membrane). These abnormalities are illustrated in Figure 6 from their study of glioblastoma (Fig. 5.4). They presented numerous other images of abnormal mitochondria in brain tumors.

The authors concluded that any tumors expressing these types of morphological abnormalities could not produce sufficient levels of ATP synthesis through OxPhos (13, 23). Their conclusion is in line with previous suggestions that OxPhos capability is closely linked to the structural integrity of mitochondrial cristae (6, 10, 11, 24, 25). Poupon, Oudard, and coworkers have also shown that the low content of normally functioning mitochondria in gliomas could underlie a shift in energy metabolism from OxPhos toward high level glycolysis (26). This shift in energy metabolism from OxPhos to glycolysis is considered to be necessary in order to generate sufficient cellular ATP to maintain viability (26, 27).

It is clear from these and numerous other studies that mitochondria in various types of tumor cells express abnormalities and are not likely capable of providing sufficient energy through OxPhos for metabolic homeostasis (11). How would it be possible for any cell to express normal OxPhos activity with multiple abnormalities in mitochondrial number, size, and shape? According to Warburg, aerobic glucose fermentation arises as a secondary consequence of irreversible injury to

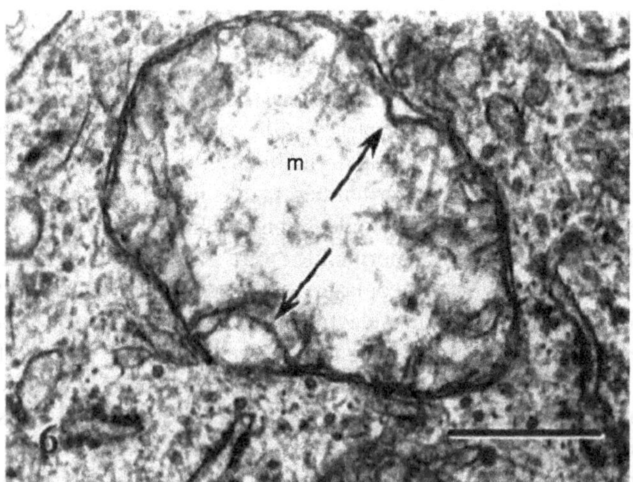

Figure 5.4 Mitochondrial abnormalities in glioblastoma multiforme. Enlarged and piriform mitochondrion (m) that shows total cristolysis and electron-lucent matrix. Note the inner membrane fold (arrows). Bar: 0.33 μm. Method of staining: uranyl acetate/lead citrate. *Source*: Reprinted with permission from Reference 13.

OxPhos (4). Bayley and Devilee (28) showed aerobic glycolysis (Warburg effect) could be directly linked to mitochondrial respiratory injury in those tumors arising from inherited mutations in the genes for succinate dehydrogenases and fumarate hydratase. Besides these defects, any abnormalities in the number of mitochondria, in their ultrastructure and morphology, and in their response to changes in growth environment would predict some degree of respiratory dysfunction.

It is hard to conceive how any rationally thinking cancer researcher would view these findings as being irrelevant to the origin of cancer. It is also inconceivable, in my mind, to think that all these types and kinds of mitochondrial structural and functional abnormalities might arise as secondary consequences of oncogenes or tumor suppressor genes. Indeed, I will later show how defects in expression of oncogenes and tumor suppressor genes can arise as a *consequence* of respiratory insufficiency. I will also show how protracted respiratory insufficiency is the origin of cancer and of genomic instability.

PROTEOMIC ABNORMALITIES IN TUMOR CELL MITOCHONDRIA

In an early biochemical study, Roskelley and coworkers (29) showed that cytochrome–oxidase activity was deficient in nearly all types of highly malignant cancers examined. The human malignancies included cancers of rectum, colon, kidney, breast, brain, prostate, stomach, skin, and testis. The same biochemical deficiencies were also found in well-established transplantable and induced tumors of rat mouse and rabbit. Moreover, carcinogenesis, both by a chemical agent and by a virus, produced the same energy defects in the animal models. Their data from a variety of human and animal cancers clearly showed that all of the normal adult tissues displayed a high oxidative response, whereas all of the frankly malignant cancers displayed a poor oxidative response (29). Pedersen (15) later provided a comprehensive review documenting the numerous protein defects in mitochondria from tumor cells. These findings have shown that mitochondrial respiratory activity is abnormal in human and animal tumors.

More recent studies from Cuezva and colleagues also provide proteomic evidence for respiratory dysfunction in cancer (30–34). These investigators evaluated the relationship of glyceraldehyde-3-phosphate dehydrogenase (GAPHD) and the β-F1 ATPase in a broad spectrum of tumors including breast, colon, lung, and esophagus (31). GAPHD and the β-F1 ATPase are key enzymes needed to drive glycolysis and OxPhos, respectively. GAPHD consumes NAD^+ and inorganic phosphate, Pi, to synthesize the energy-rich intermediate, 1,3-bisglycerophosphate with $NADH + H^+$ as by-products (Figure 4.5). Elevated GAPHD activity indicates enhanced energy production through glycolysis. The β-subunit of the F1 ATPase is required for ATP synthesis through OxPhos (35).

All cancers studied, regardless of their origin or histological grade, had significant elevations of GAPDH and reductions of the β-F1 ATPase (31, 33–38). An example of this fact is shown for colon cancer (Fig. 5.5) (31). Indeed, the

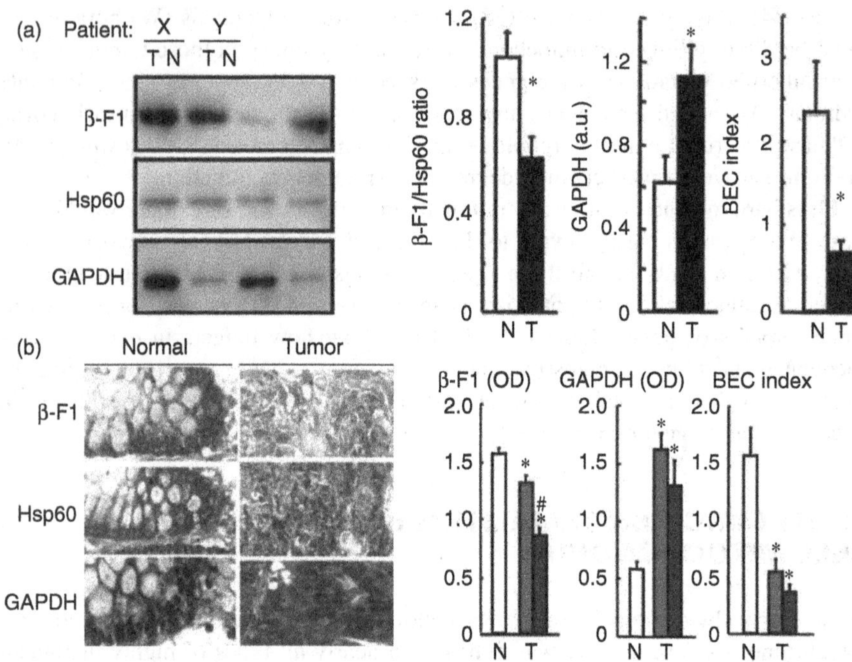

Figure 5.5 The bioenergetic signature of colon cancer. (a) Expression of β-F1-ATPase, heat shock protein 60 (Hsp60) and GAPDH in normal (N) and tumor (T) biopsies of two different colorectal cancer patients (X, Y). Histograms to the right illustrate the drop in β-F1-ATPase/Hsp60 (heat shock protein 60) ratio and the concurrent increase in the expression of the glycolytic GAPDH (glyceraldehyde 3-phosphate dehydrogenase) in colon cancer. Consistent with these changes, the bioenergetic cellular index (BEC) of the tumors was sharply diminished when compared to paired normal colon. The asterisk(*) illustrates significant differences when compared to normal. (b) Immunohistochemical analysis of the expression of β-F1-ATPase, Hsp60, and GAPDH in colorectal carcinomas using colon-tissue microarrays. Histograms to the right illustrate the absolute amount (a.u.) of the expression of β-F1-ATPase and GAPDH (OD) and of the BEC index in normal (N) and tumor (T) samples derived from patients with progressive disease (black bars) and no-evidence of disease (gray bars) after a median of 60 months clinical follow-up. The(*) and the (#) illustrate significant differences when compared to normal or no-evidence of disease group, respectively. *Source*: Reprinted with permission from Reference 31. See color insert.

GAPDH/β-F1 ATPase ratio was significantly higher in tumor tissue than in normal tissue from patients with different types of cancers including breast, colon, lung, and esophagus. Moreover, evidence was presented that the β-F1 ATPase activity is needed for initiating apoptosis (35). This would link apoptosis resistance to elevated glycolysis and reduced β-F1 ATPase activity (39).

Cuezva and colleagues concluded that all cancers regardless of their origin have a common bioenergetic signature due to dysfunction respiration. The findings from the Cuezva group were also supported by comprehensive studies from the Grammatico group in showing an altered mitochondrial structure and OxPhos function in invasive breast cancer (24). These findings, viewed together with Pedersen's extensive review of protein abnormalities in tumor mitochondria, provide substantial evidence showing that OxPhos is insufficient to maintain energy homeostasis in tumor cells.

The findings of the Cuezva and the Grammatico groups in a broad range of cancers are also supported by other proteomic studies in kidney cancer. Renal cell kidney carcinoma is the 10th most common cancer, and its incidence appears to be increasing (40). Simonnet and colleagues have shown that respiratory impairment was significantly greater in patients with clear cell or high grade renal tumors than in patients with low grade or benign renal tumors (41). Moreover, the respiratory impairment in these renal tumors was correlated with significant decreases in the content of ETC complexes II, III, and IV as well as with abnormal assembly of the complex V (the F_1F_0 ATPase).

These investigators linked their metabolic findings to defects in the von Hippel-Lindau (VHL) tumor suppressor gene and the hepatic-growth factor MET proto-oncogene. However, alterations in these genes alone were unable to account for differences in tumor aggression. Defects were found in these genes in some benign renal tumors, whereas no defects were found in these genes in some of the most aggressive and malignant renal tumors (41). It was surprising to me that these investigators tried to force their data to fit a gene defect model of renal tumor origin, but did not link their observations to Warburg's theory. Clearly, their data more strongly support an origin of cancer following respiratory dysfunction than an origin following gene dysfunction.

Unwin and coworkers (40) from the United Kingdom used a proteomic approach, based on two-dimensional gel electrophoresis and mass spectrometry, to compare the protein profiles of renal carcinoma tissue with tissue from patient-matched normal kidney cortex. The most striking findings from their study were the decreased expression of several mitochondrial enzymes implicated in OxPhos and the increased expression of enzymes for glycolysis. The increased expression of the glycolytic enzymes was also associated with a parallel decrease in three of the enzymes catalyzing the reverse reactions of gluconeogenesis (40). In addition to supporting a downregulation of mitochondrial enzymes involved in OxPhos, these investigators also found reductions in enzymes involved in other pathways including fatty acid and amino acid metabolism and the urea cycle, indicating a wider role for mitochondrial dysfunction in tumorigenesis (40). Pan and coworkers have also provided proteomic evidence for mitochondrial abnormalities in ovarian cancer (42), while Roman Eliseev and colleagues have provided credible evidence of mitochondrial dysfunction in osteosarcoma (11). I view these findings as providing evidence supporting Warburg's original theory of cancer.

LIPIDOMIC ABNORMALITIES IN TUMOR CELL MITOCHONDRIA

Besides proteomic evidence supporting the Warburg cancer theory, we have recently shown that the lipidome is also abnormal in tumor mitochondria. Although the genome and proteome have been the focus of much attention in tumorigenesis, little attention has been given to the lipidome as a potential origin of the tumorigenic phenotype. The lipidome refers to the total content and composition of all lipids in a cell or cell organelle. Lipids maintain the integrity of biomembranes. Abnormalities in lipids can compromise mitochondrial function. The functions of ETC proteins are dependent to a considerable degree on the lipid composition of the inner mitochondrial membrane. Lipid abnormalities in the inner mitochondrial membrane will therefore alter OxPhos capabilities.

Pedersen (15) has earlier reviewed a number of studies showing that mitochondrial lipid abnormalities are common in all tumors examined. A schematic diagram of the major lipids in mitochondria is shown in Figure 5.6. This diagram was generated from our extensive analysis of the mouse brain mitochondrial lipidome (43). We were also the first research group to investigate the mitochondrial lipidome in tumor cells using a multidimensional, mass-spectrometry-based shotgun lipidomic (MDMS-SL) approach (44).

It is important to see from Figure 5.6 that cholesterol is a relatively minor lipid of the inner mitochondrial membrane of normal cells. Feo and colleagues (46, 47) have previously shown that the cholesterol/phospholipid ratio was significantly higher in mitochondria from hepatomas than in mitochondria from normal liver cells. As cholesterol reduces membrane fluidity, elevated levels of cholesterol would be expected to reduce the fluidity properties of mitochondrial membranes. In contrast to mitochondrial phospholipids in normal tissues, which contain an abundance of long-chain polyunsaturated fatty acids, phospholipids in tumor mitochondria are enriched in short-chain saturated or monounsaturated species (15). We

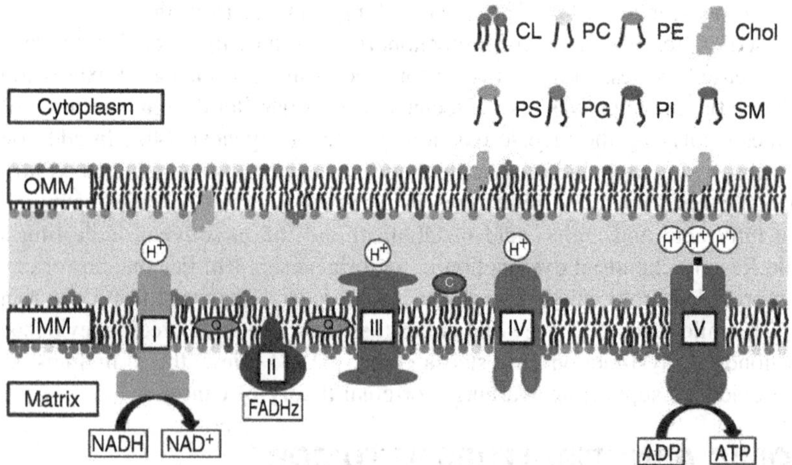

Figure 5.6 Topology of lipid distribution in mitochondrial membranes. Cardiolipin, shown containing four acyl chains, in enriched primarily in the inner mitochondrial membrane and plays an important role in maintaining the proton motive gradient and efficiency of the electron transport chain. *Source*: Reprinted with permission from Reference 45. See color insert.

have confirmed these findings in mouse brain tumors (44). Most importantly, we have found several abnormalities in the structure of cardiolipin (CL), the major lipid of the inner mitochondrial membrane (Figure 5.6).

CARDIOLIPIN: A MITOCHONDRIAL-SPECIFIC LIPID

CL (1,3-diphosphatidyl-sn-glycerol) is a complex, mitochondrial-specific phospholipid that regulates numerous enzyme activities, especially those related to OxPhos and coupled respiration (48–53). Several studies have shown that CL is essential for efficient oxidative energy production and mitochondrial function (48, 50, 53–67). CL is necessary for maintaining coupled mitochondria, and defects in CL can produce *protein independent* uncoupling (12, 32). Hence, alterations in the content or composition of CL will alter cellular respiration.

Before describing evidence linking CL abnormalities to mitochondrial dysfunction in brain tumors, it would be good to first briefly review information about the unique properties of this mitochondrial-specific lipid. CL contains two phosphate head groups, three glycerol moieties, and four fatty acyl chains and is primarily enriched in the inner mitochondrial membrane (Figs. 5.6 and 5.7). Enrichment in the inner mitochondrial membrane makes CL a pivotal molecule for regulating cristae structure and OxPhos (12). CL binds complex I, III, IV, and V and stabilizes the super complexes (I/III/IV, I/III, and III/IV), demonstrating an absolute requirement of CL for the catalytic activity of these respiratory enzyme complexes (44, 50, 51, 68, 69). CL restricts pumped protons within its head group domain, thus providing the structural basis for mitochondrial membrane potential and for supplying protons to the ATP synthase (49, 53).

Respiratory complex proteins that interact with CL form hydrophobic amino acid grooves on their surface (68, 70). These grooves accommodate the fatty acid chains of CL (Fig. 5.8). This is quite remarkable. Which biological structure evolved first, the lipids or the grooves? Since long-chain carbon molecules appeared earlier in evolution than membrane proteins, it is likely that the grooves evolved to accommodate already existing lipid fatty acids. While the amino acid sequence of electron transport proteins is highly conserved across species, considerable variability occurs for the fatty acid sequences of CL. Although the respiratory protein structure is largely invariant, the fatty acid composition of CL can be modulated through changes in nutrition and the physiological environment. Indeed, we have found that hypoxia could significantly modify brain CL fatty acid composition in VM mice (Seyfried and Ta, unpublished observation). CL can modulate ETC activities without altering the primary sequence of amino acids. Hence, changes in CL content and composition can influence electron transport and ultimately the efficiency of OxPhos.

The activity of respiratory enzymes in complex I and complex III and their linked activities are directly related to CL content (50, 63, 71). The activities of the respiratory enzyme complexes are also dependent on the composition of the CL molecular species (49). Importantly, the degree of CL unsaturation is related

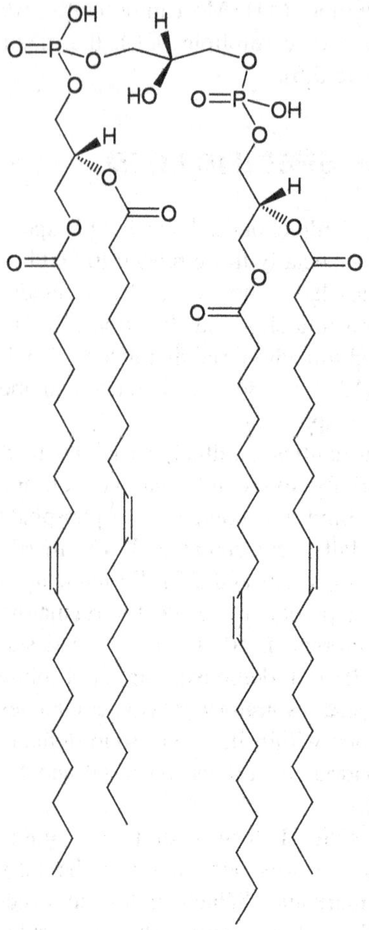

Figure 5.7 Structure of cardiolipin (1,1′,2,2′-tetraoleyl cardiolipin). Cardiolipin (CL) is a complex mitochondrial-specific phospholipid that regulates numerous enzyme activities especially those related to oxidative phosphorylation and coupled respiration (see text for details). *Source*: Reprinted with permission from Reference 44.

to states 1–3 of respiration (48, 58). Respiratory efficiency is dependent on the degree of CL remodeling. Remodeling is a complex process where immature CL is remodeled to form mature CL. This process involves the replacement of shorter chain and less unsaturated fatty acids in immature CL with longer chain and more complex (polyunsaturated) fatty acids in mature CL. In general, remodeling produces longer chain unsaturated species characteristic of CL in mature differentiated cells. The respiratory energy efficiency in tissues is therefore dependent to a large extent on the expression of mature CL.

Almost 100 molecular species of CL were recently detected in the mitochondria from mammalian brain (43, 72). Moreover, these molecular species form a beautiful symmetric pattern consisting of seven major groups when arranged according fatty acid chain length and degree of unsaturation (43) (Fig. 5.9). This unique fatty acid pattern is expressed in CL analyzed from both synaptic mitochondria (enriched in neurons) as well as in nonsynaptic mitochondria (mostly enriched in cell bodies of

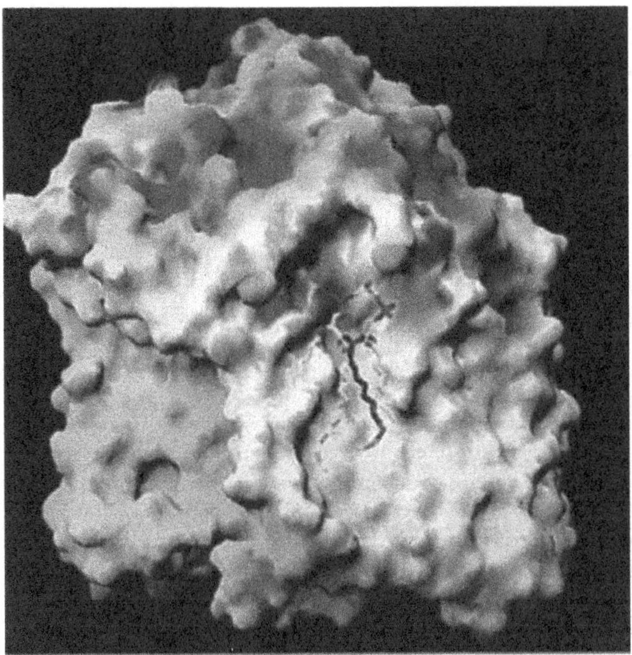

Figure 5.8 Coevolution of mitochondrial protein amino acid sequence with cardiolipin acyl chain composition. The amino acid sequence of highly conserved proteins of the electron transport chain evolved to generate hydrophobic regions to selectively mold to the structural diversity of cardiolipin molecular species. This symbiotic relationship between protein and lipid interactions generates functional regulation of enzymatic efficiency as well as emphasizes the importance of lipidomic organization of the mitochondrial membrane, thus linking the significance of the cardiolipin molecular species with enzymatic functionality. *Source*: Image reprinted with permission from Reference 70. See color insert.

neurons and glia) in mature mouse brain. CL analyzed from nonneural cells contains mostly tetra 18:2, that is, four 18-carbon chains with each chain containing two double bonds.

I consider the symmetric pattern of CL distribution as the biochemical signature of respiratory energy efficiency in brain. The pattern in the mouse brain is also seen in the human brain. Any alteration to the content or composition of CL will influence respiratory energy efficiency.

CARDIOLIPIN AND ABNORMAL ENERGY METABOLISM IN TUMOR CELLS

We have recently shown that the lipid composition and/or content in mouse brain tumor mitochondria differed markedly from that in mitochondria derived from the normal syngeneic host brain tissue (44). These brain tumors covered a spectrum of growth behaviors seen in most human malignant brain cancers. Two of the

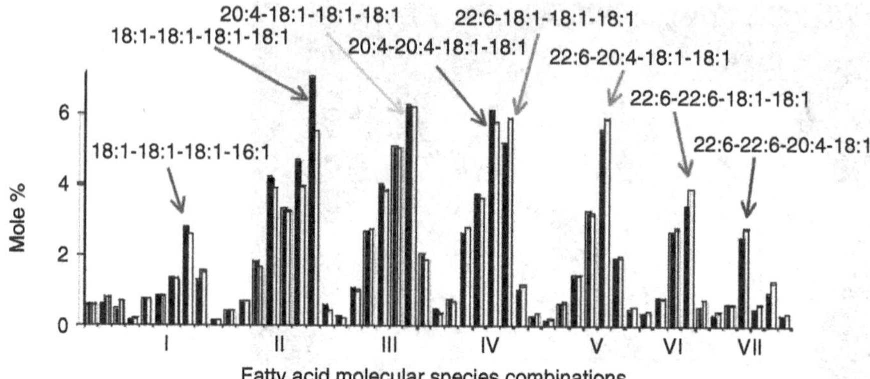

Figure 5.9 Distribution of CL molecular species in nonsynaptic (black bar) and synaptic (white bar) mitochondria of mouse brain. CL molecular species were arranged according to the mass-to-charge ratio based on percentage distribution. CL molecular species were subdivided into seven groups, which contained a predominance of oleic, arachidonic, and/or docosahexaenoic fatty acids in varying concentrations. The corresponding mass content of molecular species in nonsynaptic (NS) and synaptic (Syn) mitochondria were as described by us (43). All values are expressed as the mean of three independent samples ($n = 3$), where six mouse cerebral cortexes were pooled for each sample. *Source*: This figure is modified from its original form Reference 43. To see this figure in color please go to ftp://ftp.wiley.com/public/sci_tech_med/cancer_metabolic_disease.

tumors evaluated, an ependymoblastoma (EPEN) and an astrocytoma (CT-2A), were derived from implantation of 20-methylcholantherene into the brains of inbred C57BL/6J mice (73–75). I had received the EPEN tumor as a gift from Dr. William Sutton, as associate of Dr. Harry Zimmerman. I used Zimmerman's procedure to produce the CT-2A tumor. Three of the tumors evaluated, VM-M2, VM-M3, and VM-NM1, arose spontaneously in the brains of inbred VM mice. I have presented additional information on these tumors in Chapter 3.

The VM inbred strain is unique in developing a relatively high incidence of brain tumors (76). The VM-M2 and VM-M3 tumors express multiple properties of myeloid/mesenchymal cells and display the invasive growth behavior of human glioblastoma multiforme (77–79). The VM-NM1 is rapidly growing, but is neither highly invasive nor metastatic when grown outside the brain (78). We produced clonal cell lines from each of the five brain tumors. Each tumor was then grown subcutaneously in the syngeneic mouse host (44).

We employed both Ficoll and sucrose gradients to obtain highly purified mitochondria from normal mouse brain tissue and from mouse brain tumor tissue (43, 44). Using this isolation procedure, we were able to maintain both the structure and function of the purified mitochondria (43). Besides expressing multiple abnormalities in the major phospholipids, phosphatidylcholine, and phosphatidylethanolamine, we found that the content and composition of CL differed markedly between normal brain tissue and tumor tissue (Figs. 5.10 and 5.11a and 5.11b).

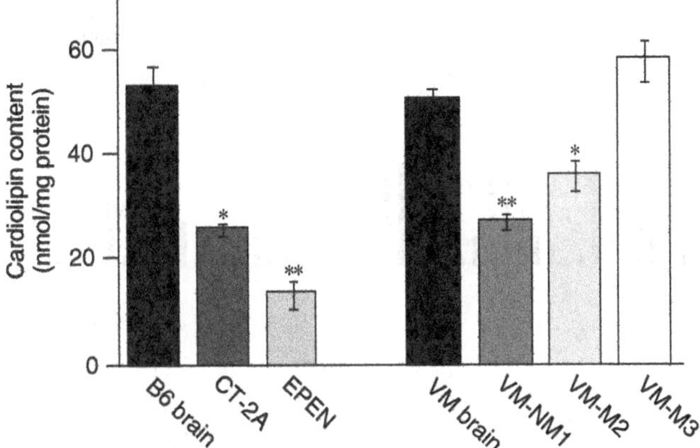

Figure 5.10 Cardiolipin content in mitochondria isolated from normal mouse brain and mouse brain tumors. Mitochondria were isolated as described by us (43). Values are represented as the mean ± standard deviation (SD) of three independent mitochondrial preparations from brain or tumor tissue. Asterisks indicate that the tumor values differ significantly from the B6 or the VM normal brain values at the $^{*}p < 0.01$ or $^{**}p < 0.001$ levels as determined by the two-tailed t-test. Reduction in CL content suggests fewer mitochondria or mitochondria with reduced amounts of inner membrane. *Source*: Reprinted with permission from Reference 44.

The CL content was significantly lower in the mitochondria from the CT-2A and the EPEN tumors than in the mitochondria from the normal control B6 mouse brain. In contrast to the B6 mouse brain, which contains about 100 molecular fatty acid species of CL distributed symmetrically over seven major groups (Fig. 5.11a), the VM mouse brain is unique in having only about 45 major CL molecular species and is missing molecular species in groups IV, V, and VII (Fig. 5.11b). We addressed the importance of the CL changes in the VM mice in relationship to the inheritance of brain tumors in this strain (80). The CL content was significantly lower in the mitochondria from the VM-NM1 and the VM-M2 tumors than in the mitochondria from the control VM mouse brain (44). We found that the CL abnormalities in these tumors were associated with significant reductions in ETC activities consistent with the pivotal role of CL in maintaining the structural integrity of the inner mitochondrial membrane (44, 48, 49, 55).

The activities of the ETC complexes I/III, II/III, and I were significantly lower in the tumors than in normal syngeneic brain tissue. As mitochondrial ETC activities depend on the content and the composition of CL, we used a *bioinformatics* approach to model ETC activities as a function of CL content and composition in the five mouse brain tumors. The two main variables included (i) the total CL content and (ii) the distribution of CL molecular species in mitochondria. The information about the molecular species distribution was simplified into a single number, which described the degree of the relationship of the CL composition of the tumor mitochondria with that of brain mitochondria from the host mouse strain

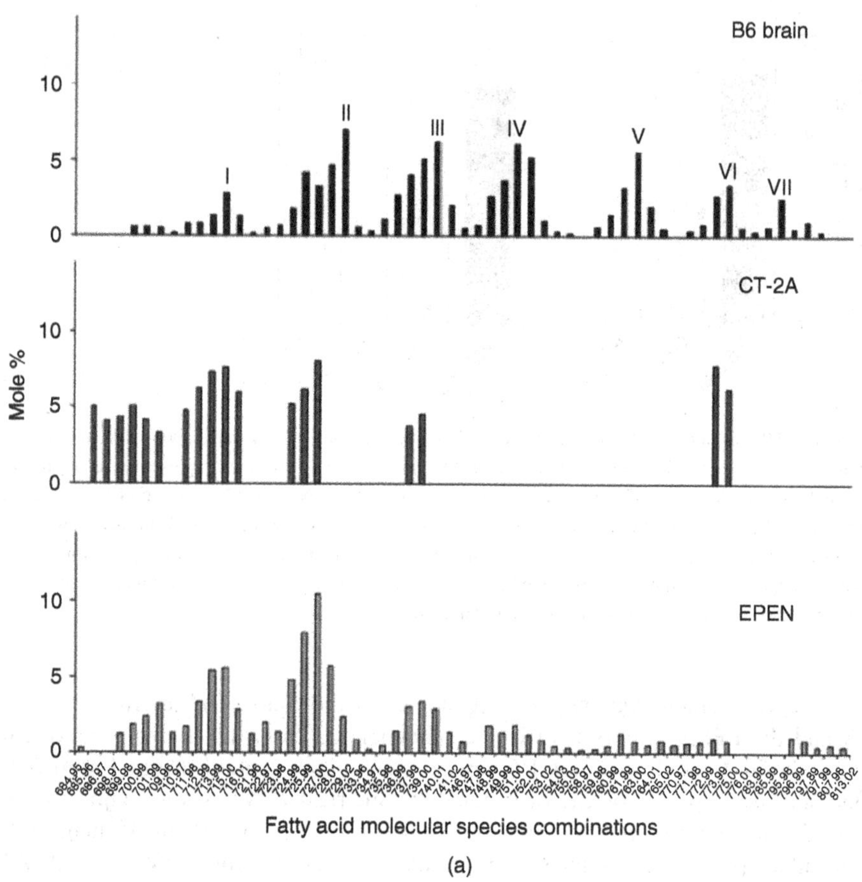

Figure 5.11 Distribution of cardiolipin molecular species in mitochondria isolated and purified from normal mouse brain and brain tumor mitochondria. (a) Distribution in a syngeneic C57BL/6J (B6) mouse brain and in the CT-2A and the EPEN tumors. (b) Distribution in syngeneic VM mouse brain and the VM-NM1, the VM-M2, and the VM-M3 tumors. Cardiolipin fatty acid molecular species are plotted on the abscissa and arranged according to the mass-to-charge ratio based on percentage distribution. The molecular species are subdivided into seven major groups (I–VII) as previously described by us and as described in Figure 5.9. Corresponding mass content of molecular species in normal brain and tumor mitochondria can be found in Table 1 of our previous study (44). It is clear that fatty acid molecular species composition differs markedly between and among tumors and their syngeneic mouse hosts brain tissue. As CL composition influences ETC activities and mitochondrial energy production, these findings indicate that mitochondrial energy efficiency differs between normal brain tissue and brain tumor tissue. All values are expressed as the mean of three independent mitochondrial preparations, where tissues from six brain cortexes or tumors were pooled for each preparation. *Source*: With permission from J. Lipid Res. (44).

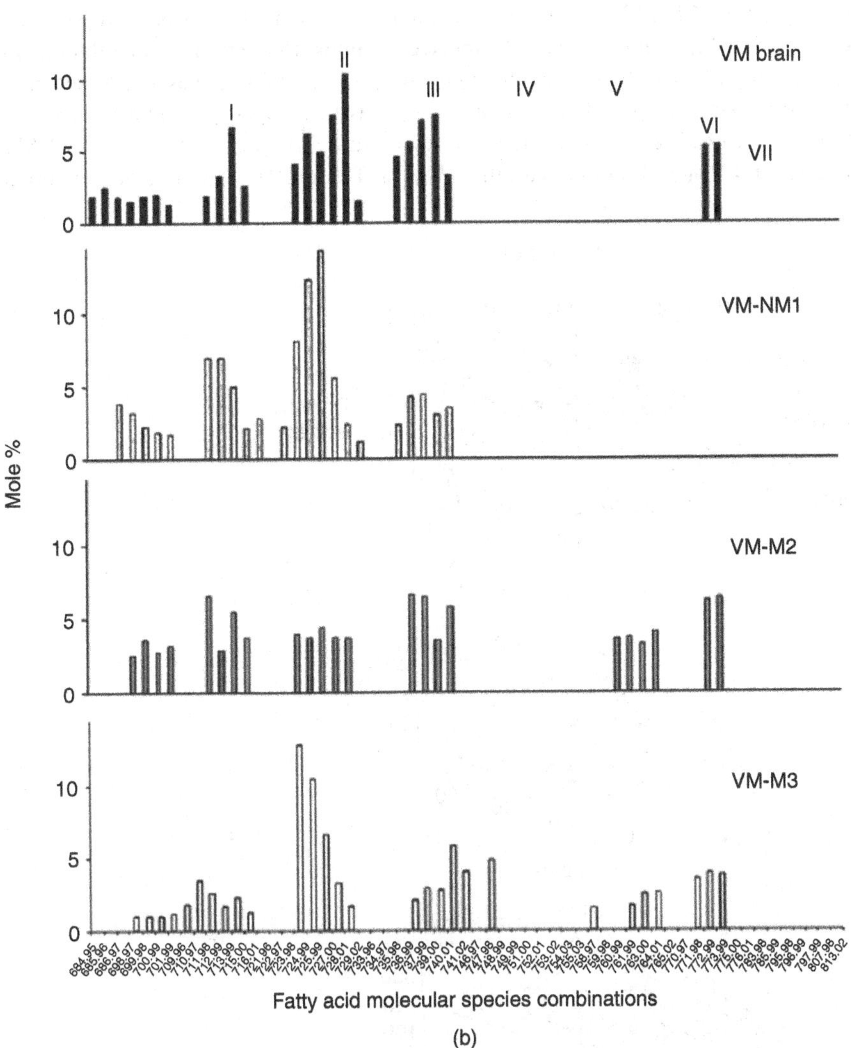

Figure 5.11 (*Continued*)

(44). This number was generated as a Pearson product–moment correlation. We used the correlation coefficient to assess the degree of "compositional similarity" of CL from the host mouse brain mitochondria with that of tumor mitochondria.

How did we interpret these data? A low coefficient indicates that CL fatty acid molecular species composition is dissimilar between the host brain mitochondria and the tumor mitochondria. A high correlation indicates that CL molecular species composition is similar between the host brain mitochondria and the tumor mitochondria. The ETC activities in each tumor were then measured using standard enzymatic procedures, whereas MDMS-SL was used to measure the content and composition of CL (43). A two-dimensional linear regression was used to fit the measured ETC activity values with the CL composition. The best-fit relationship for each complex with the content and composition of CL was expressed as a *quadratic surface* (Fig. 5.12). Our objective was to compare the data for the CT-2A and the EPEN tumors with their B6 host strain and to compare the VM-NM1, VM-M2, and VM-M3 tumors with their VM host strain. This analysis demonstrated

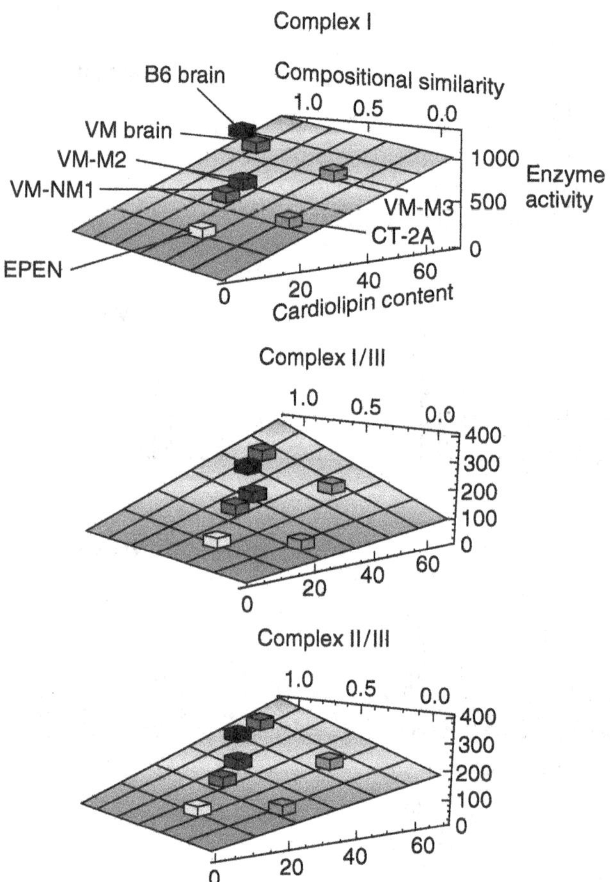

Figure 5.12 Relationship of cardiolipin abnormalities with electron transport chain activities in the B6 and the VM mouse brain tumors. The data are expressed on the best-fit three-dimensional quadratic surface for each electron transport chain complex as recently described by us (44). In order to illustrate the position of all tumors on the same graph relative to their host strain, the data for the VM strain and tumors were fit to the B6-fit quadratic surface as described (44). The data indicate that changes in tumor ETC activities can be directly related to changes in CL content and fatty acid molecular species composition. The data show that ETC complex activity differs markedly between tumors and their syngeneic B6 and VM hosts and that these differences are linked to abnormalities in the content and composition of CL. The results suggest that CL abnormalities are associated with reduced efficiency of respiration. *Source*: With permission from Reference 44. See color insert.

a direct relationship between the CL content, the distribution of molecular species, and ETC activity.

It was clear from our studies that abnormalities in the content and composition of CL could underlie the abnormal energy metabolism of these diverse brain tumors. This is the type of connection that Weinhouse (81) considered essential for establishing the credibility of the Warburg theory. Hence, our lipidomic studies in mouse brain tumors provide credibility to Warburg's original theory according to the Weinhouse argument.

Our findings are also consistent with earlier studies in rat hepatomas that show an increase in shorter chain saturated fatty acid content (palmitic and stearic) characteristic of immature CL (82, 83). Our studies are also consistent with more recent findings in rhabdomyosarcoma, a type of muscle tumor, which show that the reduction in complex I activity was associated with CL abnormalities (84). Continued expression of immature CL would reduce efficient respiratory energy production. In light of what we know about the CL structure and respiratory function, it is difficult to conceive how mitochondrial OxPhos could function normally in tumors that express CL abnormalities.

How might abnormalities in the content and composition of CL arise? Would CL abnormalities be related to the cause or effect of tumor formation? We proposed that CL abnormalities could arise from either inherited cancer risk factors as seen in the VM mice or from numerous epigenetic and environmental cancer risk factors including inflammation, viruses, hypoxia, radiation, and so on (Fig. 5.13)

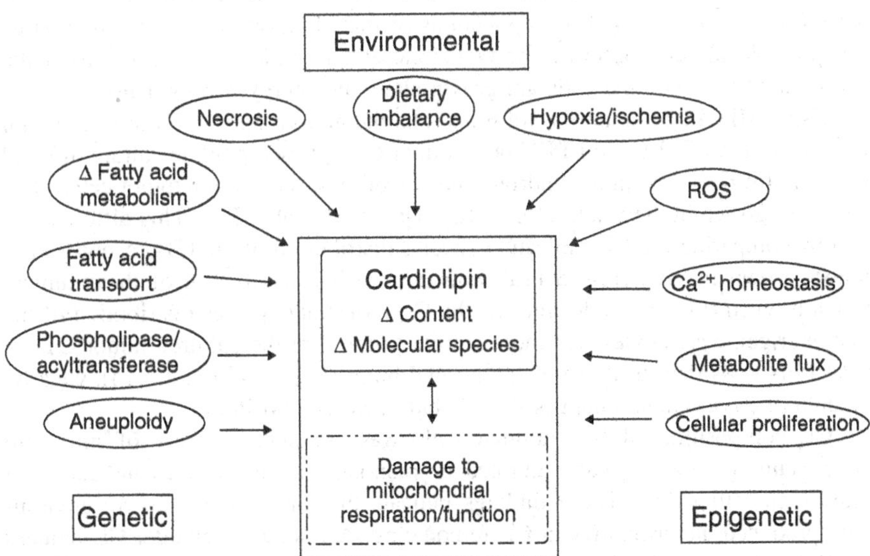

Figure 5.13 Relationship of genetic, epigenetic, and environmental factors to dysfunctional respiration associated with abnormalities in cardiolipin content and composition. ROS, reactive oxygen species. *Source*: Reprinted with permission from Reference 44.

(32, 44, 80). Indeed, γ-radiation is known to induce free radical CL fragmentation, which would compromise respiratory function (85). On the basis of these and other observations, we suggested that most tumors, regardless of cell origin, would contain abnormalities in CL composition and/or content (44). Regardless of whether the CL abnormalities are related to the cause of the tumor or arise during tumor progression, the CL abnormalities will significantly reduce the efficiency of mitochondrial OxPhos.

Our findings of immature CL molecular species in tumor mitochondria together with associated abnormalities in respiratory function are consistent with the recent findings of Eliseev and colleagues who showed enhanced replication of immature mitochondria in malignant osteosarcoma cells compared to that in normal osteoblasts (11). Mitochondrial immaturity would predict insufficient energy production through OxPhos. The association of CL abnormalities with an impaired respiratory function would be expected on the basis of the localization and role of CL in ETC activities.

COMPLICATING INFLUENCE OF THE IN VITRO GROWTH ENVIRONMENT ON CARDIOLIPIN COMPOSITION AND ENERGY METABOLISM

It is important to recognize that numerous studies of energy metabolism in tumor cells are conducted on the cells grown in tissue culture. We recently showed for the first time that the in vitro growth environment produces lipidomic and electron transport abnormalities in mitochondria from both nontumorigenic cells as well as from tumor cells (86). The implications of this observation are profound. How is it possible to fully describe the metabolic abnormalities of cancer cells if the environment in which the cells are grown alters the energy metabolism?

Using MDMS-SL analysis, we found that the non-synaptic (NS) mitochondrial lipidome of the CT-2A and EPEN brain tumor cells grown in tissue culture differed markedly from the NS mitochondrial lipidome of these same brain tumor cells when they were grown in their natural CS7BL/6 (B6) host (Table 5.1). This difference is seen by comparing the CL molecular species distribution of the CT-2A and EPEN tumors grown in vivo (Fig. 5.11a) with the species distribution of these tumors grown in vitro (Fig. 5.14). Moreover, the CL molecular species distribution of the nontumorigenic astrocytes was more similar to those of the cultured tumor CT-2A and EPEN cells than to those of the normal brain (Figs. 5.11a and 5.14). Clearly, the culture environment changes the CL fatty acid composition.

CL composition of the cultured cells was composed largely of immature CL containing shorter chain saturated or monounsaturated species indicative of failed remodeling (86). These findings indicate that the in vitro growth environment produced abnormalities in CL remodeling. A failure to remodel CL reduces efficient energy production through OxPhos. Our findings in the mouse brain tumors are consistent with numerous reports indicating that the content and composition of CL is essential for normal respiratory function (48, 50, 53–67, 84).

Table 5.1 Lipid Composition of Mitochondria Isolated from Brain, Brain Tumour and Cells

	In vivo			In vitro		
Lipid	Brain[c]	CT-2A	EPEN	Astrocyte[c]	CT-2A	EPEN
EtnGpl	187.4 ± 12.1	245.9 ± 13.7**	368.4 ± 46.4*	171.4 ± 18.6	163.0 ± 6.5	211.0 ± 16.7
PtdEtn	164.9 ± 10.0	137.3 ± 6.0*	259.4 ± 45.7	85.9 ± 3.4	69.7 ± 3.0**	98.9 ± 2.3*
PlsEtn	22.5 ± 2.2	99.3 ± 7.1**	147.8 ± 21.4**	80.5 ± 15.0	87.5 ± 9.4	106.9 ± 14.4
PakEtn	N.D.	9.3 ± 0.7**	12.4 ± 3.0*	5.0 ± 0.3	5.7 ± 0.1*	5.1 ± 1.1
ChoGpl	129.9 ± 7.7	121.2 ± 3.6	160.0 ± 29.5	168.5 ± 14.2	127.4 ± 13.2*	194.4 ± 15.7
PtdCho	119.6 ± 5.3	81.4 ± 3.4**	127.4 ± 25.4	124.6 ± 10.5	98.1 ± 12.4*	174.3 ± 13.9**
PlsCho	1.2 ± 0.1	19.4 ± 2.1***	11.6 ± 4.8	22.4 ± 4.0	15.8 ± 0.8	9.3 ± 0.8*
PakCho	9.1 ± 3.2	20.4 ± 2.6**	17.0 ± 6.2	21.5 ± 2.1	13.5 ± 0.4*	10.8 ± 1.1**
Cardiolipin	52.7 ± 4.5	26.1 ± 1.0**	13.5 ± 2.7***	28.3 ± 4.3	24.6 ± 3.7	31.1 ± 4.2
PtdIns	9.4 ± 0.8	9.5 ± 2.6	19.4 ± 2.5*	18.5 ± 2.4	18.4 ± 1.6	20.5 ± 3.3
PtdGro	7.1 ± 0.5	9.8 ± 0.5**	16.4 ± 3.6*	7.7 ± 2.6	7.6 ± 1.5	4.7 ± 0.3
CerPCho	5.3 ± 1.2	4.6 ± 0.2	5.8 ± 1.8	9.7 ± 1.4	15.9 ± 3.4	22.1 ± 1.3**
PtdSer	4.6 ± 1.5	9.1 ± 0.6*	10.4 ± 2.0*	17.8 ± 0.4	28.7 ± 5.7	24.1 ± 3.6
LysoPtdCho	2.7 ± 0.6	6.3 ± 0.6**	2.8 ± 0.4	1.5 ± 0.1	2.2 ± 0.4	2.4 ± 0.6
Cer	0.7 ± 0.2	2.3 ± 0.2***	1.7 ± 0.2**	0.9 ± 0.1	1.0 ± 0.5	2.0 ± 0.2**

Values are expressed as mean nmol/mg of protein ± S.D. ($n = 3$). *$p < 0.05$; **$p < 0.01$ and ***$p < 0.001$, significantly different values from B6 NS or astrocyte mitochondria. N.D., not detected.

[c]B6 NS brain mitochondria or astrocyte (C8-D1A) mitochondria were used as controls.

Source: Reprinted with permission from ASN Neuro 2009 May 27; 1(3).

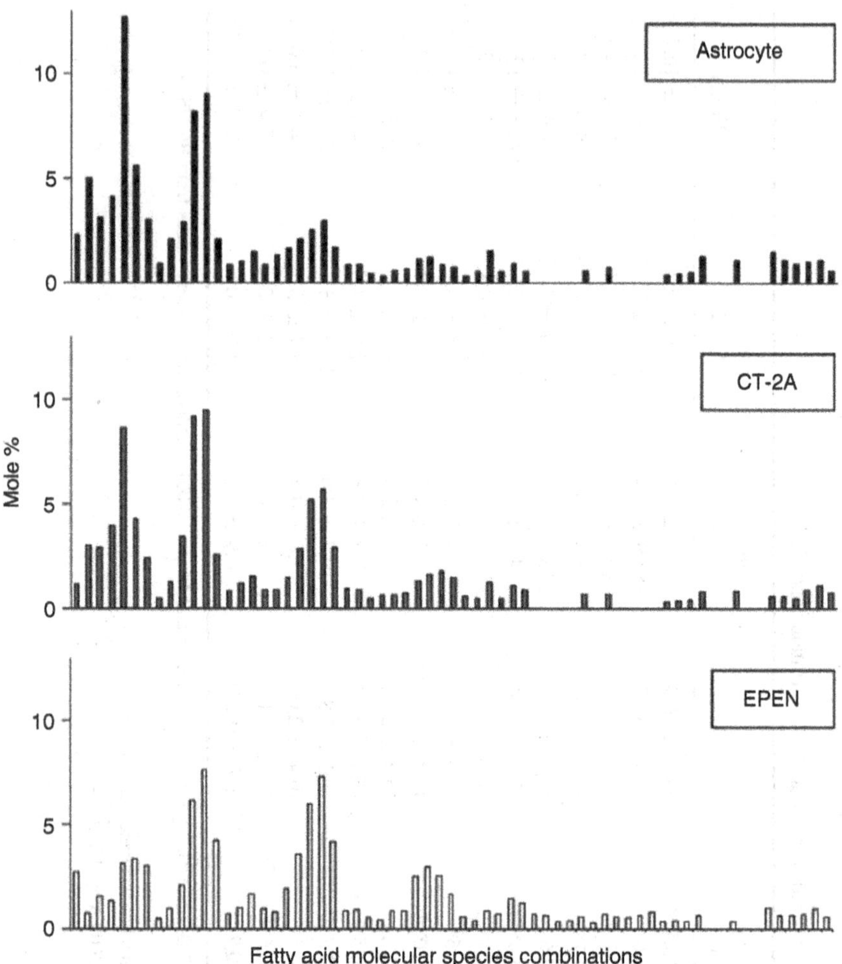

Figure 5.14 Distribution of cardiolipin molecular species in mitochondria isolated from astrocytes and brain tumor cells grown in vitro. In contrast to the unique differences seen for CL composition between normal brain and the CT-2A and EPEN brain tumors grown *in vivo* (Fig. 5.12a and 5.12b), no major differences are seen between nontumorigenic astrocytes and the CT-2A and EPEN tumors when grown as cultured cells. It appears that growth in the cell culture environment alters CL composition. *Source*: Reprinted with permission from Reference 86.

Hence, brain tumors with CL abnormalities require an alternative energy source to OxPhos in order to maintain viability.

A failure to remodel immature CL to mature CL alters the activities of respiratory enzyme activities. That such alterations occur in the cultured cells is clearly illustrated in Figure 5.15. These findings connect immature CL to reduced ETC enzyme activities. The activity of complex I was especially reduced in the cultured cells. Complex I activity is essential for the initiation of electron transport.

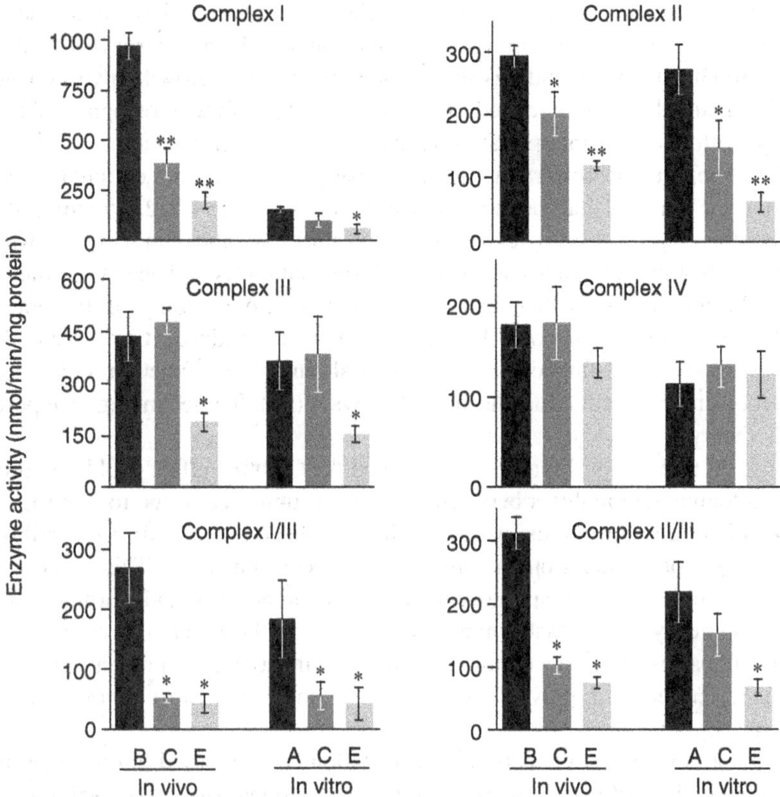

Figure 5.15 Electron transport chain (ETC) enzyme activities in purified mitochondria from mouse brain, brain tumors, and cultured tumor cells. Enzyme activities are expressed as nmol/min/mg protein as described (86). B, C, E, and A represent enzyme activities in mitochondria isolated from normal brain, CT-2A, EPEN, and astrocytes (nontumorigenic), respectively. Other conditions are as described (86). Asterisks indicate that the activities in the brain tumor samples differ from those of the control samples (either mouse brain or astrocytes) at the $^{*}p < 0.03$ or $^{**}p < 0.005$ levels as determined by the two-tailed t-test. *Source*: Reprinted with permission from Reference 86.

The linked complex I/III activities were also significantly reduced in the cultured cells. These findings indicate that growth in the in vitro environment can reduce electron transport and energy production through OxPhos. If energy through OxPhos is compromised in cultured tumor cells, how do these cells maintain their viability?

High levels of glucose and other metabolites in culture media can increase glycolysis and inhibit OxPhos. This effect was first described by Herbert Crabtree in the late 1920s and is referred to as the *Crabtree effect* (86–89). We did not exclude the possibility that the lipidomic and ETC abnormalities observed in the cultured nontumorigenic astrocytes and brain tumors cells could arise in part from the Crabtree effect. It is also interesting that several lipidomic differences found between the brain tumor mitochondria and normal mitochondria in the in vivo environment are not seen between the brain tumor cells and nontumorigenic astrocytes in the in vitro environment. These findings indicate that the in vitro growth environment obscures lipidomic differences related to tumorigenesis. A failure to recognize these facts could confound data interpretation related to energy metabolism.

Further support for respiratory energy insufficiency in the cultured cells comes from our findings that lactate production is high in the CT-2A tumor cells and the nontumorigenic astrocytes when grown under identical in vitro growth conditions, indicating that aerobic glucose fermentation is enhanced in these cells (86). Recent studies on rhabdomyosarcoma support our findings in the brain tumors (84). Freyssenet and coworkers have shown that reduced CL content in rhabdomyosarcoma was linked to mitochondrial energy dysfunction requiring compensatory energy production through glycolysis (84). These findings support Warburg's thcory.

We suggest that cell proliferation in vitro and the Crabtree effect could obscure or mask lipidomic abnormalities between normal and tumor cells due to tumorigenesis (86). In contrast to nontumorigenic cells, which do not contain irreversibly impaired respiration, respiration in tumor cells appears impaired. This could be checked by comparing the respiratory function in tumor cells and normal cells grown in respiratory media that simulate conditions in the in vivo environment. Respiratory impairment would require enhanced fermentation to prevent apoptosis. Enhanced fermentation prevents differentiation and is linked to unbridled cell proliferation.

In addition to enhancing aerobic fermentation, an impaired mitochondrial lipidome could also influence energy production through substrate-level phosphorylation in the TCA cycle itself as we have recently reported (90). It is well documented that glutamine is a necessary energy metabolite for many cells grown in culture. TCA cycle substrate-level phosphorylation, together with glycolysis could compensate for the energy lost through OxPhos in order to preserve cell viability (90–93). This could explain in large part why proliferating cultured cells, either tumorigenic or nontumorigenic, rely on glutaminolysis and glycolysis for viability. I address this issue more in Chapter 8.

As most metazoan cells did not evolve to grow as microorganisms, growth in high glucose culture media would produce a physiological state different from that of the intact tissue environment. Viewed collectively, our findings indicate that the in vitro growth environment produces lipidomic and ETC abnormalities in nontumorigenic astrocytes and in brain tumor cells, which would disrupt energy production through OxPhos and confound the relationship of altered energy metabolism to tumorigenesis. It is surprising that many researches in the cancer field appear to be unaware of this.

In summary, I have provided information documenting the role of CL in OxPhos. Abnormalities in CL can reduce ATP production through OxPhos. Abnormalities in CL can arise through the process of tumorigenesis and from the growth of mammalian cells in the in vitro environment. A take-home message from these studies is that caution should be used in comparing energy metabolism in nontumorigenic cells and tumorigenic cells grown in tissue culture environments that do not replicate the growth conditions of the in vivo environment. In light of our findings, it is surprising that many investigators still consider OxPhos as normal in cultured cancer cells (94, 95). This subject will be addressed further in the next chapter. As cancer is primarily a metabolic disease, further studies of CL in tumor cells will provide insight on the bioenergetic abnormalities of brain cancer and of other cancers. According to my hypothesis, changes in CL structure would compromise OxPhos efficiency and would precede the genomic instability seen in most cancer cells.

MITOCHONDRIAL UNCOUPLING AND CANCER

Although O_2 consumption and CO_2 production occurs in mitochondria of normal cells, oxygen uptake and CO_2 production can also occur in mitochondria that are uncoupled and/or dysfunctional (4, 32, 96). Uncoupling involves dissipation of the mitochondrial proton motive gradient. Uncoupling can produce heat rather than ATP. Mitochondrial uncoupling occurs during cold acclimation in mammals and is mediated, at least in part, by uncoupling proteins (96). Mitochondrial uncoupling in tumor cells, however, can arise from damage to the structure of the inner mitochondrial membrane (44, 96). This damage will contribute in large part to a dependence on substrate-level phosphorylation in order to maintain viability as described in Chapter 3 and in Reference 32. ATP synthesis through mitochondrial fermentation involving substrate-level phosphorylation could give the false impression that tumor mitochondria produce ATP through coupled respiration (15). The failure to recognize ATP production through nonoxidative processes in tumor mitochondria can contribute in part to the confusion surrounding the Warburg theory of cancer (32, 97).

While reduced oxygen uptake can be indicative of reduced OxPhos, increased oxygen uptake may or may not be indicative of increased OxPhos and ATP production (32, 84, 96, 98). Ramanathan and coworkers (98) have shown that oxygen consumption was greater, but oxygen-dependent (aerobic) ATP synthesis was less in cells with greater tumorigenic potential than in cells with lower tumorigenic potential. In other words, oxygen consumption was not linked to respiratory energy production in these malignant tumor cells. Hence, oxygen consumption in tumor cells could provide misinformation on the respiratory capacity of the cells.

These findings are consistent with mitochondrial uncoupling in tumor cells. Villalobo and Lehninger (99) have shown earlier that normally ejected hydrogen protons leak back into the matrix to a greater extent in tumor cells than in normal cells. Indeed, H^+ back-decay was eightfold greater in tumor mitochondria than in normal mitochondria (99). According to our hypothesis, back-decay and uncoupling arises from defects in the content or molecular species composition of CL, which is largely responsible for maintaining the proton impermeability of the inner mitochondrial membrane. I do not, however, exclude the possibility that increased expression of uncoupling proteins might also be involved in some cases (96, 100). Considered together, these findings are consistent with Warburg's views that the origin of cancer is linked to damaged or insufficient respiration.

CANCER CELL HEAT PRODUCTION AND UNCOUPLED MITOCHONDRIA

Mitochondria of brown adipose are naturally uncoupled so that oxidation of substrates in these cells produces heat rather than a proton motive gradient for ATP synthesis (101, 102). As heat production is a characteristic of uncoupled mitochondria, it would be important to know if heat production is greater in more tumorigenic cells than in less tumorigenic cells. Such evidence would support the hypothesis that mitochondrial uncoupling is expressed in tumor cells. Although few studies have been carried out on this subject, I have found evidence supporting the hypothesis that heat production is greater in less differentiated tumor cells than in more differentiated tumor cells.

For example, van Wijk and coworkers (103) have shown that glucose consumption and heat production was greater in less differentiated rat hepatoma cells (HTC) than in the more differentiated cells (H35). While CO_2 production was similar in the HTC and H35 cells, glucose consumption was fourfold greater in the HTC cells than in the H35 cells. Their data also show that heat production was threefold greater in the HTC cells than in the H35 cells (Fig. 5.16). The greater heat production in the less differentiated cells supports the hypothesis that mitochondrial uncoupling is greater in cancer cells that are more malignant than in those that are less malignant.

The findings in rat hepatoma cells were also supported with microcalorimetric investigations in patients with non-Hodgkin lymphoma (104). Monti et al. (104) have shown that the heat production rate per tumor cell was significantly greater in patients with high grade, non-Hodgkin lymphoma than in patients with lower grade malignancy. Moreover, lymphoma cell heat production was significantly greater in patients who died within two years of diagnosis than in patients who survived more than two years following diagnosis (104). Similar results were obtained in a thermal analysis of breast cancer in that prognosis was worse for patients with warmer tumors than for those with cooler tumors (105, 106). Zhao and coworkers (107) also used a thermocouple to show that tumor malignancy was positively correlated with tumor temperature. Although tumor heat production was associated with poor patient prognosis in all of these studies, none of the studies linked increased heat production with mitochondrial uncoupling.

As uncoupling leads to heat production, it is possible that the increased heat production in the more aggressive tumors is due to mitochondrial uncoupling. It is also interesting that heat production was correlated with increased glucose consumption and lactic acid production in human leukemia cells (108). This supports Warburg's theory that aerobic fermentation compensates for insufficient respiration. I suggest that these observations in tumor cells are due to greater mitochondrial

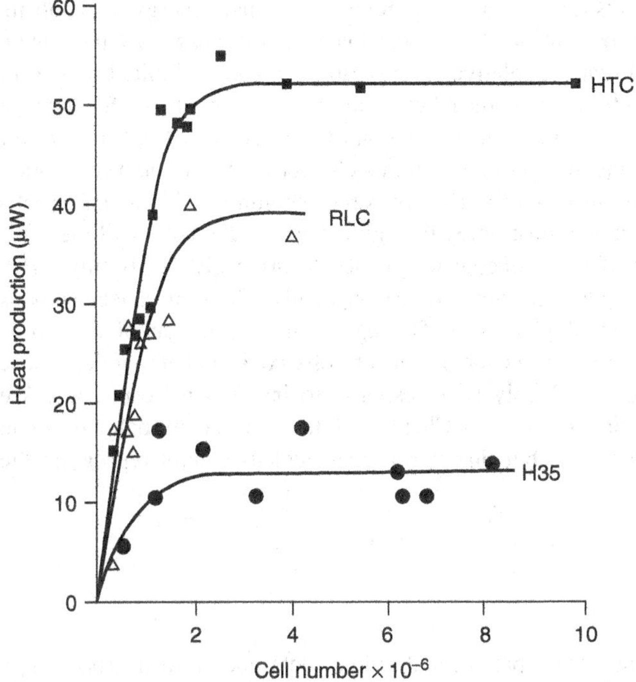

Figure 5.16 Heat production in differentiated H35 cells (dark circles), and in the poorly differentiated RLC (triangles) and HTC (dark squares) hepatoma cells. All cultures were harvested at subconfluency and prepared as described (103). Heat production was measured by diathermic isoperibol reaction calorimeter at different cell densities and at various experimental sessions. Heat production was registered for 1-h intervals; each point represents an individual measurement of initial heat production. The results show that heat production is greater in the poorly differentiated hepatoma cells than in the more differentiated H35 cells. *Source*: Modified from Chart 1 of Reference 103.

uncoupling. The greater the uncoupling, the greater will be the need to produce energy through substrate-level phosphorylation. In this case it appears to be aerobic glucose fermentation.

Considered collectively, these findings suggest that mitochondrial uncoupling, as evidenced from heat production, is greater in the more malignant tumor cells than in the less malignant tumor cells. Mitochondrial uncoupling in tumor cells is consistent with Warburg's theory that OxPhos insufficiency is the origin of cancer. Some researchers, however, might suggest that mitochondrial uncoupling is simply the result of the genetic defects in the tumor cells (96). I will address the issue of causality in later chapters.

PERSONAL PERSPECTIVE

It is not clear to me how any investigators in the cancer field would have difficulty in appreciating or comprehending the information presented in this chapter. However, most of this information is ignored in current reviews of cancer energy metabolism. Indeed, in their recent review on the subject Dang and colleagues have stated: "Today, we understand that the relative increase in glycolysis exhibited by cancer cells under aerobic conditions was mistakenly interpreted as evidence for damage to respiration instead of damage to the regulation of glycolysis." (95). It is hard for me to believe that the evidence presented in this chapter for insufficient respiration in cancer cells would be mistaken for glycolysis reprogramming. Most cancer cells suffering respiratory damage must upregulate glycolysis, or the cells will die. This will involve activation of those oncogenes needed to drive glycolysis (myc, Hif, Akt, etc.). I do not consider elevated expression of glycolysis in cancer cells as damage to the regulation of glycolysis. Damage to the regulation of glycolysis would certainly be apparent, however, if cancer cells could not upregulate those oncogenes needed to drive glycolysis in response to insufficient respiration. We know from the evidence presented in Chapter 11 that normal mitochondria can reprogram the cancer nucleus, but that the normal nucleus cannot reprogram the tumor mitochondria.

SUMMARY

The bulk of the evidence from patient and animal data and from a broad range of experimental approaches indicates that mitochondrial structure and function is abnormal in cancer cells. I think the difficulty in recognizing the evidence for OxPhos deficiency in tumor cells could arise due to several reasons. First, there are many cancer researchers and oncologists who have never heard of Warburg or his theory of cancer. This is most unsettling. I know this firsthand from having spoken to numerous cancer researchers and clinical oncologists. How will it be possible to make real advances in cancer management and prevention if many researchers and therapists working on the disease are unfamiliar with the origin of the disease? Second, there are many cancer researchers and oncologists who simply do not know what Warburg actually found in his experiments. It is surprising how many people do not read the original papers and rely on third parties for their information. Third, there are many cancer researchers and oncologists with data from their own experiments that support Warburg's theory, but this information is not recognized as supporting the theory. Finally, there are many cancer researchers and oncologists who feel that Sidney Weinhouse and others have effectively discredited Warburg's theory and that respiration functions normally in most cancer cells. I address this perspective more in Chapter 7.

On the basis of the evidence reviewed in this chapter, I am convinced that respiration differs between normal cells and cancer cells. In the next chapter, I will address the shortcomings of studies suggesting that respiration and OxPhos is normal in cancer cells. It is only through comprehensive evaluation of the evidence for insufficient respiration that we might be able to resolve this issue and move the cancer field forward.

REFERENCES

1. Ferreira LM. Cancer metabolism: the Warburg effect today. Exp Mol Pathol. 2010;89:372–80.
2. Nelson DL, Cox MM. Lehninger: Principles of Biochemistry. 5th ed. New York: W. H. Freeman and Company; 2008.
3. Manchester K. The quest by three giants of science for an understanding of cancer. Endeavour. 1997;21:72–6.
4. Warburg O. On the origin of cancer cells. Science. 1956;123:309–14.
5. Warburg O. On the respiratory impairment in cancer cells. Science. 1956;124:269–70.
6. Frey TG, Mannella CA. The internal structure of mitochondria. Trends Biochem Sci. 2000;25: 319–24.
7. Mannella CA. Introduction: our changing views of mitochondria. J Bioenerg Biomembr. 2000;32:1–4.
8. Carew JS, Huang P. Mitochondrial defects in cancer. Mol Cancer. 2002;1:9.
9. Mannella CA. The relevance of mitochondrial membrane topology to mitochondrial function. Biochim Biophys Acta. 2006;1762:140–7.
10. Benard G, Rossignol R. Ultrastructure of the mitochondrion and its bearing on function and bioenergetics. Antioxid Redox Signal. 2008;10:1313–42.
11. Shapovalov Y, Hoffman D, Zuch D, Bentley K, Eliseev RA. Mitochondrial dysfunction in cancer cells due to aberrant mitochondrial replication. J Biol Chem. 2011;286:22331–8.
12. Alirol E, Martinou JC. Mitochondria and cancer: is there a morphological connection? Oncogene. 2006;25:4706–16.
13. Arismendi-Morillo GJ, Castellano-Ramirez AV. Ultrastructural mitochondrial pathology in human astrocytic tumors: potentials implications pro-therapeutics strategies. J Electron Microsc. 2008;57:33–9.
14. Detmer SA, Chan DC. Functions and dysfunctions of mitochondrial dynamics. Nat Rev Mol Cell Biol. 2007;8:870–9.
15. Pedersen PL. Tumor mitochondria and the bioenergetics of cancer cells. Prog Exp Tumor Res. 1978;22:190–274.
16. Mates JM, Segura JA, Campos-Sandoval JA, Lobo C, Alonso L, Alonso FJ, et al. Glutamine homeostasis and mitochondrial dynamics. Int J Biochem Cell Biol. 2009;41:2051–61.
17. Potter JS, Ward EN. Mitochondria in lymphocytes of normal and leukemic mice. Cancer Res. 1942;2:655–9.
18. Springer EL. Comparative study of the cytoplasmic organelles of epithelial cell lines derived from human carcinomas and nonmalignant tissues. Cancer Res. 1980;40:803–17.
19. Kim HK, Park WS, Kang SH, Warda M, Kim N, Ko JH, et al. Mitochondrial alterations in human gastric carcinoma cell line. Am J Physiol Cell Physiol. 2007;293:C761–71.
20. Masters JR. HeLa cells 50 years on: the good, the bad and the ugly. Nat Rev. 2002;2: 315–9.
21. Piva TJ, McEvoy-Bowe E. Oxidation of glutamine in HeLa cells: role and control of truncated TCA cycles in tumour mitochondria. J Cell Biochem. 1998;68:213–25.
22. Rossignol R, Gilkerson R, Aggeler R, Yamagata K, Remington SJ, Capaldi RA. Energy substrate modulates mitochondrial structure and oxidative capacity in cancer cells. Cancer Res. 2004;64:985–93.
23. Arismendi-Morillo G. Electron microscopy morphology of the mitochondrial network in human cancer. Int J Biochem Cell Biol. 2009;41:2062–8.
24. Putignani L, Raffa S, Pescosolido R, Aimati L, Signore F, Torrisi MR, et al. Alteration of expression levels of the oxidative phosphorylation system (OXPHOS) in breast cancer cell mitochondria. Breast Cancer Res Treat. 2008;110:439–52.
25. Paumard P, Vaillier J, Coulary B, Schaeffer J, Soubannier V, Mueller DM, et al. The ATP synthase is involved in generating mitochondrial cristae morphology. EMBO J. 2002;21:221–30.

26. Oudard S, Boitier E, Miccoli L, Rousset S, Dutrillaux B, Poupon MF. Gliomas are driven by glycolysis: putative roles of hexokinase, oxidative phosphorylation and mitochondrial ultrastructure. Anticancer Res. 1997;17:1903–11.
27. Oudard S, Arvelo F, Miccoli L, Apiou F, Dutrillaux AM, Poisson M, et al. High glycolysis in gliomas despite low hexokinase transcription and activity correlated to chromosome 10 loss. Br J Cancer. 1996;74:839–45.
28. Bayley JP, Devilee P. Warburg tumours and the mechanisms of mitochondrial tumour suppressor genes. Barking up the right tree? Curr Opin Genet Dev. 2010;20:324–9.
29. Roskelley RC, Mayer N, Horwitt BN, Salter WT. Studies in Cancer Vii. Enzyme deficiency in human and experimental cancer. J Clin Invest. 1943;22:743–51.
30. Isidoro A, Casado E, Redondo A, Acebo P, Espinosa E, Alonso AM, et al. Breast carcinomas fulfill the Warburg hypothesis and provide metabolic markers of cancer prognosis. Carcinogenesis. 2005;26:2095–104.
31. Cuezva JM, Ortega AD, Willers I, Sanchez-Cenizo L, Aldea M, Sanchez-Arago M. The tumor suppressor function of mitochondria: Translation into the clinics. Biochim Biophys Acta. 2009;1792:1145–58.
32. Seyfried TN, Shelton LM. Cancer as a metabolic disease. Nutr Metab. 2010;7:7.
33. Acebo P, Giner D, Calvo P, Blanco-Rivero A, Ortega AD, Fernandez PL, et al. Cancer abolishes the tissue type-specific differences in the phenotype of energetic metabolism. Trans Oncol. 2009;2:138–45.
34. Ortega AD, Sanchez-Arago M, Giner-Sanchez D, Sanchez-Cenizo L, Willers I, Cuezva JM. Glucose avidity of carcinomas. Cancer Lett. 2009;276:125–35.
35. Cuezva JM, Krajewska M, de Heredia ML, Krajewski S, Santamaria G, Kim H, et al. The bioenergetic signature of cancer: a marker of tumor progression. Cancer Res. 2002;62:6674–81.
36. Cuezva JM, Chen G, Alonso AM, Isidoro A, Misek DE, Hanash SM, et al. The bioenergetic signature of lung adenocarcinomas is a molecular marker of cancer diagnosis and prognosis. Carcinogenesis. 2004;25:1157–63.
37. Isidoro A, Martinez M, Fernandez PL, Ortega AD, Santamaria G, Chamorro M, et al. Alteration of the bioenergetic phenotype of mitochondria is a hallmark of breast, gastric, lung and oesophageal cancer. Biochem J. 2004;378:17–20.
38. Lopez-Rios F, Sanchez-Arago M, Garcia-Garcia E, Ortega AD, Berrendero JR, Pozo-Rodriguez F, et al. Loss of the mitochondrial bioenergetic capacity underlies the glucose avidity of carcinomas. Cancer Res. 2007;67:9013–7.
39. Hanahan D, Weinberg RA. The hallmarks of cancer. Cell. 2000;100:57–70.
40. Unwin RD, Craven RA, Harnden P, Hanrahan S, Totty N, Knowles M, et al. Proteomic changes in renal cancer and co-ordinate demonstration of both the glycolytic and mitochondrial aspects of the Warburg effect. Proteomics. 2003;3:1620–32.
41. Simonnet H, Alazard N, Pfeiffer K, Gallou C, Beroud C, Demont J, et al. Low mitochondrial respiratory chain content correlates with tumor aggressiveness in renal cell carcinoma. Carcinogenesis. 2002;23:759–68.
42. Dai Z, Yin J, He H, Li W, Hou C, Qian X, et al. Mitochondrial comparative proteomics of human ovarian cancer cells and their platinum-resistant sublines. Proteomics. 2010;10:3789–99.
43. Kiebish MA, Han X, Cheng H, Lunceford A, Clarke CF, Moon H, et al. Lipidomic analysis and electron transport chain activities in C57BL/6J mouse brain mitochondria. J Neurochem. 2008;106:299–312.
44. Kiebish MA, Han X, Cheng H, Chuang JH, Seyfried TN. Cardiolipin and electron transport chain abnormalities in mouse brain tumor mitochondria: lipidomic evidence supporting the Warburg theory of cancer. J Lipid Res. 2008;49:2545–56.
45. Kiebish MA. Mitochondrial Lipidome and Genome Alterations in Mouse Brain and Experimental Murine Brain Tumors. Chestnut Hill: Boston College; 2008.
46. Feo F, Canuto RA, Bertone G, Garcea R, Pani P. Cholesterol and phospholipid composition of mitochondria and microsomes isolated from morris hepatoma 5123 and rat liver. FEBS Lett. 1973;33:229–32.

47. Feo F, Canuto RA, Garcea R, Gabriel L. Effect of cholesterol content on some physical and functional properties of mitochondria isolated from adult rat liver, fetal liver, cholesterol-enriched liver and hepatomas AH-130, 3924A and 5123. Biochim Biophys Acta. 1975;413: 116–34.
48. Hoch FL. Cardiolipins and biomembrane function. Biochim Biophys Acta. 1992;1113:71–133.
49. Chicco AJ, Sparagna GC. Role of cardiolipin alterations in mitochondrial dysfunction and disease. Am J Physiol Cell Physiol. 2007;292:C33–44.
50. Fry M, Green DE. Cardiolipin requirement for electron transfer in complex I and III of the mitochondrial respiratory chain. J Biol Chem. 1981;256:1874–80.
51. Fry M, Green DE. Cardiolipin requirement by cytochrome oxidase and the catalytic role of phospholipid. Biochem Biophys Res Commun. 1980;93:1238–46.
52. Fry M, Blondin GA, Green DE. The localization of tightly bound cardiolipin in cytochrome oxidase. J Biol Chem. 1980;255:9967–70.
53. Haines TH, Dencher NA. Cardiolipin: a proton trap for oxidative phosphorylation. FEBS Lett. 2002;528:35–9.
54. Houtkooper RH, Vaz FM. Cardiolipin, the heart of mitochondrial metabolism. Cell Mol Life Sci. 2008;65:2493–506.
55. Ordys BB, Launay S, Deighton RF, McCulloch J, Whittle IR. The role of mitochondria in glioma pathophysiology. Mol Neurobiol. 2010;42:64–75.
56. Schagger H. Respiratory chain supercomplexes of mitochondria and bacteria. Biochim Biophys Acta. 2002;1555:154–9.
57. Koshkin V, Greenberg ML. Cardiolipin prevents rate-dependent uncoupling and provides osmotic stability in yeast mitochondria. Biochem J. 2002;364:317–22.
58. Hoch FL. Cardiolipins and mitochondrial proton-selective leakage. J Bioenerg Biomembr. 1998;30:511–32.
59. Eilers M, Endo T, Schatz G. Adriamycin, a drug interacting with acidic phospholipids, blocks import of precursor proteins by isolated yeast mitochondria. J Biol Chem. 1989;264:2945–50.
60. Mileykovskaya E, Zhang M, Dowhan W. Cardiolipin in energy transducing membranes. Biochemistry. 2005;70:154–8.
61. Zhang M, Mileykovskaya E, Dowhan W. Cardiolipin is essential for organization of complexes III and IV into a supercomplex in intact yeast mitochondria. J Biol Chem. 2005;280:29403–8.
62. Shidoji Y, Hayashi K, Komura S, Ohishi N, Yagi K. Loss of molecular interaction between cytochrome c and cardiolipin due to lipid peroxidation. Biochem Biophys Res Commun. 1999;264: 343–7.
63. Pfeiffer K, Gohil V, Stuart RA, Hunte C, Brandt U, Greenberg ML, et al. Cardiolipin stabilizes respiratory chain supercomplexes. J Biol Chem. 2003;278:52873–80.
64. Ostrander DB, Zhang M, Mileykovskaya E, Rho M, Dowhan W. Lack of mitochondrial anionic phospholipids causes an inhibition of translation of protein components of the electron transport chain. A yeast genetic model system for the study of anionic phospholipid function in mitochondria. J Biol Chem. 2001;276:25262–72.
65. Gohil VM, Hayes P, Matsuyama S, Schagger H, Schlame M, Greenberg ML. Cardiolipin biosynthesis and mitochondrial respiratory chain function are interdependent. J Biol Chem. 2004;279:42612–8.
66. Kagan VE, Tyurina YY, Bayir H, Chu CT, Kapralov AA, Vlasova II, et al. The "pro-apoptotic genies" get out of mitochondria: oxidative lipidomics and redox activity of cytochrome c/cardiolipin complexes. Chem Biol Interact. 2006;163:15–28.
67. Gold VA, Robson A, Bao H, Romantsov T, Duong F, Collinson I. The action of cardiolipin on the bacterial translocon. Proc Natl Acad Sci U S A. 2010;107:10044–9.
68. Shinzawa-Itoh K, Aoyama H, Muramoto K, Terada H, Kurauchi T, Tadehara Y, et al. Structures and physiological roles of 13 integral lipids of bovine heart cytochrome c oxidase. EMBO J. 2007;26:1713–25.
69. McKenzie M, Lazarou M, Thorburn DR, Ryan MT. Mitochondrial respiratory chain supercomplexes are destabilized in Barth Syndrome patients. J Mol Biol. 2006;361:462–9.

70. McAuley KE, Fyfe PK, Ridge JP, Isaacs NW, Cogdell RJ, Jones MR. Structural details of an interaction between cardiolipin and an integral membrane protein. Proc Natl Acad Sci USA. 1999;96:14706–11.
71. Zhang M, Mileykovskaya E, Dowhan W. Gluing the respiratory chain together. Cardiolipin is required for supercomplex formation in the inner mitochondrial membrane. J Biol Chem. 2002;277:43553–6.
72. Cheng H, Mancuso DJ, Jiang X, Guan S, Yang J, Yang K, et al. Shotgun lipidomics reveals the temporally dependent, highly diversified cardiolipin profile in the mammalian brain: temporally coordinated postnatal diversification of cardiolipin molecular species with neuronal remodeling. Biochemistry. 2008;47:5869–80.
73. Mukherjee P, Abate LE, Seyfried TN. Antiangiogenic and proapoptotic effects of dietary restriction on experimental mouse and human brain tumors. Clin Cancer Res. 2004;10:5622–9.
74. Mukherjee P, El-Abbadi MM, Kasperzyk JL, Ranes MK, Seyfried TN. Dietary restriction reduces angiogenesis and growth in an orthotopic mouse brain tumour model. Br J Cancer. 2002; 86:1615–21.
75. Seyfried TN, El-Abbadi M, Roy ML. Ganglioside distribution in murine neural tumors. Mol Chem Neuropathol. 1992;17:147–67.
76. Fraser H. Brain tumours in mice, with particular reference to astrocytoma. Food Chem Toxicol. 1986;24:105–11.
77. Huysentruyt LC, Seyfried TN. Perspectives on the mesenchymal origin of metastatic cancer. Cancer Metastasis Rev. 2010;29:695–707.
78. Huysentruyt LC, Mukherjee P, Banerjee D, Shelton LM, Seyfried TN. Metastatic cancer cells with macrophage properties: evidence from a new murine tumor model. Int J Cancer. 2008;123:73–84.
79. Shelton LM, Mukherjee P, Huysentruyt LC, Urits I, Rosenberg JA, Seyfried TN. A novel pre-clinical in vivo mouse model for malignant brain tumor growth and invasion. J Neuro Oncol. 2010;99:165–76.
80. Kiebish MA, Han X, Cheng H, Chuang JH, Seyfried TN. Brain mitochondrial lipid abnormalities in mice susceptible to spontaneous gliomas. Lipids. 2008;43:951–9.
81. Weinhouse S. The Warburg hypothesis fifty years later. Z Krebsforsch Klin Onkol Cancer Res Clin Oncol. 1976;87:115–26.
82. Hartz JW, Morton RE, Waite MM, Morris HP. Correlation of fatty acyl composition of mitochondrial and microsomal phospholipid with growth rate of rat hepatomas. Lab Invest. 1982;46:73–8.
83. Canuto RA, Biocca ME, Muzio G, Dianzani MU. Fatty acid composition of phospholipids in mitochondria and microsomes during diethylnitrosamine carcinogenesis in rat liver. Cell Biochem Funct. 1989;7:11–9.
84. Jahnke VE, Sabido O, Defour A, Castells J, Lefai E, Roussel D, et al. Evidence for mitochondrial respiratory deficiency in rat rhabdomyosarcoma cells. PloS One. 2010;5:e8637.
85. Shadyro OI, Yurkova IL, Kisel MA, Brede O, Arnhold J. Radiation-induced fragmentation of cardiolipin in a model membrane. Int J Radiat Biol. 2004;80:239–45.
86. Kiebish MA, Han X, Cheng H, Seyfried TN. In vitro growth environment produces lipidomic and electron transport chain abnormalities in mitochondria from non-tumorigenic astrocytes and brain tumours. ASN Neuro. 2009;1:e00011.
87. Frezza C, Gottlieb E. Mitochondria in cancer: not just innocent bystanders. Semin Cancer Biol. 2009;19:4–11.
88. Guppy M, Greiner E, Brand K. The role of the Crabtree effect and an endogenous fuel in the energy metabolism of resting and proliferating thymocytes. Eur J Biochem. 1993;212:95–9.
89. Crabtree HG. Observations on the carbohydrate metabolism of tumors. Biochem J. 1929;23: 536–45.
90. Shelton LM, Strelko CL, Roberts MF, Seyfried NT. Krebs cycle substrate-level phosphorylation drives metastatic cancer cells. Proceedings of the 101st Annual Meeting of the American Association for Cancer Research; 2010; Washington, DC. 2010.

91. Weinberg JM, Venkatachalam MA, Roeser NF, Nissim I. Mitochondrial dysfunction during hypoxia/reoxygenation and its correction by anaerobic metabolism of citric acid cycle intermediates. Proc Natl Acad Sci U S A. 2000;97:2826–31.
92. Phillips D, Aponte AM, French SA, Chess DJ, Balaban RS. Succinyl-CoA synthetase is a phosphate target for the activation of mitochondrial metabolism. Biochemistry. 2009;48:7140–9.
93. Schwimmer C, Lefebvre-Legendre L, Rak M, Devin A, Slonimski PP, di Rago JP, et al. Increasing mitochondrial substrate-level phosphorylation can rescue respiratory growth of an ATP synthase-deficient yeast. J Biol Chem. 2005;280:30751–9.
94. Jose C, Bellance N, Rossignol R. Choosing between glycolysis and oxidative phosphorylation: a tumor's dilemma?. Biochim Biophys Acta. 2010;1807:552–61.
95. Koppenol WH, Bounds PL, Dang CV. Otto Warburg's contributions to current concepts of cancer metabolism. Nat Rev. 2011;11:325–37.
96. Samudio I, Fiegl M, Andreeff M. Mitochondrial uncoupling and the Warburg effect: molecular basis for the reprogramming of cancer cell metabolism. Cancer Res. 2009;69:2163–6.
97. Denny CA, Desplats PA, Thomas EA, Seyfried TN. Cerebellar lipid differences between R6/1 transgenic mice and humans with huntington's disease. J Neurochem. 2010;115:748–58.
98. Ramanathan A, Wang C, Schreiber SL. Perturbational profiling of a cell-line model of tumorigenesis by using metabolic measurements. Proc Natl Acad Sci U S A. 2005;102:5992–7.
99. Villalobo A, Lehninger AL. The proton stoichiometry of electron transport in Ehrlich ascites tumor mitochondria. J Biol Chem. 1979;254:4352–8.
100. Fine EJ, Miller A, Quadros EV, Sequeira JM, Feinman RD. Acetoacetate reduces growth and ATP concentration in cancer cell lines which over-express uncoupling protein 2. Cancer Cell Int. 2009;9:14.
101. Bhagavan NV. Medical Biochemistry. 4th ed. New York: Harcourt; 2002.
102. Hochachka PW, Somero GN. Biochemical Adaptation: Mechanism and Process in Physiological Evolution. New York: Oxford Press; 2002.
103. van Wijk R, Souren J, Schamhart DH, van Miltenburg JC. Comparative studies of the heat production of different rat hepatoma cells in culture. Cancer Res. 1984;44:671–3.
104. Monti M, Brandt L, Ikomi-Kumm J, Olsson H. Microcalorimetric investigation of cell metabolism in tumour cells from patients with non-Hodgkin lymphoma (NHL). Scand J Haematol. 1986;36:353–7.
105. Gautherie M. Thermopathology of breast cancer: measurement and analysis of in vivo temperature and blood flow. Ann N Y Acad Sci. 1980;335:383–415.
106. Gautherie M, Gros CM. Breast thermography and cancer risk prediction. Cancer. 1980;45:51–6.
107. Zhao Q, Zhang J, Wang R, Cong W. Use of a thermocouple for malignant tumor detection. Investigating temperature difference as a diagnostic criterion. IEEE Eng Med Biol Mag. 2008;27:64–6.
108. Nittinger J, Tejmar-Kolar L, Furst P. Microcalorimetric investigations on human leukemia cells–Molt 4. Biol Cell. 1990;70:139–42.

Chapter 6

The Warburg Dispute

Few topics have been more controversial or hotly debated in the cancer field than the role of mitochondrial function in the origin and progression of the disease. I would put this controversy on the top of Hal Hellman's "Great Feuds" list in either medicine or science (1). The controversy, as I see it, has arisen largely from Warburg's emphatic statement that cancer originates largely from injury to respiration. The mitochondrion is the key organelle responsible for cellular respiration where oxygen is consumed in the complete catabolism of organic fuel. Warburg emphasized that respiratory injury in cancer becomes irreversible since the respiration of tumor cells never returns to normal (2). Warburg went on to mention that the injury to respiration could not be so complete that the cells die, as no cancer cells could arise from dead cells.

Evidence of respiratory injury would be obvious in the cancer cells that express reduced mitochondrial adenosine triphosphate (ATP) production in association with decreased oxygen consumption, since O_2 is necessary for ATP synthesis through OxPhos. However, oxygen consumption is not reduced in some cancer cells. Indeed, O_2 consumption increases with increased malignancy in some tumor cells. Does this mean that respiration is normal or increased in such cells? Not necessarily. Warburg has attributed this phenomenon to defects in the coupling of respiration to ATP production (2). In other words, some cancer cells produce CO_2 and consume O_2, but produce insufficient energy through respiration.

Defects in the inner mitochondrial membrane of tumor cells dissipate the proton motive gradient, thus uncoupling the linkage between electron transport and ATP production through OxPhos. Uncoupling is a normal process in brown adipose tissue and is regulated by specific uncoupling proteins. Uncoupling proteins divert the proton motive gradient from ATP production to heat production (3, 4). As I have shown in the last chapter, heat is also observed in those tumor cells and tissues where thermal energy has been measured (5–7). Indeed, thermal energy is correlated with poor prognosis. The hotter is the tumor, the faster is the growth. It is my belief that heat production in the tumor tissue could arise from upregulation of uncoupling proteins or from protein-independent uncoupling of the mitochondrial

Cancer as a Metabolic Disease: On the Origin, Management and Prevention of Cancer, First Edition. Thomas Seyfried.

inner membrane. I consider heat production to be a regulated process in brown fat, but a mostly dysregulated process in cancer cells.

Disruption of the proton motive gradient, with reduced efficiency of ATP production, would naturally require a compensatory mechanism of energy production to prevent cell death. Warburg emphasized that fermentation, involving the catabolism of glucose to lactic acid, was the compensatory mechanism responsible for energy production in cancer cells. Obviously this process would need to increase substantially in tumor cells to compensate for the lost ATP production through OxPhos. If respiratory energy production is compromised, there "must" be some compensatory mechanism of energy production to maintain cell viability. Without compensation the cell would die from energy failure. Why would this hypothesis on the origin of cancer be so controversial? While most investigators have accepted the evidence that elevated glycolysis was a metabolic hallmark of almost all cancers, many investigators, especially Sidney Weinhouse, had real difficulty in accepting the idea that damaged respiration was responsible for elevated glycolysis and the origin of cancer cells.

SIDNEY WEINHOUSE'S CRITICISMS OF THE WARBURG THEORY

Sidney Weinhouse was a prominent cancer researcher of the twentieth century who served as editor of *Cancer Research* and was a member of the board of directors for the American Association of Cancer Research (8). Soon after publication of Warburg's 1956 paper in *Science*, Weinhouse (9) published a "Letter to the Editor" taking issue with Warburg's basic premise that cancer cells have impaired respiration. The debate heated with a rebuttal by Warburg and further discussion from Dean Burk and Arthur Schade (10, 11). Dean Burk appears with Warburg in Figure 2.5.

Weinhouse criticized Warburg's hypothesis even more aggressively in a guest editorial 6 years after Warburg's death in 1976 (12). It was hoped that this editorial would finally end the debate on the respiratory impairment in cancer and allow the field to move on to explore more pressing problems especially those related to the molecular genetic changes in cancer cells. While the importance of damaged respiration might have receded during the later part of the twentieth century, the issue has emerged again in the twenty-first century as a focal point of cancer research.

Why does the Warburg theory of cancer persist as a viable explanation for the disease if the key issues were supposed to have been settled in 1976 according to the evidence presented in Dr. Weinhouse's Guest Editorial? The continued debate indicates to me that the issue is not settled. If Warburg's views on the origin of cancer were correct then much of what is currently being done to manage the disease makes little sense. It is therefore important to carefully consider the evidence indicating that irreversible respiratory injury is the origin of cancer.

In reviewing the literature on respiratory impairment in cancer cells at that time, Weinhouse concluded that, "there is no sound experimental basis for the belief that oxidative metabolism in tumors is impaired" (9). Weinhouse was especially troubled by numerous findings showing that oxygen consumption and CO_2 production was high in many cancerous tissues and cells. How could cancer cells consume oxygen and produce CO_2 if their respiration were irreversibly damaged? He felt that many tumor cells could metabolize fatty acids for energy, thus producing CO_2.

The Pasteur effect also appeared to be normal in many tumors since oxygen inhibited lactate production, as would be expected if respiration were normal. He also cited evidence from metabolic tracer studies showing that the citric acid cycle operates in tumor cells. He concluded that, "*the available evidence indicates to me that high glycolysis occurs, despite quantitatively and qualitatively normal occurrence of carbon and electron transport. This can mean only that glucose catabolism is so rapid in tumors that the normal channels for disposal of pyruvic acid are over-loaded*" (12). This argument, however, does not address the fact that ATP is a Donnan active material that would destabilize ionic gradients if allowed to accumulate (13, 14). The argument also does not address the action of F_1F_0-ATPase as a hydrolase rather than as a synthase.

Weinhouse's arguments against Warburg's views were even more strongly articulated in his 1976 Guest Editorial where he wrote, "*despite massive efforts during the half-century following the Warburg proposal to find some alteration of function or structure of mitochondria, that might conceivably give some measure of support to the Warburg hypothesis, no substantial evidence has been found that would indicate a respiratory defect, either in the machinery of electron transport, or in the coupling of respiration with ATP formation, or in the unique presence or absence of mitochondrial enzymes or cofactors involved in electron transport*" (12).

This statement suggests that Weinhouse was unfamiliar with the voluminous evidence of Peter Pedersen, who had shown that the mitochondria of tumor cells express numerous abnormalities (15). Moreover, Weinhouse and colleagues actually collected data supporting Warburg's theory. In their 1968 study, "Studies on Respiration and Glycolysis in Transplanted Hepatic Tumors of the Rat", they presented evidence showing that mitochondria were quantitatively and qualitatively abnormal in hepatoma (16). In their discussion on page 8 they state: "A study of mitochondrial content, carried out in connection with assays of the mitochondrial enzyme, β-hydroxybutyrate dehydrogenase revealed that whereas liver had a mitochondrial protein content of about 50 mg/gm tissue, the range for the well-differentiated tumors was from 18 to 33 mg and for poorly differentiated tumors, 9 to 14 mg/gm tissue (16). These data reveal that, in a series of malignant hepatic neoplasms, high respiration is characteristic of the well-differentiated tumor, and that lowered respiration, coupled with loss of mitochondria, accompanies loss of differentiation." This conclusion is essentially in line with Warburg's central hypothesis (2, 10). Burk and Schade (11) also identified earlier statements made by Weinhouse regarding the misinterpretation of his data. I am not sure how the cancer field at that time was so accepting of Weinhouse's misinterpretation of his data regarding the Warburg theory.

Weinhouse went on in his 1976 review to suggest that caution is needed in comparing biochemical and physiological properties of mitochondria isolated from tumors and normal tissues, as tumor mitochondria are fragile and can be easily damaged during isolation, thus confounding data interpretation (12). While this issue might have been a concern in 1976, we clearly showed that intact and functional mitochondria could be isolated from mouse brain tumors (17, 18). We showed that the mitochondrial lipidome differs between normal brain tissue and brain tumor tissue. This was hard evidence that mitochondrial function was abnormal in these tumors. Numerous other studies have since collected evidence for mitochondrial abnormalities in cancer (14, 15). These studies will be discussed more in the next chapter.

In the end, Weinhouse said, "*the whole conception of cancer initiation or survival by 'faulty' respiration and high glycolysis seems too simplistic for serious consideration*" (9, 12). How could such a complicated disease as cancer be due to a simple replacement of respiratory energy with fermentation energy? Warburg produced a remarkably pithy and humorous response to this criticism in his 1956 rebuttal to Weinhouse when Warburg stated, "*The problem of cancer is not to explain life, but to discover the differences between cancer cells and normal growing cells. Fortunately this can be done without knowing what life really is. Imagine two engines, the one being driven by complete combustion and the other by incomplete combustion of coal. A man who knows nothing at all about engines, their structure, and their purpose, may discover the difference. He may, for example, smell it.*" In other words, when coal combustion is incomplete, sulfur is detected in the air. When glucose combustion is incomplete, lactate is detected in the microenvironment. Warburg knew that respiration was damaged or insufficient in cancer cells.

It is my opinion that many cancer researchers, through their propensity to focus on gene mutations and mechanisms of action, have made the quest for cancer management far more complicated then it actually is. We have turned the cancer problem into an exercise in cybernetics with no clear solutions. Is it necessary to fully elucidate the minutia of all cancer mechanisms before adapting therapies that exploit the Warburg theory for cancer management? The answer is "no", and this will become apparent in later chapters.

Weinhouse (12) felt that it was only natural for people and even scientists to seize upon the words of Warburg, as if he were a "prophet" promising salvation and an answer to one of humanity's greatest scourges. Weinhouse (12) mentioned in his personal reflection that the burning issues of glycolysis and respiration flickered only dimly and were no longer considered to be mainstream issues of cancer research. That these issues continue to burn brightly today suggests that Warburg might have been right all along.

ALAN AISENBERG'S CRITICISMS OF THE WARBURG THEORY

In addition to Weinhouse's arguments against Warburg's theory, other prominent investigators also raised reservations over the theory. Alan C. Aisenberg of the

Harvard Medical School devoted an entire monograph to the issue of glycolysis and respiration of tumors (19). Despite reviewing and documenting numerous studies that supported Warburg's theory, Aisenberg concluded that evidence supporting Warburg's views did not exist and that the entire concept of respiratory injury as the origin of cancer must remain just a hypothesis. I have read Aisenberg's monograph carefully and have found that much of the data presented actually supported Warburg's theory. This was also noted in Sidney Colowick's review of Aisenberg's monograph (20).

Basically, the data not supporting Warburg's theory were mostly obtained from experimental systems unrepresentative of conditions occurring in vivo. Warburg's generalization that all tumors show a high rate of anaerobic glycolysis was strongly supported by the data presented (20). Warburg's further generalization that tumor cells are unique in exhibiting substantial glycolysis under aerobic conditions was, with a few exceptions, almost entirely supported by the data presented in Aisenberg's monograph. Warburg noted that some normal tissues could also exhibit aerobic glycolysis including retina, leukocytes, kidney medulla, and jejunum mucous membrane. Warburg maintained that aerobic glycolysis in differentiated tissues would not normally occur, but that it might arise due to damage during removal or evaluation in nonphysiological environments. Aisenberg failed to mention Warburg's work showing that the leukocytes do not exhibit aerobic glycolysis until they are removed from their normal serum environment, just as he had shown for embryonic tissues. It is not clear why Aisenberg "cherry picked" data that did not support Warburg's views.

Although Aisenberg presented data showing that both aerobic and anaerobic glycolysis increased in spontaneous cancers and in chemically induced cancers compared to the tissue of origin, less emphasis was placed on identifying the cells responsible for the metabolic shift (20). According to Dean Burk, it is only when a pure clone of normal cells of one type gives rise to a tumor that the actual metabolic change accompanying the induction of malignancy can be assessed. This is why Dean Burk's work was so important in showing that a malignant cell line derived from a pure clone of heart fibroblasts in cell culture showed greater glycolytic activity than a nonmalignant cell line derived from the same clone (10, 11). Neither Aisenberg nor Weinhouse placed much emphasis on Burk's report, which strongly supported the Warburg theory (9, 11, 12, 19). Comparisons of energy metabolism in cancer cells and in normal cells that are mismatched for species, tissue, or cell type have contributed significantly to the confusion surrounding the integrity of respiratory energy sufficiency in cancer (9, 12, 19, 21, 22). Although Burk and Pedersen made reference to the importance of appropriate cellular comparisons for evaluating the Warburg theory (11, 15), few investigators of cancer energy metabolism have made the correct comparisons in their experimental design.

Considerable confusion was encountered in trying to integrate the phenomena of anaerobic glycolysis, aerobic glycolysis, and the Pasteur effect. The presence of oxygen can help distinguish the two types of glycolysis. Oxygen suppresses anaerobic glycolysis completely in most normal tissues. Oxygen, however, does not completely suppress aerobic glycolysis in tumor cells, which involves the continued expression of glycolysis in the presence of oxygen, that is, the Warburg effect. According to Warburg, aerobic glycolysis was a consequence of damaged respiration. However, both Weinhouse and Aisenberg presented data showing that the Pasteur effect was operational in tumor tissues (12, 19). As the normal respiratory capacity was assumed to underlie the Pasteur effect, why would tumor tissue show a Pasteur effect if respiration were damaged?

One problem was that much of the data presented were derived from tumors that also contained normal cells, which could express a Pasteur effect. The presence of normal cells in the tumor tissue could mask some of the metabolic abnormalities of the tumor cells. This is why the best comparisons should be between (i) tumor tissue and normal tissue from which the tumors are derived and (ii) normal cells and tumor cells derived from the same cell clone. The continued production of lactic acid in the presence of O_2, indicative of an inefficient Pasteur effect, persists in tumor cells. In contrast to the effects of O_2 in normal cells and tissues, O_2 is unable to completely suppress glycolysis in tumor cells. The incomplete O_2 suppression of glycolysis in tumor cells suggests that these cells have impaired respiration. On the other hand, some would argue that the persistent aerobic glycolysis in tumor cells results from damaged glycolysis regulation rather than damaged respiration itself (23).

Much of the confusion, in my mind, over the role of damaged respiration in cancer cells was due in large part to flaws in experimental design or to flaws in the cell systems used to evaluate respiration and fermentation. This point is made clear through examination of the data used by both Aisenberg and Weinhouse to argue against Warburg's hypothesis (11, 12, 19). Most of the data used against Warburg's hypothesis of defective respiration in cancer cells did not include adequate control cells or tissues. Caution is also required in evaluating the respiration of cells grown in culture since the culture environment itself can alter the structural integrity of the inner mitochondrial membrane, which ultimately facilitates respiration (24). Normal cells are often compared with tumor cells in growth media that is favorable to tumor cell metabolism (high glucose), but is unfavorable to cells with normal respiration.

In addition to discounting the evidence for damaged respiration in tumor cells, Aisenberg took the view of Britton Chance that tumor mitochondria were normal with respect to cytochrome content and oxidative capacity, but showed limited respiration in the cell because of an absence of the phosphate acceptor. Chance and Hess argued against impaired respiration in cancer based on their spectrophotometric studies showing mostly normal electron transfer in ascites tumor cells (22). They compared the respiration of the ascites cells with respiration in yeast cells and frog muscle cells. These studies also failed to assess the level of ATP production as a consequence of normal electron transfer and did not exclude the possibility of elevated ATP production through TCA cycle, substrate-level phosphorylation (14). Aisenberg also emphasized the findings from Monier et al. that there was no lack of *cytochrome c* relative to other members of the cytochrome system in ascites tumor cells, while minimizing their findings that other cytochromes were very low or lacking (20, 25).

I think Peter Pedersen had accurately summarized the controversy at that time in his statement: "*It is important to emphasize that before 1959 many investigators were in general agreement that the content of mitochondria in the rapidly growing, highly glycolytic class of tumors was markedly reduced. In fact, enzymatic data on this point was presented as early as 1950 by Potter et al. In 1959, however, Chance and Hess showed on the basis of spectral measurements that the cytochrome content of Ehrlich ascites cells is in the same range as that of muscle and yeast. Unfortunately, these studies were taken by some [Aisenberg, 1961] to suggest that the mitochondrial content of the rapidly growing, highly glycolytic class of tumor cells may be normal, a suggestion which may have distracted some investigators (in view of the anti-Warburg sentiment at the time) from more thoroughly investigating the metabolic significance of the marked reduction in mitochondria in rapidly growing tumors. Significantly, in preparing this review article, the author was able to find more than 20 literature references showing that the mitochondrial content of rapidly growing, highly glycolytic tumors is markedly reduced relative to their tissue of origin, but was unable to find a single literature reference showing that such tumors have a normal mitochondrial content when compared carefully with their tissue of origin. Therefore, although it may well be true that the endogenous or glucose-supported oxygen consumption rates of 'highly glycolytic' tumors are in the same range as those of a variety of normal tissues [Weinhouse, 1956, 1976], the reader should bear in mind that the total respiratory capacity of such tumors (in terms of mitochondrial content, i.e., in terms of the total DNP-stimulated or state III rates which may be exhibited by these mitochondria) may well be reduced markedly when comparisons are made directly with the known tissue of origin. This point seems to have been 'glossed over' in many writings about the respiratory properties of highly glycolytic tumor cells.*" (15). My student, Roberto Flores, has added the *italics* to emphasize key points in Pedersen's argument. It is not clear how tumor cells could have normal respiration if the number and structure of their mitochondria were abnormal as I have covered in Chapter 5.

It is my opinion that many investigators today also "gloss over" the evidence showing that cancer cells have impaired respiration or mitochondrial damage. How is this possible? Pedersen also raised an interesting point regarding an "anti-Warburg sentiment". Is it possible that this sentiment still persists?

SIDNEY COLOWICK'S ASSESSMENT OF THE AISENBERG MONOGRAPH

Sidney Colowick reviewed Aisenberg's monograph for *The Quarterly Review of Biology* and provided an alternative view to Aisenberg's conclusions (20). Colowick emphasized that, besides denying damaged respiration in tumors, Aisenberg had also attempted to refute Warburg's theory using several arguments. First, that many carcinogens were not inhibitors of respiration. However, Aisenberg also cited other work showing a correlation between the carcinogenic activity of a certain series of aminofluorene dyes and their ability to inhibit glutamate oxidation by mitochondria. Later studies have clearly shown that carcinogens damage cellular respiration (15, 26, 27).

Second, Aisenberg mentioned that the highest concentration of protein-bound carcinogen was present in cytoplasm, but admitted that mitochondria could take up carcinogenic dyes. In fact, Aisenberg cited Potter's view that the specific lesion in tumors may be due to the deletion of mitochondria rather than alteration of soluble enzymes.

Third, Aisenberg doubted that X-ray induction of tumors should be attributed to damage of mitochondrial respiration, since X rays can damage nuclear morphology. It is not clear, however, why the X rays could not also damage mitochondria as Warburg and others had suggested (10, 27). X rays can damage mitochondria.

Fourth, Aisenberg doubted the interpretation of Goldblatt and Cameron, who had shown that intermittent anoxia could transform normal fibroblasts into malignant cells. Just because malignancy could be produced in culture without anaerobiosis does not rule out the possibility that anaerobiosis, too, could cause malignancy (20). Overall, Colowick was miffed at how Aisenberg could so lightly dismiss the idea of damaged respiration in cancer cells considering the large amount of data presented in support of the hypothesis. The evidence for damaged respiration in cancer cells was even more compelling from Pedersen's later extensive review (15). I think Colowick provided an objective assessment of the information reviewed in Aisenberg's monograph.

APPLES AND ORANGES

Much of the debate involving Warburg's theory was driven by observations that apparent normal respiratory function could be found in numerous tumor cells and that respiratory function could be restored in some cancer cells (9, 12, 27–30). These and other observations challenged the notion that cancer is largely an irreversible disease of mitochondrial dysfunction (23). If many tumor cells appear to express normal respiration, how is it possible that defective respiration could be responsible for cancer as Warburg had suggested? It is also important to recognize that few investigators evaluated respiration and fermentation in tumor cells in comparison with respiration and fermentation in tissue-matched normal cells under similar growth conditions.

There is abundant evidence of flaws in experimental design or in the biological systems used that contribute to the controversy surrounding Warburg's theory. For example, many experiments compare respiration in one tumor cell with that in another tumor cell of different origin without also including nontumorigenic, cell-specific controls (12). Other experiments compare respiration of cultured tumor cells with respiration of tumor tissue or with tissue slices (21). The more informative experiments are those that compare tumor cells with nontumorigenic cells from the same tissue grown under identical in vivo and in vitro conditions (17, 24, 31). It would also be important to compare respiration in tumor cells and normal cells grown in media that favor respiration rather than glycolysis. Growth of cells in high glucose media deficient in respiratory nutrients (vitamins) will suppress normal respiration, thus potentially masking differences in respiration between cancer cells and normal cells. Burk, Schade and others have clearly shown how correct comparisons could facilitate data interpretation and highlight differences in the respiratory capacity between tumorigenic and nontumorigenic cells (11, 14, 24). These investigators have also shown how incorrect comparisons, especially those made by Weinhouse in comparing normal human tissues with mouse tumors, could confound data interpretation (11). Weinhouse made numerous errors in experimental design that obscured the differences in respiration between normal cells and cancer cells. I recommend that all those interested in this subject carefully read the Weinhouse arguments against Warburg's data and then consider Burk and Shade's rebuttal to those arguments (9, 11, 12).

It is not clear to me why Weinhouse and Aisenberg placed more emphasis on data from experiments not supporting Warburg's views than on data from numerous experiments that supported the main theory. Colowick considered the experiments supporting Warburg's findings as having better controls than the experiments not supporting Warburg's findings. Aisenberg remained unconvinced that carcinogenic agents could act simply by damaging cell respiration, or that glycolytic energy was morphologically inferior to oxidative energy. I describe in Chapter 9 how carcinogens and viruses target mitochondria to disrupt energy metabolism (14, 15, 26).

Szent-Gyorgyi had clearly described how the energy derived from glycolysis was inferior to the energy derived from respiration. The energy of respiration is associated with the intact structure of mitochondrial membranes and the differentiated state, whereas glycolysis is associated with reduced structure (mostly soluble enzymes in the cytoplasm) and the dedifferentiated state (32, 33). Inferiority was therefore linked to the disorganized organelle structure (15). It is well documented that the mitochondria in most if not all cancer cells differ in structure from those present in normal cells (Chapter 5).

I think Warburg's choice of words might have contributed to some of the confusion associated with his theory. Although Burk, Schade, Colowick, and others convincingly dispelled the main criticisms of the Warburg theory in my mind, citations to the older arguments for normal respiration in cancer cells persist in current discussions of the subject (23). The continued citation of Weinhouse's papers against the Warburg theory suggests to me that many investigators have not carefully read the arguments or evaluated the data. Although Dr. Deng and colleagues mention in their review that Warburg waffled in describing cancer cell respiration as either damaged or insufficient (23), at no time did Warburg consider respiration as being normal in cancer cells (2, 10, 34, 35). It is important for people to recognize the importance of this fact.

It will be up to each reader to judge for themselves the strength of the arguments for and against the Warburg theory. After more than one decade of research using both in vivo and in vitro tumor systems, I have concluded that respiratory insufficiency arising from any number of insults is the single most important hallmark of cancer. I challenge anyone in the cancer field to provide definitive evidence showing that the respiration of cancer cells is no different from that of their species-matched normal cell counterparts when compared in growth environments that favor respiration and not fermentation. While we were able to transition nontumorigenic astrocytes from growth in high glucose media to survival in low glucose media containing ketone bodies as a major respiratory fuel, all of our mouse tumor cells died when we attempted to transition them to this respiratory media (36, 37). Johannes Rieger and colleagues reported similar findings in a series of human brain tumor lines (38). Most cells with normal respiratory capacity can make this transition, but few if any cancer cells can make the transition. If respiration is normal in tumor cells, then the utilization of ketone bodies for energy should be similar in tumor cells and in normal cells.

It is my contention that the resolution of cancer will not likely occur until the evidence for tumor cell respiratory damage or insufficiency becomes more widely recognized and accepted (39). Once this happens, new therapeutic strategies will be developed that can significantly enhance progression-free survival for most patients with malignant cancers.

REFERENCES

1. Hellman H. Great Fueds in Medicine. New York: John Wiley & Sons, Inc; 2001.
2. Warburg O. On the origin of cancer cells. Science. 1956;123:309–14.
3. Samudio I, Fiegl M, Andreeff M. Mitochondrial uncoupling and the Warburg effect: molecular basis for the reprogramming of cancer cell metabolism. Cancer Res. 2009;69:2163–6.
4. Hochachka PW, Somero GN. Biochemical Adaptation: Mechanism and Process in Physiological Evolution. New York: Oxford Press; 2002.
5. Nittinger J, Tejmar-Kolar L, Furst P. Microcalorimetric investigations on human leukemia cells–Molt 4. Biol Cell. 1990;70:139–42.
6. van Wijk R, Souren J, Schamhart DH, van Miltenburg JC. Comparative studies of the heat production of different rat hepatoma cells in culture. Cancer Res. 1984;44:671–3.
7. Monti M, Brandt L, Ikomi-Kumm J, Olsson H. Microcalorimetric investigation of cell metabolism in tumour cells from patients with non-Hodgkin lymphoma (NHL). Scand J Haematol. 1986;36:353–7.
8. Kresge N, Hanson RW, Simoni RD, Hill RL. Sidney Weinhouse and the mechanism of ketone body synthesis from fatty acids. J Biol Chem. 2005;280:e20.
9. Weinhouse S. On respiratory impairment in cancer cells. Science. 1956;124:267–9.
10. Warburg O. On the respiratory impairment in cancer cells. Science. 1956;124:269–70.
11. Burk D, Schade AL. On respiratory impairment in cancer cells. Science. 1956;124:270–2.
12. Weinhouse S. The Warburg hypothesis fifty years later. Z Krebsforsch Klin Onkol Cancer Res Clin Oncol. 1976;87:115–26.
13. Veech RL, Kashiwaya Y, Gates DN, King MT, Clarke K. The energetics of ion distribution: the origin of the resting electric potential of cells. IUBMB Life. 2002;54:241–52.
14. Seyfried TN, Shelton LM. Cancer as a metabolic disease. Nutr Metab. 2010;7:7.
15. Pedersen PL. Tumor mitochondria and the bioenergetics of cancer cells. Prog Exp Tumor Res. 1978;22:190–274.
16. Lo C, Cristofalo VJ, Morris HP, Weinhouse S. Studies on respiration and glycolysis in transplanted hepatic tumors of the rat. Cancer Res. 1968;28:1–10.

17. Kiebish MA, Han X, Cheng H, Chuang JH, Seyfried TN. Cardiolipin and electron transport chain abnormalities in mouse brain tumor mitochondria: lipidomic evidence supporting the Warburg theory of cancer. J Lipid Res. 2008;49:2545–56.
18. Kiebish MA, Han X, Cheng H, Lunceford A, Clarke CF, Moon H, et al. Lipidomic analysis and electron transport chain activities in C57BL/6J mouse brain mitochondria. J Neurochem. 2008;106:299–312.
19. Aisenberg AC. The Glycolysis and Respiration of Tumors. New York: Academic Press; 1961.
20. Colowick SP. The status of Warburg's theory of glycolysis and respiration in tumors. Q Rev Biol. 1961;273–6.
21. Zu XL, Guppy M. Cancer metabolism: facts, fantasy, and fiction. Biochem Biophys Res Commun. 2004;313:459–65.
22. Chance B, Hess B. Spectroscopic evidence of metabolic control. Science. 1959;129:700–8.
23. Koppenol WH, Bounds PL, Dang CV. Otto Warburg's contributions to current concepts of cancer metabolism. Nat Rev. 2011;11:325–37.
24. Kiebish MA, Han X, Cheng H, Seyfried TN. In vitro growth environment produces lipidomic and electron transport chain abnormalities in mitochondria from non-tumorigenic astrocytes and brain tumours. ASN Neuro. 2009;27:1.
25. Monier R, Zajdela F, Chaix P, Petit JF. Low-temperature spectrographic study of the cytochromes in various rat and mouse tumors. Cancer Res. 1959;19:927–34.
26. Hadler HI, Daniel BG, Pratt RD. The induction of ATP energized mitochondrial volume changes by carcinogenic N-hydroxy-N-acetyl-aminofluorenes when combined with showdomycin. A unitary hypothesis for carcinogenesis. J Antibiot. 1971;24:405–17.
27. Smith AE, Kenyon DH. A unifying concept of carcinogenesis and its therapeutic implications. Oncology. 1973;27:459–79.
28. Vander Heiden MG, Cantley LC, Thompson CB. Understanding the Warburg effect: the metabolic requirements of cell proliferation. Science. 2009;324:1029–33.
29. Moreno-Sanchez R, Rodriguez-Enriquez S, Marin-Hernandez A, Saavedra E. Energy metabolism in tumor cells. FEBS J. 2007;274:1393–418.
30. Fantin VR, St-Pierre J, Leder P. Attenuation of LDH-A expression uncovers a link between glycolysis, mitochondrial physiology, and tumor maintenance. Cancer Cell. 2006;9:425–34.
31. Jahnke VE, Sabido O, Defour A, Castells J, Lefai E, Roussel D, et al. Evidence for mitochondrial respiratory deficiency in rat rhabdomyosarcoma cells. PloS One. 2010;5:e8637.
32. Szent-Gyorgyi A. The living state and cancer. Proc Natl Acad Sci USA. 1977;74:2844–7.
33. Manchester K. The quest by three giants of science for an understanding of cancer. Endeavour. 1997;21:72–6.
34. Warburg O. The Metabolism of Tumours. New York: Richard R. Smith; 1931.
35. Warburg O. Revised Lindau Lectures: The prime cause of cancer and prevention-Parts 1 & 2. In: Burk D, editor. Meeting of the Nobel-Laureates Lindau, Lake Constance, Germany: K.Triltsch; 1969. p. http://www.hopeforcancer.com/OxyPlus.htm.
36. Shelton LM. Targeting energy metabolism in brain cancer. Chestnut Hill: Boston College; 2010.
37. Seyfried TN, Mukherjee P, Adams E, Mulrooney TJ, Abate LE. Metabolic control of brain cancer: Role of glucose and ketones. Proc Amer Assoc Cancer Res. 2005;46:267.
38. Maurer GD, Brucker DP, Baehr O, Harter PN, Hattingen E, Walenta S, et al. Differential utilization of ketone bodies by neurons and glioma cell lines: a rationale for ketogenic diet as experimental glioma therapy. BMC Cancer. 2011;11:315.
39. Bambeck GS, Wolfson M. Mainstream Science's Dogma Reversal: Aerobic Glycolysis/Metabolic Alterations are finally seen as Necessary for Cancer Cell Initiation/Maintenance. 2011: http://best-resveratrol.monicathebarber.com/2011/04/29/mainstream_sciences_dogma_reversal_aerobic_glycolysis_metabolic_alterations_are_finally_seen_as_necessary_for_cancer_cell_initiation_maintena/

Chapter 7

Is Respiration Normal in Cancer Cells?

Respiration is the process by which cells use O_2 to obtain their energy through OxPhos. If cancer cells have insufficient respiration, as I described in Chapters 4–6, it is not clear why so many published studies indicate that respiration is normal or is not severely impaired in cancer cells. How is it possible for cancer cells to express normal respiration if the organelle responsible for the phenomenon is damaged? I consider this a central issue in the field of cancer metabolism. If respiration is undamaged and functional in many tumor cells then Warburg's theory of impaired respiration cannot reasonably explain the origin of cancer. The role of respiration in cancer cells festers as a persistent conundrum that must be addressed "head on." It becomes difficult for me, or anyone for that matter, to discuss Warburg's original theory if significant credible evidence exists showing that ATP is synthesized through normal OxPhos in cancer cells.

Why is it so important to establish the validity of the Warburg theory? If the Warburg theory is *correct*, then it should be possible to link all features of the disease directly or indirectly to respiratory damage or insufficiency. If the Warburg theory is *correct* in describing the nature of the disease, then the cancer field is marching in the wrong direction. On the other hand, if the Warburg theory is *incorrect*, then it becomes necessary to abandon this explanation for the origin of cancer. This would allow the cancer field to continue in the same direction. It is therefore essential for all those interested in the subject to evaluate carefully the evidence that the quantity and quality of respiration is similar in normal cells and tumor cells.

PSEUDO-RESPIRATION

Just because cultured tumor cells consume O_2, release CO_2, transport electrons through the ETC, and produce ATP in their mitochondria does not mean that

Cancer as a Metabolic Disease: On the Origin, Management and Prevention of Cancer, First Edition.
Thomas Seyfried.

the majority of this ATP arises specifically through normal OxPhos. How could this be possible? TCA cycle activity and O_2 consumption may or may not be associated with ATP synthesis through OxPhos (1–4). Many investigators have difficulty recognizing this fact, possibly because of the complex systems involved (3, 5, 6). There are a plethora of papers in the scientific literature suggesting that cancer cells respire based on data showing that they eject H^+, transport electrons, consume O_2, release CO_2, and produce ATP in mitochondria. I described before how O_2 consumption can increase in cancer cells that express uncoupled OxPhos (electron transport is not coupled to ATP synthesis). Abnormalities in cardiolipin content and composition can induce protein independent uncoupling as mentioned in Chapter 5. Tumor cells contain abnormalities in cardiolipin.

It would also be good to consider again the findings of Ramanathan regarding the role of O_2 consumption in tumor cells (7). These investigators mention the intriguing finding that cells with greater tumorigenic potential consume more oxygen and yet exhibit less oxygen-dependent (aerobic) ATP synthesis than cells with lower tumorigenic potential. They go on to suggest that such cells might use the mitochondrial electron transport chain and oxidative phosphorylation for reasons other than ATP synthesis. What other reasons are there?

Antonio Villalobo and Albert Lehninger showed that normally ejected H^+ protons leak back into the matrix to a greater extent in tumor cells than in normal cells (8). Peter Pederson also illustrated proton leak in Figure 7C, p. 215, of his review (6). Leakage of the mitochondrial membrane potential should produce heat or reactive oxygen species (Fig. 4.4). Levels of ROS are generally higher in tumor cells than in normal cells (9). I have also explained in Chapter 5 how heat production is greater in more malignant tumor cells than in the less malignant tumor cells. Consequently, evidence of O_2 consumption linked to mitochondrial ATP production is not, by itself, evidence for normal respiration.

In the next chapter, I will show how cancer cells can synthesize ATP in their mitochondria through nonoxidative processes involving amino acid fermentation. This energy is generated through the action of succinyl-CoA synthetase and a reversal of the succinate dehydrogenase, where fumarate becomes the oxidizing agent. In other words, the cells consume O_2 but do not produce ATP exclusively through OxPhos but rather also produce ATP through mitochondrial fermentation involving substrate-level phosphorylation.

This could give the impression that respiration is active when it might not be. I view this phenomenon as false or *pseudo-respiration* for lack of another term. The term *pseudo-hypoxia* was used to describe the continued expression of Hif-1α in the presence of normoxia (10). Hif-1α should only be elevated in hypoxia in order to drive glycolysis. The elevation of Hif-1α in cancer cells during normoxia arises from injured respiration. Hif-1α is needed to maintain fermentation. Without fermentation energy, most cancer cells will die.

How many of the published papers describing OxPhos activity in tumor cells have excluded mitochondrial substrate-level phosphorylation or nonoxidative phosphorylation as an alternative explanation for their findings? I am yet to find a single published paper that addresses this or includes all the necessary control experiments to distinguish energy produced through OxPhos from energy produced through mitochondrial fermentation. Without these critical experiments, it is not possible to conclude that respiration is normal in cancer cells (5). Without these critical experiments, it is not possible to reject Warburg's theory that respiratory inefficiency or injury is the origin of cancer.

Several criteria must be met to conclude that the energy derived from cancer cell mitochondria arises specifically through OxPhos and not through substrate-level phosphorylation or from an electron transfer-based ATP synthesis at the level of fumarate reductase (11). It is helpful to conduct these studies in media that is either serum free or contains dialyzed (glucose free or low glucose) serum. Serum contains many metabolites that can generate energy and confound data interpretation. On the other hand, serum contains many factors required for respiration. We know that some tumor cells can acquire lipids from serum (12). Human tumor cells grown in mice acquire unique mouse lipids, which cannot be synthesized by the human cells. Tumor cells obtain cholesterol from serum (Chapter 4). Cholesterol is needed for cell proliferation. Oxygen is needed for cholesterol synthesis. Serum can provide cholesterol in the absence of oxygen. Serum cholesterol can sustain tumor cell growth in the absence of oxygen.

It is helpful to grow the cells under at least two conditions, one under normal oxygen (normoxia) and another under hypoxia ($<0.5\%$ oxygen). But what is normoxia? Many consider that 20% O_2 is normoxia. However, normoxia in tissue is only about 5–9% O_2. Does the O_2 conditions used for cultured cells accurately mimic the in vivo environment with respect to O_2 content? In addition to using hypoxic conditions, I also think glycolytic inhibitors and antirespiration drugs should be used as additional control groups. Several of these are shown in Figure 7.1. However, some of these drugs can produce toxic effects unrelated to OxPhos inhibition (13). On the other hand, hypoxia could influence mitochondrial function independent of OxPhos inhibition. We found that absence of CO_2 can have toxic effects on cells. Cells grown in pure N_2 without CO_2 die from rapid acidity. Consequently, several observations are required as evidence that cancer cells use OxPhos for energy production.

First, that death occurs in cultured cells over a relatively short period of time when they are grown under hypoxia in serum and glucose-free minimal media not containing glutamine or any other oxidizable metabolites such as ketone bodies and fatty acids. Cells should not remain viable for very long if all energy-generating nutrients are depleted (13). Cells should not remain viable for very long if cholesterol cannot be synthesized endogenously or acquired from the environment. Membrane depolarization will occur as all energy substrates become depleted (14).

Many cancer cells can use both glucose and glutamine for maintaining viability. Energy from glucose fermentation is derived mostly through glycolysis. Energy from glutamine can be derived from OxPhos or from fermentation. Glutamine fermentation can provide energy from either substrate-level phosphorylation at the succinyl-CoA synthetase step in the TCA cycle or from the fumarate reductase reaction also in the mitochondria (15). I will describe more in the next chapter on how cancer cells can synthesize ATP through amino acid fermentation.

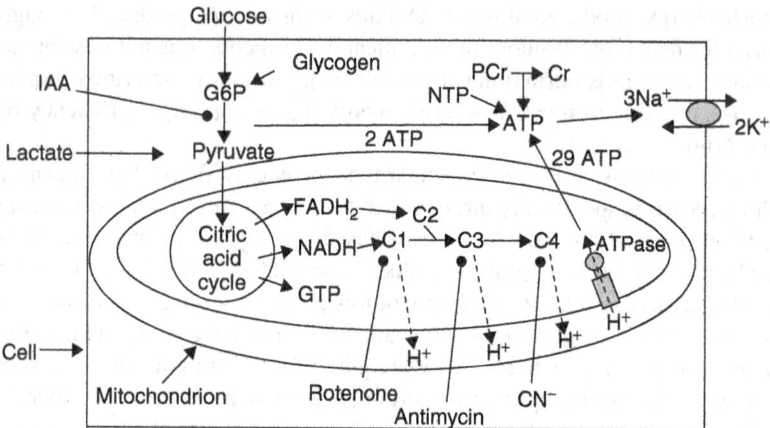

Figure 7.1 Schematic diagram of energy production pathways. Conversion of glucose (or glycogen) to pyruvate by glycolysis generates two ATP molecules and can be inhibited with iodoacetate (IAA). G6P, glucose-6-phosphate. Pyruvate is converted to acetyl-CoA to feed the citric acid cycle, which generates two GTP molecules (convertible to ATP) and exports NADH and $FADH_2$ to the electron transport chain, complexes 1, 3, and 4 (C1, C3, and C4) of which extrude protons across the mitochondrial membrane to power ATP synthesis. C1, C3, and C4 can be inhibited by rotenone, antimycin, and cyanide. ATP is used primarily to fuel ion pumping indicated here as the plasma membrane sodium pump. Short-term ATP reserves occur as phosphocreatine (PCr) and as nucleotide triphosphates (NTPs) other than ATP (UTP, GTP, and CTP). *Source*: Reprinted with permission from Ref. 14.

If glucose is not available for glycolysis or the pentose phosphate pathway, then viability cannot be maintained through these processes (Fig. 4.5). If glutamine maintains cell viability through TCA cycle metabolism and OxPhos, then hypoxia should inhibit ATP synthesis and kill cells that are grown in glutamine alone or in mixtures of glutamine or galactose. Galactose is not generally fermented well in cancer cells (16). If glutamine maintains viability in the absence of O_2, or in the presence of cyanide, then the process is not likely to involve OxPhos, which requires O_2 and cytochrome *c* for ATP production. Using potassium cyanide (KCN), Renner et al. (17) showed that ATP synthesis was similar in glioma cell lines and primary glioblastoma cells grown in the absence or presence of KCN. As KCN blocks complex IV function and respiration in normal cells, the continued viability and ATP production in glioma cells treated with KCN under glucose conditions indicate that these tumor cells were not completely using OxPhos for viability. If OxPhos was contributing to viability, then KCN would kill them. Unfortunately, these investigators did not run control experiments under low glucose in the presence of KCN and glutamine to test this. We did these control experiments in our metastatic tumor cells and found that the tumor cells maintain viability. I will describe these data in the next chapter.

Hypoxia or rapid inhibition of OxPhos should quickly kill respiring cells. The sodium pump becomes the cell's dominant energy sink during hypoxia (4, 14). If

OxPhos is shut down due to the absence of O_2, then the cell will swell and die due to depletion of energy for the sodium pump. Death will happen quickly, especially if glucose is not available as an alternative energy substrate to produce ATP through glycolysis (13, 14, 18). This was shown in the studies of the Attwell group in brain slices (14). These investigators also showed that lactate cannot substitute for glucose in maintaining viability in normal cells under hypoxia.

How long can respiring organisms survive in the absence of O_2? Survival is generally <1 h, unless the organism or cell evolved to survive for prolonged periods in hypoxia (4, 11, 19–21). If hypoxia does not kill the cell, the cell is obviously generating energy through mechanisms other than OxPhos. Lactate cannot provide energy to normal cells, if OxPhos is shut down. What about tumor cells? If lactate provides energy to tumor cells, then this is not likely to involve OxPhos, but rather fermentation. Lactate could be oxidized to pyruvate, which could then be fermented in the mitochondria through substrate-level phosphorylation. If glucose is removed from the media and the cell continues to survive in hypoxia, then OxPhos is not likely the mechanism of survival. It is important to mention, however, that Molina et al. (22) reported that lactate might serve as a metabolic fuel for subsets of breast cancer cells. Clearly, more studies are warranted to evaluate the role of lactate as an energy source for cancer cells.

I am not sure how often these complicating factors are examined when evaluating the role of OxPhos in cancer cells. I apologize to those investigators who examined these possibilities in their studies that I might have missed in my review of the literature on the subject. If our goal is to *kill* cancer cells, then we should ask serious questions about those energy systems that maintain cancer cell ATP production and viability under different physiological states. It is also to be recognized that the response to energy stress seen in cultured tumor cells might not be the same as occurs in the natural environment. If we know what cancer cells can and cannot eat, we can kill them.

Ketone bodies and fatty acids can provide alternative metabolic fuels to glutamine for mitochondrial ATP synthesis. As these alternative fuels also require O_2 for metabolism, death should occur quickly for any cell in the absence of both glucose and O_2, especially if ketone bodies and fatty acids are the only available fuels. If cells maintain viability in O_2 using either ketone bodies or fatty acids as the only energy substrates, then these cells are likely using OxPhos for survival. As far as I know, ketone bodies and fatty acids are not fermented for energy (4).

If tumor cells ferment glutamine in the presence of oxygen, as I will describe in the next chapter, then how is it possible to know for certain if glutamine-derived energy is produced from OxPhos or is derived from fermentation in the mitochondria? This is an important question that requires careful attention. If cell viability was similar for cells grown in the presence or absence of O_2, then OxPhos is not likely responsible for maintaining viability. More specifically, if viability and lactate production are similar for tumor cells grown in the presence or absence or O_2 in minimal media with glucose and glutamine as the only energy metabolites, then the cells are likely deriving their energy from fermentation rather than from OxPhos. If cancer cells are using OxPhos, then viability should be significantly lower in the absence than in the presence of O_2. If cancer cells are using OxPhos, then viability should be significantly lower in the absence than in the presence of KCN, which inhibits cytochrome *c* and OxPhos. In light of this information on the origin of energy production in tumor cells, we can now reevaluate data from a number of articles indicating that tumor cells produce ATP from OxPhos.

HOW STRONG IS THE SCIENTIFIC EVIDENCE SHOWING THAT TUMOR CELLS CAN PRODUCE ENERGY THROUGH OxPhos?

The answer to this question depends on the experimental system evaluated. In some cases, the evidence appears strong, while in other cases, the evidence is weak. I have found shortcomings in most major studies suggesting that OxPhos is normal in cancer cells. Most of these studies lack critical control experiments that can exclude mitochondrial substrate-level phosphorylation as an alternative explanation for their findings. Many experiments use respiratory poisons without also using hypoxia as an additional control group. Hypoxia can sometimes provide an alternative perspective to respiratory inhibitors. However, caution is needed when evaluating results using any respiration inhibitor since nonspecific toxic effects could complicate data interpretation. It would be best to evaluate the role of respiration in tumor cell viability using a combination of hypoxia and respiratory inhibitors. What are the most serious shortcomings in the evidence suggesting that respiration is normal in cancer cells?

OxPhos ORIGIN OF ATP IN CANCER CELLS REEVALUATED

One of the most cited papers in the scientific literature indicating that cultured tumor cells can derive energy through OxPhos is that of Reitzer et al. (16). These investigators presented evidence showing that glutamine, rather than glucose or other sugars, was the major energy source for cancerous HeLa cells. According to *Google*, this paper has been cited almost 600 times since its publication in 1979. Although the evidence presented in this paper indicates that glutamine is a major energy substrate for HeLa cells, the authors did not prove that the energy derived from glutamine actually came through coupled OxPhos, despite their suggestion that it did. It would be important therefore to carefully examine the evidence showing that glutamine provided energy through an intact TCA cycle as the authors suggest (p. 2674, first paragraph and Fig. 8 of their paper) (Fig. 7.2).

The Reitzer et al. data clearly show that internal ATP levels can be maintained in HeLa cells that are grown in serum-free media with glutamine and galactose as the only energy substrates for at least 2 h. These findings show that glutamine is a major energy metabolite for HeLa cells. They also showed in an earlier (Fig. 7 from their paper) paper that glutamine carbons appeared as CO_2, whereas only about 13% of glutamine carbons appeared as lactic acid. This is consistent with

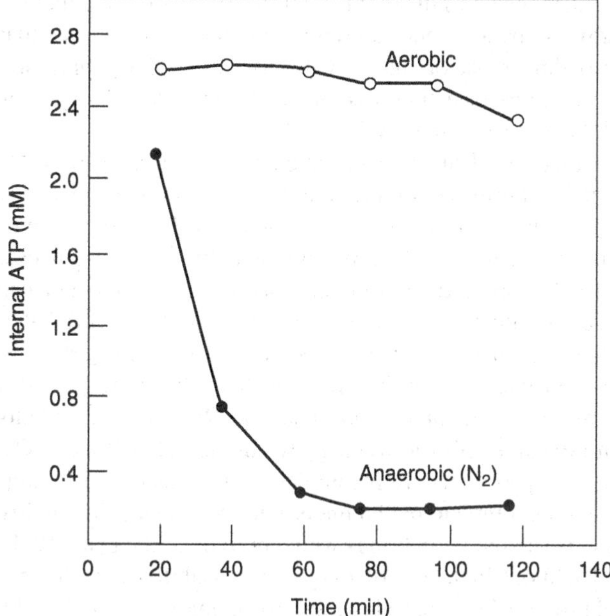

Figure 7.2 The concentration of ATP in HeLa cells incubated in the absence of sugar with or without oxygen. Cells were pregrown in Minimum Essential Medium Joklik, 5% nondialyzed fetal calf serum spinner cultures, which supports exponential growth to 10^6 cells/ml. At 3×10^5 cells/ml, 200 ml was centrifuged at 1500*g* for 15 min at room temperature and immediately resuspended in 10 ml of Minimum Essential Medium Joklik containing 20 mM 4-(2-hydroxyethyl)-L-piperazine-ethane-sulfonic acid, pH 7, without serum or glucose. The medium used for the anaerobic incubation had been deoxygenated by bubbling 100% N_2 through it for at least 10 min. The cells were resuspended gently to avoid oxygenating the medium and the gassing continued above the medium during the incubation. The aerobic culture was exposed to the atmosphere. The flasks were shaken rapidly in a water bath at 37°C. At the indicated times, 1.5 ml of culture was mixed with 0.25 ml of 2.4 N $HClO_4$ with 8 mM EDTA at 0°C and centrifuged at 12,006*g* for 20 min. The supernatant was neutralized with KOH, using the phenol red indicator present in the medium, and after removing potassium perchlorate by centrifugation, ATP was measured as described (16). *Source*: Modified from Ref. 16.

other studies indicating that little lactate is produced from glutamine in tumor cells (23–25). The Reitzer et al. findings indicate that glutamine is largely metabolized through the TCA cycle under aerobic conditions and that most of the energy derived from glutamine metabolism comes from the mitochondria and not from glycolysis in the cytoplasm. These are important findings related to energy metabolism in cancer cells.

The evidence presented showing that the cellular energy produced from glutamine metabolism was derived through OxPhos came from the influence of N_2 on ATP production. As shown in the second part of their figure, ATP production fell sharply when the HeLa cells were cultured under N_2 in the glucose/serum-free minimal media containing only glutamine (Fig. 7.2). At first glance, these observations would suggest that OxPhos must be involved in energy production since the removal of O_2 caused ATP synthesis to fall sharply. Indeed, Rietzer et al. conclude by stating, "This short term incubation under a nongrowth condition suggests that glutamine oxidation is providing most of this ATP by oxidative phosphorylation from the citric acid cycle reactions (see 'Discussion') and eliminates glycogen as a significant source of ATP energy via glycolysis" (16).

However, it is known that cell death will occur under anoxia using pure N_2 with no CO_2. CO_2 is needed to buffer acid under these conditions. We found that pure N_2 caused massive cell death in our cultured metastatic VM-M3 mouse tumor cells even when they were grown in complete media containing 25 mM glucose. We also found that HeLa cells cultured in complete medium with high glucose died rapidly when exposed to pure N_2. Papandreou et al. (26) also showed that severe anoxia caused cell death in their model tumor cells that were grown in complete media with high glucose. Rather than proving that glutamine provides energy through OxPhos in HeLa cell under normoxia, the Reitzer et al. results show that ATP production falls and cells die when grown in pure N_2 without CO_2.

Hence, the Reitzer et al. experiments are incomplete as they lack critical control experiments (cells grown in complete media under pure N_2 and a cell viability assay), showing that their experimental conditions were not toxic to the cells. Without these experiments and without inclusion of longer term viability data (longer than 2 h), it is not possible to conclude that glutamine oxidation provides ATP by OxPhos. Our findings indicate that pure N_2 and absence of CO_2 is toxic to cultured tumor cells grown in complete medium (containing serum, glutamine, and glucose) and cannot be used to assess whether OxPhos is operational. I do not exclude the possibility that glutamine provides energy to HeLa cells through OxPhos under normoxia, but I have issues with Reitzer et al.'s evidence presented to support this hypothesis.

Reitzer et al. also misinterpreted the findings of the Donnelly and Scheffler study in mentioning that these investigators found that glutamine respiration accounted for 40% of ATP production in respiration-deficient Chinese hamster fibroblasts (p. 2675). Donnelly and Scheffler concluded that glutamine respiration accounted for 40% of ATP production in wild-type "respiration competent fibroblasts," not in the respiration-deficient fibroblasts (27). How could ATP be made through respiration in cells that have no respiration?

Viewed collectively, the results from the Reitzer et al. study do not provide conclusive evidence indicating that respiration is normal in HeLa cells. Critical experiments were not done to support the conclusion that HeLa cells can produce ATP through glutamine respiration. No control experiments were included to exclude toxic effects of pure N_2 or absent CO_2. Although glutamine was metabolized through the TCA cycle with ATP production, no experiments were conducted excluding the possibility that ATP synthesis occurred through TCA cycle substrate-level phosphorylation involving amino acid fermentation. Without this evidence, it is not possible to conclude for certain that respiration is normal in HeLa cells. Moreover, we know from the findings of Rossignol et al. and Piva and McEvoy-Bowe that mitochondrial morphology is abnormal in HeLa cells (13, 28, 29). This suggests that HeLa cell mitochondrial function is not likely the same as in normal cells.

How is it possible to have normal respiration in these cells if their mitochondrial structure is abnormal? Although the Reitzer et al. study is cited heavily as evidence against the Warburg theory of cancer, it is clear that this evidence is weak at best.

WHAT ABOUT OxPhos EXPRESSION IN OTHER TUMORS?

Jose et al. (30) recently reviewed the literature on the role of OxPhos in providing energy for a broad range of tumor cell types. Besides the Reitzer et al. study, they cite numerous other studies suggesting that OxPhos provides energy to tumor cells. This information together with the previous criticisms of Sidney Weinhouse, reviewed in Chapter 6, suggests that OxPhos is functional in many tumor cells. A sampling of key reviews and studies suggesting that OxPhos in tumor cells is either functional, not seriously impaired, or can be restored to normal function include those of Guppy and coworkers (6, 31–33), Fantin et al. (34, 35), Rossignol and coworkers (28, 36–38), Moreno-Sanchez and coworkers (39–44), Griguer et al. (45), Mazurek et al. (46), Bonnet et al. (47), Funes et al. (48), Morris and coworkers (49, 50), Dang and coworkers (5, 51, 52), Fogal et al. (53), Gottlieb and Vousden (54), Thompson and coworkers (55–59), McKeehan (60), Levine and Puzio-Kuter (61), Sonveaux et al. (62), Lopez-Lazaro (63), Gatenby and Gillies (64, 65), and Weinberg and Chandel (66). I apologize to other research groups not included in this listing who hold similar views regarding the expression of OxPhos in cancer cells. As in the Reitzer et al. study, no experiments were conducted excluding the possibility that ATP synthesis occurred through TCA cycle substrate-level phosphorylation involving amino acid fermentation in any of these cited studies. Without this information, it is not possible to conclude that respiration is normal in cancer cells. While some respiration can occur in many cancer cells, it remains uncertain if respiration sufficiency is similar in tumor cells and their normal cell-matched controls.

The recent study by Wong and colleagues described evidence indicating that respiration was mostly normal in mitochondria isolated from tissues of ovarian cancer patients (67). Although these investigators did not evaluate respiration in control mitochondria isolated from normal ovarian tissue, they mentioned that ATP production and the specific activities of succinate, malate, and glutamate dehydrogenases were comparable to values reported in human skeletal muscle, heart, and liver. However, a careful examination of the data from their Table 2 shows that the rate of ATP production was markedly lower in ovarian and peritoneal cancers (mean of 37 nmol/min/mg) than in skeletal muscle (mean of 265 nmol/min/mg) when succinate was used as a substrate (67). Similar observations were obtained when TMPD+ascorbate was used as substrate.

On the basis of these data, it is not clear how these investigators could conclude that the TCA cycle was functional and that mitochondrial OxPhos was competent in these ovarian cancer tissues (67). It would be important for readers to evaluate these data for themselves in the original article. There are also reports showing that mitochondria are abnormal in ovarian cancer (68, 69). It would have been helpful if Wong and colleagues could have explained how they consider respiration normal in their tumor samples in light of these other studies and from the data in their Table 2 showing that mitochondrial function is abnormal and ATP production is reduced in ovarian cancer.

THE PEDERSEN REVIEW ON TUMOR MITOCHONDRIA AND THE BIOENERGETICS OF CANCER CELLS

Pedersen (6) presented a massive amount of evidence showing that mitochondria in tumor cells are defective when compared with mitochondria from normal cells. His review provides a comprehensive discussion of mitochondrial bioenergetics and function in cancer cells. Although a sophisticated understanding of cellular bioenergetics is needed to comprehend the information in this paper, the take-home information is clear, that is, cancer cells have defective mitochondria when compared to tissue-specific control cells. This point is important, as it is often difficult to obtain tissue-specific control cells for comparison with tumor cells. I summarize here just a few of the key conclusions from the Pedersen studies:

1. Tumor mitochondria are abnormal in morphology and ultrastructure and respond differently to changes in growth media than mitochondria from normal cells.
2. The protein and lipid composition of tumor mitochondria are markedly different from that of normal mitochondria.
3. Proton leak and uncoupling is greater in tumor mitochondria than in normal mitochondria.
4. Calcium regulation is impaired in tumor mitochondria.
5. Anion membrane transport systems are abnormal or dysregulated in mitochondria from many tumors.
6. Defective shuttle systems are not responsible for elevated glucose fermentation in tumor cells.
7. Pyruvate is not effectively oxidized in tumor mitochondria. Tumor mitochondria contain a surface-bound, fetal-like hexokinase.
8. A deficiency in some aspect of respiration could account for excessive lactic acid production in tumor cells.

Pederson indicated that it was not his intention to imply that Warburg's theory was incorrect, but only to draw attention to the underlying bioenergetic abnormalities expressed in all tumor cells. Although a generalized defect at the level of the mitochondrial electron transport chain does not exist in tumor cells, numerous other mitochondrial abnormalities do exist that would diminish respiratory function. Indeed, Warburg never stated that a generalized defect in electron transport was responsible for the origin of cancer. Rather, Warburg stated that insufficient respiration was responsible for the origin of cancer (70–73). We know from the work of numerous investigators that electron transport may not be coupled to ATP synthesis in cancer cells. Any mitochondrial defect that would uncouple electron transport from OxPhos could reduce respiratory sufficiency.

Although there are a variety of reasons why many investigators rejected Warburg's central hypothesis that damage to respiration is the origin of cancer, most of the reasons are unsupported by evidence. In none of the cited works arguing against the Warburg theory has the investigators excluded mitochondrial amino acid fermentation and substrate-level phosphorylation as an alternative to OxPhos for mitochondrial energy production (5). Indeed, I am yet to find any study, other than those we conducted, that actually addresses this issue in tumor cells (74, 75). Without this information, it is not possible to state with certainty that respiration is normal or can be restored in cancer cells. Without this information, it is not possible to reject the Warburg theory of cancer. It is therefore imperative that researchers who claim that OxPhos is normal in tumor cells design experiments that exclude mitochondrial ATP production through amino acid fermentation as an alternative explanation for their findings. This is important, as amino acid fermentation mostly involving glutamine oxidation can easily be mistaken for OxPhos since significant ATP is synthesized within the mitochondria whether or not O_2 is present.

On the basis of the data presented over many years by numerous investigators, I am convinced that OxPhos is universally impaired to some degree in all cancer cells. While the damage might be profound in some cancer cells and less profound in other cancer cells, most if not all cancer cells will express some degree of OxPhos insufficiency. Hence, the answer to the question posed in the title of this chapter is "not likely."

REFERENCES

1. Vaupel P. Strikingly high respiratory quotients: a further characteristic of the tumor pathophysiome. Adv Exp Med Biol. 2008;614:121–5.
2. Herst PM, Berridge MV. Cell surface oxygen consumption: a major contributor to cellular oxygen consumption in glycolytic cancer cell lines. Biochim Biophys Acta. 2007;1767:170–7.
3. Ferreira LM. Cancer metabolism: the Warburg effect today. Exp Mol Pathol. 2010;89:372–80.
4. Hochachka PW, Somero GN. Biochemical Adaptation: Mechanism and Process in Physiological Evolution. New York: Oxford Press; 2002.
5. Koppenol WH, Bounds PL, Dang CV. Otto Warburg's contributions to current concepts of cancer metabolism. Nat Rev. 2011;11:325–37.
6. Pedersen PL. Tumor mitochondria and the bioenergetics of cancer cells. Prog Exp Tumor Res. 1978;22:190–274.
7. Ramanathan A, Wang C, Schreiber SL. Perturbational profiling of a cell-line model of tumorigenesis by using metabolic measurements. Proc Natl Acad Sci USA. 2005;102:5992–7.
8. Villalobo A, Lehninger AL. The proton stoichiometry of electron transport in Ehrlich ascites tumor mitochondria. J Biol Chem. 1979;254:4352–8.
9. Seoane M, Mosquera-Miguel A, Gonzalez T, Fraga M, Salas A, Costoya JA. The mitochondrial genome is a "genetic sanctuary" during the oncogenic process. PloS One. 2011;6:e23327.
10. Tennant DA, Duran RV, Boulahbel H, Gottlieb E. Metabolic transformation in cancer. Carcinogenesis. 2009;30:1269–80.
11. Hochachka PW, Owen TG, Allen JF, Whittow GC. Multiple end products of anaerobiosis in diving vertebrates. Comp Biochem Physiol B. 1975;50:17–22.

12. Ecsedy JA, Holthaus KA, Yohe HC, Seyfried TN. Expression of mouse sialic acid on gangliosides of a human glioma grown as a xenograft in SCID mice. J Neurochem. 1999;73:254–9.
13. Lyamzaev KG, Izyumov DS, Avetisyan AV, Yang F, Pletjushkina OY, Chernyak BV. Inhibition of mitochondrial bioenergetics: the effects on structure of mitochondria in the cell and on apoptosis. Acta Biochim Pol. 2004;51:553–62.
14. Allen NJ, Karadottir R, Attwell D. A preferential role for glycolysis in preventing the anoxic depolarization of rat hippocampal area CA1 pyramidal cells. J Neurosci. 2005;25:848–59.
15. Tomitsuka E, Kita K, Esumi H. The NADH-fumarate reductase system, a novel mitochondrial energy metabolism, is a new target for anticancer therapy in tumor microenvironments. Ann N Y Acad Sci. 2010;1201:44–9.
16. Reitzer LJ, Wice BM, Kennell D. Evidence that glutamine, not sugar, is the major energy source for cultured HeLa cells. J Biol Chem. 1979;254:2669–76.
17. Renner C, Asperger A, Seyffarth A, Meixensberger J, Gebhardt R, Gaunitz F. Carnosine inhibits ATP production in cells from malignant glioma. Neurol Res. 2010;32:101–5.
18. Guzy RD, Schumacker PT. Oxygen sensing by mitochondria at complex III: the paradox of increased reactive oxygen species during hypoxia. Exp Physiol. 2006;91:807–19.
19. Niitsu Y, Hori O, Yamaguchi A, Bando Y, Ozawa K, Tamatani M, et al. Exposure of cultured primary rat astrocytes to hypoxia results in intracellular glucose depletion and induction of glycolytic enzymes. Brain Res Mol Brain Res. 1999;74:26–34.
20. Murdoch C, Muthana M, Lewis CE. Hypoxia regulates macrophage functions in inflammation. J Immunol. 2005;175:6257–63.
21. Gulliksson M, Carvalho RF, Ulleras E, Nilsson G. Mast cell survival and mediator secretion in response to hypoxia. PloS One. 2010;5:e12360.
22. Molina JR, Dennison JB, Mills GB. Breast cancer cells utilize lactate as energy during nutrient stress. Metabolism and Cancer; 2011 Oct 16–19; Baltimore (MD). 2011. p. 75.
23. Moreadith RW, Lehninger AL. The pathways of glutamate and glutamine oxidation by tumor cell mitochondria. Role of mitochondrial NAD(P)+-dependent malic enzyme. J Biol Chem. 1984;259:6215–21.
24. Lanks KW. End products of glucose and glutamine metabolism by L929 cells. J Biol Chem. 1987;262:10093–7.
25. Scott DA, Richardson AD, Filipp FV, Knutzen CA, Chiang GG, Ronai ZA, et al. Comparative metabolic flux profiling of melanoma cell lines: beyond the Warburg effect. J Biol Chem. 2011; 286:42626–34.
26. Papandreou I, Krishna C, Kaper F, Cai D, Giaccia AJ, Denko NC. Anoxia is necessary for tumor cell toxicity caused by a low-oxygen environment. Cancer Res. 2005;65:3171–8.
27. Donnelly M, Scheffler IE. Energy metabolism in respiration-deficient and wild type Chinese hamster fibroblasts in culture. J Cell Physiol. 1976;89:39–51.
28. Rossignol R, Gilkerson R, Aggeler R, Yamagata K, Remington SJ, Capaldi RA. Energy substrate modulates mitochondrial structure and oxidative capacity in cancer cells. Cancer Res. 2004;64:985–93.
29. Piva TJ, McEvoy-Bowe E. Oxidation of glutamine in HeLa cells: role and control of truncated TCA cycles in tumour mitochondria. J Cell Biochem. 1998;68:213–25.
30. Jose C, Bellance N, Rossignol R. Choosing between glycolysis and oxidative phosphorylation: a tumor's dilemma? Biochim Biophys Acta. 2010;1807:552–61.
31. Alirol E, Martinou JC. Mitochondria and cancer: is there a morphological connection? Oncogene. 2006;25:4706–16.
32. Zu XL, Guppy M. Cancer metabolism: facts, fantasy, and fiction. Biochem Biophys Res Commun. 2004;313:459–65.
33. Guppy M, Leedman P, Zu X, Russell V. Contribution by different fuels and metabolic pathways to the total ATP turnover of proliferating MCF-7 breast cancer cells. Biochem J. 2002;364:309–15.
34. Fantin VR, St-Pierre J, Leder P. Attenuation of LDH-A expression uncovers a link between glycolysis, mitochondrial physiology, and tumor maintenance. Cancer Cell. 2006;9:425–34.
35. Bui T, Thompson CB. Cancer's sweet tooth. Cancer Cell. 2006;9:419–20.

36. Smolkova K, Bellance N, Scandurra F, Genot E, Gnaiger E, Plecita-Hlavata L, et al. Mitochondrial bioenergetic adaptations of breast cancer cells to aglycemia and hypoxia. J Bioenerg Biomembr. 2010;42:55–67.
37. Smolkova K, Plecita-Hlavata L, Bellance N, Benard G, Rossignol R, Jezek P. Waves of gene regulation suppress and then restore oxidative phosphorylation in cancer cells. Int J Biochem Cell Biol. 2010;43:950–68.
38. Jezek P, Plecita-Hlavata L, Smolkova K, Rossignol R. Distinctions and similarities of cell bioenergetics and the role of mitochondria in hypoxia, cancer, and embryonic development. Int J Biochem Cell Biol. 2010;42:604–22.
39. Moreno-Sanchez R, Rodriguez-Enriquez S, Marin-Hernandez A, Saavedra E. Energy metabolism in tumor cells. FEBS J. 2007;274:1393–418.
40. Marin-Hernandez A, Rodriguez-Enriquez S, Vital-Gonzalez PA, Flores-Rodriguez FL, Macias-Silva M, Sosa-Garrocho M, et al. Determining and understanding the control of glycolysis in fast-growth tumor cells. Flux control by an over-expressed but strongly product-inhibited hexokinase. FEBS J. 2006;273:1975–88.
41. Rodriguez-Enriquez S, Vital-Gonzalez PA, Flores-Rodriguez FL, Marin-Hernandez A, Ruiz-Azuara L, Moreno-Sanchez R. Control of cellular proliferation by modulation of oxidative phosphorylation in human and rodent fast-growing tumor cells. Toxicol Appl Pharmacol. 2006;215: 208–17.
42. Rodriguez-Enriquez S, Marin-Hernandez A, Gallardo-Perez JC, Carreno-Fuentes L, Moreno-Sanchez R. Targeting of cancer energy metabolism. Mol Nutr Food Res. 2009;53:29–48.
43. Moreno-Sanchez R, Rodriguez-Enriquez S, Saavedra E, Marin-Hernandez A, Gallardo-Perez JC. The bioenergetics of cancer: is glycolysis the main ATP supplier in all tumor cells? Biofactors. 2009;35:209–25.
44. Rodriguez-Enriquez S, Carreno-Fuentes L, Gallardo-Perez JC, Saavedra E, Quezada H, Vega A, et al. Oxidative phosphorylation is impaired by prolonged hypoxia in breast and possibly in cervix carcinoma. Int J Biochem Cell Biol. 2010;42:1744–51.
45. Griguer CE, Oliva CR, Gillespie GY. Glucose metabolism heterogeneity in human and mouse malignant glioma cell lines. J Neurooncol. 2005;74:123–33.
46. Mazurek S, Grimm H, Boschek CB, Vaupel P, Eigenbrodt E. Pyruvate kinase type M2: a crossroad in the tumor metabolome. Br J Nutr. 2002;87(Suppl 1): S23–S29.
47. Bonnet S, Archer SL, Allalunis-Turner J, Haromy A, Beaulieu C, Thompson R, et al. A mitochondria-K+ channel axis is suppressed in cancer and its normalization promotes apoptosis and inhibits cancer growth. Cancer Cell. 2007;11:37–51.
48. Funes JM, Quintero M, Henderson S, Martinez D, Qureshi U, Westwood C, et al. Transformation of human mesenchymal stem cells increases their dependency on oxidative phosphorylation for energy production. Proc Natl Acad Sci USA. 2007;104:6223–8.
49. Regan DH, Lavietes BB, Regan MG, Demopoulos HB, Morris HP. Glutamate-mediated respiration in tumors. J Natl Cancer Inst. 1973;51:1013–7.
50. Kovacevic Z, Morris HP. The role of glutamine in the oxidative metabolism of malignant cells. Cancer Res. 1972;32:326–33.
51. Dang CV, Semenza GL. Oncogenic alterations of metabolism. Trends Biochem Sci. 1999;24: 68–72.
52. Dang CV. p32 (C1QBP) and cancer cell metabolism: is the Warburg effect a lot of hot air? Mol Cell Biol. 2010;30:1300–2.
53. Fogal V, Richardson AD, Karmali PP, Scheffler IE, Smith JW, Ruoslahti E. Mitochondrial p32 protein is a critical regulator of tumor metabolism via maintenance of oxidative phosphorylation. Mol Cell Biol. 2010;30:1303–18.
54. Gottlieb E, Vousden KH. p53 regulation of metabolic pathways. Cold Spring Harb Perspect Biol. 2010;2:a001040.
55. Wise DR, DeBerardinis RJ, Mancuso A, Sayed N, Zhang XY, Pfeiffer HK, et al. Myc regulates a transcriptional program that stimulates mitochondrial glutaminolysis and leads to glutamine addiction. Proc Natl Acad Sci USA. 2008;105:18782–7.

56. Jones RG, Thompson CB. Tumor suppressors and cell metabolism: a recipe for cancer growth. Genes Dev. 2009;23:537–48.
57. Vander Heiden MG, Cantley LC, Thompson CB. Understanding the Warburg effect: the metabolic requirements of cell proliferation. Science. 2009;324:1029–33.
58. Elstrom RL, Bauer DE, Buzzai M, Karnauskas R, Harris MH, Plas DR, et al. Akt stimulates aerobic glycolysis in cancer cells. Cancer Res. 2004;64:3892–9.
59. Buzzai M, Bauer DE, Jones RG, DeBerardinis RJ, Hatzivassiliou G, Elstrom RL, et al. The glucose dependence of Akt-transformed cells can be reversed by pharmacologic activation of fatty acid beta-oxidation. Oncogene. 2005;24:4165–73.
60. McKeehan WL. Glycolysis, glutaminolysis and cell proliferation. Cell Biol Int Rep. 1982;6: 635–50.
61. Levine AJ, Puzio-Kuter AM. The control of the metabolic switch in cancers by oncogenes and tumor suppressor genes. Science. 2010;330:1340–4.
62. Sonveaux P, Vegran F, Schroeder T, Wergin MC, Verrax J, Rabbani ZN, et al. Targeting lactate-fueled respiration selectively kills hypoxic tumor cells in mice. J Clin Invest. 2008;118: 3930–42.
63. Lopez-Lazaro M. The warburg effect: why and how do cancer cells activate glycolysis in the presence of oxygen? Anticancer Agents Med Chem. 2008;8:305–12.
64. Gatenby RA, Gillies RJ. Why do cancers have high aerobic glycolysis? Nat Rev. 2004;4:891–9.
65. Gillies RJ, Gatenby RA. Adaptive landscapes and emergent phenotypes: why do cancers have high glycolysis? J Bioenerg Biomembr. 2007;39:251–7.
66. Weinberg F, Chandel NS. Mitochondrial metabolism and cancer. Ann N Y Acad Sci. 2009;1177: 66–73.
67. Lim HY, Ho QS, Low J, Choolani M, Wong KP. Respiratory competent mitochondria in human ovarian and peritoneal cancer. Mitochondrion. 2011;11:437–43.
68. Carew JS, Huang P. Mitochondrial defects in cancer. Mol Cancer. 2002;1:9.
69. Dai Z, Yin J, He H, Li W, Hou C, Qian X, et al. Mitochondrial comparative proteomics of human ovarian cancer cells and their platinum-resistant sublines. Proteomics. 2010;10:3789–99.
70. Warburg O. The Metabolism of Tumours. New York: Richard R. Smith; 1931.
71. Warburg O. On the origin of cancer cells. Science. 1956;123:309–14.
72. Warburg O. On the respiratory impairment in cancer cells. Science. 1956;124:269–70.
73. Warburg O. Revidsed Lindau Lectures: The prime cause of cancer and prevention - Parts 1 & 2. In: Burk D, editor. Meeting of the Nobel-Laureates Lindau, Lake Constance, Germany: K.Triltsch; 1969. p. http://www.hopeforcancer.com/OxyPlus.htm.
74. Seyfried TN. Mitochondrial glutamine fermentation enhances ATP synthesis in murine glioblastoma cells. Proceedings of the 102nd Annual Meeting of the American Association Cancer Research, Orlando (FL). 2011.
75. Shelton LM, Strelko CL, Roberts MF, Seyfried NT. Krebs cycle substrate-level phosphorylation drives metastatic cancer cells. Proceedings of the 101st Annual Meeting of the American Association for Cancer Research, Washington (DC). 2010.

Chapter 8

Is Mitochondrial Glutamine Fermentation a Missing Link in the Metabolic Theory of Cancer?

AMINO ACID FERMENTATION CAN MAINTAIN CELLULAR ENERGY HOMEOSTASIS DURING ANOXIA

Mitochondrial amino acid fermentation is known to maintain metabolic homeostasis under hypoxia in several species of diving animals (1, 2). Mitochondrial amino acid fermentation can also maintain metabolic homeostasis in the heart and kidney under low glucose and low O_2 conditions (1–5). The possibility that tumor cells might also obtain energy through amino acid fermentation has not been considered previously as an alternative energy source to OxPhos. Although Warburg considered respiration and glucose fermentation as the sole producers of energy within cells, amino acid fermentation in the mitochondria can also produce energy through substrate-level phosphorylation (1).

Schwimmer et al. (6) showed that the energy derived from TCA cycle substrate phosphorylation (succinyl-CoA synthetase step) (Fig. 4.6) was sufficient to compensate for F1-ATPase deficiency in yeast cells. It is unclear if Warburg was aware of energy that could be derived through this step, as he did not to my knowledge discuss this in his writings (7–11). Indeed, we were the first group to report that Krebs cycle substrate-level phosphorylation might compensate for insufficient respiration in metastatic cancer cells (12). On the basis of preliminary studies, I suggest that energy through glutamine fermentation could compensate for

Cancer as a Metabolic Disease: On the Origin, Management and Prevention of Cancer, First Edition. Thomas Seyfried.

insufficient or suppressed respiration in those tumor cells that can use glutamine for energy.

While it is well known that glucose can be fermented, less is known about amino acid fermentation. Lactate is the by-product of glucose fermentation, whereas succinate, alanine, and aspartate are by-products of glutamine or amino acid fermentation under hypoxia (1–5). The expression of lactate in the presence of O_2 is abnormal and would indicate that the cells are fermenting. The degree of fermentation (lactate production) is positively correlated with the degree of malignant growth (10). Also, the less is the respiration, the greater is the fermentation. Under anoxia, fumarate can replace O_2 as an electron acceptor. If the cells consume oxygen, it is unlikely that succinate would accumulate. Under high glucose, amino acid fermentation can occur whether or not succinate accumulates. Hence, it is important to account for the multiple variables required to assure that cells are actually using OxPhos alone or are using some combination of OxPhos and mitochondrial substrate-level phosphorylation to maintain their viability.

EVIDENCE SUGGESTING THAT METASTATIC MOUSE CELLS DERIVE ENERGY FROM GLUTAMINE FERMENTATION

My graduate student, Roberto Flores, and I propose that glutamine and its metabolites (glutamate and α-ketoglutarate) could be fermented for energy in cancer cells under certain metabolic conditions, for example, under hypoxic or in high glucose under normoxia. High glucose levels suppress respiration through a Crabtree effect, thus producing increased fermentation. We presented evidence for this possibility at the 2011 meetings of the American Association of Cancer Research (13). We examined the influence of glucose and glutamine on ATP synthesis and viability in cultured mouse VM-M3 cells, a model for invasive human glioblastoma and systemic metastasis. These cells are known to have abnormalities in the content and composition of cardiolipin, which is linked to abnormal respiration (14).

Using a bioluminescent-based in vitro ATP assay, we found that ATP production and cell viability were similar in the metastatic cells grown in media containing either glutamine alone or glucose alone (Fig. 8.1). Shelton (15) also showed that lactate production was significantly lower in the metastatic cells grown in glutamine than in the cells grown in glucose, indicating that these cells produce little lactate from glutamine alone (Fig. 4.10). Recent findings from my graduate student, Linh Ta, showed that only trace levels of ^{14}C-labeled glutamine carbons are found in lactate when the VM-M3 cells are grown in 25 mM glucose (unlabeled) and 4 mM glutamine (radiolabeled). Significant radiolabeled lactate was found, however, when ^{14}C-labeled glucose was added. These findings indicate that very few glutamine carbons are present in lactate when high glucose is also present in the media.

However, ATP synthesis and lactate production were significantly greater in the VM-M3 tumor cells grown in glucose and glutamine than that for the tumor

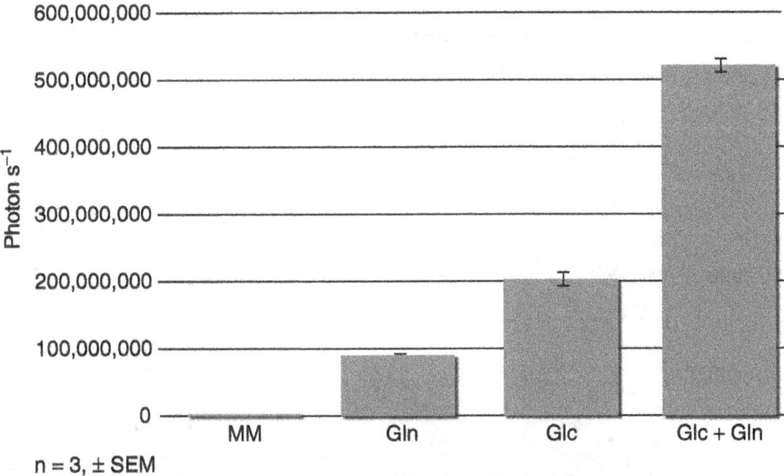

Figure 8.1 Effect of glucose and glutamine on viability of metastatic VM-M3 glioblastoma cells. In total, 5×10^4 cells were seeded in 96-well plates in 100 μl of DMEM plus 5% FBS and allowed to settle for 6 h before washing with 1 × phosphate buffered saline (PBS) with subsequent addition of minimal DMEM media containing either 25 mM glucose alone, 4 mM glutamine alone, or both metabolites together. After the cells were incubated for 24 h in 95% air and 5% CO_2, the Promega CellTiter Glo ATP assay was performed. Values represent the mean ± SEM of three independent samples per group. The results show that glucose and glutamine work synergistically compared to each metabolite alone (MM = minimal media). These data were presented at the 2011 meeting of the American Association of Cancer Research (13).

cells grown in either glutamine alone or glucose alone (Figs. 4.10 and 8.1). These findings show that glucose and glutamine work synergistically to enhance ATP synthesis, lactate production, and growth.

The synergy we found in the VM-M3 tumor cells was due to glutamine, as neither aspartate nor alanine (alternative nitrogen sources) could replace glutamine for the effect (Fig. 8.2). Previous results from Reitzer et al. (16) showed that the nonfermentable sugars, galactose and fructose, could replace glucose as a driver of energy metabolism in HeLa cells. Synergy for ATP synthesis and growth in the metastatic mouse cells arises from a specific interaction between glucose and glutamine metabolism. However, many tumor cells such as A549, HepG2, HeLa, U-87, U-251, and MDA-MB-453 can grow with minimal glucose. Many of these lines have a low glycolytic capacity relative to highly glycolytic tumors such as VM-M3, MCF-7, D-54 MG, GL 261, and 143B. All of the highly glycolytic cell lines are unable to grow without glucose.

We also collaborated with Cheryl Strelko and Mary Roberts in the Boston College Chemistry Department to further examine mitochondrial function in the metastatic mouse cells. Strelko and Roberts identified succinate, aspartate, alanine, and citrate in the tumor cells grown in pan-labeled glutamine using [C^{13}] NMR analysis (Fig. 8.3). These data were presented to support mitochondrial energy production through Krebs cycle substrate-level phosphorylation (12) and indicate

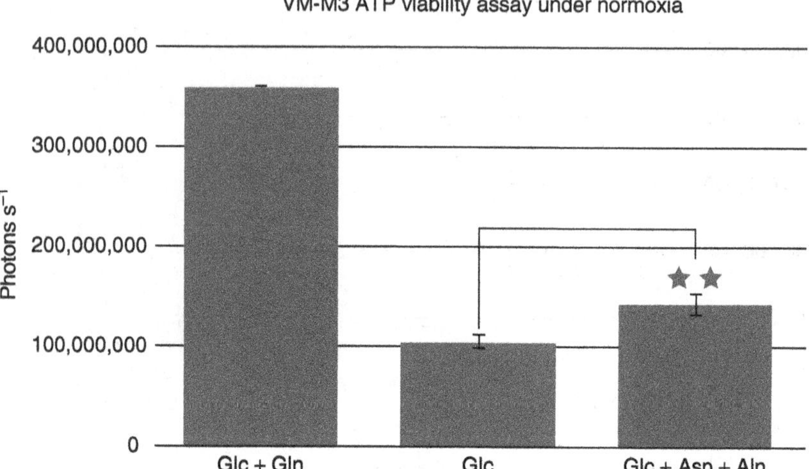

Figure 8.2 Glutamine is a better metabolic fuel for VM-M3 glioblastoma cells than are aspartate and alanine. Cells were grown in minimal DMEM media containing 25 mM glucose plus 4 mM glutamine, 25 mM glucose, or 25 mM glucose plus 4 mM aspartate and alanine. After the cells were incubated for 24 h in 95% air and 5% CO_2, the respective media were removed from each well and 100 μl DMEM plus 5% FBS were added followed by a 30-min equilibration to room temperature. ATP synthesis was measured as in Figure 8.1. Values represent the mean ± SEM of six independent samples per group. The asterisks indicate that the Glc+Asp+Aln values differ significantly from the Glc values at $p < 0.01$. The results show that neither aspartate nor alanine can substitute for glutamine as an energy metabolite in the VM-M3 cells. Other conditions are described in Figure 8.1. These data were presented at the 2011 meeting of the American Association of Cancer Research (13).

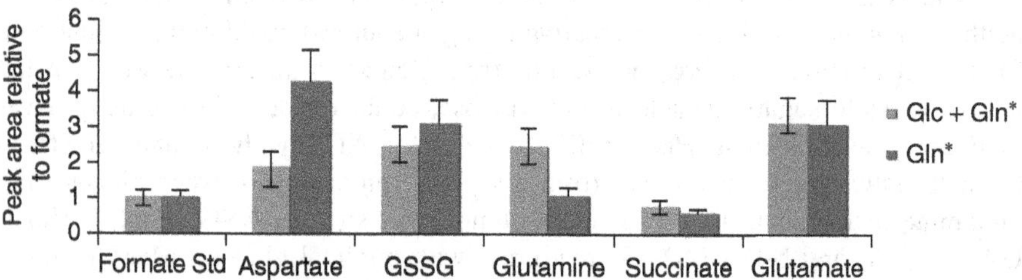

Figure 8.3 [C^{13}] NMR analysis of glutamine-labeled metabolites in VM-M3 glioblastoma cells. VM-M3 cells were grown in the presence of 4 mM C^{13} glutamine ± unlabeled 25 mM glucose. Cell extracts were collected via ethanol extraction after 12 h. Lyophilized extracts were redissolved in D_2O with 2 mM sodium formate standard and pH adjusted to 7.4. 1D-g HSQC (Heteronuclear Multiple Quantum Correlation) spectra were analyzed, and peaks were integrated with respect to the formate standard. Data are expressed as the average peak area relative to the formate standard of three independent samples ± the average % error. Asterisk indicates C^{13}. These data were presented at the 2010 meeting of the American Association of Cancer Research (12).

that glutamine is metabolized through the TCA cycle in these cells. In other words, mitochondria are capable of metabolizing glutamine in these tumor cells. The question arose as to whether these cells were using the glutamine to produce energy through OxPhos or through mitochondrial fermentation. The role for glutamine in energy production would be in addition to the known role of glutamine in replenishing TCA cycle metabolites (anapleurosis) (17, 18).

We showed that tumor cell viability and ATP production were robust in either anoxia or cyanide as long as both glucose and glutamine were present in the media (Figs 8.4 and 8.5). Since anoxia (95% N_2, 5% CO_2) or cyanide (an inhibitor of complex IV respiration) inhibits OxPhos, the robust synergy seen for glucose and glutamine is unlikely due to significant energy from OxPhos. Scott et al. (18) also found significant ATP production from glutamine under hypoxia in human melanoma cells but did not describe how ATP was formed from glutamine in the absence of O_2. We propose that the glucose/glutamine energy synergy observed in our metastatic mouse cells arises from linked fermentation redox couples in the cytoplasm and mitochondria that synthesize ATP largely through nonoxidative substrate-level phosphorylations.

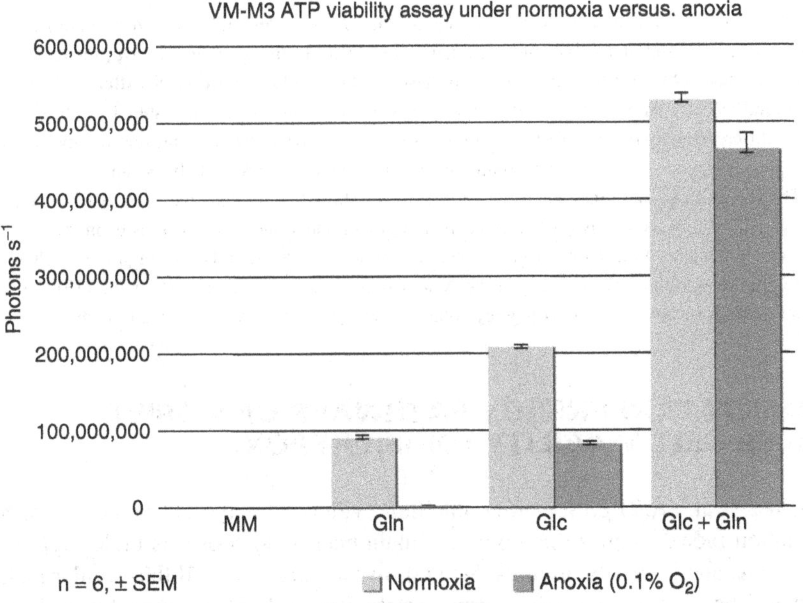

Figure 8.4 Influence of anoxia on viability of VM-M3 glioblastoma cells. Cells were grown in minimal DMEM media containing 25 mM glucose alone, 4 mM glutamine alone, or both metabolites together. After 24-h incubation of one 96-well plate in 95% air and 5% CO_2 and another in 95% nitrogen and 5% CO_2 (Biospherix Chamber), the ATP assay was performed. Values represent the mean SEM of three independent samples per group. Other conditions were as described in Figure 8.1. These data were presented at the 2011 meeting of the American Association of Cancer Research (13). To see this figure in color please go to ftp://ftp.wiley.com/public/sci_tech_med/cancer_metabolic_disease.

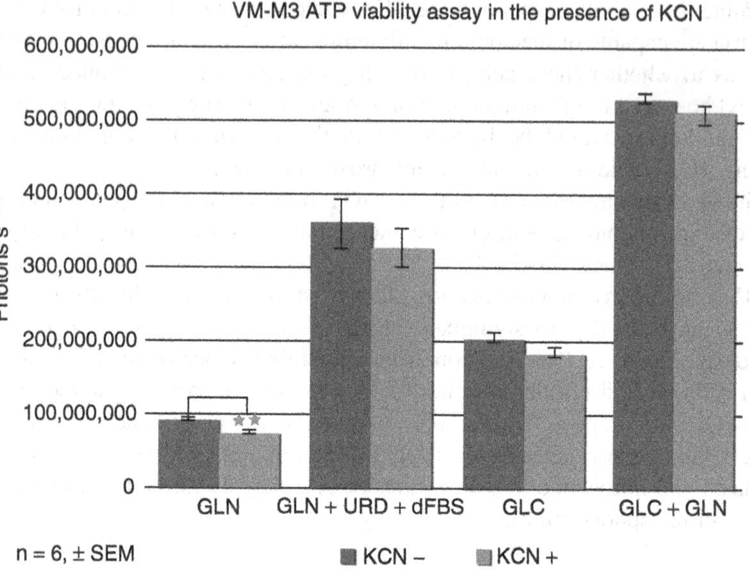

Figure 8.5 Potassium cyanide influence on viability of VM-M3. Cells were grown in minimal DMEM (Dulbecco's Modified Eagle Medium) containing 25 mM glucose alone, 4 mM glutamine alone, or both metabolites together as well as all these conditions plus 1 mM KCN. After 24 h of incubation in 95% air and 5% CO_2, ATP synthesis was measured as above. The asterisks indicate significant difference between Gln+KCN and Gln−KCN at $p < 0.01$. Values represent the mean ± SEM of three independent samples per group. URD = uridine and dFBS = dialyzed fetal bovine serum. The results from these data and those in Figure 8.5 show that OxPhos plays an insignificant role in VM-M3 energy metabolism when the cells are grown under anoxic conditions or in the presence of the complex IV inhibitor KCN. Other conditions were as described in Figure 8.1. These data were presented at the 2011 meeting of the American Association of Cancer Research (13). To see this figure in color please go to ftp://ftp.wiley.com/public/sci_tech_med/cancer_metabolic_disease.

FERMENTATION ENERGY PATHWAYS CAN DRIVE CANCER CELL VIABILITY UNDER HYPOXIA

Hochachka et al. (1, 2) presented compelling evidence for the existence of linked fermentation redox couples that could maintain energy homeostasis under hypoxia in metazoans and in diving animals. Several investigators showed that similar pathways could be used to maintain energy metabolism and cellular viability in heart and kidney under periodic hypoxia (3–5, 19–22). Tomitsuka et al. (23) recently provided the first evidence for the existence of this type of energy metabolism in cancer cells. Hence, cytoplasmic and mitochondrial amino acid fermentation could compensate for OxPhos under hypoxia. Many cancer cells can grow in hypoxic environments. Is it possible that cancer cells use energy through these pathways to compensate for respiratory injury? We suggest that they might, but this would occur only under specific conditions, for example, hypoxia or high glucose.

Our new concept to explain cancer cell energy metabolism is illustrated in Figure 8.6 and is a modification of the concept of Hochachka. We presented these pathways for the first time at the 2011 meetings of the American Association of Cancer Research (13). The malate–aspartate and glycerol 3-phosphate shuttles can link the redox couples in cytoplasm and mitochondria. This linkage is consistent with evidence showing high expression of these shuttle systems in various cancer cells (24–26). Shuttle expression in tumor cells, however, depends in part on whether cells can grow in the presence or absence of glucose.

In addition to the shuttles, the mitochondrial fumarate reductase pathway is also thought to produce ATP under certain hypoxic conditions (1, 23). NADH serves as the electron and proton donor, whereas fumarate serves as the ultimate electron and proton acceptor with succinate as an end product. Our model would be most relevant in those cancers that proliferate when using both glucose and glutamine to drive energy metabolism. The model would require modification to explain energy metabolism for those tumors that express defects in the TCA cycle and depend more heavily on glucose than glutamine for energy metabolism (28, 29).

According to our model, simultaneous glutamine and glucose fermentation would maintain cancer cell viability in those environments where oxygen is limited (hypoxia). It remains to be determined, however, if glutamine can also be fermented in tumor cells in the presence of oxygen. The Warburg effect involves the continued fermentation of glucose in oxygen. Aerobic lactate production provides this evidence. Succinate accumulation is indicative of amino acid fermentation under hypoxia. It is not yet clear if the succinate detected in the tumor cells under aerobic conditions from the NMR experiments results from glutamine fermentation. Succinate should not accumulate in cells that respire (1). It is also possible that glutamine is oxidized under aerobic conditions but is fermented under hypoxia.

Glutamine could also be metabolized under hypoxia through anaerobic respiration involving uncoupled electron transport. Elevated glucose levels would suppress OxPhos through a Crabtree effect, thus allowing the possibility of glutamine fermentation under normoxia. It would be difficult to distinguish glutamine respiration from glutamine fermentation under normoxia since both processes would involve electron transfer and TCA cycle activity.

Glutamine fermentation, occurring under high glucose conditions, will generate considerable energy through substrate-level phosphorylation and possibly through the fumarate reductase reaction (1, 2). Neither process involves OxPhos but would still require uncoupled electron transport. ATP uptake into the mitochondria from the cytoplasm and electron transport would be needed to drive the F1-F0-ATPase in reverse in order to maintain a proton motive gradient (2, 30). We think this situation would be present in those highly glycolytic tumor cells where the hexokinase-2 becomes attached to the outer mitochondrial membrane as described by Pedersen (31). The ATP needed to drive the ATP synthase in reverse under hypoxia would come almost exclusively from glucose and glutamine fermentation. *Hence, targeting glucose and glutamine could effectively shutdown energy metabolism in many cancers that depend on these metabolites for energy.*

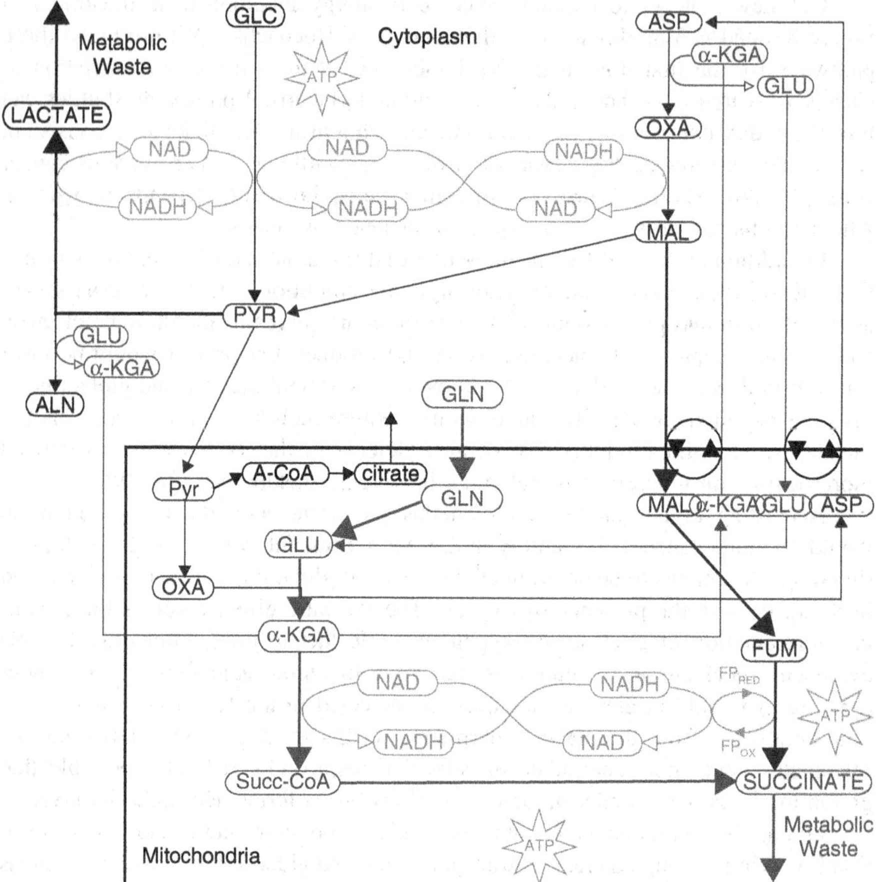

Figure 8.6 Proposed pathways for fermentation energy metabolism in VM-M3 glioblastoma cells. Fermentation redox couples formed in the mitochondria and cytoplasm can generate ATP under hypoxia. Since cancer cells are known to be in a state of pseudo-hypoxia, the proposed schematic is logical. In the mitochondrial fermentation scheme, the fumarate reductase (FRD) system presides, as fumarate rather than oxygen becomes the final electron acceptor. The activity of the aspartate–malate shuttle can link the cytoplasmic and mitochondrial redox couples under hypoxia. Glucose-derived pyruvate is considered the sole source of lactate and alanine. For each mole of alanine "bled off" the glycolytic pathway, 1 mole of NAD+ must be generated from a source other than lactate dehydrogenase. The redox imbalance in the glycolytic pathway can be corrected by reduction of aspartate-derived oxaloacetate to malate. The malate and α-ketoglutarate formed in the cytoplasm can then be transported into the mitochondria in exchange for other anions (27). We believe that this mechanism also provides cancer cells energy under normoxia when high glucose is also present in the media. Under normoxia and high glucose, oxygen would replace fumarate as the electron acceptor and the F1F0-ATPase would run in reverse. This would be linked to attachment of hexokinase-2 to the mitochondria (refer to text for more detail). This metabolic pathway was presented at the 2011 meeting of the American Association of Cancer Research (13). See color insert.

Warburg was aware of the difficulty in attempting to shutdown tumor energy metabolism in the body (11). Restricting availability of glucose and glutamine becomes a simple and effective therapeutic strategy for cancer management. I address in Chapter 17 how we can shutdown tumor energy metabolism in vivo using combinations of energy-restricted ketogenic diets and drugs that target glucose and glutamine metabolism.

Tumor cells survive in hypoxia "not" because they have a growth advantage over normal cells but because they can ferment organic molecules. Organic molecules become O_2 surrogates in accepting electrons. Cancer cells not only ferment glucose, as Warburg first showed, but they might also ferment glutamine and possibly other amino acids in the mitochondria under hypoxia and when glucose levels are high under normoxia. Unlike normal cells that can switch back to OxPhos when O_2 becomes available, most tumor cells depend on fermentation metabolism whether or not O_2 is present in the environment. Tumor cells adapt to fermentation because their OxPhos is insufficient to maintain energy homeostasis. Fermentation adaptation underlies the pathology of cancer.

The failure to consider amino acid fermentation as an alternative energy source for tumor cells can cause confusion regarding energy metabolism in cancer. It can be difficult to distinguish the effects of glutamine oxidation from glutamine fermentation since both processes occur in the mitochondria. The difference between glutamine oxidation and glutamine fermentation is that the latter does not couple the proton motive gradient to ATP production. Warburg was also unaware of this energy source in tumor cells, as he considered residual OxPhos activity as the likely origin of the low aerobic ATP production in cancer cells (7, 8). We also do not exclude this possibility, as it remains to be determined if glutamine is fermented or oxidized under normoxic when glucose levels are low. Residual glutamine oxidation coupled with detectable but low glycolysis could occur in low glycolytic tumor cells. As Warburg mentioned, however, no tumor cells are known that do not ferment at least some glucose indicative of respiratory insufficiency (10).

COMPETING EXPLANATIONS FOR THE METABOLIC ORIGIN OF CANCER

Currently, I consider that there are three major hypotheses regarding the role of energy metabolism in the origin cancer cells. The first hypothesis is that of Weinhouse, which considers that cancer cells express aerobic glycolysis despite having normal respiratory function. The evidence for this view was presented in Chapter 6. This view is also consistent with the gene theory of cancer in that abnormalities in oncogenes and tumor suppressor genes are ultimately responsible for aerobic glycolysis. More specifically, gene defects cause aerobic glycolysis and the metabolic defects seen in cancer cells. Dang and colleagues summarized this view in their recent paper where they stated: "*Today, we understand that the relative increase in glycolysis exhibited by cancer cells under aerobic conditions was mistakenly interpreted as evidence for damage to respiration instead of damage to the regulation of glycolysis*" (32). According to this view, abnormal expression of oncogenes and tumor suppressor genes are ultimately responsible for glycolytic damage and the metabolic reprogramming of cancer cells.

This view of cancer origin is at odds with the metabolic theory that insufficient respiration is the origin of cancer. Warburg argued that damaged respiration is more common in cancer cells than is damaged fermentation. Respiration is more complicated than fermentation because it requires mitochondrial structure and many more enzymatic steps than glycolysis (9, 33). Warburg stated, "*it is one of the fundamental facts of present-day biochemistry that adenosine triphosphate can be synthesized in homogeneous solutions with crystallized fermentation enzymes, whereas no one has succeeded in synthesizing adenosine triphosphate in homogeneous solutions with dissolved respiratory enzymes, and the structure always goes with oxidative phosphorylation*" (8). Simply put, respiratory damage is more likely in cancer than is damage to fermentation (glycolysis).

In order to accept the Weinhouse hypothesis, one would need to overlook or discount the massive data of Pedersen and others (presented in Chapters 5–7) showing that mitochondrial structure and respiration are damaged in cancer cells. In addition, one would need to ignore or overlook the evidence from the nuclear/cytoplasmic transfer experiments showing that normal mitochondria can reprogram cancer nuclei to form normal tissues (covered in Chapter 11). However, normal nuclei are unable to reprogram the tumor cytoplasm to form normal cells. These experiments rule out a chromosomal (somatic mutation) origin of cancer and strongly implicate the importance of extrachromosomal, nonnuclear systems (mitochondria).

The second hypothesis suggests that elevated glycolysis suppresses respiration in cancer cells. Under this hypothesis, cancer respiration is considered repressed, but the repression arises secondary to the appearance of aerobic glycolysis. In other words, many of the abnormalities seen in tumor mitochondria structure and function would arise as effects rather than the cause of aerobic glycolysis. The findings of the Cuezva, Mazurek, and Rossignol groups seem to support variations of this hypothesis (34–37). While this hypothesis is consistent with many of Warburg's findings, this hypothesis also seems in line with the genetic origin of cancer, as abnormalities in oncogenes and tumor suppressor genes are thought responsible for elevated tumor glycolysis. To accept this hypothesis, one would also need to overlook the evidence from the nuclear/cytoplasmic transfer experiments showing that extrachromosomal processes, rather than nuclear mutations, drive tumorigenesis.

In contrast to the first two hypotheses, we favor Warburg's original hypothesis with the caveat that tumor cells can also use mitochondrial fermentation in addition to glycolysis to compensate for insufficient respiration. While the evidence supporting our hypothesis is still preliminary, I believe that this will help clarify the metabolic origin of cancer. It is my opinion that the view of cancer as a nuclear gene-driven process has stymied investigations into the mitochondrial origin of the disease. Our hypothesis can accommodate most characteristics of cancer, once the nuclear gene origin of the disease is rejected. Consequently, a critical reevaluation of the gene theory of cancer is necessary before the metabolic theory of cancer can be fully appreciated. I cover this reevaluation in Chapters 9–11 and 15.

CHAPTER SUMMARY

Cancer is a disease of abnormal energy metabolism. In order to survive with insufficient respiration, tumor cells have adapted to energy production through fermentation. Powerful synergy is established between fermentation redox couples in the cytoplasm and mitochondria. These redox couples are linked through shuttle systems that drive tumor cell energy metabolism using glucose and glutamine as fermentable metabolic fuels. Adaptation to fermentation allows tumor cells to survive and grow in hypoxic environments. The information covered in this chapter raises the specter of mitochondrial glutamine fermentation as an energy source for tumor metabolism under certain conditions.

REFERENCES

1. Hochachka PW, Owen TG, Allen JF, Whittow GC. Multiple end products of anaerobiosis in diving vertebrates. Comp Biochem Physiol B. 1975;50:17–22.
2. Hochachka PW, Somero GN. Biochemical Adaptation: Mechanism and Process in Physiological Evolution. New York: Oxford Press; 2002.
3. Pisarenko OI, Solomatina ES, Ivanov VE, Studneva IM, Kapelko VI, Smirnov VN. On the mechanism of enhanced ATP formation in hypoxic myocardium caused by glutamic acid. Basic Res Cardiol. 1985;80:126–34.
4. Weinberg JM, Venkatachalam MA, Roeser NF, Nissim I. Mitochondrial dysfunction during hypoxia/reoxygenation and its correction by anaerobic metabolism of citric acid cycle intermediates. Proc Natl Acad Sci USA. 2000;97:2826–31.
5. Weinberg JM, Venkatachalam MA, Roeser NF, Saikumar P, Dong Z, Senter RA, et al. Anaerobic and aerobic pathways for salvage of proximal tubules from hypoxia-induced mitochondrial injury. Am J Physiol Renal Physiol. 2000;279:F927–43.
6. Schwimmer C, Lefebvre-Legendre L, Rak M, Devin A, Slonimski PP, di Rago JP, et al. Increasing mitochondrial substrate-level phosphorylation can rescue respiratory growth of an ATP synthase-deficient yeast. J Biol Chem. 2005;280:30751–9.
7. Warburg O. The Metabolism of Tumours. New York: Richard R. Smith; 1931.
8. Warburg O. On the origin of cancer cells. Science. 1956;123:309–14.
9. Warburg O. On the respiratory impairment in cancer cells. Science. 1956;124:269–70.
10. Warburg O. Revidsed Lindau Lectures: The prime cause of cancer and prevention - Parts 1 & 2. In: Burk D, editor. Meeting of the Nobel-Laureates Lindau, Lake Constance, Germany: K.Triltsch; 1969. http://www.hopeforcancer.com/OxyPlus.htm.
11. Warburg O, Wind F, Negelein E. The metabolism of tumors in the body. J Gen Physiol. 1927;8:519–30.
12. Shelton LM, Strelko CL, Roberts MF, Seyfried NT. Krebs cycle substrate-level phosphorylation drives metastatic cancer cells. Proceedings of the 101st Annual Meeting of the American Association for Cancer Research; Washington (DC). 2010.
13. Seyfried TN. Mitochondrial glutamine fermentation enhances ATP synthesis in murine glioblastoma cells. Proceedings of the 102nd Annual Meeting of the American Association of Cancer Research; Orlando (FL). 2011.
14. Kiebish MA, Han X, Cheng H, Chuang JH, Seyfried TN. Cardiolipin and electron transport chain abnormalities in mouse brain tumor mitochondria: lipidomic evidence supporting the Warburg theory of cancer. J Lipid Res. 2008;49:2545–56.

15. Shelton LM. Targeting Energy Metabolism in Brain Cancer. Chestnut Hill: Boston College; 2010.
16. Reitzer LJ, Wice BM, Kennell D. Evidence that glutamine, not sugar, is the major energy source for cultured HeLa cells. J Biol Chem. 1979;254:2669–76.
17. DeBerardinis RJ, Cheng T. Q's next: the diverse functions of glutamine in metabolism, cell biology and cancer. Oncogene. 2010;29:313–24.
18. Scott DA, Richardson AD, Filipp FV, Knutzen CA, Chiang GG, Ronai ZA, et al. Comparative metabolic flux profiling of melanoma cell lines: beyond the Warburg effect. J Biol Chem. 2011;286:42626–34.
19. Phillips D, Aponte AM, French SA, Chess DJ, Balaban RS. Succinyl-CoA synthetase is a phosphate target for the activation of mitochondrial metabolism. Biochemistry. 2009;48:7140–9.
20. Penney DG, Cascarano J. Anaerobic rat heart. Effects of glucose and tricarboxylic acid-cycle metabolites on metabolism and physiological performance. Biochem J. 1970;118:221–7.
21. Gronow GH, Cohen JJ. Substrate support for renal functions during hypoxia in the perfused rat kidney. Am J Physiol. 1984;247:F618–31.
22. Rumsey WL, Abbott B, Bertelsen D, Mallamaci M, Hagan K, Nelson D, et al. Adaptation to hypoxia alters energy metabolism in rat heart. Am J Physiol. 1999;276:H71–80.
23. Tomitsuka E, Kita K, Esumi H. The NADH-fumarate reductase system, a novel mitochondrial energy metabolism, is a new target for anticancer therapy in tumor microenvironments. Ann N Y Acad Sci. 2010;1201:44–9.
24. Greenhouse WV, Lehninger AL. Occurrence of the malate-aspartate shuttle in various tumor types. Cancer Res. 1976;36:1392–6.
25. Greenhouse WV, Lehninger AL. Magnitude of malate-aspartate reduced nicotinamide adenine dinucleotide shuttle activity in intact respiring tumor cells. Cancer Res. 1977;37:4173–81.
26. Mazurek S, Michel A, Eigenbrodt E. Effect of extracellular AMP on cell proliferation and metabolism of breast cancer cell lines with high and low glycolytic rates. J Biol Chem. 1997;272: 4941–52.
27. Klingenberg M. Metabolite Transport in Mitochondria: An Example of Intracellular Membrane Function. In: Campbell PN, Dickins F, editors. Essays in Biochemistry. New York/London: Academic Press; 1970. p.119–59.
28. Sandulache VC, Ow TJ, Pickering CR, Frederick MJ, Zhou G, Fokt I, et al. Glucose, not glutamine, is the dominant energy source required for proliferation and survival of head and neck squamous carcinoma cells. Cancer. 2011;117:2926–38.
29. Pollard PJ, Wortham NC, Tomlinson IP. The TCA cycle and tumorigenesis: the examples of fumarate hydratase and succinate dehydrogenase. Ann Med. 2003;35:632–9.
30. Chevrollier A, Loiseau D, Chabi B, Renier G, Douay O, Malthiery Y, et al. ANT2 isoform required for cancer cell glycolysis. J Bioenerg Biomembr. 2005;37:307–16.
31. Pedersen PL. Warburg, me and hexokinase 2: multiple discoveries of key molecular events underlying one of cancers' most common phenotypes, the "Warburg effect", i.e., elevated glycolysis in the presence of oxygen. J Bioenerg Biomembr. 2007;39:211–22.
32. Koppenol WH, Bounds PL, Dang CV. Otto Warburg's contributions to current concepts of cancer metabolism. Nat Rev. 2011;11:325–37.
33. Fosslien E. Cancer morphogenesis: role of mitochondrial failure. Ann Clin Lab Sci. 2008;38: 307–29.
34. Jose C, Bellance N, Rossignol R. Choosing between glycolysis and oxidative phosphorylation: a tumor's dilemma? Biochim Biophys Acta. 2010;1807:552–61.
35. Rossignol R, Gilkerson R, Aggeler R, Yamagata K, Remington SJ, Capaldi RA. Energy substrate modulates mitochondrial structure and oxidative capacity in cancer cells. Cancer Res. 2004;64:985–93.
36. Ortega AD, Sanchez-Arago M, Giner-Sanchez D, Sanchez-Cenizo L, Willers I, Cuezva JM. Glucose avidity of carcinomas. Cancer Lett. 2009;276:125–35.
37. Ristow M, Cuezva JM. Oxidative Phosphorylation and Cancer: The Ongoing Warburg Hypothesis. In: Apte SP, Sarangarajan R, editors. Cellular Respiration and Carcinogenesis. New York: Humana Press; 2009. p.1–18.

Chapter 9

Genes, Respiration, Viruses, and Cancer

DOES CANCER HAVE A GENETIC ORIGIN?

Despite overwhelming evidence showing that cancer is a metabolic disease in line with Warburg's original theory, most investigators today view cancer as a genetic disease where mutations and chromosomal abnormalities underlie most aspects of tumor initiation and progression. The view of cancer as a genetic disease is the dogma driving the academic pursuit for resolution and is what currently underlies the pharmaceutical industry's approach to new therapies. Each person's tumor contains mutations unique to that tumor and to that person. Consequently, tailored or personalized molecular therapies are considered to be the future for cancer treatment. This therapeutic strategy has emerged from the widely held view that cancer is a *genetic* disease (1–5). How sure are we that cancer really is a genetic disease?

What if most cancers are not of genetic origin and that the multitude of gene and chromosomal defects seen in cancers are effects rather than causes of the disease? Despite the hype given to molecular targets and cancer therapeutics at the 2010 and 2011 meetings of the American Association of Cancer Research, the evidence supporting this therapeutic approach is weak at best. With the exception of imatinib (Gleevec), which targets the Abelson (ABL) proto-oncogene receptor tyrosine kinase, little success has been found to date for other targeted therapies (6–9). In view of the investment already made in the development of targeted molecular therapies, I consider the momentum for this personalized therapy as a type of escalation situation where good money is thrown after bad and where *heard mentality* trumps rational thinking in decision making (10). I am afraid that the cancer body count will need to go much higher before the medical establishment,

Cancer as a Metabolic Disease: On the Origin, Management and Prevention of Cancer, First Edition. Thomas Seyfried.

the NCI (National Cancer Institute), and the cancer industry come to recognize the futility of gene-based molecular therapies as a "primary" course of cancer management.

While the metabolic defects in tumor cells are receiving renewed attention in the cancer field, many investigators view the Warburg effect and other metabolic defects in cancer cells as a consequence of genomic instability (11–15). Many investigators attempt to force the metabolic abnormalities seen in cancer cells into preexisting ideas on the gene theory, where activation of oncogenes and inactivation of tumor suppressor genes underlie the origin of the metabolic abnormalities. We recently reviewed new evidence, however, showing how chromosomal abnormalities and somatic gene changes in cancer cells can arise as secondary effects rather than as primary causes of abnormal energy metabolism (16). How was it possible for the gene theory to gain precedence over Warburg's metabolic theory for the origin of cancer? As with most man-made fiascos, there is usually a convergence of several mishaps. The same can be said for why the gene theory displaced the Warburg metabolic theory for the origin of cancer.

First, the appearance of normal respiratory function in cancer cells leads many to question Warburg's central hypothesis that injury to OxPhos was the origin of cancer. As discussed in Chapter 4, the attacks of Weinhouse and other investigators were especially effective in discouraging investigation into the respiratory origin of cancer. Moreover, how could cancer cells arise from injured respiration if so many investigators working in the cancer metabolism field have reported that OxPhos is normal in many tumor cell types? I have addressed the shortcomings of these arguments in Chapters 4, 5, and 8. The experimental evidence linking the origin of cancer to defective energy metabolism appeared to be confused to many investigators working both within and outside the metabolism field. It was also difficult to see how defective respiration could cause gene mutations or metastasis (17). The failure to craft a cohesive cancer theory based on defective energy metabolism raised the possibility that other explanations of cancer might be more credible than any metabolic hypothesis.

The gene theory gained momentum over the viral theory of cancer once the perceived molecular mechanisms of viral action were revealed. A mechanistic linkage between gene defects and viruses was convenient, as viruses had long been recognized as the origin of cancer (18–21). It gradually became recognized that viruses might cause cancer by turning on certain cancer-causing genes called *oncogenes*, or by turning off other genes that prevented cancer, that is, tumor suppressor genes (4, 22–24). Oncogenes are those that are assumed to cause cancer. This accounts for the attention given to these kinds of genes in the cancer field. According to James German, a pioneer in cytogenetics, 1981 was the turning point when scientific evidence overwhelmingly supported the mutational origin of human cancer (20). Stratton and colleagues have considered 1982 as this turning point with the seminal discovery that the human *HRAS* oncogene could transform normal mouse NIH3T3 cells into cancer cells (25). In 1994, Harold Varmus was quoted as saying "there's incontrovertible evidence that cancer is a genetic disease" (22). Dr. Varmus now heads the NCI.

The Nobel Prize to Michael Bishop and Harold Varmus for their discovery of cellular oncogenes together with Peter Nowell's evidence that acquired genetic liability underlies tumor progression has solidified the idea that cancer is primarily a disease of genetic origin (5, 22, 26–28). From their work on colorectal cancer, Fearon and Vogelstein considered cancer to be primarily a genetic disease that arises from accumulation of mutations. These mutations were assumed to promote clonal selection of cells with increasingly aggressive behavior (29). The genetic origin of cancer is now considered to be dogma in major reviews on the subject (2–4, 30). Even those evaluating the metabolism of cancer consider gene defects as the drivers of the metabolic abnormalities (31, 32). The genetic dogma was further solidified in Robert Weinberg's textbook, *The Biology of Cancer* (33).

Problems with the Gene Theory

Although there is incontrovertible evidence that genomic instability is found in most cancers, this does not mean that cancer is primarily a genetic disease. According to Gibbs, "No one questions that cancer is ultimately a disease of the DNA" (26). I must apologize to Dr. Gibbs, but I seriously question this notion. I consider the majority of gene defects described in tumor cells as downstream epiphenomena of insufficient or damaged respiration. This includes the majority of recognized oncogenes and tumor suppressor genes. Alterations in these genes are required in order to enhance *nonoxidative* energy metabolism. In other words, the genetic damage seen in cancer arises as an effect of damaged respiration with compensatory fermentation rather than as the direct cause of cancer. If oncogene upregulation does not follow respiratory injury, the cell will die. Oncogenes are needed to maintain cellular viability following protracted respiratory insufficiency. There is growing evidence supporting this concept (16, 34).

How would the genomic instability theory of cancer be viewed if there were evidence showing that nuclear genomic stability is dependent on normal respiratory function? How would the genomic instability theory of cancer be viewed if there were evidence showing that oncogene upregulation and suppressor gene downregulation are required for maintaining cell viability following respiratory damage? How would the genomic instability theory of cancer be viewed if there were evidence showing that tumor suppressor gene mutations and viruses damage respiration?

I will review evidence showing that genomic instability, DNA damage, and abnormal expression of many oncogenes and tumor suppressor genes arise as secondary downstream effects of abnormal respiration rather than as primary causes of most cancers. I will review evidence showing that inherited cancer genes damage respiration, which then produces cancer. Once genomic defects become established in the tumor cell, they can contribute to the irreversibility of the disease. The persistent view of cancer as a DNA disease is largely responsible for the failure to develop effective cancer therapies. It is difficult to develop an effective therapy for a disease when the origin of the disease is misunderstood.

Theodor Boveri, Aneuploidy, and the Genetic Origin of Cancer

Where did the idea arise that gene defects cause cancer? The gene theory of cancer originated with Theodor Boveri's suggestion in 1914 that cancer could arise from defects in the segregation of chromosomes during cell division (26, 35–37). Boveri is best recognized for showing that Gregor Mendel's inherited traits in pea plants had their origin in chromosomes (37). This observation along with the work of Walter Sutton established the field of cytogenetics. As chromosomal instability in the form of aneuploidy (extra chromosomes, missing chromosomes, or broken chromosomes) is present in many tumor tissues (5, 28, 38, 39), it was not too much of a reach to extend these observations to somatic mutations within individual genes including oncogenes and tumor suppressor genes (40–43).

According to Ulrich Wolf, however, Boveri did not examine chromosome behavior in tumor cells (37). Boveri's hypothesis on the role of chromosomes in the origin of malignancy was based primarily on his observations of chromosome behavior in nematodes (*Ascaris*) and sea urchins (*Paracentrotus*) and on von Hansemann's earlier observations of chromosome behavior in tumors (37). Hence, the founder of the genetic theory of cancer appears not to have directly studied the disease.

In his 2002 review, Knudson stated that, "considerable evidence has been amassed in support of Boveri's early hypothesis that cancer is a somatic genetic disease" (3). The seeds of the somatic mutation theory (SMT) of cancer might have been sown even before Boveri's work. Virchow considered that cancer cells arise from other cancer cells (44). Robert Wagner provides a good overview of those early studies, leading to the idea that somatic mutations give rise to cancer (44). It gradually became clear that almost every kind of genomic defect could be found in tumor cells whether or not the mutations were connected to carcinogenesis (26, 28).

Inconsistencies with the Genetic Origin of Cancer

As I have mentioned in Chapter 2, Sonnenschein and Soto highlighted numerous inconsistencies in the SMT of cancer. David Tarin has also highlighted similar inconsistencies, while Duesberg and coworkers outright rejected the role of somatic mutations and oncogenes in the origin of cancer (18, 26, 34, 38, 45–49). It is important for readers to carefully consider the multiple inconsistencies supporting the gene theory of cancer. Soto and Sonnenschein state: "the emergence of conflicting data within the SMT (somatic mutation theory) did not result in the rejection of premises and hypotheses. For example, an oncogene could be 'dominant' and express a gain of function with respect to the non-mutated homologue, and its biological effect could be contextual at the same time. That is, a mutation that should have produced uncontrolled cell proliferation resulted in cell death or arrest of cell proliferation. Again, ad hoc explanations were proposed to resolve conflicting evidence, leading to a situation whereby any possible conclusion is valid because no alternative concept is ever disproved and abandoned. The lack of fit is attributed to the unfathomable complexity of nature/biology. In short, something can be anything and its opposite" (47).

Support for the Soto and Sonnenschein argument was recently highlighted regarding mutations in the gene for isocitrate dehydrogenase 1 (IDH1) (50). Some investigators suggest that the *IDH1* gene acts as a tumor-provoking oncogene, whereas others suggest that *IDH1* acts as a tumor-inhibiting suppressor gene. The problem becomes even more confusing with suggestions that *IDH1* can act simultaneously as an oncogene and as a tumor suppressor gene (50). In other words, when it comes to the SMT of cancer, "something can be anything and its opposite."

Rous may have hit the nail on the head regarding the SMT as early as 1959 when he stated: "*Most serious of all the results of the somatic mutation hypothesis has been its effect on research workers. It acts as a tranquilizer on those who believe in it*" (19). The concerns raised over the years regarding the SMT as a rational explanation for the origin of cancer are so profound that it is remarkable that this theory has persisted for as long as it has. How many more patients must die before the cancer field abandons the failed therapies based on the SMT of cancer?

There are also issues regarding the role of aneuploidy in the origin of cancer. I view gene mutations and aneuploidy as opposite sides of the same coin. Both arguments are based on a DNA origin of the disease. Knudson considered the linkage of the Philadelphia (Ph1) chromosome to chronic myelocytic leukemia (CML) as evidence for the genetic origin of cancer. The Ph1 chromosome involves a translocation between chromosomes 9 and 22, which then activates the ABL oncogene. The chronic phase of the disease invariably progresses into an acute blastic phase in which one of the main events has been characterized as a second Ph1, which further increases the activity of the ABL oncogene (3). However, the Ph1 chromosome and mutations in the ABL oncogene have been found in some people who do not have CML or any cancers for that matter (51). These findings indicate that mutation in the ABL oncogene is insufficient alone to cause CML.

There are simply too many inconsistencies with the hypothesis that most cancers arise specifically from gene or chromosomal defects. The most damming evidence against the gene theory comes from the nuclear/cytoplasmic transfer experiments (Chapter 11). Gene and chromosomal defects can, however, contribute to the respiratory insufficiency in tumor cells, thus solidifying the insufficiency once it occurs. Aneuploidy can disrupt respiratory function, thus forcing cells to rely more heavily on fermentation for energy (52). This would be consistent with the Warburg theory. I hope to make it clear that respiratory insufficiency precedes and induces both the somatic mutations and the aneuploidy widely found in cancer cells.

Just because the majority of cancer researchers do not question the theory that guides their work does not mean that the theory is correct. Indeed, it appears that the average cancer researcher is not guided by any grand theory (46), rather they formulate restricted hypotheses for the next few experiments and tend to go on collecting data without reference to the problem of carcinogenesis (Ponten J. In: Iversen OH, editor. *New Frontiers in Cancer Causation*. Washington, DC: Taylor & Francis; 1992. p. 59) (46). More disturbingly, many investigators pursue their research in areas considered to be "hot" simply because well-known researchers have defined the area as such. Many correctly surmise that it is easier to get papers published and grants funded in hot areas than in areas not considered hot. Cancer is one of the few fields where research areas are consistently hot, but progress toward the cure is consistently cold.

The cancer research field has drifted off course for too long in my opinion. It is now time for all cancer researchers to pause, and to reconsider the foundation upon which their views rest. In light of the compelling counterarguments against the gene-based theories of cancer together with our extensive in vivo studies in brain cancer (53–55), it has become clear to me that genetic theories are wanting in their ability to explain the origin of cancer. I do not dispute the overwhelming evidence that defects in DNA, genes, and chromosomes occur in all cancers. The evidence is massive. What I do question, however, is whether these defects actually cause the disease. I will review evidence showing that most of the genomic defects seen in tumor cells can be linked directly or indirectly to insufficient respiration.

RESPIRATORY INSUFFICIENCY AS THE ORIGIN OF CANCER

Is it genomic instability or is it insufficient respiration that is primarily responsible for the origin of cancer? As we have recently mentioned, this is more than an academic question, as the answer will impact approaches to cancer management and prevention (16). Metabolic studies in a variety of human cancers have previously shown that loss of respiratory function precedes the appearance of malignancy and aerobic glycolysis (the Warburg effect). Besides the evidence obtained from Warburg (56, 57), Roskelley and coworkers also illustrated this fact in their studies of various animal and human tumor tissues (58). They used two chemical systems to assess the respiratory function in tumor tissue and in the normal host tissue from which the tumor arose. These systems included the following:

1. O^2 → cytochrome oxidase → cytochrome-*c*-*p*-phenylenediamine.
2. O^2 → cytochrome oxidase → cytochrome-*c*-succinic dehydrase → succinate.

These enzyme systems provide the main pathway by which oxygen is fed to the vital combustive processes occurring in most normal cells. These pathways therefore represented a physiologic unit for evaluating the likelihood that a given tissue is neoplastic (58). It was clear from their findings that the respiratory function was seriously impaired in human cancer tissue in comparison to the respiratory function in normal nondiseased host tissues (58). A representative figure for metastatic rectal cancer is shown in Figure 9.1. This observation in rectal cancer was replicated in a broad range of human cancers including breast, brain, kidney, and stomach.

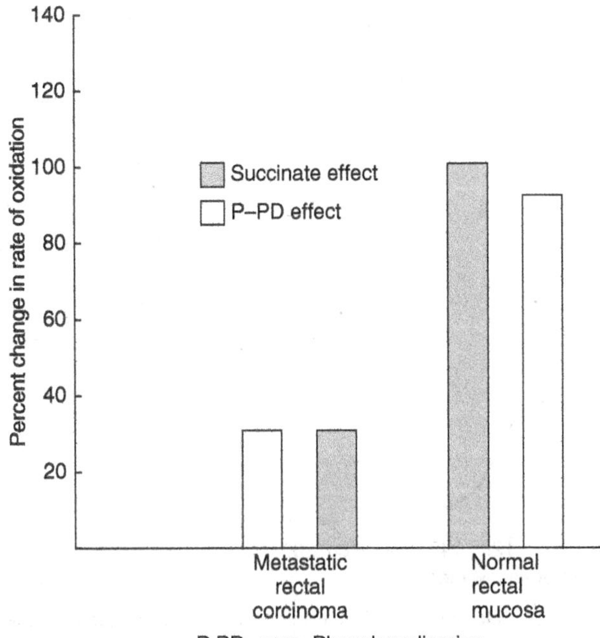

Figure 9.1 Oxidative behavior is less in human rectal tumor tissue than in normal rectal mucosa from the same patient. The authors concluded that the loss of oxidative activity occurs only in the neoplastic tissue, but not in the adjacent normal tissue (58). *Source*: Reprinted with permission from Ref. 58. To see this figure in color please go to ftp://ftp.wiley.com/public/sci_tech_med/cancer_metabolic_disease.

To further investigate the mechanisms involved, these investigators have followed the development of liver cancer in rats fed with butter yellow (a chemical carcinogen), and the development of skin cancer in rabbits treated with the Shope papilloma virus. Remarkably, the respiratory activity increased for several weeks in the treated tissues of both experimental groups, but then rapidly decreased until little or no respiratory activity remained in the treated liver or skin tissues (Figs. 9.2 and 9.3). Histological evidence of "frank neoplasia" did not appear for some time later in the treated tissues and was correlated with the onset of aerobic glycolysis (58).

Several profound insights regarding the origin of cancer emerged from this study. First, no human cancer tissues expressed normal respiratory capacity. Despite differences in the tissue of origin and in histological heterogeneity within tumors, all neoplastic cells express insufficient respiration. Second, the induction of carcinogenesis in animal tissues using either a chemical agent or a virus altered respiration in a similar way. It is now recognized that carcinogenic hydrocarbons, aflatoxin, viruses, and X rays all damage mitochondrial function and energy metabolism in similar a way (17, 34, 57, 59–62). This is interesting since chemical carcinogens and viruses also activate oncogenes in a similar way (23, 63),

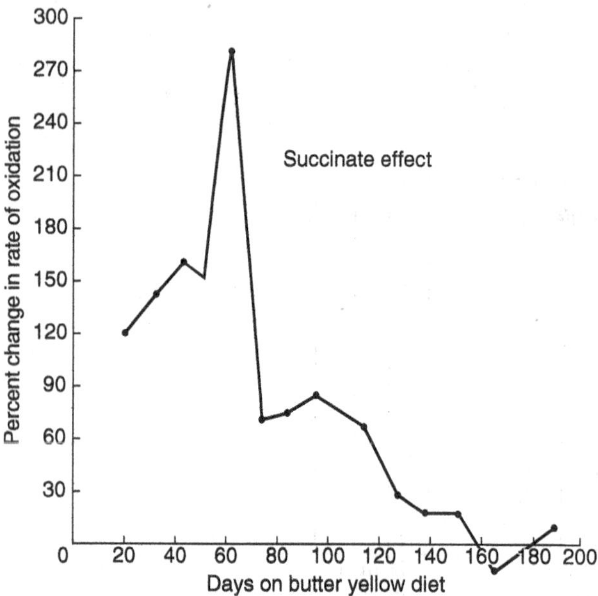

Figure 9.2 Time to neoplasia in rat liver after tissue oxidation is destroyed. Oxidative behavior was examined in rats fed with the butter yellow carcinogen in the diet for 190 days. Oxidative behavior was evaluated using the O^2 → cytochrome oxidase → cytochrome-*c*-succinic dehydrase → succinate reaction. At day 70, liver cells showed pronounced cytoplasmic degeneration with changes in nuclear size and shape (58). These characteristics were even more pronounced at day 137. Defined neoplasia (cell proliferation, etc.) was not detected until day 163. It was clear from these studies that oxidative damage preceded the onset of neoplasia and carcinogen-induced liver cancer. *Source*: Reprinted with permission from Ref. 58. To see this figure in color please go to ftp://ftp.wiley.com/public/sci_tech_med/cancer_metabolic_disease.

suggesting that oncogene activation follows mitochondrial damage. While γ-radiation causes mutations, it is the effect of radiation on mitochondrial respiration that causes cancer (17, 57). The spike in the respiratory activity prior to neoplasia is consistent with the recent findings of Seoane and coworkers showing that a spike in oncogene-induced cytochrome *c* activity and ROS production preceded tumorigenesis in glioma cells (64). Their study has shown that mitochondrial ROS production was ultimately responsible for the nuclear genomic instability seen in these tumor cells. The loss of the respiratory function "preceded" the histological appearance of any precancerous growth, frank neoplasia, or even aerobic glycolysis. de Groof et al. also showed that H-RasV12/E1A transformation of cells causes an instantaneous and dramatic upregulation of mitochondrial OxPhos activity that "precedes" the upregulation of glycolysis (65). This observation is similar to what Roskelley and coworkers had observed almost 70 years earlier. Respiratory insufficiency and tumorigenic transformation emerges after a transient OxPhos upregulation. Considered together, these findings support the Warburg theory in showing that respiratory insufficiency occurs prior to the

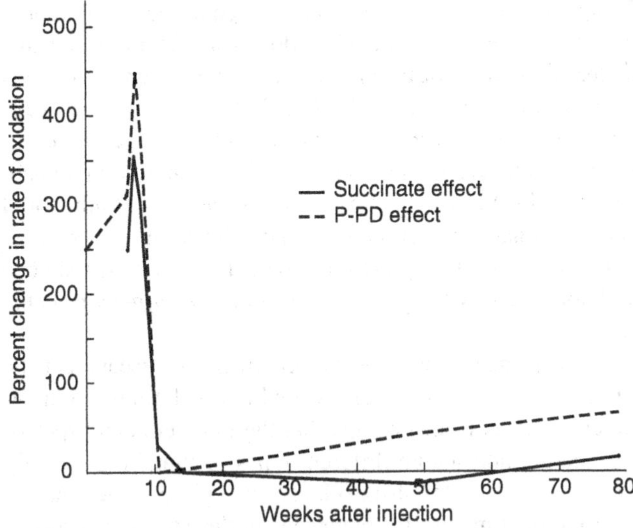

Figure 9.3 Time to neoplasia in rabbit skin after tissue oxidation is destroyed. Domestic rabbits developed cutaneous papilloma after inoculation with the carcinogenic Shope virus (58). Oxidative behavior was evaluated using the succinate reaction, as in Figure 9.2, and using $O^2 \rightarrow$ cytochrome oxidase $\rightarrow$ cytochrome-*c*-*p*-phenylenediamine reaction. Defined neoplasia (cell proliferation, etc.) and metastatic cancer was not detected until week 79. As seen for carcinogen-induced liver cancer in rats (above), oxidative damage preceded the onset of viral-induced skin cancer in rabbits. *Source*: Reprinted with permission from Reference 58. To see this figure in color please go to ftp://ftp.wiley.com/public/sci_tech_med/cancer_metabolic_disease.

onset of aerobic glucose fermentation, mutations, and neoplastic transformation. The data supporting the mitochondrial origin of cancer are too strong to dismiss.

None of the current integrated gene/chromosome theories can account for such observations (26). As neoplastic transformation is linked to genomic changes, one could argue that injury to respiration also precedes genomic instability. This will be discussed more in the next chapter. The Roskelley findings are consistent with the vast evidence presented in Pederson's review (discussed in Chapter 7), showing that respiration is defective or insufficient in all cancer cells. The Roskelley findings are also consistent with the findings of the Costoya and Singh groups in showing that the Warburg effect and tumorigenesis arise from mitochondrial damage and insufficient respiration (64, 66). Viewed collectively, these findings indicate that cancer is not a collection of many different diseases, but is rather a singular disease of respiratory insufficiency regardless of the tissue origin or cellular composition. While cancer cells arising in one organ will look morphologically different from cancer cells arising in another organ, they all suffer from a common malady, that is, respiratory insufficiency with compensatory fermentation.

Despite the evidence presented in support of the Warburg theory, the general view over the last 50 years has been that gene mutations and chromosomal abnormalities underlie most aspects of tumor initiation and progression including the Warburg effect and impaired respiratory function. The gene theory of cancer would argue that mitochondrial and respiratory dysfunction is an effect rather than a cause of cancer. It is not clear how many of the chromosomal defects and mutations found in tumor cells arise prior to cell division (38). Respiratory damage can occur prior to cell division. It is important to recognize that most of the gene and chromosome defects seen in tumor cells are found after the cells become premalignant or malignant. We recently described how somatic mutations and aneuploidy could arise following mitochondrial damage, as normal respiratory function is needed for DNA repair and mitotic fidelity (16). I will present more evidence against the gene theory of cancer in later chapters and will show that *respiratory insufficiency is the harbinger of tumorigenicity!*

If gene mutations are the primary cause of cancer, then the disease can be considered to be etiologically complicated requiring multiple solutions for management and prevention. This comes from findings that the numbers and types of mutations differ markedly among and within different types of tumors, especially the metastatic cancers (26, 28, 67–69). This also explains why "personalized molecular therapy" is considered the new mantra for management despite the failures of this approach to significantly impact cancer diagnosis or management (Chapter 16). If, on the other hand, impaired energy metabolism is primarily responsible for cancer, then most cancers can be considered to be a type of metabolic disease requiring fewer and less complicated solutions. How long will it take before this concept becomes widely recognized? How many more cancer patients must die before this concept is recognized?

GERMLINE MUTATIONS, DAMAGED RESPIRATION, AND CANCER

In general, cancer-causing germline mutations are rare and contribute to only about 5–7% of all cancers (40, 47). Although mitochondrial function is impaired in all tumor cells, it remains unclear how these impairments relate to the large number of somatic mutations and chromosomal abnormalities found in tumors (40, 70–72). I will show later how somatic mutations and aneuploidy can be linked to respiratory insufficiency with compensatory fermentation. But what can be said about germline mutations and cancer? Most inherited "inborn errors of metabolism" do not specifically compromise mitochondrial function or cause cancer in mammals (16). There are several exceptions, however, as rare germline mutations in genes encoding proteins of the TCA cycle can increase the risk to certain human cancers (73, 74).

For example, the risk for paraganglioma involves mutations in the succinate dehydrogenase (*SDH*) gene, whereas risk for leiomyomatosis and renal cell carcinoma involves mutations in the fumarate hydratase (*FH*, fumarase) gene (74–77). Mutations in the von Hippel–Lindau (VHL) tumor suppressor gene enhance risk for the VHL syndrome involving a predisposition to renal clear cell cancer (RCC), retinal and central nervous system hemangioblastomas, pancreatic cysts, and adrenal tumors (pheochromocytomas) (78, 79). The VHL tumor suppressor gene targets the mitochondria (80). It is important to recognize that these and similar mutations directly impair mitochondrial energy production leading to increased glycolysis and the Warburg effect (78, 81). Bayley and Devilee recently described how the inherited defects in these genes provide direct support for Warburg's original hypothesis that impaired respiration can be the origin of some cancers (81). Hence, respiratory damage sufficient to cause cancer can arise from mutations in these genes.

Inherited Mutations in p53 and Damaged Respiration

Besides these cancers, other inherited genes impact mitochondrial function and enhance cancer risk. It is well documented that rare inherited mutations in the *p*53 tumor suppressor gene increase risk for cancers of the Li Fraumeni syndrome (82). The spectrum of cancers identified in this syndrome includes breast carcinomas, soft tissue sarcomas, brain tumors, osteosarcoma, leukemia, and adrenocortical carcinoma. Tumor incidence is also greater in p53 knockout mice than in normal mice, although lymphomas appear to be more common than other tumor types (83). While many investigators view *p*53 as regulating tumorigenesis through its effects on mitochondria apoptotic signaling pathways or by influencing transcriptional factor response elements (33, 84, 85), recent evidence also indicates that p53 directly influences mitochondrial energy production.

Hwang and coworkers have shown that *p*53 regulates mitochondrial respiration through its transcriptional target gene *synthesis of cytochrome c oxidase* 2 (*SCO2*) (86–88). Most importantly, these investigators have shown that genome stability is dependent on OxPhos. These findings are also consistent with the earlier findings of Singh and coworkers showing that mitochondrial energy metabolism is impaired in human cancer cells containing defects in *p*53 (89). Genome stability is dependent on OxPhos, while mutations in *p*53 influence cancer susceptibility through disturbance in mitochondrial OxPhos. Werner syndrome, a disease of rapid aging and cancer predisposition, can also be linked to abnormalities in *p*53 and defective mitochondrial function (90, 91). Hence, the guardian function of *p*53 appears to reside in its ability to maintain sufficient OxPhos activity. This evidence would support Warburg's original theory.

A recent commentary in *Science* has suggested that the tumor suppressor genes *p*73 and *p*63 could serve along with *p*53 as "brothers in arms against cancer" (92). Although *p*73 also appears to function in the mitochondria like *p*53 (93), germline mutations in *p*73 are not associated with increased cancer risk (94). As no effective cancer therapies have yet emerged from attempts to manipulate *p*53 in tumor cells, it is unlikely in my opinion that effective therapies will emerge from attempts to manipulate either *p*73 or *p*63 in tumor cells (84, 92, 95–97).

Inherited Mutations in BRCA1 and Damaged Respiration?

Individuals carrying germline mutations in the *BRCA1* tumor suppressor gene are at increased risk of developing breast and ovarian tumors (98). *BRCA1* encodes a protein that is part of the large DNA repair complex (99). Recent findings from Coene and coworkers show that several anti-BRCA1 antibodies colocalize with mitochondrial staining in a variety of normal and cancer cell lines (100). BRCA1 was mostly localized to the mitochondrial matrix possibly in association with mtDNA (mitochondrial DNA). About 20% of BRCA1 staining was also found in the inner mitochondrial membrane, suggesting involvement in multiple mitochondrial functions (100) (Figure 9.4). Like the BRCA1 tumor suppressor protein, a mitochondrial location was also reported for the adenomatous polyposis coli (APC) tumor suppressor protein, which is mutated in most of the colon cancers (101, 102). These findings raise the possibility that mutations in the *BRCA1* and *APC* genes influence cancer susceptibility through alterations of mitochondrial function and OxPhos efficiency.

Inherited Mutations in RB and Damaged Respiration

The tumor suppressor protein, RB (retinoblastoma), regulates cell cycle exit and is dysregulated in numerous cancers (103). Germline mutations in the *RB* gene cause familial forms of retinal tumors. Recent studies indicate that the RB protein regulates mitochondrial biogenesis and the control of cell differentiation (104). In other words, defects in RB alter mitochondrial function, thus sustaining cell proliferation while preventing differentiation. Normal mitochondrial function is required for maintaining cellular differentiation and quiescence (16). These findings also link the action of oncogene-induced, cell senescence through effects on mitochondria and RB activity (105). Abnormalities in ATP production through OxPhos are linked to aerobic glycolysis in tumors with RB abnormalities (104, 105). Hence, abnormalities in the RB gene expression can underlie cancer susceptibility through alterations in OxPhos.

Xeroderma Pigmentosum and Damaged Respiration

Enhanced susceptibility to skin cancers is seen in patients who inherit the autosomal recessive gene for xeroderma pigmentosum (XP) (39). XP involves defects in nuclear DNA repair, thus enhancing susceptibility to cancer in skin cells and neural defects in brain cells (33). This disease is often used to support the hypothesis that cancer is a genetic disease since defects in genomic stability is the linchpin for the gene theory of cancer (33). However, studies from Rothe and coworkers have shown that XP mutations alter mitochondrial energy production (106). Altered mitochondrial ATP production is consistent with other studies showing that mitochondrial morphology and structure is also abnormal in XP patients and fibroblasts (106). These findings support my hypothesis that abnormalities in

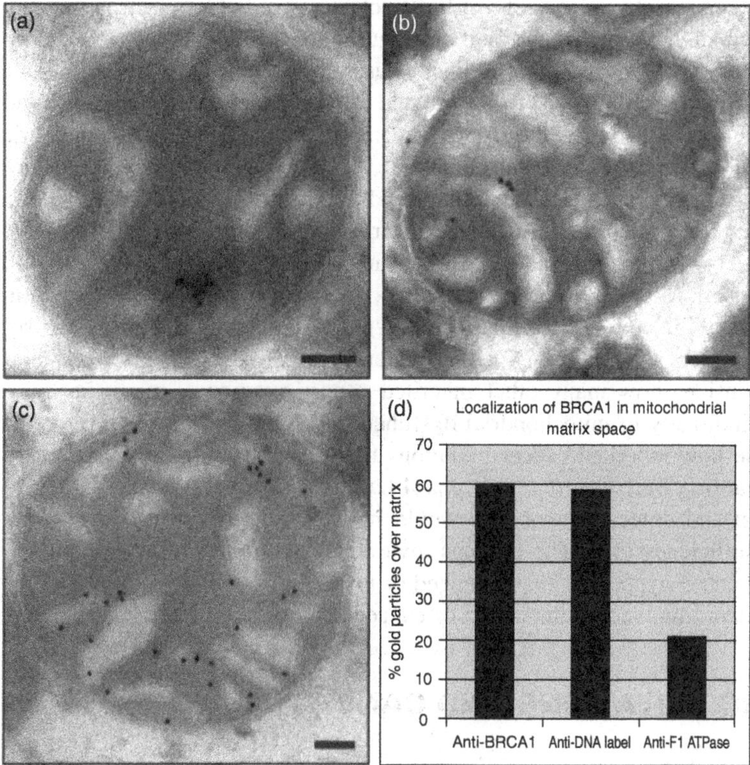

Figure 9.4 Intramitochondrial localization of BRCA1. (a) EM (electron microscopic) analysis of rat liver mitochondria with anti-BRCA1 Ab-1 shows BRCA1 gold clusters in the matrix. (b) EM analysis of rat liver mitochondria with anti-DNA IgM shows IgM signal in the matrix. (c) EM analysis of rat liver mitochondria with anti-F1 ATPase shows that F1 ATPase is associated with the mitochondrial membrane. Bars, 100 nm. (d) Table shows localization of BRCA1 in mitochondrial matrix space. BRCA1 (60%) and IgM (59%) are both predominantly located in the mitochondrial matrix space; F1 ATPase is predominantly associated with the cristae and therefore only a minority (20%) is located over the matrix space. *Source*: Reprinted with permission from Ref. 100. To see this figure in color please go to ftp://ftp.wiley.com/public/sci_tech_med/cancer_metabolic_disease.

genomic stability and DNA repair mechanisms can be attributed to defects in mitochondrial energy metabolism (16). In other words, the abnormal phenotypes seen in XP patients can be linked to mitochondrial dysfunction.

Friedrich's Ataxia and Damaged Respiration

Enhanced susceptibility to certain malignancies and neurological defects are seen in patients who inherit the gene for Friedrich's ataxia (107, 108). Friedrich's ataxia involves reduced expression of the mitochondrial protein frataxin, which regulates OxPhos and mitochondrial ATP production (109). It appears that frataxin directs the intramitochondrial synthesis of iron/sulfur clusters needed for electron transport (110). Recent studies from Ristow and colleagues have shown that targeted disruption of hepatic frataxin expression in mice causes impaired mitochondrial function and tumor growth (108). These studies provide further evidence showing that tumor formation can be linked directly to inherited mutations that disrupt mitochondrial function and energy production.

Viewed collectively, these findings provide both direct and indirect evidence that mitochondrial abnormalities can arise from inherited mutations that target various aspects of mitochondrial respiration. Inherited mutations that disturb mitochondrial function and respiratory energy production can produce abnormalities in genomic stability through compensatory fermentation, thus increasing the risk of various cancers. On the basis of the evidence reviewed here, it will be interesting to consider how many other inherited cancer syndromes might be linked directly or indirectly to mitochondrial dysfunction. Moreover, it will be important to determine how inherited cancer mutations link disturbed respiration with aerobic fermentation (Warburg effect), a central hallmark of cancer. Increased aerobic fermentation and defective apoptosis would be an expected consequence of respiratory insufficiency (16). The findings reviewed here provide strong support for Warburg's central hypothesis that damaged or insufficient respiration, regardless of the mechanism involved, is the origin of cancer.

SOMATIC MUTATIONS AND CANCER

Most of the gene defects found in cancers are not inherited, but arise sporadically, as do most mutations in the *p53* gene (4, 40, 42, 85, 111, 112). While germline mutations can increase the risk of some rare cancers as I have mentioned above, most cancer mutations are somatic and will contribute more to the progression than to the origin of most cancers (16). It is interesting to note, however, that somatic mutations occur only rarely in cells and tissues. Rous and Duesberg have considered this to be a major shortcoming of the somatic mutation theory of cancer (19, 38). If somatic mutations rarely occur in normal tissues, why are somatic mutations so common in tumor tissues?

Loeb and colleagues had initially proposed that the multiple mutations found in tumor cells resulted from mutations in genes responsible for maintaining the fidelity of DNA synthesis or the adequacy of DNA repair (40). More specifically, mutations in genomic caretakers underlie genomic stability and the large number of somatic mutations found in cancer. Mutations in these genes would then trigger an explosion of new mutations throughout the genome. As I mentioned in Chapter 2, however, it is not clear how mutations would occur so frequently in caretaker genes that had supposedly evolved to maintain the fidelity of DNA synthesis and repair.

If the spontaneous mutation rate in normal cells were as low as Loeb suggests, then why would the mutation rate be so high in the supposed genome guardians? Indeed, Loeb mentions that the mutator phenotype hypothesis does not address the sources of mutations that initiate carcinogenesis (40). According to my hypothesis, the large number of mutations and aneuploidy seen in tumor cells arises as a consequence of insufficient respiration with compensatory fermentation, as I will make clear in the next chapter. As the integrity of the nuclear genome is dependent on the fidelity of OxPhos, *respiratory insufficiency* becomes the *real* mutator phenotype. On the other hand, a normal respiratory function can suppress tumorigenicity as will be made clear in Chapter 11.

It is also important to recognize that mutations in the *p53* caretaker gene are not expressed in all common human malignancies, suggesting a more complicated involvement of this and other genome guardians in carcinogenesis (40, 42, 46, 85, 113–115). Although *p53* mutations are considered to be common in human glioblastoma multiforme, no defects in the *p53* gene are found in about 60% of these tumors (71). While numerous genetic abnormalities have been described in most human cancers, no specific mutation is reliably diagnostic of any specific type of tumor (16, 25–27, 40, 116–119). I find this very unsettling. How can mutations be relevant to the origin of cancer if the complement of mutations differs from one neoplastic cell to the next within most tumors of non-germline origin (28, 40, 71, 120)? These findings indicate that most of the tumor-associated somatic mutations are neither necessary nor sufficient to cause cancer.

Although common somatic mutations occur in some tumors, it is unlikely that these mutations are expressed in every individual cell of the tumor due to cellular and genetic heterogeneity. The data from Loeb and others make this fact clear (28, 40, 69). It is, nevertheless, interesting that progression of malignant gliomas is generally slower in patients with chromosome 1p/19q co-deletions, promoter hypermethylation of the O^6-methylguanine methyltransferase (MGMT) gene, or with mutations in the gene for IDH1 (71, 121–123). Should we consider these as "good" mutations since tumors containing these mutations grow slower than tumors not containing these mutations? GBM patients with IDH1 mutations live slightly longer than patients without this mutation (71). Mutations in this gene could inhibit mitochondrial amino acid fermentation, thus disrupting the glucose/glutamine synergy (Chapter 8). It is unclear if targeting these genes or their pathways would reduce or enhance patient survival. It is my opinion that it would be easier and more therapeutic to simply target glucose and glutamine availability than to target IDH1 mutations in GBM patients. I hope others interested in cancer management will come to share this view.

Considering the complexity of metabolic flux, genetic heterogeneity, and gene–environmental interactions (124–128), caution should be used in assuming that targeting any specific mutation or signaling pathway will have a major effect on tumor growth or patient survival (69). Sandra Yin's piece in Medscape Medical News made this point clear (6). It should be no surprise that attempts to restore p53 guardian function in cancer patients have met with little success (96, 129–131). The promise has not been realized (6, 132). Do those directing the NCI know about this?

Dr. Brad Ozenberger, Director of the Cancer Genome Atlas, predicts that 10 years from now, each cancer patient will want to get a genomic analysis of their cancer (133). If Dr. Ozenberger's prediction becomes a reality, I can predict that the cost of cancer care will be more expensive than it is today, while the number of cancer deaths per year will remain largely unchanged. When will we come to our senses? *Until we abandon the idea that cancer is a genetic disease and recognize that the mutations are downstream epiphenomena of the disease, there will be little progress in defeating cancer*.

REVISITING THE ONCOGENE THEORY

In light of the information presented above, it is difficult to see how cancer-associated mutations could be viewed as being the origin of cancer. It would therefore be important to revisit the key evidence suggesting that oncogenes cause cancer. According to Michael Stratton, the key evidence supporting the oncogene theory arose from studies showing that the introduction of total genomic DNA from human cancers into normal NIH3T3 cells could transform them into cancer cells (25). He cited the paper from Krontriris and Cooper as providing this evidence (134). However, high molecular weight DNA from only 2 out of 24 cancers, both bladder cancers, was able to transform the NIH3T3 cells into cancer cells. The authors were unable to exclude the possibility that the transformation resulted form viral infection.

Viral infection can damage mitochondria. No information was presented showing that mitochondria were normal or unaffected in the transformed cells. This might be difficult, however, since fermentation is elevated in the NIH3T3, suggesting that they suffer from some type of respiratory insufficiency (135). This led Rubin to agree with Leslie Foulds' conclusions that epigenetic phenomena contribute in part to the transformation of normal cells including NIH3T3 cells (136). Mitochondria represent an extrachromosomal epigenetic system (Chapter 10).

It would have been better for Krontriris and Cooper to have demonstrated tumorigenic transformation in nonglycolytic cells than in glycolytic cells, which are already on the path to tumorigenesis. According to the data of Moiseeva et al., however, Ras transfection (Ha-RasV12) of normal cells causes senescence rather than tumorigenic transformation (105). As Ras transfection damages OxPhos (137), it is not clear if Ras causes cancer through an effect in the nucleus or through an effect on the mitochondrial function. Warburg noted that acute damage to respiration is more likely to cause cell death than to cause cancer (57). Only those cells capable of upregulating fermentation to compensate for chronic mitochondrial damage can become tumor cells. Moiseeva et al. have shown that mitochondrial dysfunction and not a defect in glucose consumption is the underlying cause of the bioenergetic defects of Ras-senescent cells (105). A defect in tumor cell glucose consumption would be expected if tumorigenesis were dependent on the damage to the regulation of glycolysis as suggested by Koppenol and Dang (31). The findings from Moiseeva et al. and Hu et al. provide compelling evidence for the mitochondrial origin of cancer (105, 137).

We also found that mouse BV2 microglial cells, immortalized by the v-raf/v-myc carrying J2 retrovirus (138), were unable to form tumors when implanted into the brains of the syngeneic host C57BL/6J host (Michael Kiebish, unpublished observation). Although these cells are highly dependent on glucose and glutamine for energy, they could survive and grow when transitioned to media containing low glucose/glutamine (3 mM and 2 mM, respectively) and elevated ketone bodies (7.0 mM β-hydroxybutyrate). As OxPhos is necessary for ketone energy metabolism, the findings suggest that respiration is not impaired in these cells. In contrast to the BV2 cells, our VM-M3 glioblastoma cells also express characteristics of microglia, but are unable to survive in this low glucose/glutamine, high ketone media. The VM-M3 cells express impaired respiration and are highly invasive and metastatic (139–142). Our findings in the BV2 cells suggest that OxPhos function is maintained despite transfection with the *raf* and *myc* oncogenes. In other words, apparent damage to the regulation of glycolysis in these cells does not cause them to form tumors. It is damage to the mitochondria and the resulting insufficient respiration that causes tumorigenesis, not the reverse.

In a more extensive series of studies, Weinberg and coworkers have shown that DNA isolated from a variety of interspecific tumors could also transform NIH3T3 cells (143). It was clear that the DNA from the donor tumors was present in the transfected cells, but it was not determined if mitochondrial function was also altered following the DNA transfection. This is important since Moiseeva and colleagues have shown that transfection of cells with the Ha-RasV12 oncogene damages OxPhos in human fibroblasts (105). Moreover, Huang and colleagues have also shown that K-Ras damage to the mitochondria was the origin of the Warburg effect and that the damage preceded glycolytic upregulation and tumorigenesis (137). This point was highlighted in the Neuzil et al. timeline of events leading to malignant transformation (144). Aerobic glycolysis or the Warburg effect arises from damaged or insufficient respiration, just as Warburg had shown in his metabolic experiments. Just like genomic instability, aerobic glycolysis is a downstream effect rather than a cause of respiratory insufficiency (16, 137). Perhaps this is why Warburg placed little emphasis on aerobic glycolysis considering it to be a labile epiphenomenon of respiratory injury (57).

Elevated fermentation allows cells to bypass senescence, thus enhancing the likelihood of oncogenic transformation. Several research groups provide compelling evidence showing that oncogene transformation increases ROS expression and damages mitochondria (145–148). Lee and coworkers have shown that transfection of human diploid cells with V12Ras significantly increased the damage to oxygen species in mitochondria (145), whereas Weinberg and colleagues have shown that mitochondrial ROS generation and damage to complex III was essential for K-Ras-induced cell proliferation and tumorigenesis (146). Moreover, Yang and colleagues have shown that H-Ras transformation of mouse fibroblasts damaged respiration, thus forcing the cells into a glycolytic metabolism (147). This is notable since activated Ras has been proposed to induce MYC activity and to enhance non-hypoxic levels of HIF-1α (31). As MYC and HIF-1 drive glycolysis, their upregulation would be necessary to prevent senescence following respiratory damage.

Similar findings to those seen with Ras transfection were also observed when MYC expression was elevated in normal cardiomyocytes (149). MYC-induced damage to mitochondrial structure and respiratory function caused some cardiomyocytes to die, but caused other cardiomyocytes to reenter the cell cycle and proliferate. There is also emerging evidence that the c-RAF oncogene targets the mitochondria, which produces ROS and damages mitochondrial physiology (64). Elevated Hif-1α expression follows RAF-c-induced mitochondrial damage (148). Hif-1α upregulates glucose transport and multiple glycolytic pathways. Hence, oncogenes can sometimes target and damage mitochondrial function.

Viewed together, these findings indicate that respiratory dysfunction is an effector pathway of oncogene-induced senescence. An upregulation of glycolysis following respiratory insufficiency will prevent senescence leading to cell-cycle reentry and proliferation, that is, the initiating events in tumorigenesis. Hello, is anyone out there listening? Do people really think oncogenes are the specific cause of cancer in light of this information?

The evidence presented in this treatise makes a compelling argument that oncogenic transformation can act through the mitochondria and that cancer can arise from damaged or insufficient respiration. Neoplastic transformation can arise from oncogene-induced damage to mitochondrial function and OxPhos. The evidence supporting this statement is strong (105, 137, 145–147). However, not all oncogenes cause cancer as mentioned above for the immortalized mouse BV2 microglia. It would therefore be important to characterize the differences between those oncogenes that transform cells without damaging mitochondrial respiration or causing tumors from those that damage OxPhos leading to tumorigenesis. These issues are far from settled.

Nuclear genomic instability would arise as a downstream consequence of damaged mitochondrial respiration with compensatory fermentation according to evidence presented in the next chapter. Michael Stratton and others who consider cancer to be a genetic disease might want to reconsider the foundation upon which their beliefs rest. The cancer field will not move forward in my opinion until it is recognized that inefficient respiration with compensatory fermentation underlies the origin and progression of the disease.

While no mutation is known that causes a single type of cancer, few if any cancers are known that express normal respiration (16, 150). The importance of this fact cannot be overemphasized. Gibbs mentions that neither the standard genetic dogma nor any of the new theories can explain the 100-odd diseases we call cancer as variations of a single principle (26). It appears that Dr. Gibbs is unfamiliar with Warburg's theory, which explains cancer as a singular disease of impaired respiration with compensatory fermentation. I am perplexed that so many investigators in the cancer field focus on the highly capricious genetic minutia of tumors, while paying little attention to the most consistent metabolic phenotype of all tumors, that is, dysfunctional or insufficient respiration. Might the origin of cancer as a simple metabolic disease create anxiety in those who assume that the disease must be infinitely complex?

MITOCHONDRIAL MUTATIONS AND THE ABSENCE OR PRESENCE OF CANCER

If defective mitochondrial respiratory function is the origin of all cancers, why are cancers rare in those persons that inherit mutations damaging mitochondrial respiration? For example, mutations in Cu/Zn superoxide dismutase (*SOD*) gene, which disturbs respiratory function, are associated with familial amyotrophic lateral sclerosis (151–153). However, cancer is rare in patients with ALS (amyotrophic lateral sclerosis) (154). Eng and colleagues have addressed the issue of mitochondrial mutations and cancer in a comprehensive review on the subject (155). First, most of the inherited mutations that affect respiratory chain function and the TCA cycle are homozygous and cause profound damage to multiple organ systems (156). Inherited mutations are found in all cells, whereas the mitochondrial defects in cancer cells are found only in the cancer cells. Also, some individuals that inherit mitochondrial mutations do not live long enough to get cancer, for example, those with Barth syndrome involving abnormalities in cardiolipin remodeling. Second, those mutations that alter the TCA cycle function and cause cancer, that is, mutations in the *SDH* and *FH* genes, are generally heterozygous and do not affect the physiology of multiple organ systems. Homozygous mutations in these genes are associated with neurodegeneration rather than cancer (157). Neurodegeneration is also seen in those individuals with heterozygous mutations in the *SOD* gene (153). What role might differential mitochondrial damage and fermentation play in these inflammatory diseases?

As Warburg had originally mentioned, cells that die can never become tumorigenic (57). There are intriguing differences between genes that damage mitochondria and cause, or do not cause, cancer. Douglas Wallace suggests that mutations producing mitochondrial ROS rather than energy impairment is the missing link to cancer (157). However, mitochondrial ROS kills dopaminergic cells in Parkinson's disease without producing cancer. I agree with the view of Eng and colleagues that further research is needed into the genetic, cellular, and clinical aspects of mitochondrial function in relationship to cancer risk (155).

Critical Evaluation of Pathogenic Mitochondrial DNA Mutations in Tumors

There is substantial scientific literature suggesting that mtDNA mutations contribute to the origin of human brain tumors and to various other cancers (157–160). mtDNA changes can also alter cellular energy metabolism (73, 157, 161–163). To determine if mtDNA mutations might contribute to defects in brain tumor energy metabolism, we evaluated mtDNA for pathogenic mutations in five independently derived mouse brain tumors (160). These tumors covered a spectrum of growth behaviors seen in most malignant brain cancers. Two of the tumors evaluated, an ependymoblastoma (EPEN) and an astrocytoma (CT-2A), were derived from 20-methylcholantherene implantation into the brains of inbred C57BL/6J mice (164–166). Three of the tumors evaluated, VM-M2, VM-M3, and VM-NM1, arose spontaneously in the brains of inbred VM mice. I previously presented information on these tumors and the VM inbred strain in Chapter 3.

The VM inbred strain is unique in developing a relatively high incidence of spontaneous brain tumors (167). The VM-M2 and VM-M3 tumors express multiple properties of myeloid/mesenchymal cells and display the invasive growth behavior of human glioblastoma multiforme (139–141). The VM-NM1 is rapidly growing, but is neither highly invasive nor metastatic when grown outside the brain (140). We produced clonal cell lines from each of the five brain tumors. Each tumor was then grown subcutaneously in the syngeneic mouse host (142). We did this in order to obtain enough tumor tissue from which mitochondria could be isolated and purified according to our established procedures (168).

The mtDNA of each tumor was compared to that of mtDNA in purified brain mitochondrial populations from the corresponding normal syngeneic mouse host strain. Direct sequencing of the entire mitochondrial genome in each tumor and in the normal brain tissue from each host mouse strain revealed few genetic alterations. Most of the mutations found were in regions of mononucleotide repeats, but no mutations were found in protein coding genes. Remarkably, none of the genetic changes in the tumors were considered to be pathogenic (160). It was clear that the high glycolytic phenotype and rapid growth in these mouse brain tumors were not due to pathogenic mtDNA mutations.

These findings were surprising considering the vast literature which suggests that pathogenic mtDNA lesions can cause cancer (157). If mtDNA mutations are considered to be so important for the origin of cancer, why would there be no pathogenic mutations present in any of the five independently derived mouse brain tumors? We found, however, that all of the tumors expressed a robust Warburg effect suggestive of respiratory insufficiency.

Our failure to find pathogenic mtDNA mutations in the five independently derived mouse brain tumors does not support suggestions that mtDNA mutations contributed significantly to carcinogenesis, at least in the tumors from these mice. It is not clear why mtDNA mutations are so common in human tumors, but do not occur in our mouse tumors. Wallace considered it unlikely that sequencing errors could contribute to the high mtDNA mutation rate reported in human tumors (157). However, Salas and coworkers have shown that much of the evidence for pathogenic mtDNA mutations in human tumor cells was largely due to artifacts of data interpretation or methodologies used in mtDNA analysis (64, 169). In order to prove that mtDNA mutations contribute to the OxPhos deficiency in tumor cells, it is necessary to isolate and purify mitochondria from the tumor tissue and from the normal tissue of the patient, and then sequence the entire genome of the purified mtDNA of the tumor and normal tissue. Many studies of mtDNA mutations in human tumor tissue fail to include all of the necessary controls to exclude misinformation (160, 169). We included all the necessary controls and showed that none of the gene changes in the mouse brain tumors were pathogenic. Is it possible that pathogenic mtDNA mutations are more common in human tumors than in mouse tumors? Human and mouse are similar, however, in expressing respiratory insufficiency with compensatory fermentation.

Our carefully executed and comprehensive experiments in the mouse brain tumors clearly showed that mtDNA mutations were not involved in the origin or in the metabolic abnormalities present in these diverse mouse brain tumors (160). As the mitochondrial genome is highly redundant, it is unlikely that many cancers arise directly from mtDNA mutations due to the multiple copies of normal alleles in the mitochondrial genome (162). However, some cancers could arise if mutations are expressed in all copies of the circular mitochondrial genome or where the entire mitochondrial genome is depleted as Singh and coworkers have recently described (66, 73). Our studies evaluated only the mtDNA sequence, but not the mtDNA content. It is possible that the mtDNA content is lower in the tumor cells than in normal cells. Depletion of mtDNA increases the expression of uncoupling proteins (UCPs) (66). Activation of mitochondrial UCPs, especially uncoupling protein 2 (UCP-2), has been detected in a broad range of tumor cells (66, 170–172). Normal cells activate UCP-2 in response to elevated glucose levels in order to reduce mitochondrial membrane hyperpolarization (173). UCP activation can also help reduce ROS, which arises from elevated glucose levels (174). UCP-2 activation in tumor cells can be an attempt to regulate oxidative stress following OxPhos damage and mtDNA depletion (66).

It is also interesting that mtDNA polymorphisms can explain the risk for some maternally inherited cancers (73). Could this also be related to the maternal inheritance of some viral-derived cancers that Rous described (19)? On the other hand, direct evidence of mtDNA involvement in the origin of cancer comes from the recent studies of Rebbeck and colleagues who have shown that numerous pathogenic mtDNA mutations were present in canine transmissible venereal cancer (175). These mutations would disrupt OxPhos, thus causing the disease. Hence, mtDNA deficiency or pathogenic mtDNA mutations can cause cancer as long as the defects induce respiratory insufficiency (161).

VIRAL INFECTION, DAMAGED RESPIRATION, AND THE ORIGIN OF CANCER

Viruses have long been recognized as the cause of some cancers (21, 176, 177). About 15% of human cancers are caused by tumor viruses (178). Kofman and colleagues recently reviewed substantial information linking viral infections to the origin of malignant gliomas (177). It is interesting that several cancer-associated viruses or their protein products localize to, or accumulate in, the mitochondria (16, 17, 178, 179). Viral alteration of mitochondrial function could potentially disrupt energy metabolism, thus altering the expression of tumor suppressor genes and oncogenes over time. Viruses that affect mitochondrial function and increase cancer risk include the Rous sarcoma virus (src), Epstein–Barr virus (EBV), Kaposi's sarcoma-associated herpes virus (KSHV), human papilloma virus (HPV), hepatitis B virus (HBV), hepatitis C virus (HCV), human immunodeficiency virus (HIV), human cytomegalovirus (HCMV), and human T-cell leukemia virus type 1 (HTLV-1) (17, 177–186). Although viral disruption of mitochondrial function will kill many cells through apoptosis following an acute infection (180), those infected cells that can upregulate fermentation through substrate-level phosphorylation will survive and potentially produce a neoplasm following chronic infection.

Studies from Duensing and Munger show that the HPV type 16 E7 oncoprotein induces abnormal centrosome duplication, thereby increasing the propensity for multipolar mitoses, which can cause chromosome missegregation and aneuploidy (187). Although the mechanism is independent of RB protein inactivation, it can, nevertheless, involve mitochondrial damage. This comes from findings that E7 binds tightly to the human DNA polymerase pol interacting protein 38 (PDIP38), which localizes to the mitochondria (188). On the basis of the mitochondrial localization of other HPV oncoproteins (E1–E4) (189), Xie et al. suggest that PDIP38 would move from the mitochondria to the nucleus after HPV infection, possibly through either structural links between mitochondrial and nuclear membranes or release from mitochondria after the reduction of the mitochondrial membrane potential (188). According to my hypothesis, the PDIP38/E7 interaction would damage mitochondrial function prior to PDIP38 nuclear localization and initiation of genomic instability. More specifically, the oncogenic action of HPV originates with respiratory damage.

Siddiqui and colleagues have shown that the HBV-encoded protein HBx, which enhances the risk of hepatocellular carcinoma, disrupts the mitochondrial proton motive gradient (181). The HBx protein also blocks ubiquitination of HIF-1α thus increasing HIF-1α stability and activity in a hypoxia-independent manner (190). Alterations in calcium homeostasis, ROS production, and expression of NF-kB and HIF-1α are also expected to alter the metabolic state as was previously found for some viral infections (182, 183). HIF-1α stability is essential for upregulation of glycolysis following mitochondrial dysfunction (16). Thus, viruses can potentially cause cancer through displacement of respiration with substrate-level phosphorylation in the infected cells. Alterations in the expression of tumor suppressor genes and oncogenes will follow this energy transformation as we have previously described (16).

It is not known how many transforming retroviruses cause cancer by disrupting mitochondrial function and OxPhos in the infected cells. This appears to be the case for the KSHV, HPV, HIV, and HCMV (179, 184–186). Table 3.3 from the *Biology of Cancer* lists many of the known retroviruses and their acquired oncogenes (33). Although it is assumed that these retroviruses cause cancer by nuclear DNA insertion and oncogene upregulation (33), considerable evidence indicates that either the viruses themselves or their protein products damage OxPhos, leading to respiratory insufficiency. Viral infections can cause cancer through damage to cellular respiration.

It is my contention that the viral association with oncogenesis could arise more from their damage to respiration than from their influence on the nuclear genome. It will be up to those working in the viral oncology field to prove me wrong on this point. Viruses can also enhance cell fusions (191, 192), which could further damage mitochondrial function leading to genomic instability. I will discuss this more in the next chapter. Hence, it is naive to assume that retroviruses cause cancer solely through nuclear-DNA-based mechanisms (33). According to the new generalization proposed here, retroviruses disrupt mitochondrial energy metabolism, thus initiating the path to carcinogenesis.

HIV and Cancer Risk

The failure to appreciate the mechanism by which viral infection increases the risk for cancer is evidenced from the NCI "Fact Sheet" on HIV Infection and Cancer Risk (www.cancer.gov/images/documents/45cf39f5-569f-4c7f-a9e9-c0941765bc73/Fs3_97.pdf). The fact sheet mentions that people infected with HIV have a substantially higher risk for contracting some types of cancers than uninfected people of the same age. Three of these cancers are known as acquired immunodeficiency syndrome (AIDS)-defining cancers or AIDS-defining malignancies: Kaposi sarcoma, non-Hodgkin lymphoma, and cervical cancer. People infected with HIV are about 800 times more likely than uninfected people to be diagnosed with Kaposi sarcoma, at least seven times more likely to be diagnosed with non-Hodgkin lymphoma, and, among women, at least three times more likely to be diagnosed with cervical cancer. In addition, people infected with HIV are also at higher risk for anal cancer, Hodgkin lymphoma, liver cancer, and lung cancer.

The explanation given for increased cancer risk in HIV infection is that the infection weakens the immune system and reduces the body's ability to destroy cancer cells and fight infections that may lead to cancer. This explanation does not address where all these cancer cells came from in the HIV-infected patients. This explanation is also not connected to a molecular mechanism. Chronic viral infections can cause inflammation.

Inflammation damages OxPhos, thus shifting energy metabolism to fermentation. OxPhos insufficiency is the origin of cancer regardless of the involved tissue. Nothing is mentioned in the "Fact Sheet," however, on how the HIV infection might disrupt mitochondrial function and respiration, thus altering energy metabolism and cancer risk according to the metabolic cancer theory. On the basis of their explanation regarding the origin of HIV-associated cancers, it is not clear if those working at the NCI know about the Warburg's theory.

Equally disturbing are attempts to target human lung tumors using recombinant adenoviral vectors expressing *p*53 under control of the constitutively active src (www.genetherapyreview.com/gene-therapy-education/technology-overview/56-p53-gene-therapy.html). It is unlikely that this therapeutic approach will benefit patients significantly and could actually produce new kinds of tumors from accumulation of viruses or their products in mitochondria of normal cells. These types of therapies are guided more by lack of knowledge than by an understanding of what cancer actually is.

I find it amazing that so few people know about the Warburg theory and how it can explain many observations associated with the origin of cancer. It is interesting in this regard that carcinogenesis, whether arising from viral infection or from chemical agent, produces similar impairment in the respiratory enzyme activity and mitochondrial function as Roskelley and others have shown (17, 58). It is therefore imperative that all investigators working on the viral origin of cancer know about the Warburg cancer theory and the molecular mechanisms by which this theory can explain their observations.

SUMMARY

In this chapter, I take a hard and critical look at the data suggesting that cancer is a genetic disease. There is overwhelming inconsistency with the data supporting this hypothesis. While genomic instability is a common hallmark of nearly all cells within tumors, the evidence that genomic instability actually causes the disease is marginal at best. Is it nuclear genomic instability or is it insufficient respiration that ultimately gives rise to neoplasia? Little attention has been given to the possibility that tumorigenic transformation also damages OxPhos. Many germline cancer mutations damage OxPhos. Many known carcinogenic agents damage cellular respiration while also producing nuclear genomic instability. Oncogene activation and tumor suppressor gene inactivation are necessary changes in order to drive fermentation when OxPhos is insufficient. These changes are effects rather than causes of the disease. The data reviewed in this chapter raise the likelihood that respiratory insufficiency precedes the onset of genomic instability. The centrist might argue that damage to both organelles is required for the initiation and progression of the disease. However, the nuclear-cytoplasmic transfer experiments described in Chapter 11 show that the tumor cell nucleus can direct normal development when delivered to normal cytoplasm, but the normal cell nucleus cannot direct normal development when delivered to the tumor cytoplasm. Such data argue against the hypothesis that cancer arises from defects in the nuclear genome. In the next chapter, I will present additional information showing how respiratory insufficiency can give rise to nuclear instability.

REFERENCES

1. Hayden EC. Personalized cancer therapy gets closer. Nature. 2009;458:131–2.
2. Stratton MR. Exploring the genomes of cancer cells: progress and promise. Science. 2011; 331:1553–8.
3. Knudson AG. Cancer genetics. Am J Med Genet. 2002;111:96–102.
4. Hanahan D, Weinberg RA. The hallmarks of cancer. Cell. 2000;100:57–70.
5. Nowell PC. The clonal evolution of tumor cell populations. Science. 1976;194:23–28.
6. Yin S. Experts question benefits of high-cost cancer care. Medscape Today. 2011. http://www.medscape.com/viewarticle/754808?src=iphone.
7. Kaiser J. Combining targeted drugs to stop resistant tumors. Science. 2011;331:1542–5.
8. Fojo T, Parkinson DR. Biologically targeted cancer therapy and marginal benefits: are we making too much of too little or are we achieving too little by giving too much? Clin Cancer Res. 2010; 16:5972–80.

9. Kolata G. How bright promise in cancer testing fell apart. New York Times. 2011 July 7.
10. Staw BM, Ross J. Understanding behavior in escalation situations. Science. 1989;246:216–20.
11. Vander Heiden MG, Cantley LC, Thompson CB. Understanding the Warburg effect: the metabolic requirements of cell proliferation. Science. 2009;324:1029–33.
12. Kaelin WG, Thompson CB Jr Q&A: Cancer: clues from cell metabolism. Nature. 2010;465: 562–4.
13. Levine AJ, Puzio-Kuter AM. The control of the metabolic switch in cancers by oncogenes and tumor suppressor genes. Science. 2010;330:1340–4.
14. Bui T, Thompson CB. Cancer's sweet tooth. Cancer Cell. 2006;9:419–20.
15. Dang CV, Semenza GL. Oncogenic alterations of metabolism. Trends Biochem Sci. 1999; 24:68–72.
16. Seyfried TN, Shelton LM. Cancer as a metabolic disease. Nutr Metabol. 2010;7:7.
17. Smith AE, Kenyon DH. A unifying concept of carcinogenesis and its therapeutic implications. Oncology. 1973;27:459–79.
18. Sonnenschein C, Soto AM The Society of Cells: Cancer and the Control of Cell Proliferation. New York: Springer; 1999.
19. Rous P. Surmise and fact on the nature of cancer. Nature. 1959;183:1357–61.
20. German J. Constitutional hyperrecombinability and its consequences. Genetics. 2004;168:1–8.
21. Parkin DM. The global health burden of infection-associated cancers in the year 2002. Int J Cancer. 2006;118:3030–44.
22. Marx J. Oncogenes reach a milestone. Science. 1994;266:1942–4.
23. Parada LF, Tabin CJ, Shih C, Weinberg RA. Human EJ bladder carcinoma oncogene is homologue of Harvey sarcoma virus ras gene. Nature. 1982;297:474–8.
24. Hayward WS, Neel BG, Astrin SM. Activation of a cellular onc gene by promoter insertion in ALV-induced lymphoid leukosis. Nature. 1981;290:475–80.
25. Stratton MR, Campbell PJ, Futreal PA. The cancer genome. Nature. 2009;458:719–24.
26. Gibbs WW. Untangling the roots of cancer. Sci Am. 2003;289:56–65.
27. Nowell PC. Tumor progression: a brief historical perspective. Semin Cancer Biol. 2002; 12:261–6.
28. Salk JJ, Fox EJ, Loeb LA. Mutational heterogeneity in human cancers: origin and consequences. Annu Rev Pathol. 2010;5:51–75.
29. Fearon ER. Human cancer syndromes: clues to the origin and nature of cancer. Science. 1997;278:1043–50.
30. Vogelstein B, Kinzler KW. Cancer genes and the pathways they control. Nat Med. 2004; 10:789–99.
31. Koppenol WH, Bounds PL, Dang CV. Otto Warburg's contributions to current concepts of cancer metabolism. Nat Rev. 2011;11:325–37.
32. Nakajima EC, Van Houten B. Metabolic symbiosis in cancer: Refocusing the Warburg lens. Molecular carcinogenesis. 2012;doi: 10.1002/mc.21863. [Epub ahead of print].
33. Weinberg RA. The Biology of Cancer. New York: Garland Science; 2007.
34. Tarin D. Cell and tissue interactions in carcinogenesis and metastasis and their clinical significance. Semin Cancer Biol. 2011;21:72–82.
35. Hameroff SR. A new theory of the origin of cancer: quantum coherent entanglement, centrioles, mitosis, and differentiation. Biosystems. 2004;77:119–36.
36. Manchester K. The quest by three giants of science for an understanding of cancer. Endeavour. 1997;21:72–6.
37. Wolf U. Theodor Boveri and His Book, On the Problem of the Origin of Malignant Tumors. In: German J, editor. Chromosomes and Cancer. New York: John Wiley & Sons, Inc.; 1974. p.1–20.
38. Duesberg P, Rasnick D. Aneuploidy, the somatic mutation that makes cancer a species of its own. Cell Motil Cytoskeleton. 2000;47:81–107.
39. Cairns J. The origin of human cancers. Nature. 1981;289:353–7.
40. Loeb LA. A mutator phenotype in cancer. Cancer Res. 2001;61:3230–9.

41. Whitman RC. Somatic mutations as a factor in the production of cancer; a critical review of von Hansemanns's theory of anaplasia in light of modern knowledge of genetics. J Cancer Res. 1919;4:181–202.
42. Nigro JM, Baker SJ, Preisinger AC, Jessup JM, Hostetter R, Cleary K, et al. Mutations in the p53 gene occur in diverse human tumour types. Nature. 1989;342:705–8.
43. Fearon ER, Vogelstein B. A genetic model for colorectal tumorigenesis. Cell. 1990;61:759–67.
44. Wagner RP. Anecdotal, historical and critical commentaries on genetics. Rudolph Virchow and the genetic basis of somatic ecology. Genetics. 1999;151:917–20.
45. Duesberg PH, Schwartz JR. Latent viruses and mutated oncogenes: no evidence for pathogenicity. Prog Nucleic Acid Res Mol Biol. 1992;43:135–204.
46. Sonnenschein C, Soto AM. Somatic mutation theory of carcinogenesis: why it should be dropped and replaced. Mol Carcinog. 2000;29:205–11.
47. Soto AM, Sonnenschein C. The somatic mutation theory of cancer: growing problems with the paradigm?. Bioessays. 2004;26:1097–107.
48. Sonnenschein C, Soto AM. Theories of carcinogenesis: an emerging perspective. Semin Cancer Biol. 2008;18:372–7.
49. Duesberg PH. Oncogenes and cancer. Science. 1995;267:1407–8.
50. Garber K. Oncometabolite? IDH1 discoveries raise possibility of new metabolism targets in brain cancers and leukemia. J Natl Cancer Inst. 2010;102:926–8.
51. Bose S, Deininger M, Gora-Tybor J, Goldman JM, Melo JV. The presence of typical and atypical BCR-ABL fusion genes in leukocytes of normal individuals: biologic significance and implications for the assessment of minimal residual disease. Blood. 1998;92:3362–7.
52. Torres EM, Sokolsky T, Tucker CM, Chan LY, Boselli M, Dunham MJ, et al. Effects of aneuploidy on cellular physiology and cell division in haploid yeast. Science. 2007;317: 916–24.
53. Seyfried TN, Marsh J, Shelton LM, Huysentruyt LC, Mukherjee P. Is the restricted ketogenic diet a viable alternative to the standard of care for managing malignant brain cancer?. Epilepsy Res. 2011.
54. Marsh J, Mukherjee P, Seyfried TN. Akt-dependent proapoptotic effects of dietary restriction on late-stage management of a phosphatase and tensin homologue/tuberous sclerosis complex 2-deficient mouse astrocytoma. Clin Cancer Res. 2008;14:7751–62.
55. Seyfried TN, Kiebish MA, Marsh J, Shelton LM, Huysentruyt LC, Mukherjee P. Metabolic management of brain cancer. Biochim Biophys Acta. 2010;1807:577–94.
56. Warburg O The Metabolism of Tumours. New York: Richard R. Smith; 1931.
57. Warburg O. On the origin of cancer cells. Science. 1956;123:309–14.
58. Roskelley RC, Mayer N, Horwitt BN, Salter WT. Studies in cancer. Vii. Enzyme deficiency in human and experimental cancer. J Clin Invest. 1943;22:743–51.
59. Sajan MP, Satav JG, Bhattacharya RK. Effect of aflatoxin B in vitro on rat liver mitochondrial respiratory functions. Indian J Exp Biol. 1997;35:1187–90.
60. Bhat NK, Emeh JK, Niranjan BG, Avadhani NG. Inhibition of mitochondrial protein synthesis during early stages of aflatoxin B1-induced hepatocarcinogenesis. Cancer Res. 1982;42:1876–80.
61. Warburg O. On the respiratory impairment in cancer cells. Science. 1956;124:269–70.
62. Hadler HI, Daniel BG, Pratt RD. The induction of ATP energized mitochondrial volume changes by carcinogenic N-hydroxy-N-acetyl-aminofluorenes when combined with showdomycin. A unitary hypothesis for carcinogenesis. J Antibiot. 1971;24:405–17.
63. Parada LF, Weinberg RA. Presence of a Kirsten murine sarcoma virus ras oncogene in cells transformed by 3-methylcholanthrene. Mol Cell Biol. 1983;3:2298–301.
64. Seoane M, Mosquera-Miguel A, Gonzalez T, Fraga M, Salas A, Costoya JA. The mitochondrial genome is a "Genetic Sanctuary" during the oncogenic process. PloS One. 2011;6:e23327.
65. de Groof AJ, te Lindert MM, van Dommelen MM, Wu M, Willemse M, Smift AL, et al. Increased OXPHOS activity precedes rise in glycolytic rate in H-RasV12/E1A transformed fibroblasts that develop a Warburg phenotype. Mol Cancer. 2009;8:54.

66. Ayyasamy V, Owens KM, Desouki MM, Liang P, Bakin A, Thangaraj K, et al. Cellular model of Warburg effect identifies tumor promoting function of UCP2 in breast cancer and its suppression by genipin. PloS One. 2011;6:e24792.
67. Steeg PS. Heterogeneity of drug target expression among metastatic lesions: lessons from a breast cancer autopsy program. Clin Cancer Res. 2008;14:3643–5.
68. Wu JM, Fackler MJ, Halushka MK, Molavi DW, Taylor ME, Teo WW, et al. Heterogeneity of breast cancer metastases: comparison of therapeutic target expression and promoter methylation between primary tumors and their multifocal metastases. Clin Cancer Res. 2008;14:1938–46.
69. Gabor Miklos GL. The human cancer genome project–one more misstep in the war on cancer. Nat Biotechnol. 2005;23:535–7.
70. Rasnick D, Duesberg PH. How aneuploidy affects metabolic control and causes cancer. Biochem J. 1999;340(Pt 3):621–30.
71. Parsons DW, Jones S, Zhang X, Lin JC, Leary RJ, Angenendt P, et al. An integrated genomic analysis of human glioblastoma multiforme. Science. 2008;321:1807–12.
72. Jones S, Zhang X, Parsons DW, Lin JC, Leary RJ, Angenendt P, et al. Core signaling pathways in human pancreatic cancers revealed by global genomic analyses. Science. 2008;321: 1801–6.
73. Chandra D, Singh KK. Genetic insights into OXPHOS defect and its role in cancer. Biochim Biophys Acta. 2010;1807:620–5.
74. Pollard PJ, Wortham NC, Tomlinson IP. The TCA cycle and tumorigenesis: the examples of fumarate hydratase and succinate dehydrogenase. Ann Med. 2003;35:632–9.
75. Hao HX, Khalimonchuk O, Schraders M, Dephoure N, Bayley JP, Kunst H, et al. SDH5, a gene required for flavination of succinate dehydrogenase, is mutated in paraganglioma. Science. 2009;325:1139–42.
76. Baysal BE, Ferrell RE, Willett-Brozick JE, Lawrence EC, Myssiorek D, Bosch A, et al. Mutations in SDHD, a mitochondrial complex II gene, in hereditary paraganglioma. Science. 2000;287:848–51.
77. Alam NA, Rowan AJ, Wortham NC, Pollard PJ, Mitchell M, Tyrer JP, et al. Genetic and functional analyses of FH mutations in multiple cutaneous and uterine leiomyomatosis, hereditary leiomyomatosis and renal cancer, and fumarate hydratase deficiency. Hum Mol Genet. 2003;12: 1241–52.
78. Favier J, Briere JJ, Burnichon N, Riviere J, Vescovo L, Benit P, et al. The Warburg effect is genetically determined in inherited pheochromocytomas. PloS One. 2009;4:e7094.
79. Kim WY, Kaelin WG. Role of VHL gene mutation in human cancer. J Clin Oncol. 2004; 22:4991–5004.
80. Shiao YH, Resau JH, Nagashima K, Anderson LM, Ramakrishna G. The von Hippel-Lindau tumor suppressor targets to mitochondria. Cancer Res. 2000;60:2816–9.
81. Bayley JP, Devilee P. Warburg tumours and the mechanisms of mitochondrial tumour suppressor genes. Barking up the right tree?. Curr Opin Genet Dev. 2010;20:324–9.
82. Malkin D, Li FP, Strong LC, Fraumeni JF Jr, Nelson CE, Kim DH, et al. Germ line p53 mutations in a familial syndrome of breast cancer, sarcomas, and other neoplasms. Science. 1990;250:1233–8.
83. Donehower LA, Harvey M, Slagle BL, McArthur MJ, Montgomery CA Jr, Butel JS, et al. Mice deficient for p53 are developmentally normal but susceptible to spontaneous tumours. Nature. 1992;356:215–21.
84. Lane D, Levine A. p53 Research: the past thirty years and the next thirty years. Cold Spring Harb Perspect Biol. 2010;2:a000893.
85. Levine AJ. p53, the cellular gatekeeper for growth and division. Cell. 1997;88:323–31.
86. Sung HJ, Ma W, Wang PY, Hynes J, O'Riordan TC, Combs CA, et al. Mitochondrial respiration protects against oxygen-associated DNA damage. Nat Commun. 2011;1:1–8.
87. Lago CU, Sung HJ, Ma W, Wang PY, Hwang PM. p53, Aerobic metabolism and cancer. Antioxid Redox Signal. 2010;15:1739–48.
88. Matoba S, Kang JG, Patino WD, Wragg A, Boehm M, Gavrilova O, et al. p53 regulates mitochondrial respiration. Science. 2006;312:1650–3.

89. Zhou S, Kachhap S, Singh KK. Mitochondrial impairment in p53-deficient human cancer cells. Mutagenesis. 2003;18:287–92.
90. Pallardo FV, Lloret A, Lebel M, d'Ischia M, Cogger VC, Le Couteur DG, et al. Mitochondrial dysfunction in some oxidative stress-related genetic diseases: Ataxia-telangiectasia, Down syndrome, Fanconi anaemia and Werner syndrome. Biogerontology. 2010;11:401–19.
91. Blander G, Kipnis J, Leal JF, Yu CE, Schellenberg GD, Oren M. Physical and functional interaction between p53 and the Werner's syndrome protein. J Biol Chem. 1999;274:29463–9.
92. Leslie M. Brothers in arms against cancer. Science. 2011;331:1551–2.
93. Sayan AE, Sayan BS, Gogvadze V, Dinsdale D, Nyman U, Hansen TM, et al. P73 and caspase-cleaved p73 fragments localize to mitochondria and augment TRAIL-induced apoptosis. Oncogene. 2008;27:4363–72.
94. Peters MA, Janer M, Kolb S, Jarvik GP, Ostrander EA, Stanford JL. Germline mutations in the p73 gene do not predispose to familial prostate-brain cancer. Prostate. 2001;48:292–6.
95. McGill G, Fisher DE. p53 and cancer therapy: a double-edged sword. J Clin Invest. 1999;104:223–5.
96. Swisher SG, Roth JA, Nemunaitis J, Lawrence DD, Kemp BL, Carrasco CH, et al. Adenovirus-mediated p53 gene transfer in advanced non-small-cell lung cancer. J Natl Cancer Inst. 1999;91:763–71.
97. Lowe SW, Bodis S, McClatchey A, Remington L, Ruley HE, Fisher DE, et al. p53 status and the efficacy of cancer therapy in vivo. Science. 1994;266:807–10.
98. Miki Y, Swensen J, Shattuck-Eidens D, Futreal PA, Harshman K, Tavtigian S, et al. A strong candidate for the breast and ovarian cancer susceptibility gene BRCA1. Science. 1994;266: 66–71.
99. Wang Y, Cortez D, Yazdi P, Neff N, Elledge SJ, Qin J. BASC, a super complex of BRCA1-associated proteins involved in the recognition and repair of aberrant DNA structures. Genes Dev. 2000;14:927–39.
100. Coene ED, Hollinshead MS, Waeytens AA, Schelfhout VR, Eechaute WP, Shaw MK, et al. Phosphorylated BRCA1 is predominantly located in the nucleus and mitochondria. Mol Biol Cell. 2005;16:997–1010.
101. Brocardo M, Henderson BR. APC shuttling to the membrane, nucleus and beyond. Trends Cell Biol. 2008;18:587–96.
102. Brocardo M, Lei Y, Tighe A, Taylor SS, Mok MT, Henderson BR. Mitochondrial targeting of adenomatous polyposis coli protein is stimulated by truncating cancer mutations: regulation of Bcl-2 and implications for cell survival. J Biol Chem. 2008;283:5950–9.
103. Classon M, Harlow E. The retinoblastoma tumour suppressor in development and cancer. Nat Rev. 2002;2:910–17.
104. Sankaran VG, Orkin SH, Walkley CR. Rb intrinsically promotes erythropoiesis by coupling cell cycle exit with mitochondrial biogenesis. Genes Dev. 2008;22:463–75.
105. Moiseeva O, Bourdeau V, Roux A, Deschenes-Simard X, Ferbeyre G. Mitochondrial dysfunction contributes to oncogene-induced senescence. Mol Cell Biol. 2009;29:4495–507.
106. Rothe M, Werner D, Thielmann HW. Enhanced expression of mitochondrial genes in xeroderma pigmentosum fibroblast strains from various complementation groups. J Cancer Res Clin Oncol. 1993;119:675–84.
107. Thierbach R, Drewes G, Fusser M, Voigt A, Kuhlow D, Blume U, et al. The Friedreich's ataxia protein frataxin modulates DNA base excision repair in prokaryotes and mammals. Biochem J. 2010;432:165–72.
108. Thierbach R, Schulz TJ, Isken F, Voigt A, Mietzner B, Drewes G, et al. Targeted disruption of hepatic frataxin expression causes impaired mitochondrial function, decreased life span and tumor growth in mice. Hum Mol Genet. 2005;14:3857–64.
109. Ristow M. Oxidative metabolism in cancer growth. Curr Opin Clin Nutr Metab Care. 2006;9: 339–45.
110. Rotig A, de Lonlay P, Chretien D, Foury F, Koenig M, Sidi D, et al. Aconitase and mitochondrial iron-sulphur protein deficiency in Friedreich ataxia. Nat Genet. 1997;17:215–7.

111. Fearon ER, Cho KR, Nigro JM, Kern SE, Simons JW, Ruppert JM, et al. Identification of a chromosome 18q gene that is altered in colorectal cancers. Science. 1990;247:49–56.
112. Yokota J. Tumor progression and metastasis. Carcinogenesis. 2000;21:497–503.
113. Duesberg P, Rasnick D, Li R, Winters L, Rausch C, Hehlmann R. How aneuploidy may cause cancer and genetic instability. Anticancer Res. 1999;19:4887–906.
114. Kruse JP, Gu W. Modes of p53 regulation. Cell. 2009;137:609–22.
115. Olovnikov IA, Kravchenko JE, Chumakov PM. Homeostatic functions of the p53 tumor suppressor: regulation of energy metabolism and antioxidant defense. Semin Cancer Biol. 2009;19: 32–41.
116. Mandinova A, Lee SW. The p53 pathway as a target in cancer therapeutics: obstacles and promise. Sci Transl Med. 2011;3:64rv1.
117. Gravendeel LA, Kouwenhoven MC, Gevaert O, de Rooi JJ, Stubbs AP, Duijm JE, et al. Intrinsic gene expression profiles of gliomas are a better predictor of survival than histology. Cancer Res. 2009;69:9065–72.
118. Network TCGAR. Comprehensive genomic characterization defines human glioblastoma genes and core pathways. Nature. 2008;455:1061–8.
119. Dang L, White DW, Gross S, Bennett BD, Bittinger MA, Driggers EM, et al. Cancer-associated IDH1 mutations produce 2-hydroxyglutarate. Nature. 2009;462:739–44.
120. Greenman C, Stephens P, Smith R, Dalgliesh GL, Hunter C, Bignell G, et al. Patterns of somatic mutation in human cancer genomes. Nature. 2007;446:153–8.
121. van den Bent MJ, Dubbink HJ, Marie Y, Brandes AA, Taphoorn MJ, Wesseling P, et al. IDH1 and IDH2 mutations are prognostic but not predictive for outcome in anaplastic oligodendroglial tumors: a report of the European Organization for Research and Treatment of Cancer Brain Tumor Group. Clin Cancer Res. 2010;16:1597–1604.
122. Stupp R, Mason WP, van den Bent MJ, Weller M, Fisher B, Taphoorn MJ, et al. Radiotherapy plus concomitant and adjuvant temozolomide for glioblastoma. N Engl J Med. 2005;352:987–96.
123. Krex D, Klink B, Hartmann C, von Deimling A, Pietsch T, Simon M, et al. Long-term survival with glioblastoma multiforme. Brain. 2007;130:2596–606.
124. Saad AG, Sachs J, Turner CD, Proctor M, Marcus KJ, Wang L, et al. Extracranial metastases of glioblastoma in a child: case report and review of the literature. J Pediatr Hematol Oncol. 2007;29:190–4.
125. Ohgaki H, Kleihues P. Genetic alterations and signaling pathways in the evolution of gliomas. Cancer Sci. 2009;100:2235–41.
126. Greenspan RJ. The flexible genome. Nat Rev Genet. 2001;2:383–7.
127. Strohman R. Maneuvering in the complex path from genotype to phenotype. Science. 2002;296:701–3.
128. Strohman R. Thermodynamics–old laws in medicine and complex disease. Nat Biotechnol. 2003;21:477–9.
129. Lang FF, Bruner JM, Fuller GN, Aldape K, Prados MD, Chang S, et al. Phase I trial of adenovirus-mediated p53 gene therapy for recurrent glioma: biological and clinical results. J Clin Oncol. 2003;21:2508–18.
130. Makower D, Rozenblit A, Kaufman H, Edelman M, Lane ME, Zwiebel J, et al. Phase II clinical trial of intralesional administration of the oncolytic adenovirus ONYX-015 in patients with hepatobiliary tumors with correlative p53 studies. Clin Cancer Res. 2003;9:693–702.
131. Schuler M, Herrmann R, De Greve JL, Stewart AK, Gatzemeier U, Stewart DJ, et al. Adenovirus-mediated wild-type p53 gene transfer in patients receiving chemotherapy for advanced non-small-cell lung cancer: results of a multicenter phase II study. J Clin Oncol. 2001;19:1750–8.
132. Gibbs JB. Mechanism-based target identification and drug discovery in cancer research. Science. 2000;287:1969–73.
133. Park HR. Cracking cancer's code: tumor DNA holds key to beating the disease. TIME. 2011;177:69–71.
134. Krontiris TG, Cooper GM. Transforming activity of human tumor DNAs. Proc Natl Acad Sci USA. 1981;78:1181–4.

135. Lanks KW, Li PW. End products of glucose and glutamine metabolism by cultured cell lines. J Cell Physiol. 1988;135:151–5.
136. Rubin H. Experimental control of neoplastic progression in cell populations: Foulds' rules revisited. Proc Natl Acad Sci USA. 1994;91:6619–23.
137. Hu Y, Lu W, Chen G, Wang P, Chen Z, Zhou Y, et al. K-ras(G12V) transformation leads to mitochondrial dysfunction and a metabolic switch from oxidative phosphorylation to glycolysis. Cell res. 201222:399–412.
138. Blasi E, Barluzzi R, Bocchini V, Mazzolla R, Bistoni F. Immortalization of murine microglial cells by a v-raf/v-myc carrying retrovirus. J Neuroimmunol. 1990;27:229–37.
139. Shelton LM, Mukherjee P, Huysentruyt LC, Urits I, Rosenberg JA, Seyfried TN. A novel pre-clinical in vivo mouse model for malignant brain tumor growth and invasion. J Neuro Oncol. 2010;99:165–76.
140. Huysentruyt LC, Mukherjee P, Banerjee D, Shelton LM, Seyfried TN. Metastatic cancer cells with macrophage properties: evidence from a new murine tumor model. Int J Cancer. 2008;123:73–84.
141. Huysentruyt LC, Seyfried TN. Perspectives on the mesenchymal origin of metastatic cancer. Cancer Metastasis Rev. 2010;29:695–707.
142. Kiebish MA, Han X, Cheng H, Chuang JH, Seyfried TN. Cardiolipin and electron transport chain abnormalities in mouse brain tumor mitochondria: lipidomic evidence supporting the Warburg theory of cancer. J Lipid Res. 2008;49:2545–56.
143. Shih C, Padhy LC, Murray M, Weinberg RA. Transforming genes of carcinomas and neuroblastomas introduced into mouse fibroblasts. Nature. 1981;290:261–4.
144. Neuzil J, Rohlena J, Dong LF. K-Ras and mitochondria: Dangerous liaisons. Cell res. 2012;22:285–287.
145. Lee AC, Fenster BE, Ito H, Takeda K, Bae NS, Hirai T, et al. Ras proteins induce senescence by altering the intracellular levels of reactive oxygen species. J Biol Chem. 1999;274:7936–40.
146. Weinberg F, Hamanaka R, Wheaton WW, Weinberg S, Joseph J, Lopez M, et al. Mitochondrial metabolism and ROS generation are essential for Kras-mediated tumorigenicity. Proc Natl Acad Sci USA. 2010;107:8788–93.
147. Yang D, Wang MT, Tang Y, Chen Y, Jiang H, Jones TT, et al. Impairment of mitochondrial respiration in mouse fibroblasts by oncogenic H-RAS(Q61L). Cancer Biol Ther. 2010;9:122–33.
148. Galmiche A, Fueller J. RAF kinases and mitochondria. Biochim Biophys Acta. 2007;1773: 1256–62.
149. Lee HG, Chen Q, Wolfram JA, Richardson SL, Liner A, Siedlak SL, et al. Cell cycle reentry and mitochondrial defects in myc-mediated hypertrophic cardiomyopathy and heart failure. PloS One. 2009;4:e7172.
150. Lopez-Lazaro M. A new view of carcinogenesis and an alternative approach to cancer therapy. Mol Med. 2010;16:144–53.
151. Rosen DR. Mutations in Cu/Zn superoxide dismutase gene are associated with familial amyotrophic lateral sclerosis. Nature. 1993;364:362.
152. Dupuis L, Gonzalez de Aguilar JL, Oudart H, de Tapia M, Barbeito L, Loeffler JP. Mitochondria in amyotrophic lateral sclerosis: a trigger and a target. Neurodegener Dis. 2004;1:245–54.
153. Dupuis L, Oudart H, Rene F, Gonzalez de Aguilar JL, Loeffler JP. Evidence for defective energy homeostasis in amyotrophic lateral sclerosis: benefit of a high-energy diet in a transgenic mouse model. Proc Natl Acad Sci USA. 2004;101:11159–64.
154. Vigliani MC, Polo P, Chio A, Giometto B, Mazzini L, Schiffer D. Patients with amyotrophic lateral sclerosis and cancer do not differ clinically from patients with sporadic amyotrophic lateral sclerosis. J Neurol. 2000;247:778–82.
155. Eng C, Kiuru M, Fernandez MJ, Aaltonen LA. A role for mitochondrial enzymes in inherited neoplasia and beyond. Nat Rev. 2003;3:193–202.
156. Schoffner JM, Wallace DC. Oxidative Phosphorylation Diseases. In: Scriver CR, Beaudet AL, Sly WS, Valle D, editors. The Metabolic and Molecular Bases of Inherited Diseases. New York: McGraw-Hill Inc.; 1995. p.1535–609.

157. Wallace DC. Mitochondria and cancer: Warburg addressed. Cold Spring Harb Symp Quant Biol. 2005;70:363–74.
158. Kirches E, Krause G, Warich-Kirches M, Weis S, Schneider T, Meyer-Puttlitz B, et al. High frequency of mitochondrial DNA mutations in glioblastoma multiforme identified by direct sequence comparison to blood samples. Int J Cancer. 2001;93:534–8.
159. Lueth M, von Deimling A, Pietsch T, Wong LJ, Kurtz A, Henze G, et al. Medulloblastoma harbor somatic mitochondrial DNA mutations in the D-loop region. J Pediatr Hematol Oncol. 2010;32:156–9.
160. Kiebish MA, Seyfried TN. Absence of pathogenic mitochondrial DNA mutations in mouse brain tumors. BMC Cancer. 2005;5:102.
161. Lu J, Sharma LK, Bai Y. Implications of mitochondrial DNA mutations and mitochondrial dysfunction in tumorigenesis. Cell Res. 2009;19:802–15.
162. Carew JS, Huang P. Mitochondrial defects in cancer. Mol Cancer. 2002;1:9.
163. Singh KK, Kulawiec M, Still I, Desouki MM, Geradts J, Matsui S. Inter-genomic cross talk between mitochondria and the nucleus plays an important role in tumorigenesis. Gene. 2005; 354:140–6.
164. Mukherjee P, Abate LE, Seyfried TN. Antiangiogenic and proapoptotic effects of dietary restriction on experimental mouse and human brain tumors. Clin Cancer Res. 2004;10:5622–9.
165. Mukherjee P, El-Abbadi MM, Kasperzyk JL, Ranes MK, Seyfried TN. Dietary restriction reduces angiogenesis and growth in an orthotopic mouse brain tumour model. Br J Cancer. 2002;86:1615–21.
166. Seyfried TN, el-Abbadi M, Roy ML. Ganglioside distribution in murine neural tumors. Mol Chem Neuropathol. 1992;17:147–67.
167. Fraser H. Brain tumours in mice, with particular reference to astrocytoma. Food Chem Toxicol. 1986;24:105–11.
168. Kiebish MA, Han X, Cheng H, Lunceford A, Clarke CF, Moon H, et al. Lipidomic analysis and electron transport chain activities in C57BL/6J mouse brain mitochondria. J Neurochem. 2008;106:299–312.
169. Salas A, Yao YG, Macaulay V, Vega A, Carracedo A, Bandelt HJ. A critical reassessment of the role of mitochondria in tumorigenesis. PLoS Med. 2005;2:e296.
170. Fine EJ, Miller A, Quadros EV, Sequeira JM, Feinman RD. Acetoacetate reduces growth and ATP concentration in cancer cell lines which over-express uncoupling protein 2. Cancer Cell Int. 2009;9:14.
171. Harper ME, Antoniou A, Villalobos-Menuey E, Russo A, Trauger R, Vendemelio M, et al. Characterization of a novel metabolic strategy used by drug-resistant tumor cells. FASEB J. 2002;16:1550–7.
172. Samudio I, Fiegl M, Andreeff M. Mitochondrial uncoupling and the Warburg effect: molecular basis for the reprogramming of cancer cell metabolism. Cancer Res. 2009;69:2163–6.
173. Chan CB, De Leo D, Joseph JW, McQuaid TS, Ha XF, Xu F, et al. Increased uncoupling protein-2 levels in beta-cells are associated with impaired glucose-stimulated insulin secretion: mechanism of action. Diabetes. 2001;50:1302–10.
174. Affourtit C, Jastroch M, Brand MD. Uncoupling protein-2 attenuates glucose-stimulated insulin secretion in INS-1E insulinoma cells by lowering mitochondrial reactive oxygen species. Free Radical Biol Med. 2011;50:609–16.
175. Rebbeck CA, Leroi AM, Burt A. Mitochondrial capture by a transmissible cancer. Science. 2011;331:303.
176. Boccardo E, Villa LL. Viral origins of human cancer. Curr Med Chem. 2007;14:2526–39.
177. Kofman A, Marcinkiewicz L, Dupart E, Lyshchev A, Martynov B, Ryndin A, et al. The roles of viruses in brain tumor initiation and oncomodulation. J Neurooncol. 2011;105:451–66.
178. D'Agostino DM, Bernardi P, Chieco-Bianchi L, Ciminale V. Mitochondria as functional targets of proteins coded by human tumor viruses. Adv Cancer Res. 2005;94:87–142.
179. Ackermann WW, Kurtz H. The relation of herpes virus to host cell mitochondria. J Exp Med. 1952;96:151–7.

180. Macho A, Castedo M, Marchetti P, Aguilar JJ, Decaudin D, Zamzami N, et al. Mitochondrial dysfunctions in circulating T lymphocytes from human immunodeficiency virus-1 carriers. Blood. 1995;86:2481–7.
181. Rahmani Z, Huh KW, Lasher R, Siddiqui A. Hepatitis B virus X protein colocalizes to mitochondria with a human voltage-dependent anion channel, HVDAC3, and alters its transmembrane potential. J Virol. 2000;74:2840–6.
182. Koike K. Hepatitis B virus X gene is implicated in liver carcinogenesis. Cancer Lett. 2009; 286:60–8.
183. Clippinger AJ, Bouchard MJ. Hepatitis B virus HBx protein localizes to mitochondria in primary rat hepatocytes and modulates mitochondrial membrane potential. J Virol. 2008;82:6798–811.
184. Yu Y, Clippinger AJ, Alwine JC. Viral effects on metabolism: changes in glucose and glutamine utilization during human cytomegalovirus infection. Trends Microbiol. 2011;19:360–7.
185. Yu Y, Maguire TG, Alwine JC. Human cytomegalovirus activates glucose transporter 4 expression to increase glucose uptake during infection. J Virol. 2011;85:1573–80.
186. Miro O, Lopez S, Martinez E, Pedrol E, Milinkovic A, Deig E, et al. Mitochondrial effects of HIV infection on the peripheral blood mononuclear cells of HIV-infected patients who were never treated with antirctrovirals. Clin Infect Dis. 2004;39:710–6.
187. Duensing S, Munger K. Human papillomavirus type 16 E7 oncoprotein can induce abnormal centrosome duplication through a mechanism independent of inactivation of retinoblastoma protein family members. J Virol. 2003;77:12331–5.
188. Xie B, Li H, Wang Q, Xie S, Rahmeh A, Dai W, et al. Further characterization of human DNA polymerase delta interacting protein 38. J Biol Chem. 2005;280:22375–84.
189. Raj K, Berguerand S, Southern S, Doorbar J, Beard P. E1 empty set E4 protein of human papillomavirus type 16 associates with mitochondria. J Virol. 2004;78:7199–207.
190. Moon EJ, Jeong CH, Jeong JW, Kim KR, Yu DY, Murakami S, et al. Hepatitis B virus X protein induces angiogenesis by stabilizing hypoxia-inducible factor-1alpha. FASEB J. 2004;18:382–4.
191. Duelli D, Lazebnik Y. Cell-to-cell fusion as a link between viruses and cancer. Nat Rev. 2007;7:968–76.
192. Duelli DM, Padilla-Nash HM, Berman D, Murphy KM, Ried T, Lazebnik Y. A virus causes cancer by inducing massive chromosomal instability through cell fusion. Curr Biol. 2007;17: 431–7.

Chapter 10

Respiratory Insufficiency, the Retrograde Response, and the Origin of Cancer

Although respiratory insufficiency can explain most of the observations associated with the origin and progression of cancer, it is interesting that this concept or Warburg's theory is not mentioned in a popular textbook on the biology of cancer (1). The failure to discuss the role of mitochondria in the origin of cancer would be like failing to discuss the role of the sun in the origin of the solar system. Many in the cancer field attribute the origin of cancer to mutations in genes. This theory, however, is fraught with inconsistencies as I have described in the last chapter. A resolution to the origin of cancer becomes possible only when we replace any number of supposed origins (genes, viruses, aneuploidy, etc.) with respiratory insufficiency. This would be similar to the replacement of the Earth with the Sun in order to explain the orbits of the planets (2, 3). In light of this synopsis, it would be important to consider how respiratory insufficiency can be linked to the origin of cancer.

THE RETROGRADE (RTG) RESPONSE: AN EPIGENETIC SYSTEM RESPONSIBLE FOR NUCLEAR GENOMIC STABILITY

A good hypothesis is one that can explain most of the observations associated with a phenomenon. If the hypothesis cannot be rejected and is supported by a broad range of experimental observations, then it becomes a theory (4). Although Warburg's observations have generated controversy, they have never been disproved as I have described in Chapters 7 and 8. While the cumulative data more strongly support a respiratory origin than a gene origin of cancer, it is not clear to many people how

Cancer as a Metabolic Disease: On the Origin, Management and Prevention of Cancer, First Edition. Thomas Seyfried.

mitochondrial damage and respiratory insufficiency relate to the genetic defects observed in tumor cells. How is it possible that respiratory insufficiency could underlie the genomic instability seen in most of the tumor cells?

Emerging evidence indicates that a persistent retrograde response can link respiratory injury to the genomic instability seen in tumor cells (5–7). The RTG response is the general term used for mitochondria-to-nuclear signaling and involves cellular responses to changes in respiration and the functional state of mitochondria (6, 8–14). The RTG response is initiated following interruption in the respiratory energy production. Genomic stability is dependent on the integrity of the mitochondrial function. If respiratory insufficiency is not corrected, the RTG response will persist, thus producing the Warburg effect, genomic instability, and the path to tumorigenesis.

The RTG response can be viewed as a classic extrachromosomal epigenetic control system (15, 16). Although DNA methylation and histone modification are considered to be one type of epigenetic mechanism (17, 18), the mitochondrion as an extrachromosomal element is the predominant driver of epigenetic control within the cell (15). Mitochondria maintain cellular differentiation through well-established nuclear cytoplasmic interactions. What is the evidence supporting the role of the RTG response in genomic instability and the epigenetic origin of cancer?

The RTG response has been mostly studied in yeast, but mitochondrial stress signaling is an analogous response in mammalian cells (5, 8, 10, 14, 19). Jazwinski and colleagues (8) have recently shown that the RTG metabolic stress response in yeast is similar to the NF-kB metabolic stress response in humans. Expression of multiple nuclear genes controlling energy metabolism is profoundly altered following impairment in mitochondrial energy homeostasis (10, 20, 21). Respiratory insufficiency can arise from abnormalities in mtDNA, the TCA cycle, the electron transport chain, or in the proton motive gradient ($\Delta\Psi_m$) of the inner membrane. In other words, any interruption in mitochondrial respiration can trigger an RTG response (7). How does this relate to the origin of cancer?

The RTG response evolved in eukaryotic microorganisms to maintain cell viability following periodic disruption of respiratory ATP production (8, 14, 22). This mostly involves an energy transition from OxPhos to substrate-level phosphorylation including glycolysis and amino acid fermentation. Similar systems are also expressed in mammalian cells (8, 14, 19–21). According to our hypothesis, the RTG response would include upregulation of networks needed for nonoxidative energy metabolism. This is supported by findings showing that respiratory damage upregulates the expression of the *Myc* and *Ras* oncogenes (6, 20). MYC enhances ROS production while mitigating p53 function (23, 24). ROS production also stimulates the RTG response and produces genomic instability (9, 25). MYC also upregulates genes needed for both glycolysis and glutamine metabolism, which drive nonoxidative energy metabolism as I have described in Chapter 8. The upregulation of oncogenes becomes necessary for maintaining nonoxidative energy metabolism when energy production through respiration is insufficient for maintaining energy homeostasis. RTG signaling coordinates oncogene upregulation in order to prevent cell death. Oncogene upregulation is a genetic hallmark of cancer.

Besides upregulating oncogene expression, prolonged or continued activation of the RTG response can have dire consequences on nuclear genome stability and function. Warburg was also aware of the linkage between respiration and the maintenance of cell structure and the linkage between fermentation and the loss of cell structure (26). The structural organization of the cell including its morphology and genome integrity is dependent on sufficient respiration (25). The maintenance of structure and genome integrity is dependent on the regulatory elements of the RTG response. Although the RTG response evolved to protect cell viability following transient disruption of respiration, a prolonged RTG response will lead to genomic instability and disorder.

Three main regulatory elements define the RTG response in yeast, including the Rtg2 signaling protein and the Rtg1/Rtg3 transcriptional factor complex (both are basic helix-loop-helix-leucine zippers) (7, 14). Rtg2 contains an N-terminal, ATP-binding motif that senses changes in mitochondrial ATP production. Rtg2 also regulates the function and cellular localization of the heterodimeric Rtg1/Rtg3 complex (Fig. 10.1).

The RTG response is "off" in healthy cells with sufficient respiratory energy production. In the off state, the Rtg1/Rtg3 complex is sequestered in the cytoplasm with Rtg1 attached (dimerized) to a highly phosphorylated form of Rtg3

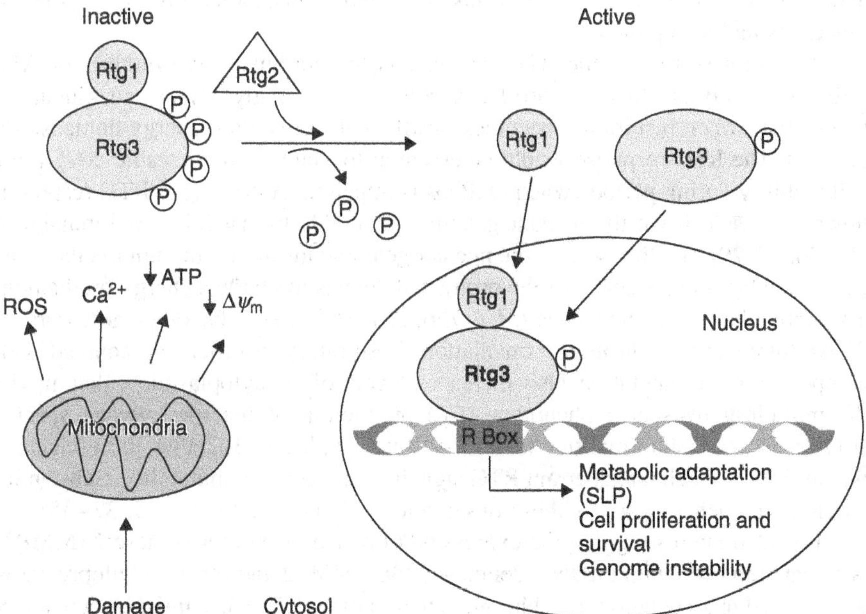

Figure 10.1 Activation of the retrograde response (RTG) in yeast cells. The RTG response in yeast is mechanistically similar to the mitochondrial stress response in mammalian cells. The circled Ps are phosphate groups. SLP, substrate-level phosphorylation. The RTG response can upregulate genes needed for fermentation when energy through OxPhos becomes compromised. See text for further details. *Source*: Reprinted with permission from Ref. 7. See color insert.

(14). Besides its role in the cytoplasm as an energy sensor, Rtg2 also functions in the nucleus as a regulator of chromosomal integrity (10, 27). The RTG response also maintains the function of the chromosome stability gene, *SMC4* (10). The RTG response reduces the expression of this gene following damage to OxPhos (10). Moreover, a prolonged RTG response disrupts DNA repair mechanisms, thus producing a plethora of random DNA mutations and chromosomal defects (7, 12, 25, 27).

The RTG response is turned "on" following insufficient energy production through OxPhos. In the on state, cytoplasmic Rtg2 disengages the Rtg1/Rtg3 complex through a dephosphorylation of Rtg3 (14). The Rtg1 and Rtg3 proteins then individually enter the nucleus where Rtg3 binds to R box sites, Rtg1 reengages Rtg3, and transcription and signaling commences for multiple energy and anti-apoptotic related genes and proteins to include MYC, TOR, Ras, CREB, NF-kB, and CHOP (14, 20, 21, 28–30). CHOP, also known as GADD153, is a member of the C/EBP transcription factor family that forms heterodimers with other C/EBPs (31). Increased expression of these genes and proteins is linked to tumor inflammation, proliferation, and progression, that is, the key hallmarks of cancer. The RTG response also involves the participation of multiple negative and positive regulators, which facilitate the bioenergetic transition from respiration to fermentation involving substrate-level phosphorylations (14, 20). Most importantly, persistent activation of the RTG response leads to genomic instability involving somatic mutations and aneuploidy.

The primary role of the RTG response is to coordinate the synthesis of ATP through glycolysis alone or through a combination of glycolysis and glutamine metabolism when respiration becomes insufficient to maintain energy homeostasis (14, 19). The RTG response would be essential for maintaining a stable $\Delta G'_{\text{ATP}}$ for cell viability during periods when OxPhos is impaired. A prolonged RTG response, however, would leave the nuclear genome vulnerable to instability and mutability (10, 20, 27, 29). In other words, the nuclear genomic instability in tumor cells arises as a secondary consequence of the protracted defects in OxPhos energy production. The upregulation of oncogenes (*Myc, Ras, Akt, Hif-1*, etc.) becomes necessary to derive tumor energy through fermentation. Respiratory insufficiency coupled with compensatory fermentation also increases levels of (i) cytoplasmic calcium; (ii) the multidrug resistance phenotype; (iii) production of reactive oxygen species (ROS); and (iv) abnormalities in iron–sulfur complexes. Together these changes would further accelerate aberrant RTG signaling and genome mutability. Substantial empirical evidence supports these observations (5, 11, 12, 14, 19, 25, 32–35).

It is also interesting that the expression of matrix metalloproteinase 2 (MMP2) is elevated in cells with mtDNA deficiency (36). MMP2 and other metalloproteases are elevated in association with chronic inflammation (37). Since mtDNA deficiency and ROS production reduces mitochondrial respiration, it is not unreasonable to speculate that MMP2 expression would also be elevated following any number of insults to mitochondrial respiration. Elevated MMP expression is the phenotype seen in activated macrophages, which hybridize with neoplastic epithelial to form cancer cells with a high metastatic potential (Chapter 13). ROS production associated with inflammation would also activate the RTG response as previously seen in other systems (9). Hence, the RTG response can be linked to both the initiation and the progression of carcinogenesis.

It is also interesting that the human Myc/Max transcription factor complex shows interesting homologies to the yeast Rtg1/Rtg3 proteins (6, 8, 20). MYC is a member of the basic, helix-loop-helix-leucine zipper family of transcription factors as are Rtg1/Rtg3. MYC upregulation is also necessary for the induction of genes needed for glycolysis and glutamine metabolism (38). Although there is currently no known counterpart in higher eukaryotes for the Rtg2 protein as a sensor of mitochondrial dysfunction and a transducer of mitochondrial signals that activate Rtg1/3-like transcription factors, there is recognized conservation of the NF-kB stress response in humans and the RTG stress response in yeast (8). The yeast RTG response also shares interesting functional homologies with the mTor, Akt, and RAS signaling pathways. Jazwinsk and colleagues have prepared an excellent review linking the similarities between the yeast and human stress responses (8).

Considered collectively, these findings indicate that the integrity of the nuclear genome is dependent to a large extent on normal respiratory function (5, 7, 25). The mitochondrial–nuclear interaction is an example of a classic epigenetic system as David Nanney had first described in 1958 (16, 39). Although the concept of epigenetics originates with the work of Waddington (40), Nanney's views of epigenetic control systems are quite relevant to the role of mitochondria in cancer (15, 41). It would therefore be helpful for students of cancer epigenetics to carefully consider the information presented in Dr. Nanny's review.

Epigenetics involves more than just DNA methylation, genomic imprinting, and histone modification (41–43). Mitochondrial function is also epigenetic. It is interesting that inherited defects in p53 can damage OxPhos, leading to genomic instability (44, 45). Moreover, Hwang and colleagues (46) recently showed that efficient mitochondrial respiration is essential for maintaining genomic stability in environments where oxygen is present. As mentioned above, the multiple carcinogenic effects of the *Myc* oncogene can also be linked to OxPhos damage. While the RTG response evolved to protect cells from acute energy failure, a persistent RTG response associated with insufficient respiration can eventually initiate genomic instability and tumorigenesis. Hence, chronic respiratory insufficiency together with an activated RTG response is the gateway to cellular disorder and the origin of neoplasia regardless of whether genetic or environmental factors initiate the response.

INFLAMMATION INJURES CELLULAR RESPIRATION

Although chronic inflammation has long been linked to carcinogenesis, it is not clear how inflammation specifically causes cancer (47–51). It is known that inflammation associated with sepsis or LPS impairs mitochondrial respiration (52–55). Sepsis is an acute inflammatory condition that can lead to systemic organ failure and death. In contrast to inflammation from sepsis and LPS, which induce acute mitochondrial failure and cell death, the inflammation that causes many cancers is chronic. Chronic inflammation will produce protracted mitochondrial damage (48, 50, 54). Injury or damage to the mitochondrial ETC can arise from persistent nitric oxide expression in the inflamed microenvironment (54). Nel and colleagues (56) have shown how ultrafine particles could exacerbate oxidative stress and mitochondrial damage while depleting intracellular glutathione levels in macrophage and epithelial cell lines. Bissell and colleagues together with Bierie and Moses have reviewed information showing how chronic inflammation in the microenvironment activates the expression of transforming growth factor beta (TGF-β) (37, 57, 58). Yoon and colleagues (59) showed that TGF-β induces protracted mitochondrial ROS production, which damages respiratory control and enhances senescence in lung epithelial cells. Seoane et al. (25) showed that nuclear genomic instability was directly correlated with mitochondrial ROS production. Fosslien (60) described how gradients of TGF-β could alter mitochondrial ATP generation in the morphogenetic field.

Viewed collectively, these findings indicate that respiratory damage links inflammation to carcinogenesis. Chronic inflammation, which enhances the expression of nitric oxide and TGF-β, damages respiration. Most cells that suffer respiratory damage die. According to Warburg's theory, tumors arise only from those cells that are capable of increasing fermentation in order to compensate for insufficient respiration. Enhanced fermentation prevents senescence (61, 62). Although it is clear that damaged respiration links inflammation to the origin of cancer, further studies are necessary to better define the molecular mechanisms responsible for this linkage.

HYPOXIA-INDUCIBLE FACTOR (HIF) STABILITY IS REQUIRED FOR THE ORIGIN OF CANCER

While respiratory insufficiency is the initiating event in carcinogenesis, enhanced fermentation is required to maintain cell viability following damaged respiration. Interesting analogies exist between yeast and mammalian cells for the physiological response to impaired respiration (7, 20, 29, 63–65). Mammalian cells increase the expression of HIF-1α in response to transient hypoxia (66). HIF-1α is rapidly degraded under normoxia, but becomes stabilized under hypoxia. This is a conserved physiological response that has evolved to protect mammalian mitochondria from hypoxic damage and to provide an alternative source of energy to respiration. HIF-1α induces the expression of genes that are involved in glucose uptake, glycolysis, and lactic acid production (66, 67). It remains controversial whether the HIF-1α expression also activates pyruvate dehydrogenase kinase 1, thus blocking entry of pyruvate into the mitochondria since pyruvate metabolism to citrate is considered to be essential for fatty acid synthesis (68). However, HIF-1α expression remains elevated in most tumor cells whether or not oxygen is present and could largely mediate aerobic glycolysis (67, 69–74).

The continued stability of HIF-1α in the presence of oxygen is sometimes referred to as *pseudo-hypoxia* (70, 75). As cancer originates from insufficient OxPhos, HIF-1α stability would be essential for maintaining glycolytic substrate-level phosphorylation, which compensates for respiratory insufficiency whether or not oxygen is present. Although the mechanisms of HIF-1α stabilization under hypoxic conditions are well defined, the mechanisms by which HIF-1α is stabilized under aerobic or normoxic conditions are less clear (69, 73, 75). HIF-1α is generally unstable in cells under normal aerobic conditions through its interaction with the VHL tumor suppressor protein, which facilitates HIF-1α hydroxylation, ubiquitination, and proteasomal degradation (71). It appears that several factors contribute to HIF-1α stability in cancer cells.

The rapid degradation of HIF-1α in oxygen is regulated by oxygen-dependent prolyl hydroxylases (PHDs). PHDs hydroxylate prolyl residues in an oxygen-dependent degradation domain (67, 75). The inhibition of PHDs stabilizes Hif-1α even in the presence of oxygen. HIF-1α stabilization under aerobic conditions can be linked to respiratory insufficiency through abnormalities in calcium homeostasis, ROS generation, NF-kB signaling, accumulation of TCA cycle metabolites (succinate and fumarate), and oncogenic viral infections (25, 69, 76–80). Genomic instability arises, in part, through ROS production and "prolonged" HIF-1α stabilization under aerobic conditions. This process would be linked to the RTG system as described above.

Studies from Gottlieb and colleagues (75) indicate that certain energy metabolites, that is, succinate, α-ketoglutarate, and fumarate, can stabilize HIF-1α in the presence of oxygen. As I have described in Chapter 7, succinate and fumarate are also products of amino acid fermentation. Hence, succinate and fumarate, arising through glutamine fermentation, could contribute to inhibited PHDs and the stabilization of Hif-1α. As Hif-1α expression regulates multiple genes needed for glycolysis, Hif-1α stabilization would be important for maintaining fermentation energy production through substrate-level phosphorylation following deficiency in OxPhos. It is important to recognize that respiratory insufficiency is ultimately responsible for Hif-1α stabilization in cancer cells.

MITOCHONDRIA AND THE MUTATOR PHENOTYPE

Mitochondria maintain cellular differentiation through an interaction with the nucleus in a classical nuclear/epigenetic homeostasis (7, 16). Most human cancer cells display genome instability involving elevated mutation rates, gross chromosomal rearrangements, and alterations in chromosome number (81–87). The studies of the Singh and the Jazwinski groups provide compelling evidence that mitochondrial dysfunction, operating largely through the epigenetic RTG response (mitochondrial stress signaling), can underlie the mutator phenotype of tumor cells (10, 21, 29, 36, 88, 89). Chromosomal instability, expression of gene mutations, and the tumorigenic phenotype are significantly greater in human cells with mtDNA depletion than in cells with normal mtDNA. Although mitochondrial mutations are not found in all tumors as described in Chapter 9, the depletion of mtDNA as occurs in rho^0 cells will compromise OxPhos.

Singh and coworkers have shown that mtDNA depletion downregulates the expression of the apurinic/apyrimidinic endonuclease (APE1). APE1 is a redox-sensitive multifunctional endonuclease that regulates DNA transcription and repair (10, 21, 90). In other words, the function of this DNA repair enzyme is dependent on the mitochondrial function. Any protracted disruption of mitochondrial respiration would be expected to compromise mechanisms involved with DNA transcription and repair. Damage to respiration increases ROS, which enhances mutation rates (25). ROS induce nuclear genomic instability in tumor cells. APE1 expression is significantly decreased in most of the tumors examined (Fig. 10.2). The risk of mutations and genomic instability will increase if APE1 expression is reduced. Regardless of the process by which mitochondria get damaged, respiratory insufficiency in tumor cells as an initial event in carcinogenesis can account for the eventual genomic instability seen in cancer cells. Hence, the elevated mutation rates, gross chromosomal rearrangements, and alterations in chromosome number observed in tumor cells can be linked to impaired mitochondrial respiration.

Besides APE1, other DNA repair proteins are downregulated in association with mtDNA depletion and OxPhos insufficiency, including p53 and SMC4 (10). It is well documented that genomic mutability increases when p53 expression is reduced. Since gene expression is different in different tissues, it is expected that disturbed energy metabolism would produce different kinds of mutations in different types of cancers (7). Genetic heterogeneity can be even more complex, as many metastatic cancer cells arise from fusions of macrophages and neoplastic epithelial cells (Chapter 13). Even different tumors within the same cancer type could appear to represent different diseases when evaluated at the genomic level. When evaluated at the metabolic level, however, most cancers and tumors are alike in expressing respiratory insufficiency and elevated fermentation. Impaired mitochondrial function can induce abnormalities in tumor suppressor genes and oncogenes. For example, an impaired mitochondrial function can induce abnormalities in p53 activation, while abnormalities in p53 expression and regulation can further impair mitochondrial respiration (21, 28, 35, 45, 46, 91–95). Viewed together, these findings indicate that insufficient respiration underlies the mutator phenotype of tumor cells. Hello, is anyone listening?

Has anyone connected these observations to what I am saying? I think Lu and coworkers (5) might have recognized some of these connections. The findings of Seoane and coworkers make these connections. The function of the pRB tumor suppressor protein, which controls the cell cycle, is also sensitive to ROS production through the redox state of the cell (96). Elevated expression of the *MYC* and *Ras* oncogenes can be linked to the requirements of fermentation energy in order to maintain tumor cell viability (62). The numerous gene defects found in various cancers can arise as secondary consequences of mitochondrial dysfunction and respiratory insufficiency. Do those directing the cancer genome projects know about this?

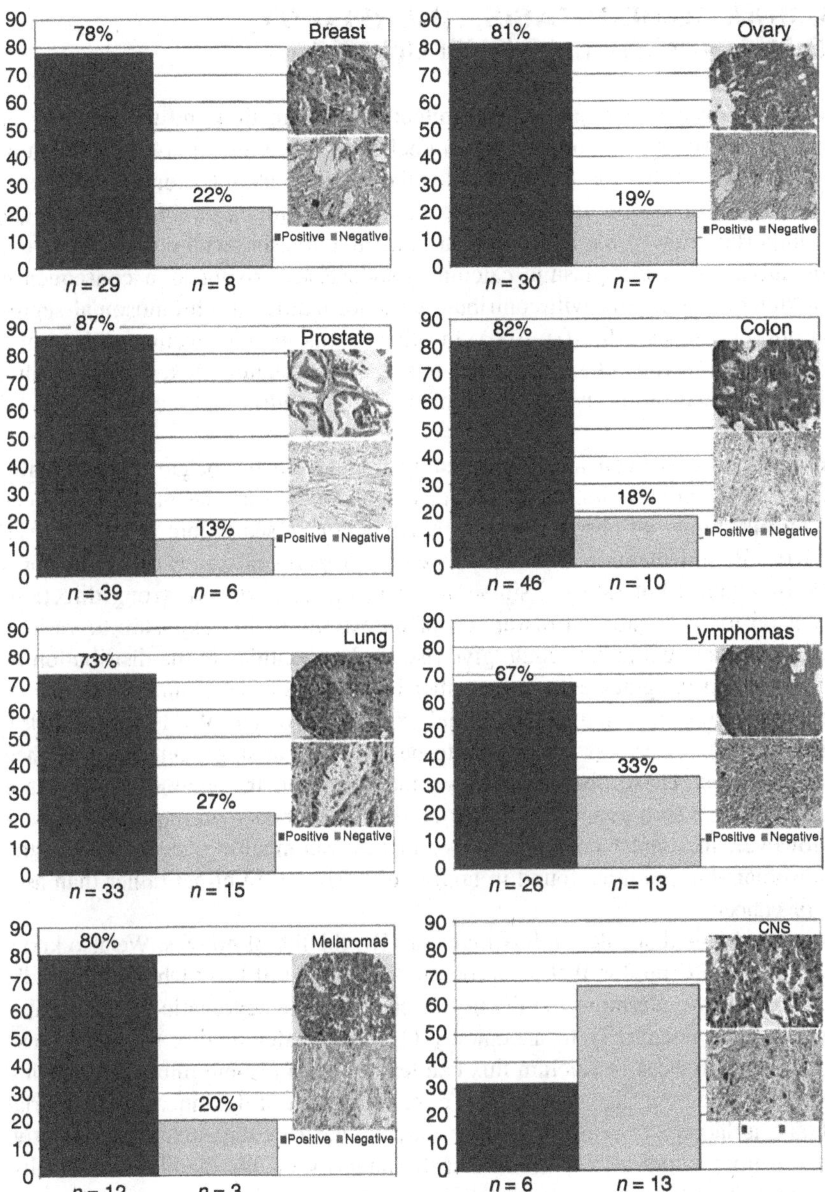

Figure 10.2 Involvement of the APE1 DNA repair gene in tumorigenesis. APE1 expression was analyzed in normal and carcinoma tissues. Immunohistochemical (IHC) analysis was carried out on a variety of cancer tissues using the Tissue Array Research Program (TARP2) of the National Cancer Institute, National Institutes of Health. The bar graph shows the percent of positive and negative carcinoma cases as a whole. Each panel shows a representative positive and negative carcinoma case as well as expression in the normal tissue. APE1 protein was visualized using DAB with hematoxylin counterstain. The results show that APE1 expression is significantly decreased in most of the tumors examined. Bar = 50 μm. *Source*: Reprinted with permission from Ref. (21). See color insert.

CALCIUM HOMEOSTASIS, ANEUPLOIDY, AND MITOCHONDRIAL DYSFUNCTION

Calcium homeostasis is dependent on mitochondrial function and the integrity of the proton motive gradient of the inner mitochondrial membrane (14, 97). Calcium homeostasis is essential for the fidelity of cell division to include spindle and microtubule assemblies, sister chromosome separation, and cytokinesis (98–103). In light of the important role of the mitochondria in maintaining intracellular calcium flux, any disturbances in cytoplasmic calcium homeostasis, arising as a consequence of insufficient respiration, will contribute to abnormalities in chromosomal segregation during mitosis (19, 104–106). In other words, nondisjunction and mitotic defects can arise from changes in the intracellular calcium flux, which is ultimately determined by the health status of the mitochondria and the sufficiency of OxPhos.

It is important to consider these findings in light of the origin of aneuploidy in cancer cells. Boveri first suggested that aneuploidy was the origin of cancer based on his studies of chromosomal nondisjunction in sea urchin embryogenesis (107, 108). Boveri mentioned that the essential property of tumor cells was not a disease of vitality, but rather a situation where the cell takes a wrong direction. He went on to speculate that disruption of mitosis following exposure to various physical and chemical insults could give rise to abnormalities in the distribution of chromosomes in daughter cells. In a rather bold position, he claimed that disturbances in chromosomal segregation could explain the origin of all cancers (108). In other words, cancer was considered to be a disease arising from an imbalance of chromosomes. These observations eventually lead to the view of cancer as a genetic disease, which persists today as the dominant theory on the origin of cancer. However, the data I presented above makes a compelling argument that the chromosomal abnormalities found in tumor cells arise as an effect rather than as a cause of cancer.

We now know that calcium flux maintains the fidelity of mitosis. We also know from the work of Compton that inappropriate attachment of kinetochores to spindle microtubules can undermine the fidelity of chromosome segregation during mitosis, which leads eventually to aneuploidy (109). As calcium flux regulates these processes, disturbances in calcium flux can lead to chromosome imbalances during cell division. The integrity of the proton motive gradient of the inner mitochondrial membrane is largely responsible for the intracellular calcium flux. Consequently, damage to mitochondrial respiration and the integrity of this membrane will ultimately contribute to chromosome imbalances and aneuploidy.

Duesberg and coworkers (110, 111) have also argued that it is aneuploidy, rather than somatic mutations, that underlies the origin of cancer. They suggested a two-stage mechanism of carcinogenesis. During stage one, carcinogens would cause aneuploidy, either through chromosome fragmentation or by damaging the spindle apparatus. During stage two, tumorigenic karyotypes would evolve autocatalytically because aneuploidy destabilizes the karyotype, that is, causes genetic instability.

In light of my hypothesis, protracted respiratory insufficiency would account for the origin of aneuploidy. The integrity of the mitotic spindle assembly is based on calcium flux, which is linked to the mitochondrial proton motive gradient. We also know that ROS, arising from any number of environmental insults including tissue inflammation, damage the proton motive gradient. These insults include X rays, chemicals, or viruses (112). Samper (113) and colleagues and Seoane et al. (25) have clearly shown that mitochondrial stress, arising through ROS, causes genomic instability including aneuploidy. The findings of Lu and coworkers also support these findings (5). Once aneuploidy occurs, it would facilitate energy production through fermentation, causing further disorder to nuclear genome stability and to mitochondrial respiration. The work of Amon and colleagues has shown that aneuploidy destabilizes cellular physiology and energy homeostasis (114–116). If aneuploidy is deleterious to cell viability and inhibits cell proliferation, how is it possible for tumor cells to grow if they are aneuploid? The answer is fermentation.

Fermentation is the mechanism that leads to tolerance of aneuploidy in tumor cells. Thompson and Compton (117) have shown that loss of the p53 tumor suppressor could promote growth of aneuploid cells. We know that p53 is required for normal mitochondrial function and that defects in p53 will enhance energy through fermentation (44–46). Hence, damage to respiration with compensation through fermentation can permit viability in the presence of genomic instability.

The aneuploidy–cancer mechanism of Duesberg and colleagues could play a role in cancer progression once the origin or aneuploidy becomes linked to respiratory damage and aerobic glycolysis. I therefore propose that aneuploidy and the numerous somatic mutations and other genomic aberrations found in cancer cells ultimately arise as a consequence of damage to mitochondrial proteins, lipids, and mtDNA. This damage dissipates the proton motive gradient, leading to elevated fermentation and an imbalance in the cellular calcium flux. The vast number of genomic changes identified in sporadic cancers ultimately arises as a consequence of mitochondrial dysfunction and respiratory insufficiency. The evidence amassed in support of this hypothesis is compelling and will be difficult to disprove or, worse yet, to ignore.

MITOCHONDRIAL DYSFUNCTION AND LOSS OF HETEROZYGOSITY (LOH)

Most normal genes on autosomes (nonsex chromosomes) contain two alleles (alternate forms of the gene), which produce a normal protein product. Abnormal phenotypes do not normally arise from allelic loss in recessive genes since the product of the single normal allele is usually sufficient to prevent pathology. A normal phenotype can be maintained in the heterozygous state for most recessive genes. Loss of function or deletion of the single normal allele, however, will prevent production of any normal product from that gene. The loss of heterozygosity (LOH) in relationship to the origin of cancer is often referred to as the *Knudson hypothesis* and originates with Alfred Knudson, who had first developed the idea that LOH in critical genes such as *p53* and *RB* could predispose individuals to cancer. This concept is now a well-accepted mechanism for the somatic mutation theory of cancer (118).

However, recent studies in yeast indicate that damage to the inner mitochondrial membrane potential ($\Delta\Psi_m$), following mtDNA depletion, induces mitochondrial dysfunction and LOH in the nuclear genome (12). Yeast colonies formed following mtDNA depletion varied in size, but eventually expressed improved growth following repeated passaging despite the continued absence of their mtDNA. Remarkably, these clones were unable to respire and had a slower growth than cells with intact mtDNA. After 30 h of growth, however, they formed colonies that grew faster, and displayed fewer nuclear LOH events than cells within the first 30 h following loss of mtDNA.

These findings show that LOH is an early event following mtDNA depletion in these cells. Moreover, the function of several nuclear iron–sulfur-dependent DNA repair enzymes involving the Rad3 helicase, the Pri2 primase, and the Ntg2 glycase were defective in the mtDNA-depleted cells (12). Abnormalities in these DNA repair enzymes contribute to the LOH phenotype in specific genes. These findings indicate that LOH, which is common for many genes of cancer cells (85), is linked to mitochondrial dysfunction and respiratory insufficiency.

The findings of Veatch et al. (12) are consistent with the earlier findings of Roskelley et al. (119) in showing that mitochondrial dysfunction and damaged respiration is the initiating event in the origin of cancer. The findings of Veatch et al. are also consistent with the findings of the Singh, Jazwinski, and Seoane et al. groups in showing that genomic instability is a consequence of mitochondrial dysfunction. When considered together, these observations indicate that the bulk of the genetic abnormalities found in cancer cells, ranging from point mutations to gross chromosomal rearrangements, arise following damage to the structure and function of mitochondria.

TISSUE INFLAMMATION, DAMAGED RESPIRATION, AND CANCER

Impairment of mitochondrial function can occur following prolonged injury or irritation to tissues including disruption of morphogenetic fields (37, 48, 60). Sonnenschein and Soto argued persuasively that disruption of tissue organization and structure, rather than random somatic mutations, could give rise to cancers (3, 49, 120, 121). They describe this process as the *tissue organizational field theory* (TOFT) of carcinogenesis. This concept is based on evidence that cancer-provoking agents disturb the tridimensional organization of tissue architecture, thus disturbing positional and historical information embodied in the morphogenetic field. These views are closely aligned with those of Dr. Mina Bissell, who has long considered disturbances in the microenvironment as being a driver of carcinogenesis (37). David Tarin (122) also considers disturbances in the microenvironment, rather than gene defects, as being the origin of cancer. While the mechanisms by which abnormalities in the tissue organization underlie carcinogenesis are multiple, the TOFT can be incorporated into the mitochondrial theory.

For example, it is the accumulation of respiratory damage over time that ultimately leads to malignant tumor formation. ROS damage mitochondrial proteins, lipids, and nuclear DNA. ROS arise from chronic inflammation, which also disrupts tissue morphogenetic fields. Chronic disruption of the tissue morphogenetic field (microenvironment) would ultimately impair respiratory function in cells within the field. Acquired abnormalities in mitochondrial function would produce a type of vicious cycle where insufficient mitochondrial energy production initiates genome instability and mutability, which then promotes further mitochondrial dysfunction and energy impairment, and so on, in a cumulative way. This would ultimately be seen as a gross disturbance in the structural organization of the local tissue and eventually as a carcinoma (49). An increased dependency on fermentation energy for cell survival would follow each round of metabolic and genetic damage, thus initiating uncontrolled cell growth with the eventual formation of a malignant neoplasm. Hence, the well-documented, tumor-associated abnormalities and genomic instability seen in cancer can arise as a consequence of the progressive impairment of OxPhos. It is my view that *chronic OxPhos insufficiency, arising from any number of genetic or environmental insults, is the origin of cancer*.

REFERENCES

1. Weinberg RA. The Biology of Cancer. New York: Garland Science; 2007.
2. Dobzhansky T. Nothing in biology makes sense except in the light of evolution. Am Biol Teach. 1973;35:125–9.
3. Sonnenschein C, Soto AM. Somatic mutation theory of carcinogenesis: why it should be dropped and replaced. Mol Carcinog. 2000;29:205–11.
4. Lands B. A critique of paradoxes in current advice on dietary lipids. Prog Lipid Res. 2008;47:77–106.
5. Lu J, Sharma LK, Bai Y. Implications of mitochondrial DNA mutations and mitochondrial dysfunction in tumorigenesis. Cell Res. 2009;19:802–15.
6. Erol A. Retrograde regulation due to mitochondrial dysfunction may be an important mechanism for carcinogenesis. Med Hypotheses. 2005;65:525–9.
7. Seyfried TN, Shelton LM. Cancer as a metabolic disease. Nutr Metabol. 2010;7:7.
8. Srinivasan V, Kriete A, Sacan A, Jazwinski SM. Comparing the yeast retrograde response and NF-kappaB stress responses: implications for aging. Aging Cell. 2010;9:933–41.
9. Woodson JD, Chory J. Coordination of gene expression between organellar and nuclear genomes. Nat Rev Genet. 2008;9:383–95.
10. Chandra D, Singh KK. Genetic insights into OXPHOS defect and its role in cancer. Biochim Biophys Acta. 2010;1807:620–5.
11. Traven A, Wong JM, Xu D, Sopta M, Ingles CJ. Interorganellar communication. Altered nuclear gene expression profiles in a yeast mitochondrial dna mutant. J Biol Chem. 2001;276:4020–7.
12. Veatch JR, McMurray MA, Nelson ZW, Gottschling DE. Mitochondrial dysfunction leads to nuclear genome instability via an iron-sulfur cluster defect. Cell. 2009;137:1247–58.
13. Jazwinski SM. The retrograde response links metabolism with stress responses, chromatin-dependent gene activation, and genome stability in yeast aging. Gene. 2005;354:22–7.
14. Butow RA, Avadhani NG. Mitochondrial signaling: the retrograde response. Mol Cell. 2004;14: 1–15.

15. Serb AM, Owen RD, Edgar RS. Extrachromosomal and Epigenetic Systems. General Genetics. San Francisco (CA): W.H. Freeman; 1965. p.315–351.
16. Nanney DL. Epigenetic control systems. Proc Natl Acad Sci USA. 1958;44:712–7.
17. Bonasio R, Tu S, Reinberg D. Molecular signals of epigenetic states. Science. 2011;330:612–6.
18. Riddihough G, Zahn LM. Epigenetics. What is epigenetics? Introduction. Science. 2011; 330:611.
19. Amuthan G, Biswas G, Ananadatheerthavarada HK, Vijayasarathy C, Shephard HM, Avadhani NG. Mitochondrial stress-induced calcium signaling, phenotypic changes and invasive behavior in human lung carcinoma A549 cells. Oncogene. 2002;21:7839–49.
20. Miceli MV, Jazwinski SM. Common and cell type-specific responses of human cells to mitochondrial dysfunction. Exp Cell Res. 2005;302:270–80.
21. Singh KK, Kulawiec M, Still I, Desouki MM, Geradts J, Matsui S. Inter-genomic cross talk between mitochondria and the nucleus plays an important role in tumorigenesis. Gene. 2005;354:140–6.
22. Liu Z, Butow RA. Mitochondrial retrograde signaling. Annu Rev Genet. 2006;40:159–85.
23. Chung YM, Kim JS, Yoo YD. A novel protein, Romo1, induces ROS production in the mitochondria. Biochem Biophys Res Commun. 2006;347:649–55.
24. Vafa O, Wade M, Kern S, Beeche M, Pandita TK, Hampton GM, et al. c-Myc can induce DNA damage, increase reactive oxygen species, and mitigate p53 function: a mechanism for oncogene-induced genetic instability. Mol Cell. 2002;9:1031–44.
25. Seoane M, Mosquera-Miguel A, Gonzalez T, Fraga M, Salas A, Costoya JA. The mitochondrial genome Is a "Genetic Sanctuary" during the oncogenic process. PloS One. 2011;6:e23327.
26. Warburg O. On the origin of cancer cells. Science. 1956;123:309–14.
27. Borghouts C, Benguria A, Wawryn J, Jazwinski SM. Rtg2 protein links metabolism and genome stability in yeast longevity. Genetics. 2004;166:765–77.
28. Kulawiec M, Ayyasamy V, Singh KK. p53 regulates mtDNA copy number and mitocheckpoint pathway. J Carcinog. 2009;8:8.
29. Kulawiec M, Safina A, Desouki MM, Still I, Matsui SI, Bakin A, et al. Tumorigenic transformation of human breast epithelial cells induced by mitochondrial DNA depletion. Cancer Biol Ther. 2008;7:1732–43.
30. Wolfman JC, Planchon SM, Liao J, Wolfman A. Structural and functional consequences of c-N-Ras constitutively associated with intact mitochondria. Biochim Biophys Acta. 2006;1763:1108–24.
31. Endo M, Oyadomari S, Suga M, Mori M, Gotoh T. The ER stress pathway involving CHOP is activated in the lungs of LPS-treated mice. J Biochem. 2005;138:501–7.
32. Simbula G, Glascott PA Jr, Akita S, Hoek JB, Farber JL. Two mechanisms by which ATP depletion potentiates induction of the mitochondrial permeability transition. Am J Physiol. 1997;273:C479–C488.
33. Arnould T, Vankoningsloo S, Renard P, Houbion A, Ninane N, Demazy C, et al. CREB activation induced by mitochondrial dysfunction is a new signaling pathway that impairs cell proliferation. EMBO J. 2002;21:53–63.
34. Whitfield JF. Calcium, calcium-sensing receptor and colon cancer. Cancer Lett. 2009;275:9–16.
35. Trachootham D, Alexandre J, Huang P. Targeting cancer cells by ROS-mediated mechanisms: a radical therapeutic approach? Nat Rev Drug Discov. 2009;8:579–91.
36. Miceli MV, Jazwinski SM. Nuclear gene expression changes due to mitochondrial dysfunction in ARPE-19 cells: implications for age-related macular degeneration. Invest Ophthalmol Vis Sci. 2005;46:1765–73.
37. Bissell MJ, Hines WC. Why don't we get more cancer? A proposed role of the microenvironment in restraining cancer progression. Nat Med. 2011;17:320–9.
38. Dang CV, Le A, Gao P. MYC-induced cancer cell energy metabolism and therapeutic opportunities. Clin Cancer Res. 2009;15:6479–83.
39. Haig D. The (dual) origin of epigenetics. Cold Spring Harbor Symp Quant Biol. 2004;69:67–70.
40. Holliday R. Epigenetics: a historical overview. Epigenetics. 2006;1:76–80.

41. Holliday R. A new theory of carcinogenesis. Br J Cancer. 1979;40:513–22.
42. Smiraglia DJ, Kulawiec M, Bistulfi GL, Gupta SG, Singh KK. A novel role for mitochondria in regulating epigenetic modification in the nucleus. Cancer Biol Ther. 2008;7:1182–90.
43. Feinberg AP, Tycko B. The history of cancer epigenetics. Nat Rev. 2004;4:143–53.
44. Matoba S, Kang JG, Patino WD, Wragg A, Boehm M, Gavrilova O, et al. p53 regulates mitochondrial respiration. Science. 2006;312:1650–3.
45. Lago CU, Sung HJ, Ma W, Wang PY, Hwang PM. p53, aerobic metabolism and cancer. Antioxid Redox Signal. 2010;15:1739–48.
46. Sung HJ, Ma W, Wang PY, Hynes J, O'Riordan TC, Combs CA, et al. Mitochondrial respiration protects against oxygen-associated DNA damage. Nat Commun. 2011;1:1–8.
47. Hanahan D, Weinberg RA. Hallmarks of cancer: the next generation. Cell. 2011;144:646–74.
48. Coussens LM, Werb Z. Inflammation and cancer. Nature. 2002;420:860–7.
49. Sonnenschein C, Soto AM. The Society of Cells: Cancer and the Control of Cell Proliferation. New York: Springer-Verlag; 1999.
50. Colotta F, Allavena P, Sica A, Garlanda C, Mantovani A. Cancer-related inflammation, the seventh hallmark of cancer: links to genetic instability. Carcinogenesis. 2009;30:1073–81.
51. Ohshima H, Bartsch H. Chronic infections and inflammatory processes as cancer risk factors: possible role of nitric oxide in carcinogenesis. Mutat Res. 1994;305:253–64.
52. Brealey D, Karyampudi S, Jacques TS, Novelli M, Stidwill R, Taylor V, et al. Mitochondrial dysfunction in a long-term rodent model of sepsis and organ failure. Am J Physiol Regul Integr Comp Physiol. 2004;286:R491–7.
53. Hunter RL, Dragicevic N, Seifert K, Choi DY, Liu M, Kim HC, et al. Inflammation induces mitochondrial dysfunction and dopaminergic neurodegeneration in the nigrostriatal system. J Neurochem. 2007;100:1375–86.
54. Frost MT, Wang Q, Moncada S, Singer M. Hypoxia accelerates nitric oxide-dependent inhibition of mitochondrial complex I in activated macrophages. Am J Physiol Regul Integr Comp Physiol. 2005;288:R394–R400.
55. Navarro A, Boveris A. Hypoxia exacerbates macrophage mitochondrial damage in endotoxic shock. Am J Physiol Regul Integr Comp Physiol. 2005;288:R354–5.
56. Li N, Sioutas C, Cho A, Schmitz D, Misra C, Sempf J, et al. Ultrafine particulate pollutants induce oxidative stress and mitochondrial damage. Environ Health Perspect. 2003;111:455–60.
57. Bierie B, Moses HL. Tumour microenvironment: TGFbeta: the molecular Jekyll and Hyde of cancer. Nat Rev Cancer. 2006;6:506–20.
58. Bierie B, Moses HL. TGF-beta and cancer. Cytokine Growth Factor Rev. 2006;17:29–40.
59. Yoon YS, Lee JH, Hwang SC, Choi KS, Yoon G. TGF beta1 induces prolonged mitochondrial ROS generation through decreased complex IV activity with senescent arrest in Mv1Lu cells. Oncogene. 2005;24:1895–903.
60. Fosslien E. Cancer morphogenesis: role of mitochondrial failure. Ann Clin Lab Sci. 2008;38: 307–29.
61. Ortega AD, Sanchez-Arago M, Giner-Sanchez D, Sanchez-Cenizo L, Willers I, Cuezva JM. Glucose avidity of carcinomas. Cancer Lett. 2009;276:125–35.
62. Moiseeva O, Bourdeau V, Roux A, Deschenes-Simard X, Ferbeyre G. Mitochondrial dysfunction contributes to oncogene-induced senescence. Mol Cell Biol. 2009;29:4495–507.
63. Diaz-Ruiz R, Uribe-Carvajal S, Devin A, Rigoulet M. Tumor cell energy metabolism and its common features with yeast metabolism. Biochim Biophys Acta. 2009;1796:252–65.
64. Amuthan G, Biswas G, Zhang SY, Klein-Szanto A, Vijayasarathy C, Avadhani NG. Mitochondria-to-nucleus stress signaling induces phenotypic changes, tumor progression and cell invasion. EMBO J. 2001;20:1910–20.
65. Biswas G, Guha M, Avadhani NG. Mitochondria-to-nucleus stress signaling in mammalian cells: nature of nuclear gene targets, transcription regulation, and induced resistance to apoptosis. Gene. 2005;354:132–9.
66. Semenza GL. Oxygen-dependent regulation of mitochondrial respiration by hypoxia-inducible factor 1. Biochem J. 2007;405:1–9.

67. PORPORATO PE, DHUP S, DADHICH RK, COPETTI T, SONVEAUX P. Anticancer targets in the glycolytic metabolism of tumors: a comprehensive review. Front Pharmacol. 2011;2:49.
68. JOSE C, BELLANCE N, ROSSIGNOL R. Choosing between glycolysis and oxidative phosphorylation: A tumor's dilemma? Biochim Biophys Acta. 2010;1807:552–61.
69. GUZY RD, SCHUMACKER PT. Oxygen sensing by mitochondria at complex III: the paradox of increased reactive oxygen species during hypoxia. Exp Physiol. 2006;91:807–19.
70. FAVIER J, BRIERE JJ, BURNICHON N, RIVIERE J, VESCOVO L, BENIT P, et al. The warburg effect is genetically determined in inherited pheochromocytomas. PloS One. 2009;4:e7094.
71. SEMENZA GL. HIF-1 mediates the Warburg effect in clear cell renal carcinoma. J Bioenerg Biomembr. 2007;39:231–4.
72. DANG CV, SEMENZA GL. Oncogenic alterations of metabolism. Trends Biochem Sci. 1999;24: 68–72.
73. DENKO NC. Hypoxia, HIF1 and glucose metabolism in the solid tumour. Nat Rev. 2008;8:705–13.
74. VANDER HEIDEN MG, CANTLEY LC, THOMPSON CB. Understanding the Warburg effect: the metabolic requirements of cell proliferation. Science. 2009;324:1029–33.
75. TENNANT DA, DURAN RV, BOULAHBEL H, GOTTLIEB E. Metabolic transformation in cancer. Carcinogenesis. 2009;30:1269–80.
76. KING A, SELAK MA, GOTTLIEB E. Succinate dehydrogenase and fumarate hydratase: linking mitochondrial dysfunction and cancer. Oncogene. 2006;25:4675–82.
77. RIUS J, GUMA M, SCHACHTRUP C, AKASSOGLOU K, ZINKERNAGEL AS, NIZET V, et al. NF-kappaB links innate immunity to the hypoxic response through transcriptional regulation of HIF-1alpha. Nature. 2008;453:807–11.
78. ZHANG L, LI L, LIU H, PRABHAKARAN K, ZHANG X, BOROWITZ JL, et al. HIF-1alpha activation by a redox-sensitive pathway mediates cyanide-induced BNIP3 upregulation and mitochondrial-dependent cell death. Free Radical Biol Med. 2007;43:117–27.
79. HAEBERLE HA, DURRSTEIN C, ROSENBERGER P, HOSAKOTE YM, KUHLICKE J, KEMPF VA, et al. Oxygen-independent stabilization of hypoxia inducible factor (HIF)-1 during RSV infection. PloS One. 2008;3:e3352.
80. MOON EJ, JEONG CH, JEONG JW, KIM KR, YU DY, MURAKAMI S, et al. Hepatitis B virus X protein induces angiogenesis by stabilizing hypoxia-inducible factor-1alpha. FASEB J. 2004;18:382–4.
81. CAMPBELL PJ, YACHIDA S, MUDIE LJ, STEPHENS PJ, PLEASANCE ED, STEBBINGS LA, et al. The patterns and dynamics of genomic instability in metastatic pancreatic cancer. Nature. 2010;467:1109–13.
82. LOEB LA. A mutator phenotype in cancer. Cancer Res. 2001;61:3230–39.
83. SALK JJ, FOX EJ, LOEB LA. Mutational heterogeneity in human cancers: origin and consequences. Annu Rev Pathol. 2010;5:51–75.
84. LENGAUER C, KINZLER KW, VOGELSTEIN B. Genetic instabilities in human cancers. Nature. 1998;396:643–9.
85. YOKOTA J. Tumor progression and metastasis. Carcinogenesis. 2000;21:497–503.
86. NOWELL PC. Tumor progression: a brief historical perspective. Semin Cancer Biol. 2002;12: 261–6.
87. KOLODNER RD, PUTNAM CD, MYUNG K. Maintenance of genome stability in Saccharomyces cerevisiae. Science. 2002;297:552–7.
88. DELSITE R, KACHHAP S, ANBAZHAGAN R, GABRIELSON E, SINGH KK. Nuclear genes involved in mitochondria-to-nucleus communication in breast cancer cells. Mol Cancer. 2002;1:6.
89. RASMUSSEN AK, CHATTERJEE A, RASMUSSEN LJ, SINGH KK. Mitochondria-mediated nuclear mutator phenotype in Saccharomyces cerevisiae. Nucleic Acids Res. 2003;31:3909–17.
90. EVANS AR, LIMP-FOSTER M, KELLEY MR. Going APE over ref-1. Mutat Res. 2000;461:83–108.
91. MA Y, BAI RK, TRIEU R, WONG LJ. Mitochondrial dysfunction in human breast cancer cells and their transmitochondrial cybrids. Biochim Biophys Acta. 2010;1797:29–37.
92. LEBEDEVA MA, EATON JS, SHADEL GS. Loss of p53 causes mitochondrial DNA depletion and altered mitochondrial reactive oxygen species homeostasis. Biochim Biophys Acta. 2009;1787: 328–34.

93. Holley AK, St Clair DK. Watching the watcher: regulation of p53 by mitochondria. Future Oncol. 2009;5:117–30.
94. Olovnikov IA, Kravchenko JE, Chumakov PM. Homeostatic functions of the p53 tumor suppressor: regulation of energy metabolism and antioxidant defense. Semin Cancer Biol. 2009;19:32–41.
95. Busso CS, Iwakuma T, Izumi T. Ubiquitination of mammalian AP endonuclease (APE1) regulated by the p53-MDM2 signaling pathway. Oncogene. 2009;28:1616–25.
96. Burhans WC, Heintz NH. The cell cycle is a redox cycle: linking phase-specific targets to cell fate. Free Radical Biol Med. 2009;47:1282–93.
97. Gunter TE, Yule DI, Gunter KK, Eliseev RA, Salter JD. Calcium and mitochondria. FEBS Lett. 2004;567:96–102.
98. Whitaker M. Calcium microdomains and cell cycle control. Cell Calcium. 2006;40:585–92.
99. Liu Y, Malureanu L, Jeganathan KB, Tran DD, Lindquist LD, van Deursen JM, et al. CAML loss causes anaphase failure and chromosome missegregation. Cell Cycle. 2009;8: 940–9.
100. Marx J. Cell biology: do centrosome abnormalities lead to cancer? Science. 2001;292:426–9.
101. Chang DC, Meng C. A localized elevation of cytosolic free calcium is associated with cytokinesis in the zebrafish embryo. J Cell Biol. 1995;131:1539–45.
102. Salmon ED, Segall RR. Calcium-labile mitotic spindles isolated from sea urchin eggs (Lytechinus variegatus). J Cell Biol. 1980;86:355–65.
103. Anghileri LJ. Warburg's cancer theory revisited: a fundamentally new approach. Arch Geschwulstforsch. 1983;53:1–8.
104. Keith CH. Effect of microinjected calcium-calmodulin on mitosis in PtK2 cells. Cell Motil Cytoskeleton. 1987;7:1–9.
105. Schon EA, Kim SH, Ferreira JC, Magalhaes P, Grace M, Warburton D, et al. Chromosomal non-disjunction in human oocytes: is there a mitochondrial connection? Hum Reprod. 2000;15 Suppl 2:160–72.
106. Chen RH, Waters JC, Salmon ED, Murray AW. Association of spindle assembly checkpoint component XMAD2 with unattached kinetochores. Science. 1996;274:242–6.
107. Wolf U. Theodor Boveri and his book, on the problem of the origin of malignant tumors. In: German J, editor. Chromosomes and Cancer. New York: John Wiley & Sons, Inc; 1974. p.1–20.
108. Manchester K. The quest by three giants of science for an understanding of cancer. Endeavour. 1997;21:72–6.
109. Compton DA. Mechanisms of aneuploidy. Curr Opin Cell Biol. 2011;23:109–13.
110. Duesberg PH. Oncogenes and cancer. Science. 1995;267:1407–8.
111. Duesberg P, Rasnick D. Aneuploidy, the somatic mutation that makes cancer a species of its own. Cell Motil Cytoskeleton. 2000;47:81–107.
112. Smith AE, Kenyon DH. A unifying concept of carcinogenesis and its therapeutic implications. Oncology. 1973;27:459–79.
113. Samper E, Nicholls DG, Melov S. Mitochondrial oxidative stress causes chromosomal instability of mouse embryonic fibroblasts. Aging Cell. 2003;2:277–85.
114. Torres EM, Sokolsky T, Tucker CM, Chan LY, Boselli M, Dunham MJ, et al. Effects of aneuploidy on cellular physiology and cell division in haploid yeast. Science. 2007;317:916–24.
115. Torres EM, Williams BR, Amon A. Aneuploidy: cells losing their balance. Genetics. 2008;179: 737–46.
116. Torres EM, Williams BR, Tang YC, Amon A. Thoughts on aneuploidy. Cold Spring Harb Symp Quant Biol. 2011;75:445–51.
117. Thompson SL, Compton DA. Chromosomes and cancer cells. Chromosome Res. 2010;19: 433–44.
118. Knudson AG. Cancer genetics. Am J Med Genet. 2002;111:96–102.
119. Roskelley RC, Mayer N, Horwitt BN, Salter WT. Studies in cancer. Vii. Enzyme deficiency in human and experimental cancer. J Clin Invest. 1943;22:743–51.

120. Soto AM, Sonnenschein C. The somatic mutation theory of cancer: growing problems with the paradigm? Bioessays. 2004;26:1097–107.
121. Sonnenschein C, Soto AM. Theories of carcinogenesis: an emerging perspective. Semin Cancer Biol. 2008;18:372–7.
122. Tarin D. Cell and tissue interactions in carcinogenesis and metastasis and their clinical significance. Semin Cancer Biol. 2011;21:72–82.

Chapter 11

Mitochondria: The Ultimate Tumor Suppressor

MITOCHONDRIAL SUPPRESSION OF TUMORIGENICITY

According to Warburg's theory, respiratory insufficiency is the origin of cancer. All other characteristics of cancer arise either directly or indirectly from insufficient respiration. Up to this point, I have amassed substantial evidence from a variety of fields that strongly supports the theory. It is also clear that genomic instability and the vast numbers of gene and chromosome defects seen in tumor cells can arise as secondary consequences of protracted respiratory insufficiency. Genome instability is linked to mitochondrial dysfunction through the retrograde signaling system. If all cancer arises from mitochondrial dysfunction, then replacement of damaged mitochondria with normal mitochondria should prevent cancer. In other words, mitochondria producing sufficient respiration should suppress tumor growth regardless of the numbers and types of mutations or aneuploidy present.

Energy derived from substrate level phosphorylation (including the Warburg effect and amino acid fermentation) will persist in the presence of insufficient respiration. Up-regulation of oncogenes and down regulation of tumor suppressor genes is necessary to maintain fermentation when mitochondria fail to produce sufficient energy through respiration. While the mutator phenotype of cancer can be linked to impaired mitochondrial function as I described in the last chapter, substantial evidence also exists showing that normal mitochondrial function suppresses tumorigenesis. Further support for the Warburg theory would come from evidence showing that normal mitochondria can suppress malignant growth in tumor cells. If respiratory insufficiency is the origin of cancer, then tumor nuclei should not induce malignancy when placed in cytoplasm containing respiration competent normal mitochondria. Alternatively, if mitochondrial dysfunction

Cancer as a Metabolic Disease: On the Origin, Management and Prevention of Cancer, First Edition.
Thomas Seyfried.

is the origin of cancer, normal nuclei should be unable to prevent tumorigenesis when placed into the tumor cytoplasm. I refer to these types of experiments as *nuclear–cytoplasm transfer studies*. What is the evidence from these types of studies that support the metabolic origin of cancer?

NORMAL MITOCHONDRIA SUPPRESS TUMORIGENESIS IN CYBRIDS

It is well documented that tumorigenicity can be suppressed when cytoplasm from enucleated normal cells is fused with nucleated tumor cells to form cybrids. Cybrids contain a single nucleus and mixtures of cytoplasm from two different cells. To examine the effect of cytoplasm on the expression of tumorigenicity in cybrids, Koura formed fusions between intact B16 mouse melanoma cancer cells with cytoplasts (absent nucleus) from nontumorigenic rat myoblasts (1). The reconstituted clones and cybrids showed unique morphology and cellular arrangements. Tumorigenicity was suppressed in all the reconstituted clones and cybrids soon after their isolation, but tumorigenicity reappeared in some clones after prolonged cultivation of the cells. The adverse effects of the cell culture environment on mitochondrial respiration could account in part for the tumorigenic reversion of some clones (2). Koura's findings showed that cytoplasm containing normal mitochondria could suppress the malignant phenotype of tumor cells. Unfortunately, Koura did not link these observations to the Warburg theory of cancer.

In a more extensive series of experiments, Israel and Schaeffer showed that suppression of the malignant state could reach 100% in cybrids containing normal cytoplasm and tumorigenic nuclei (3). The unique aspect of their study was that all of the cells utilized, both normal and transformed, were derived from an original cloned progenitor (4). They also showed that nuclear/cytoplasmic hybrids derived by fusion of cytoplasts from malignant cells (nucleus absent) with karyoplasts from normal cells (nucleus present) produced tumors in 97% of the animals injected. *These findings showed that normal cell nuclei could not suppress tumorigenesis when placed in tumor cell cytoplasm*. In other words, normal nuclear gene expression was unable to suppress malignancy. These findings showed that it was the cytoplasm, rather than the nucleus, that dictated the malignant state of the cells. Although these investigators did not define the molecular basis for the cytoplasmic mediation of tumorigenesis, they suggested that epigenetic alterations of nuclear gene expression might be responsible. It is clear that the findings of Israel and Schaeffer strongly supported the concepts of Warburg's theory. However, these investigators also did not link their observations to Warburg's theory.

The findings and conclusions of Israel and Schaeffer that cytoplasmic factors suppress tumorigenicity were also strongly supported by the findings of Shay and Werbin (5, 6). These investigators also discussed the various factors that could influence the success or failure of cybrid experiments designed to uncover cytoplasmic suppressors of tumorigenicity. These factors included, (i) the relative amounts of tumorigenic and nontumorigenic cytoplasm in cybrids; (ii) the time interval that cybrids are passaged prior to testing their tumorigenicity; (iii) whether mutagenesis with carcinogens was used to introduce genetic markers on the cells; and (iv) the specific cell combinations used. They were not surprised that some investigators could obtain varying results if the various factors were not monitored carefully. Nevertheless, their results with mouse tumor cells were consistent with the conclusion of the Israel and Schaeffer experiments mentioned above. While Shay and Werbin discussed the role of the mitochondria in the suppressive effects of the cytoplasm on tumorigenesis, they did not discuss their results in light of Warburg's theory.

Howell and Sager (7), however, were cognizant of the relationship between Warburg's theory and the findings from the various cybrid studies. These investigators knew that analysis of cybrids could help distinguish whether it was the nucleus or the cytoplasm that determined tumorigenicity. Their results showed that cytoplasm from nontumorigenic normal cells suppressed the rate and extent of tumor formation in nude mice when fused with their nucleated tumorigenic counterparts. They concluded "*if tumor cell mitochondria are defective, as Warburg postulated, then suppression could result from the introduction of mitochondria from normal cells into cybrids*" (7). These findings like those of Koura, Israel and Schaeffer, and Shay and Werbin supported Warburg's theory. How was it possible that so many investigators in the cancer field failed to link their findings to those of Warburg's theory?

Jonasson and Harris conducted one of the more interesting studies in human mouse hybrids to evaluate the role of cytoplasm and the nucleus in the control of malignancy. They evaluated in vivo tumor malignancy in a range of hybrid clones derived from fusions of diploid human fibroblasts and lymphocytes with the cells of a malignant mouse melanoma (8). They showed that the human diploid cells were as effective as mouse diploid cells in suppressing the malignancy of the mouse melanoma cells even though human chromosomes were preferentially eliminated in the hybrid clones. Malignancy was also suppressed in a hybrid clone in which a single human X chromosome was present. They went on to show that this clone continued to produce few tumors even after back selection was used to eliminate this remaining X chromosome. It was clear that no human nuclear genetic material was responsible for suppression.

They also made hybrids between the melanoma cells and diploid human fibroblasts that were irradiated before cell fusion. Interestingly, the incidence of tumor take was substantially higher in crosses between the mouse melanoma cells and the irradiated human fibroblasts than in crosses between the melanoma cells and the unirradiated human fibroblasts (8). They concluded that the suppression of malignancy involves the activity of a radiosensitive extrachromosomal element.

The findings from the Jonasson and Harris study were remarkable for several reasons. First, their findings were consistent with those of many other cybrid studies indicating that something in normal cytoplasm suppresses tumorigenicity in malignant cells. Second, no human chromosome or nuclear genetic material was responsible for the suppressive effect. Finally, radiation could destroy the cytoplasmic factor responsible for tumor suppression. This last fact is consistent with

Warburg findings that radiation destroys mitochondrial respiration. Surprisingly, Jonasson and Harris (8) excluded a mitochondrial origin in preference to a centrosome origin for the suppression effect. This decision was based on the findings of others who showed that no human mitochondrial DNA or proteins were found in human mouse cybrids. However, new studies in transmissible cancers show that tumor mitochondria can integrate with normal mitochondria in some tumors (9). I suggest that this integration could reduce or correct in part the respiratory damage in the tumor cell mitochondria thus suppressing the malignant phenotype. This possibility is also supported further from the work of King and Attardi (10, 11), showing that exogenous mtDNA enhances respiration in cells lacking functional mtDNA. Such a possibility would be consistent with Warburg's original theory.

Paul Saxon and colleagues showed that the microcell transfer of chromosome 11 could suppress tumorigenicity in HeLa cells (12). They conclude that chromosome 11 contained a tumor suppressor gene. These findings are interesting and also suggest an interaction between chromosome 11 and the mitochondria. It is possible that a gene on chromosome 11 facilitates mitochondrial respiration thus suppressing tumorigenicity in the HeLa cells. It is also interesting that neuroblastoma and Wilms tumor are associated with defects on chromosome 11. Further studies are needed to determine if tumorigenic suppression involves specific interactions between chromosome 11-encoded genes and mitochondrial respiration efficiency.

EVIDENCE FROM rho^0 CELLS

Singh and coworkers also provided evidence for the role of mitochondria in the suppression of tumorigenicity by showing that exogenous transfer of wild-type mitochondria to cells with depleted mitochondria DNA (rho^0 cells) could reverse the altered expression of the APE1 DNA repair protein and the tumorigenic phenotype (13). The efficiency of APE1-mediated DNA repair is dependent on the sufficiency of mitochondrial respiration. The rho^0 cells have deficient respiration because they lack mtDNA, which is necessary for normal respiration. Consequently, transfer of normal mtDNA to rho^0 cells will restore respiration, turn off the RTG response, and prevent genomic instability. Again, it is the integrity of mitochondrial respiration that prevents cancer. Cancer arises from respiratory insufficiency just as Warburg predicted.

Further support for the importance of respiration in the origin of cancer comes from the findings of Petros, Wallace, and colleagues with prostate cancer. To determine whether mutant pancreatic tumors had increased ROS and tumor growth rates, these investigators introduced the T8993G pathogenic mtDNA mutation into PC3 prostate cancer cells through cybrid transfer. They then tested the cells for tumor growth in nude mice. The resulting T8993G mutant cybrids generated tumors that were seven times larger than those seen in the wild-type cybrids. Moreover, the wild-type cybrids barely grew in the mice. The tumors derived from T8993G mutant cybrids also generated significantly more ROS than tumors without this mutation. ROS generated in mitochondria damages respiration thus producing genomic instability (14). Additional experiments showed that introduction of mtDNA mutations could reverse the antitumorigenic effect of normal mitochondria in cybrids (15). *The authors concluded that mtDNA mutations play an important role in the etiology of prostate cancer and that cancer can be best defined as a type of mitochondrial disease*. These findings provide direct support for Warburg's theory.

NORMAL MITOCHONDRIA SUPPRESS TUMORIGENESIS IN VIVO

It is also well documented that nuclei from cancer cells can be reprogrammed to form normal tissues when transplanted into normal cytoplasm despite the continued presence of the tumor-associated genomic defects in the cells of the derived tissues. Dramatic evidence for this fact was obtained from studies in neoplastic tissue from frogs and mice. McKinnell et al. (16) provided some of the first evidence showing that tumor cell nuclei could direct normal vertebrate development following transplantation of the tumor cell nucleus into an enucleated normal egg cell. Triploid nuclei isolated from Lucke frog renal cell tumors were surgically implanted into fertilized enucleated eggs from normal diploid frogs (Fig. 11.1 from their study shows a large rapidly growing tumor on the left kidney of a Lucke frog). All cells of the tumor were triploid in containing three copies of all chromosomes. Triploid tadpoles developed from the triploid tumor cell nuclei. Remarkably, the living triploid tadpoles revealed functional tissues of many types.

The transplantation of nuclei from triploid tumor cells into an enucleated eggs makes it possible to distinguish development initiated by the transplanted nucleus from that guided by an inadvertently retained maternal diploid nucleus (16). The investigators noted that ciliated epithelium propelled the tadpoles in the culture dishes. The tadpoles swam when stimulated. The tadpoles had functional receptors, nerve tissue, and striated muscle necessary for swimming. Cardiac muscle pumped blood cells through the gills. Suckers secreted abundant mucus. Clearly seen were a pronephric ridge, eye anlage, nasal pit, and open mouth, as was the differentiation of the head, body, and the tail. The tail fin regenerated after being clipped for chromosome study. Moreover, sections of embryos developed from transplanted triploid tumor nuclei revealed apparent normal development of brain, spinal cord, optic cup with lens, auditory vesicle, somites, pronephric tubules, pharynx, midgut, and notochord. These findings indicate that nuclei derived from tumor cells can direct normal developmental processes.

It is difficult to reconcile these observations based on the somatic mutation theory of cancer, but the findings are consistent with the tenets of the Warburg theory that cancer is a disease of respiration. Normal mitochondria derived from the fertilized egg suppress tumorigenesis because their OxPhos is sufficient for maintaining energy homeostasis. Later studies suggested that the loss of tumorigenicity was associated with the loss of Lucke tumor herpes virus. This virus was considered the etiological agent for the origin of this tumor (18). As I discussed in Chapter 9, however, herpes virus can interfere with mitochondrial function to induce tumorigenesis (19). Indeed, herpes viruses have an intimate attachment with

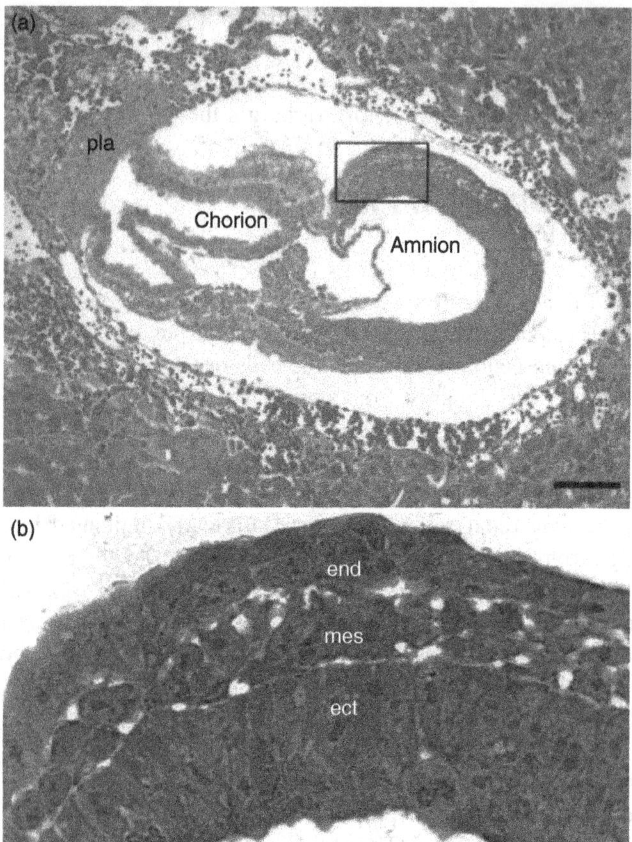

Figure 11.1 Brain tumor nucleus can support normal mouse embryonic development. (a) An E-7.5 mouse embryo derived from a transplanted medulloblastoma nucleus stained with H&E. (b) A higher magnification of the boxed area in (a) to show the three distinguishable germ layers: pla, ecto-placental cone; end, embryonic endoderm; mes, embryonic mesoderm; ect, embryonic ectoderm. Normal mitochondria will be present in the cytoplasm. Scale bar, 20 μm. The results show that a nucleus from a brain tumor implanted into normal cytoplasm can direct normal embryonic development. *Source*: Reprinted with permission from Reference 17. To see this figure in color please go to ftp://ftp.wiley.com/public/sci_tech_med/cancer_metabolic_disease.

mitochondria that causes dysfunctional respiration (20). Hence, the suppression of tumorigenesis in the Lucke frog tumor is likely due to the replacement of virus-damaged mitochondria with normal mitochondria. These results are similar to those described above with cell cybrids.

NORMAL MOUSE CYTOPLASM SUPPRESSES TUMORIGENIC PHENOTYPES

Results similar to those with the Lucke frog tumor were also obtained following nuclear transfer in mouse tumors. Morgan and colleagues also showed that nuclei from mouse medulloblastoma (a brain tumor thought to arise from cerebellar granule cells) could direct normal development when transplanted into enucleated somatic cells (17). Figure 11.1 shows that normal embryonic tissue and germ cell layers were formed from medulloblastoma nuclei. These investigators concluded that somatic nuclear transfer into normal cytoplasm suppressed the tumorigenic phenotype (17). Moreover transplanted medulloblastoma nuclei gave rise to postimplantation embryos that underwent tissue differentiation and early stages of organogenesis. Remarkably, no malignancies were observed in any of the recipient mice, and normal proliferation control was observed in cultured blastocysts (17).

These investigators went on to suggest that the tumorigenic mutations causing medulloblastoma must act within the context of the cerebellar granule cell lineage, and these changes did not support malignant cell proliferation. Although an epigenetic reprogramming of medulloblastoma nuclei was offered as an explanation for their findings, it is more likely that their findings resulted from the replacement of abnormal mitochondria with normal mitochondria similar to that seen with the Lucke frog experiments. The findings also supported the earlier work of Mintz and Illmensee (21) showing that normal mice could be produced from tumor cells and that *structural mutations in the nuclear genome could not be responsible for tumor formation*. Considered together, these findings indicate that nuclear gene mutations alone cannot account for the origin of cancer and further highlight the dynamic role of mitochondria in the epigenetic origin of carcinogenesis.

The findings from the Lucke frog and mouse medulloblastoma experiments were further supported from the work of Konrad Hochedlinger, Rudy Jaenisch and colleagues at MIT (22). These investigators showed that the nuclei of many cancer cells including pancreatic cancer and melanoma were able to support preimplantation mouse development into normal-appearing blastocysts without signs of abnormal proliferation (Fig. 11.2). They also showed that normal blastocysts could be formed from p53 –/– breast cancer cells, and that normal blastocysts and embryonic cell lines could be formed from melanoma nuclei. These investigators concluded that the oocyte environment suppressed the malignant phenotype of the various tumor types and that tumor nuclei could direct normal development in early mouse embryos. The oocyte cytoplasm would, of course, contain normal mitochondria. It is the respiratory competent normal mitochondria that would suppress tumorigenicity. The tumor nuclei direct normal development as long as normal mitochondria are present in the cytoplasm.

These studies also showed that embryonic stem (ES) cells derived from one of the cloned melanoma cells were able to differentiate into most if not all somatic cell lineages in chimeras including fibroblasts, lymphocytes, and melanocytes (22). *Remarkably, normal development occurred despite the persistence of severe chromosomal changes and mutations documented by array-comparative genome hybridization (CHG)*. The investigators went on to conclude that secondary chromosomal changes associated with malignancy do not necessarily interfere with preimplantation development; ES cell derivation, and a broad nuclear differentiation potential (22). Nevertheless, these observations indicate that nuclear gene mutations cannot

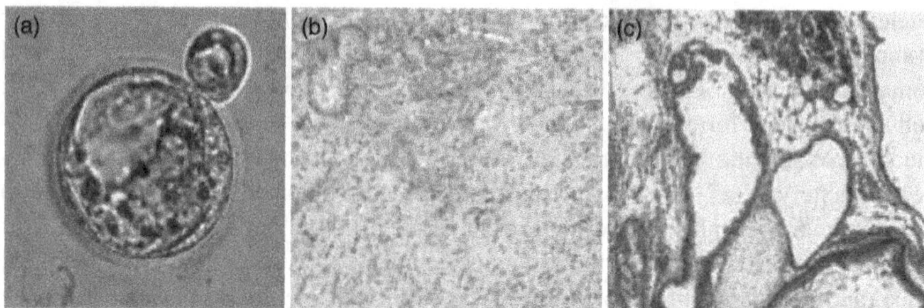

Figure 11.2 Cancer nuclei can support mouse development. (a) Analysis of the developmental potential of embryonic stem cell with melanoma nucleus. A hatching blastocyst derived from a breast cancer cell by nuclear transfer shows a blastocoel cavity, trophectoderm layer, and an inner cell mass. (b and c) H&E staining of teratoma sections produced from R545-1 ES cells shows differentiation into mature neurons, mesenchymal cells, and squamous epithelium (b), and columnar epithelium, chondrocytes, and adipocytes (c). The results show that nuclei from various tumor cells can direct normal mouse development when placed into normal cytoplasm containing normal mitochondria. *Source*: Reprinted with permission from Reference 22. To see this figure in color please go to ftp://ftp.wiley.com/public/sci_tech_med/cancer_metabolic_disease.

account for the origin of cancer and further highlight the dynamic role of mitochondria in the epigenetic regulation of carcinogenesis. Unfortunately, Hochedlinger et al. did not link their findings to the Warburg theory despite showing strong evidence supporting this theory.

ENHANCED DIFFERENTIATION AND SUPPRESSED TUMORIGENICITY IN THE LIVER MICROENVIRONMENT

Grisham and colleagues reported that two aneuploid liver tumor cell lines, which formed aggressive tumors when grown subcutaneously, did not form tumors when grown in the liver (23). The tumors became morphologically differentiated when they were transplanted and grown in liver. The authors concluded that close cell contact or factors in the liver microenvironment suppressed tumorigenicity. It is well documented, however, that cell–cell fusion is a common physiological process in murine liver (24). I suggest that fusion between normal liver cells and neoplastic liver cells in the unique liver microenvironment would suppress tumorigenicity in a manner similar to that seen in the cybrid experiments mentioned above. It is likely the normal mitochondria in the fused cell hybrids are responsible for enhanced differentiation and suppressed tumorigenicity.

The suppressive effect of normal mitochondria on tumorigenesis links mitochondrial respiration to the long-standing controversy on cellular differentiation and tumorigenicity (25–27). *Respiration is required for the emergence and maintenance of differentiation, while loss of respiration leads to glycolysis, dedifferentiation, and unbridled proliferation* (28, 29). These observations are consistent with the general hypothesis presented here, that prolonged impairment of mitochondrial energy metabolism underlies carcinogenesis. This would represent an epigenetic origin of the disease in the classic sense (30, 31). Replacement of damaged mitochondria with normal mitochondria, which will produce sufficient energy through respiration, restores the differentiated state. Hello! Is anyone listening?

SUMMARY OF NUCLEAR-CYTOPLASMIC TRANSFER EXPERIMENTS

Viewed collectively, these findings provide compelling evidence showing that normal mitochondria can suppress tumorigenicity. The evidence reviewed supports the Warburg theory of cancer as a disease of respiratory insufficiency. Normal mitochondria will reverse the Warburg effect because this effect is due to insufficient respiration. Statements about a "reverse Warburg effect," which do not involve restored respiration, are hard to reconcile in light of the information presented here (32). Normal mitochondria suppress respiratory dysfunction and tumorigenicity, whereas abnormal mitochondria cannot suppress respiratory dysfunction or tumorigenicity.

According to Warburg's theory, it would be expected that the presence of normal mitochondria in tumor cells would restore the cellular redox status, turn off the mitochondrial stress response, and reduce or eliminate the need for fermentation to maintain viability. In rephrasing, *normal mitochondrial function maintains*

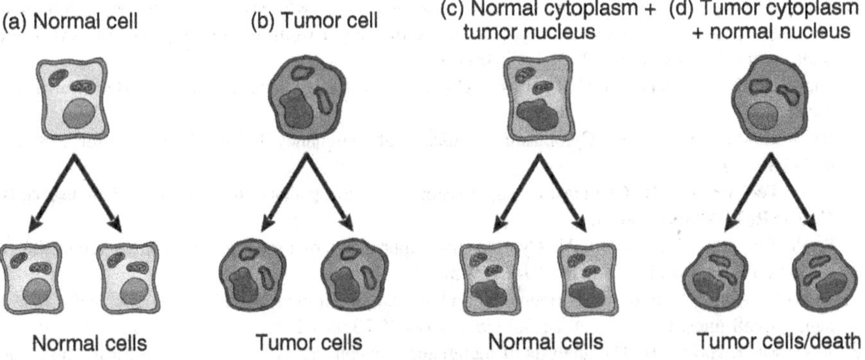

Figure 11.3 Summary of nuclear/cytoplasmic transfer experiments and the origin of tumors. This image summarizes the experimental evidence presented in this chapter. Normal cells are depicted in lighter shade with mitochondrial and nuclear morphology indicative of normal respiration and nuclear gene expression, respectively. Tumor cells are depicted in darker shade with abnormal mitochondrial and nuclear morphology indicative of abnormal respiration and genomic instability. (a) Normal cells beget normal cells. (b) Tumor cells beget tumor cells. (c) Delivery of a tumor cell nucleus into a normal cell cytoplasm begets normal cells despite the persistence of tumor-associated genomic abnormalities. (d) Delivery of a normal cell nucleus into a tumor cell cytoplasm begets tumor cells or dead cells, but not normal cells. The results show that nuclear genomic defects alone cannot cause tumors and that normal mitochondria can suppress tumorigenesis. *Source*: Original diagram from Jeffrey Ling and Thomas N. Seyfried, with permission. See color insert.

the differentiated state thereby suppressing carcinogenesis, whereas dysfunctional mitochondria can enhance dedifferentiation thereby facilitating carcinogenesis. The biochemical studies of Cuezva and Ristow also indicate that normal mitochondrial respiration suppresses tumorigenesis (33–35). The evidence for the mitochondrial origin of cancer is supported from a broad range of experimental data. Figure 11.3 summarizes the phenomenon.

In summary, the origin of carcinogenesis resides with the *mitochondria* in the cytoplasm, not with the *genome* in the nucleus. How is it possible that so many in the cancer field seem unaware of the evidence supporting this concept? How is it possible that so many in the cancer field have ignored these findings while embracing the flawed gene theory? Perhaps Payton Rous was correct when he mentioned "*the somatic mutation theory acts like a tranquilizer on those who believe in it*" (36). I attribute the absence of any real progress in the war on cancer over the last 40 years to the flawed concepts of the somatic mutation theory, and to the failure in recognizing mitochondrial dysfunction as a credible scientific explanation for the origin of the disease. This failure is an inexcusable tragedy ultimately responsible for the deaths of millions of cancer patients.

REFERENCES

1. Koura M, Isaka H, Yoshida MC, Tosu M, Sekiguchi T. Suppression of tumorigenicity in interspecific reconstituted cells and cybrids. Gann. 1982;73:574–80.
2. Kiebish MA, Han X, Cheng H, Seyfried TN. In vitro growth environment produces lipidomic and electron transport chain abnormalities in mitochondria from non-tumorigenic astrocytes and brain tumours. ASN Neuro. 2009;1:pii:e00011.
3. Israel BA, Schaeffer WI. Cytoplasmic suppression of malignancy. In Vitro Cell Dev Biol. 1987; 23:627–32.
4. Israel BA, Schaeffer WI. Cytoplasmic mediation of malignancy. In Vitro Cell Dev Biol. 1988;24: 487–90.
5. Shay JW, Werbin H. Cytoplasmic suppression of tumorigenicity in reconstructed mouse cells. Cancer Res. 1988;48:830–3.
6. Shay JW, Liu YN, Werbin H. Cytoplasmic suppression of tumor progression in reconstituted cells. Somat Cell Mol Genet. 1988;14:345–50.
7. Howell AN, Sager R. Tumorigenicity and its suppression in cybrids of mouse and Chinese hamster cell lines. Proc Natl Acad Sci USA. 1978;75:2358–62.
8. Jonasson J, Harris H. The analysis of malignancy by cell fusion. VIII. Evidence for the intervention of an extra-chromosomal element. J Cell Sci. 1977;24:255–63.
9. Rebbeck CA, Leroi AM, Burt A. Mitochondrial capture by a transmissible cancer. Science. 2011;331:303.
10. King MP, Attardi G. Injection of mitochondria into human cells leads to a rapid replacement of the endogenous mitochondrial DNA. Cell. 1988;52:811–9.
11. King MP, Attardi G. Human cells lacking mtDNA: repopulation with exogenous mitochondria by complementation. Science. 1989;246:500–3.
12. Saxon PJ, Srivatsan ES, Stanbridge EJ. Introduction of human chromosome 11 via microcell transfer controls tumorigenic expression of HeLa cells. EMBO J. 1986;5:3461–6.
13. Singh KK, Kulawiec M, Still I, Desouki MM, Geradts J, Matsui S. Inter-genomic cross talk between mitochondria and the nucleus plays an important role in tumorigenesis. Gene. 2005;354:140–6.

14. Seoane M, Mosquera-Miguel A, Gonzalez T, Fraga M, Salas A, Costoya JA. The mitochondrial genome is a "Genetic Sanctuary" during the oncogenic process. PloS One. 2011;6:e23327.
15. Petros JA, Baumann AK, Ruiz-Pesini E, Amin MB, Sun CQ, Hall J, et al. mtDNA mutations increase tumorigenicity in prostate cancer. Proc Natl Acad Sci USA. 2005;102:719–24.
16. McKinnell RG, Deggins BA, Labat DD. Transplantation of pluripotential nuclei from triploid frog tumors. Science. 1969;165:394–6.
17. Li L, Connelly MC, Wetmore C, Curran T, Morgan JI. Mouse embryos cloned from brain tumors. Cancer Res. 2003;63:2733–6.
18. Carlson DL, Sauerbier W, Rollins-Smith LA, McKinnell RG. Fate of herpesvirus DNA in embryos and tadpoles cloned from Lucke renal carcinoma nuclei. J Comp Pathol. 1994;111: 197–204.
19. D'Agostino DM, Bernardi P, Chieco-Bianchi L, Ciminale V. Mitochondria as functional targets of proteins coded by human tumor viruses. Adv Cancer Res. 2005;94:87–142.
20. Ackermann WW, Kurtz H. The relation of herpes virus to host cell mitochondria. J Exp Med. 1952;96:151–7.
21. Mintz B, Illmensee K. Normal genetically mosaic mice produced from malignant teratocarcinoma cells. Proc Natl Acad Sci USA. 1975;72:3585–9.
22. Hochedlinger K, Blelloch R, Brennan C, Yamada Y, Kim M, Chin L, et al. Reprogramming of a melanoma genome by nuclear transplantation. Genes Dev. 2004;18:1875–85.
23. Coleman WB, Wennerberg AE, Smith GJ, Grisham JW. Regulation of the differentiation of diploid and some aneuploid rat liver epithelial (stemlike) cells by the hepatic microenvironment. Am J Pathol. 1993;142:1373–82.
24. Faggioli F, Sacco MG, Susani L, Montagna C, Vezzoni P. Cell fusion is a physiological process in mouse liver. Hepatology. 2008;48:1655–64.
25. Seyfried TN, Shelton LM. Cancer as a metabolic disease. Nutr Metab. 2010;7:7.
26. Harris H. The analysis of malignancy by cell fusion: the position in 1988. Cancer Res. 1988;48: 3302–6.
27. Soto AM, Sonnenschein C. The somatic mutation theory of cancer: growing problems with the paradigm. Bioessays. 2004;26:1097–107.
28. Warburg O. Revidsed Lindau Lectures: The prime cause of cancer and prevention - Parts 1 & 2. In: Burk D, editor. Meeting of the Nobel-Laureates Lindau, Lake Constance, Germany: K.Triltsch; 1969. http://www.hopeforcancer.com/OxyPlus.htm.
29. Szent-Gyorgyi A. The living state and cancer. Proc Natl Acad Sci USA. 1977;74:2844–7.
30. Nanney DL. Epigenetic Control Systems. Proc Natl Acad Sci USA. 1958;44:712–7.
31. Holliday R. Epigenetics: a historical overview. Epigenetics. 2006;1:76–80.
32. Pavlides S, Whitaker-Menezes D, Castello-Cros R, Flomenberg N, Witkiewicz AK, Frank PG, et al. The reverse Warburg effect: aerobic glycolysis in cancer associated fibroblasts and the tumor stroma. Cell Cycle. 2009;8:3984–4001.
33. Ristow M, Cuezva JM. Oxidative Phosphorylation and Cancer: The Ongoing Warburg Hypothesis. In: Apte SP, Sarangarajan R, editors. Cellular Respiration and Carcinogenesis. New York: Humana Press; 2009. p.1–18.
34. Ristow M. Oxidative metabolism in cancer growth. Curr Opin Clin Nutr Metab Care. 2006;9: 339–45.
35. Cuezva JM, Ortega AD, Willers I, Sanchez-Cenizo L, Aldea M, Sanchez-Arago M. The tumor suppressor function of mitochondria: translation into the clinics. Biochim Biophys Acta. 2009; 1792:1145–58.
36. Rous P. Surmise and fact on the nature of cancer. Nature. 1959;183:1357–61.

Chapter 12

Abnormalities in Growth Control, Telomerase Activity, Apoptosis, and Angiogenesis Linked to Mitochondrial Dysfunction

Hanahan and Weinberg (1, 2) argued that genomic instability was an essential enabling characteristic for manifesting the hallmarks of cancer. However, I suggest that OxPhos insufficiency is the essential enabling characteristic in the origin of cancer. Our recent hypothesis defined how the acquired capabilities of cancer could be linked specifically to impaired energy metabolism (3). Kroemer and Pouyssegur (4) also provided a nice overview on how the Hanahan and Weinberg hallmarks of cancer could be linked to signaling cascades and to the metabolic reprogramming of cancer cells. Although similar topics are considered in the Kroemer/Pouyssegur review and in my treatise, my view differs from theirs on the role of respiratory damage in the origin of cancer. Kroemer and Pouyssegur mentioned that Warburg's respiratory damage hypothesis is not universally applicable to all cancers and cited the study of Funes et al. (5) to support their contention. I consider insufficient respiration as universal phenotype of all cancers. I described in Chapter 7 how none of the studies suggesting normal respiration in cancer cells have excluded amino acid fermentation as an alternative explanation for their findings. Funes and coworkers also did not consider the possibility of mitochondrial amino acid fermentation as a source of tumor cell ATP. I also disagree with the view of

Cancer as a Metabolic Disease: On the Origin, Management and Prevention of Cancer, First Edition.
Thomas Seyfried.

Kroemer and Pouyssegur that tumor cells have a growth advantage over normal cells. I will address the growth advantage topic more in Chapter 17.

GROWTH SIGNALING ABNORMALITIES AND LIMITLESS REPLICATIVE POTENTIAL

A central concept in linking abnormalities of growth signaling and replicative potential to impaired energy metabolism is in recognizing that proliferation, rather than quiescence, is the default state of both microorganisms and metazoans (3, 6–9). The default state of the cell is the condition under which cells are found when they are freed from any active control. Respiring cells in mature organ systems are largely quiescent because their replicative potential is under negative control through the action of normal mitochondrial function. Respiration maintains differentiation and quiescence. In addition, tumor suppressor genes such as *p*53 and the retinoblastoma protein, pRB, can also help to maintain quiescence (6, 10). As *p*53 function is linked to cellular respiration, prolonged respiratory insufficiency will gradually reduce *p*53 function, thus inactivating the negative control of *p*53 and other tumor suppressor genes on cell proliferation. In contrast to the view of Hanahan and Weinberg that quiescence is the default state of cells, I believe that proliferation is the default state of cells. Sonnenschein and Soto (7) provide an excellent review describing how proliferation is the default state of metazoan cells.

A persistent impairment in respiratory function will trigger the RTG response, which is necessary for upregulating fermentation pathways needed to maintain the $\Delta G'_{ATP}$ for viability. The RTG response will activate numerous oncogenes such as *MYC*, *Ras*, *HIF-1α*, *Akt*, and *m-Tor*, which are required to sustain fermentation as the prime energy source in the presence of insufficient respiration (11–15). Glucose and glutamine become the main energy substrates for driving fermentation through glycolysis and TCA cycle substrate-level phosphorylation. Salvadore Moncada and colleagues showed how glucose and glutamine are linked to the cell cycle and proliferation (16). This is important, as glucose and glutamine are the main energy metabolites needed for fermentation and tumor growth.

In addition to facilitating the metabolism of glucose and glutamine through fermentation and substrate-level phosphorylation, MYC and Ras also stimulate cell proliferation (17–19). Part of this mechanism also includes inactivation of pRB, the function of which is dependent on mitochondrial activities and the cellular redox state (10). Disruption of the pRB signaling pathway will contribute to cell proliferation and neoplasia (2). Cell proliferation is linked to fermentation, whereas quiescence is linked to respiration (9, 20). Unlike normal cells, which engage respiration following proliferation, tumor cells remain dependent on glucose and glutamine for fermentation because their OxPhos is insufficient to maintain homeostasis. Hence, the growth signaling abnormalities and limitless replicative potential of tumor cells can be linked directly to the requirements of fermentation energy, which originates ultimately from impaired or insufficient respiration.

It is also interesting that the RTG response underlies replicative life span extension in budding yeast. Yeast longevity is linked to the number of buds that a mother cell produces before it dies (14). The greater the loss of mitochondrial function, the greater the induction of the RTG response and the greater the longevity (bud production) in yeast (21). As mitochondrial energy efficiency declines with age, fermentation energy becomes necessary to compensate for insufficient respiration. A reliance on fermentation becomes essential if a cell is to remain alive. A greater dependency on substrate-level phosphorylation will induce oncogene expression and unbridled proliferation, which could underlie in part the enhanced longevity in yeast (14, 22, 23). When this process occurs in mammalian cells, however, the phenomenon is referred to as *neoplasia* or *new growth*. We proposed that replicative life span extension in yeast and the limitless replicative potential in tumor cells can be linked through common bioenergetic mechanisms involving impaired mitochondrial function (3). It is insufficient respiration that underlies yeast longevity as well as the origin of cancer.

LINKING TELOMERASE ACTIVITY TO CELLULAR ENERGY AND CANCER

Telomerase is a ribonucleoprotein enzyme complex associated with cellular immortality through telomere maintenance. Telomerase is activated in about 90% of human cancers, suggesting a role in tumorigenesis (24–26). Emerging evidence suggests that mitochondrial dysfunction could underlie the relocation of telomerase from the mitochondria, where it seems to have a protective role, to the nucleus, where it maintains telomere integrity necessary for limitless replicative potential (27–29). Interestingly, telomerase activity is high during early embryonic development when cell proliferation is high, but telomerase activity is low in adult tissues where most cells are differentiated and quiescent (30, 31).

These findings suggest a linkage of telomerase activity with energy metabolism. Telomerase activity is high in normal cells or tumor cells that primarily use fermentation for energy, but the activity is low or nonexistent in nontumorigenic differentiated cells that primarily use OxPhos for energy. These findings suggest that the energy state of the cells dictates the activity level of telomerase. Elevated telomerase activity in tumor cells would therefore be an effect rather than a cause of cancer. Further studies will be necessary to determine how changes in telomerase expression and subcellular localization could be related to mitochondrial dysfunction, elevated fermentation, and the limitless replication of tumor cells.

EVASION OF PROGRAMMED CELL DEATH (APOPTOSIS)

Apoptosis is a coordinated process that initiates cell death following a variety of cellular insults. Damage to mitochondrial energy production is one type of insult that can trigger the apoptotic cascade, which ultimately involves release of mitochondrial cytochrome *c*, activation of intracellular caspases, and death (2, 3). In contrast to normal cells, acquired resistance to apoptosis is a hallmark of most types of cancer cells (2). The evasion of apoptosis is a predictable physiological response for tumor cells that primarily use fermentation and substrate-level phosphorylation for energy production following respiratory damage during the protracted process of carcinogenesis (32). Only those cells capable of making the gradual energy transition from respiration to fermentation in response to respiratory insufficiency will be able to evade apoptosis. Cells unable to make this energy transition will die and thus never become tumor cells. The ability of a cell to replace respiration with fermentation is a central tenet of the Warburg theory (20, 33–35).

Numerous findings indicate that the genes and signaling pathways needed to upregulate and sustain fermentation are themselves antiapoptotic. For example, sustained glycolysis or glutamine fermentation requires participation of mTOR, MYC, Ras, HIF-1a, and the IGF-1/PI3K/Akt signaling pathways (13–15, 22, 32, 36, 37). The upregulation of these genes and pathways together with inactivation of tumor suppressor genes such as *p*53, which is required to initiate apoptosis, will disengage the apoptotic signaling cascade thus preventing programmed cell death (3, 38).

Abnormalities in the outer mitochondrial membrane and inner membrane potential ($\Delta\Psi_m$) can also induce expression of known antiapoptotic genes (*Bcl2* and *Ccl-X$_L$*) (32, 39). Tumor cells will continue to evade apoptosis as long as they have access to glucose and glutamine, which are required to maintain fermentation energy. Glycolytic tumor cells, however, can readily express a robust apoptotic phenotype if their glucose supply is targeted. We clearly showed that dietary or calorie restriction could significantly increase the number of apoptotic cells in experimental brain tumors (15, 40, 41). This is also thought to contribute in part to the therapeutic action of dietary energy reduction in managing human glioblastoma (42–44). Hence, the evasion of apoptosis in tumor cells is linked directly to a dependency on fermentation and substrate-level phosphorylation for energy, which is itself a consequence of impaired respiratory function.

SUSTAINED VASCULARITY (ANGIOGENESIS)

Angiogenesis involves neovascularization or the formation of new capillaries from existing blood vessels and is associated with the processes of tissue inflammation, wound healing, and tumorigenesis (45–48). Figure 1.3 highlights the role of angiogenesis in tumor progression. Angiogenesis is required for most tumors to grow beyond an approximate size of 0.2–2.0 mm (49). This is necessary to provide the tumor with essential energy nutrients, including glucose and glutamine, and to remove toxic tumor waste products such as lactic acid and ammonia (3, 50).

In addition to its role in upregulating glycolysis in response to hypoxia, HIF-1α is also the main transcription factor for vascular endothelial growth factor (VEGF), which stimulates angiogenesis (15, 51–53) (see also Figs. 17.11 and 17.16). HIF-1α is part of the IGF-1/PI3K/Akt signaling pathway that also indirectly influences the expression of βfibroblast growth factor (FGF), another key angiogenesis growth factor (15, 54). Many of the genes and metabolites that drive angiogenesis arise as secondary consequences of tumor cell fermentation. Hence, the sustained vascularity of tumors can be linked mechanistically to the metabolic requirements of fermentation and substrate-level phosphorylation necessary for tumor cell survival.

When viewed collectively, the information presented in this chapter provides compelling evidence linking several of the Hanahan and Weinberg hallmarks of cancer directly to insufficient respiration in tumor cells. It is also interesting that the cancer hallmarks discussed in this chapter are not unique to malignant cancers but are also present in benign tumors (55). Indeed, abnormalities in growth control, telomerase activity, apoptosis, and angiogenesis are present in many tumors that do not invade or metastasize. Nevertheless, I think it is easy to see how respiratory insufficiency can account for these characteristics in tumors.

REFERENCES

1. Hanahan D, Weinberg RA. Hallmarks of cancer: the next generation. Cell. 2011;144:646–74.
2. Hanahan D, Weinberg RA. The hallmarks of cancer. Cell. 2000;100:57–70.
3. Seyfried TN, Shelton LM. Cancer as a metabolic disease. Nutr Metab. 2010;7:7.
4. Kroemer G, Pouyssegur J. Tumor cell metabolism: cancer's Achilles' heel. Cancer Cell. 2008;13:472–82.
5. Funes JM, Quintero M, Henderson S, Martinez D, Qureshi U, Westwood C, et al. Transformation of human mesenchymal stem cells increases their dependency on oxidative phosphorylation for energy production. Proc Natl Acad Sci USA. 2007;104:6223–8.
6. Tzachanis D, Boussiotis VA. Tob, a member of the APRO family, regulates immunological quiescence and tumor suppression. Cell Cycle. 2009;8:1019–25.
7. Sonnenschein C, Soto AM. The Society of Cells: Cancer and the Control of Cell Proliferation. New York: Springer-Verlag; 1999.
8. Soto AM, Sonnenschein C. The somatic mutation theory of cancer: growing problems with the paradigm? Bioessays. 2004;26:1097–107.
9. Szent-Gyorgyi A. The living state and cancer. Proc Natl Acad Sci USA. 1977;74:2844–7.
10. Burhans WC, Heintz NH. The cell cycle is a redox cycle: linking phase-specific targets to cell fate. Free Radic Biol Med. 2009;47:1282–93.
11. Ramanathan A, Wang C, Schreiber SL. Perturbational profiling of a cell-line model of tumorigenesis by using metabolic measurements. Proc Natl Acad Sci USA. 2005;102:5992–7.
12. Godinot C, de Laplanche E, Hervouet E, Simonnet H. Actuality of Warburg's views in our understanding of renal cancer metabolism. J Bioenerg Biomembr. 2007;39:235–41.
13. Singh KK, Kulawiec M, Still I, Desouki MM, Geradts J, Matsui S. Inter-genomic cross talk between mitochondria and the nucleus plays an important role in tumorigenesis. Gene. 2005;354:140–6.
14. Butow RA, Avadhani NG. Mitochondrial signaling: the retrograde response. Mol Cell. 2004;14:1–15.
15. Marsh J, Mukherjee P, Seyfried TN. Akt-dependent proapoptotic effects of dietary restriction on late-stage management of a phosphatase and tensin homologue/tuberous sclerosis complex 2-deficient mouse astrocytoma. Clin Cancer Res. 2008;14:7751–62.
16. Colombo SL, Palacios-Callender M, Frakich N, De Leon J, Schmitt CA, Boorn L, et al. Anaphase-promoting complex/cyclosome-Cdh1 coordinates glycolysis and glutaminolysis with transition to S phase in human T lymphocytes. Proc Natl Acad Sci USA. 2010;107:18868–73.
17. Dang CV, Le A, Gao P. MYC-Induced cancer cell energy metabolism and therapeutic opportunities. Clin Cancer Res. 2009;15:6479–83.
18. Gao P, Tchernyshyov I, Chang TC, Lee YS, Kita K, Ochi T, et al. c-Myc suppression of miR-23a/b enhances mitochondrial glutaminase expression and glutamine metabolism. Nature. 2009;458:762–5.

19. Wise DR, DeBerardinis RJ, Mancuso A, Sayed N, Zhang XY, Pfeiffer HK, et al. Myc regulates a transcriptional program that stimulates mitochondrial glutaminolysis and leads to glutamine addiction. Proc Natl Acad Sci USA. 2008;105:18782–7.
20. Warburg O. On the origin of cancer cells. Science. 1956;123:309–14.
21. Jazwinski SM. The retrograde response links metabolism with stress responses, chromatin-dependent gene activation, and genome stability in yeast aging. Gene. 2005;354:22–7.
22. Miceli MV, Jazwinski SM. Common and cell type-specific responses of human cells to mitochondrial dysfunction. Exp Cell Res. 2005;302:270–80.
23. Borghouts C, Benguria A, Wawryn J, Jazwinski SM. Rtg2 protein links metabolism and genome stability in yeast longevity. Genetics. 2004;166:765–77.
24. Kovalenko OA, Caron MJ, Ulema P, Medrano C, Thomas AP, Kimura M, et al. A mutant telomerase defective in nuclear-cytoplasmic shuttling fails to immortalize cells and is associated with mitochondrial dysfunction. Aging Cell. 2010;9:203–19.
25. Kovalenko OA, Kaplunov J, Herbig U, Detoledo S, Azzam EI, Santos JH. Expression of (NES-)hTERT in cancer cells delays cell cycle progression and increases sensitivity to genotoxic stress. PloS One. 2010;5:e10812.
26. Bagheri S, Nosrati M, Li S, Fong S, Torabian S, Rangel J, et al. Genes and pathways downstream of telomerase in melanoma metastasis. Proc Natl Acad Sci USA. 2006;103:11306–11.
27. Saretzki G. Telomerase, mitochondria and oxidative stress. Exp Gerontol. 2009;44:485–92.
28. Santos JH, Meyer JN, Van Houten B. Mitochondrial localization of telomerase as a determinant for hydrogen peroxide-induced mitochondrial DNA damage and apoptosis. Hum Mol Genet. 2006;15:1757–68.
29. Ahmed S, Passos JF, Birket MJ, Beckmann T, Brings S, Peters H, et al. Telomerase does not counteract telomere shortening but protects mitochondrial function under oxidative stress. J Cell Sci. 2008;121:1046–53.
30. Fu W, Begley JG, Killen MW, Mattson MP. Anti-apoptotic role of telomerase in pheochromocytoma cells. J Biol Chem. 1999;274:7264–71.
31. Forsyth NR, Wright WE, Shay JW. Telomerase and differentiation in multicellular organisms: turn it off, turn it on, and turn it off again. Differentiation. 2002;69:188–97.
32. Kroemer G. Mitochondria in cancer. Oncogene. 2006;25:4630–2.
33. Warburg O. The Metabolism of Tumours. New York: Richard R. Smith; 1931.
34. Warburg O. On the respiratory impairment in cancer cells. Science. 1956;124:269–70.
35. Warburg O. Revidsed Lindau Lectures: The prime cause of cancer and prevention-Parts 1 & 2. In: Burk D, editor. Meeting of the Nobel-Laureates Lindau, Lake Constance, Germany: K.Triltsch; 1969. p. http://www.hopeforcancer.com/OxyPlus.htm.
36. Dang CV, Semenza GL. Oncogenic alterations of metabolism. Trends Biochem Sci. 1999;24:68–72.
37. Semenza GL. HIF-1 mediates the Warburg effect in clear cell renal carcinoma. J Bioenerg Biomembr. 2007;39:231–4.
38. Holley AK, St Clair DK. Watching the watcher: regulation of p53 by mitochondria. Future Oncol. 2009;5:117–30.
39. Amuthan G, Biswas G, Ananadatheerthavarada HK, Vijayasarathy C, Shephard HM, Avadhani NG. Mitochondrial stress-induced calcium signaling, phenotypic changes and invasive behavior in human lung carcinoma A549 cells. Oncogene. 2002;21:7839–49.
40. Mukherjee P, Abate LE, Seyfried TN. Antiangiogenic and proapoptotic effects of dietary restriction on experimental mouse and human brain tumors. Clin Cancer Res. 2004;10:5622–9.
41. Mukherjee P, El-Abbadi MM, Kasperzyk JL, Ranes MK, Seyfried TN. Dietary restriction reduces angiogenesis and growth in an orthotopic mouse brain tumour model. Br J Cancer. 2002;86:1615–21.
42. Zuccoli G, Marcello N, Pisanello A, Servadei F, Vaccaro S, Mukherjee P, et al. Metabolic management of glioblastoma multiforme using standard therapy together with a restricted ketogenic diet: Case Report. Nutr Metab. 2010;7:33.
43. Seyfried TN, Shelton LM, Mukherjee P. Does the existing standard of care increase glioblastoma energy metabolism? Lancet Oncol. 2010;11:811–3.

44. Seyfried TN, Kiebish MA, Marsh J, Shelton LM, Huysentruyt LC, Mukherjee P. Metabolic management of brain cancer. Biochim Biophys Acta. 2010;1807:577–94.
45. Coussens LM, Werb Z. Inflammation and cancer. Nature. 2002;420:860–7.
46. Colotta F, Allavena P, Sica A, Garlanda C, Mantovani A. Cancer-related inflammation, the seventh hallmark of cancer: links to genetic instability. Carcinogenesis. 2009;30:1073–81.
47. Iruela-Arispe ML, Dvorak HF. Angiogenesis: a dynamic balance of stimulators and inhibitors. Thromb Haemost. 1997;78:672–7.
48. Folkman J. The role of angiogenesis in tumor growth. Semin Cancer Biol. 1992;3:65–71.
49. Folkman J. Incipient angiogenesis. J Natl Cancer Inst. 2000;92:94–5.
50. DeBerardinis RJ. Is cancer a disease of abnormal cellular metabolism? New angles on an old idea. Genet Med. 2008;10:767–77.
51. Greenberg JI, Cheresh DA. VEGF as an inhibitor of tumor vessel maturation: implications for cancer therapy. Expert Opin Biol Ther. 2009;9:1347–56.
52. Claffey KP, Brown LF, del Aguila LF, Tognazzi K, Yeo KT, Manseau EJ, et al. Expression of vascular permeability factor/vascular endothelial growth factor by melanoma cells increases tumor growth, angiogenesis, and experimental metastasis. Cancer Res. 1996;56:172–81.
53. Ferrara N, Gerber HP, LeCouter J. The biology of VEGF and its receptors. Nat Med. 2003;9:669–76.
54. Bos R, van Diest PJ, de Jong JS, van der Groep P, van der Valk P, van der Wall E. Hypoxia-inducible factor-1alpha is associated with angiogenesis, and expression of bFGF, PDGF-BB, and EGFR in invasive breast cancer. Histopathology. 2005;46:31–6.
55. Lazebnik Y. What are the hallmarks of cancer? Nat Rev Cancer. 2010;10:232–3.

Chapter 13

Metastasis

Because metastasis is the single most important phenomenon of cancer, I am devoting considerable attention to this subject. My knowledge of metastasis comes from our extensive studies of spontaneous brain tumors in the VM mouse strain. The VM strain is unique in developing a relatively high incidence of spontaneous brain tumors as I have described in Chapter 3. The cells derived from some of these spontaneous tumors are highly metastatic when grown outside the brain. The metastatic behavior of these tumor cells is remarkably similar to that seen in the cells of many human systemic metastatic cancers.

For many years, I studied chemically induced brain tumors that expressed numerous hallmarks of cancer, but none showed the highly invasive properties seen in human brain tumors. Indeed, most mouse cancer models rarely express the invasive and metastatic behaviors seen in the human disease (Chapter 3). It was only after we isolated and characterized the tumor cells in several independent VM brain tumors that we recognized the importance of these tumors in explaining the cellular origin and characteristics of metastatic cancers. It gradually became clear to me that these VM brain tumors not only expressed the most salient features of human glioblastoma (the most common malignant brain cancer) but also expressed the most salient features of most human metastatic cancers. Our studies of the spontaneous VM mouse brain tumors changed our views on the origin of metastatic cancer.

METASTASIS OVERVIEW

Metastasis is the general term used to describe the spread of cancer cells from the primary tumor to surrounding tissues and to distant organs and is the primary cause of cancer morbidity and mortality (1–8). It is estimated that metastasis is responsible for about 90% of cancer deaths (9). This estimate has not changed significantly in more than 50 years (10, 11). Although systemic metastasis is responsible for 90% of cancer deaths, most research in cancer does not involve metastasis in the

Cancer as a Metabolic Disease: On the Origin, Management and Prevention of Cancer, First Edition.
Thomas Seyfried.

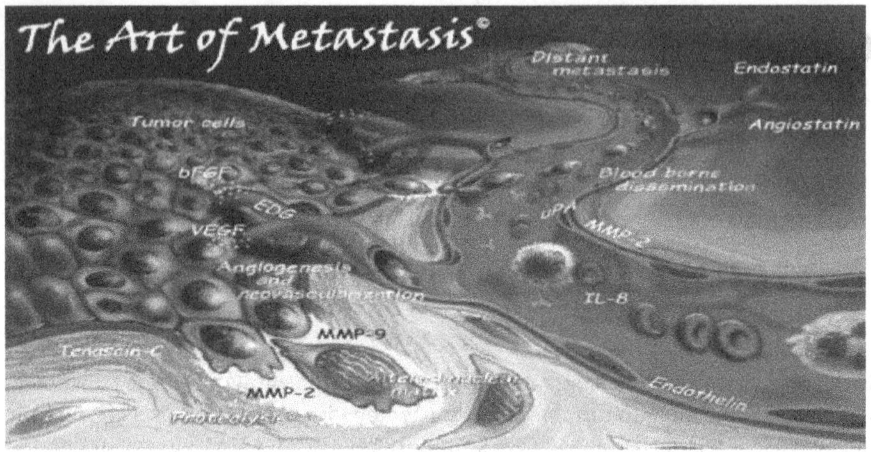

Figure 13.1 The art of metastasis depicting metastatic skin cancer (melanoma). *Source:* Reprinted with permission from *Oncogene*. See color insert.

in vivo state (5). That about 1500 people continue to die each day from cancer further attests to the failure in managing the disease once it spreads to other organs.

Metastasis involves a series of sequential and interrelated steps. In order to complete the metastatic cascade, cancer cells must detach from the primary tumor, intravasate into the circulatory and lymphatic systems, evade immune attack, extravasate at distant capillary beds, and invade and proliferate in distant organs (1–4, 7, 12, 13). Metastatic cells also establish a microenvironment that facilitates angiogenesis and proliferation, resulting in macroscopic, malignant secondary tumors. Figure 13.1 provides an image of metastatic skin cancer (melanoma). There are also illustrative videos from the web, which can help describe the phenomenon of metastasis (http://www.youtube.com/watch?v=rrMq8uA_6iA). One misconception with such simplified overviews of metastasis is that metastatic cancer cells have difficulties surviving multiple hazards in the circulation. I address this misconception later in the chapter.

A difficulty in characterizing the cellular origin of metastasis comes in large part from the lack of animal models that show *systemic* metastasis. As I have mentioned in Chapter 3, tumor cells that are naturally metastatic should not require intravenous injection to initiate the metastatic phenotype. The key phenotype of metastasis is that the tumor cells spread *naturally* from the primary tumor site to secondary locations. Nevertheless, numerous investigators use intravenous tumor cell injection models to study metastasis. While these models can provide information on tumor cell survival in the circulation, it is not clear if this information is relevant to survival of naturally metastatic tumor cells.

If the tumor cells evaluated are not naturally metastatic, it is not clear why they would be used as models of metastasis. According to Lazebnik (5), much of what is known about metastasis comes from model systems that have more in

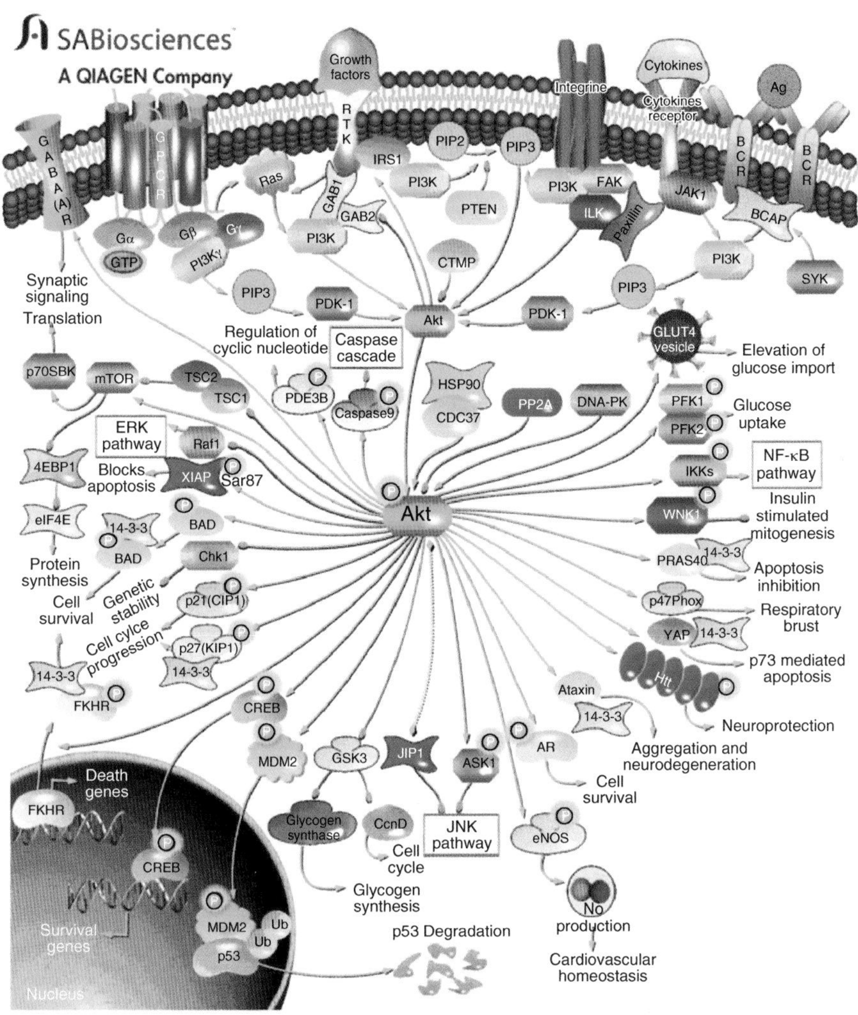

Figure 1.2 See full caption on page 4.

Cancer as a Metabolic Disease: On the Origin, Management and Prevention of Cancer, First Edition. Thomas Seyfried.

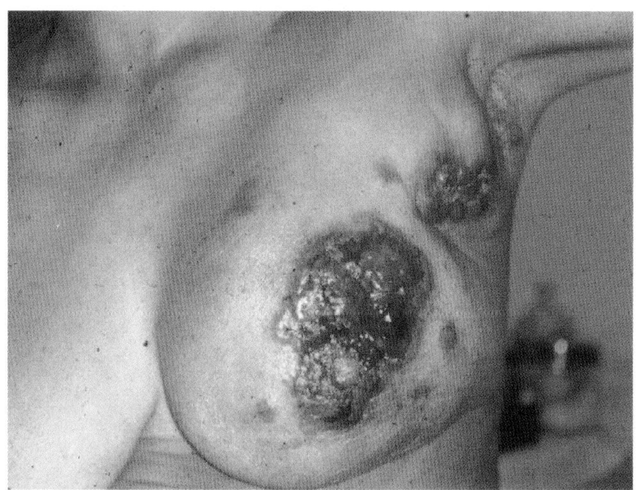

Figure 1.5a See full caption on page 7.

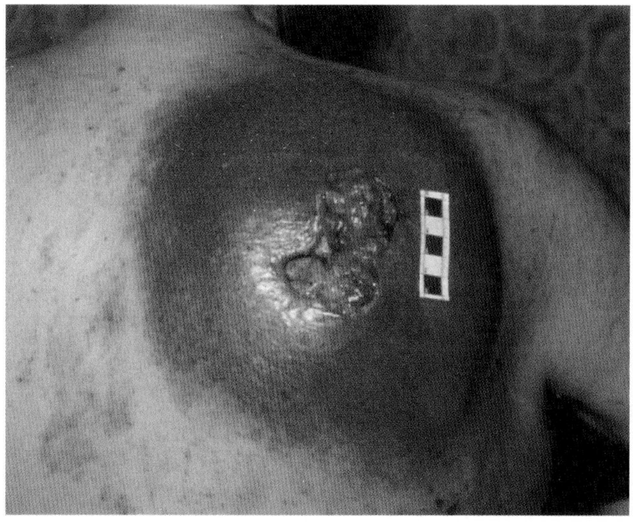

Figure 1.5d See full caption on page 7.

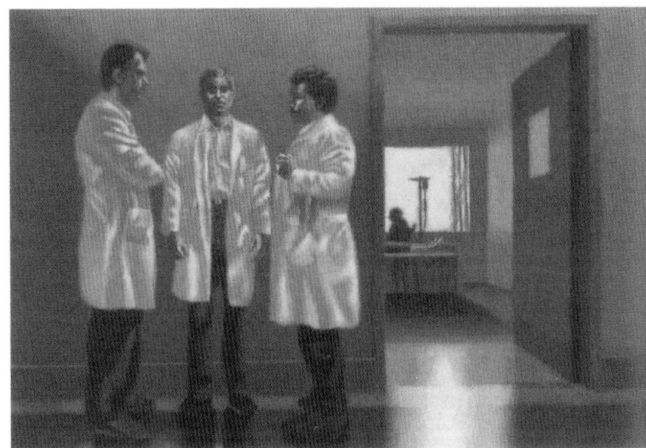

Figure 1.6 See full caption on page 7.

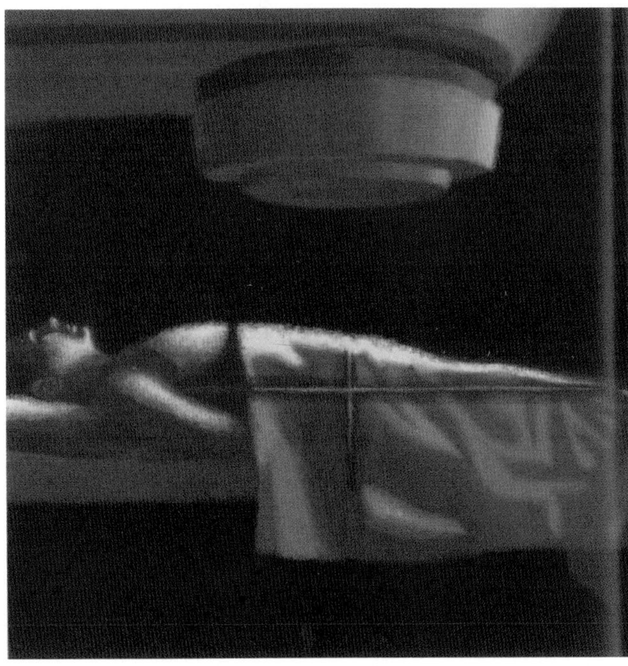

Figure 1.7 See full caption on page 8.

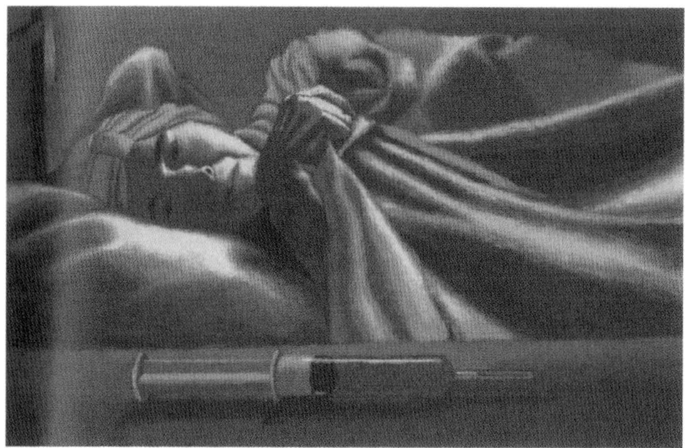

Figure 1.8 See full caption on page 8.

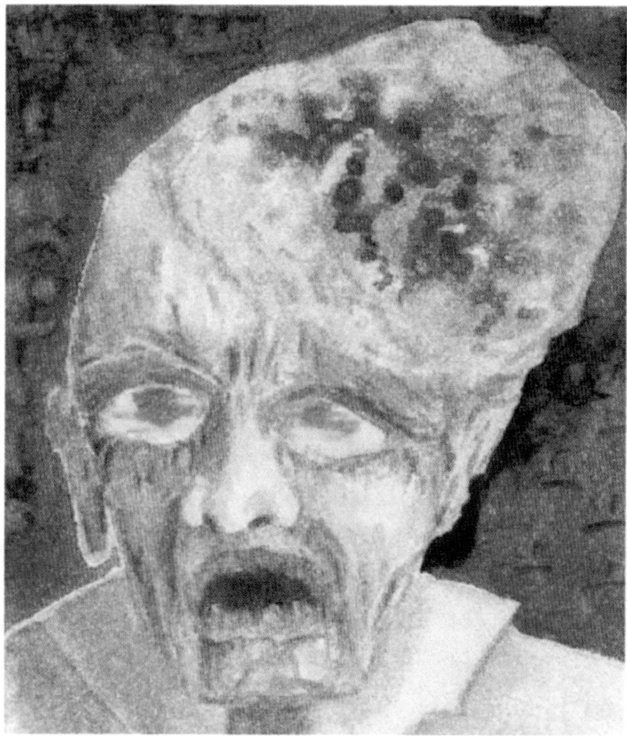

Figure 1.13 See full caption on page 11.

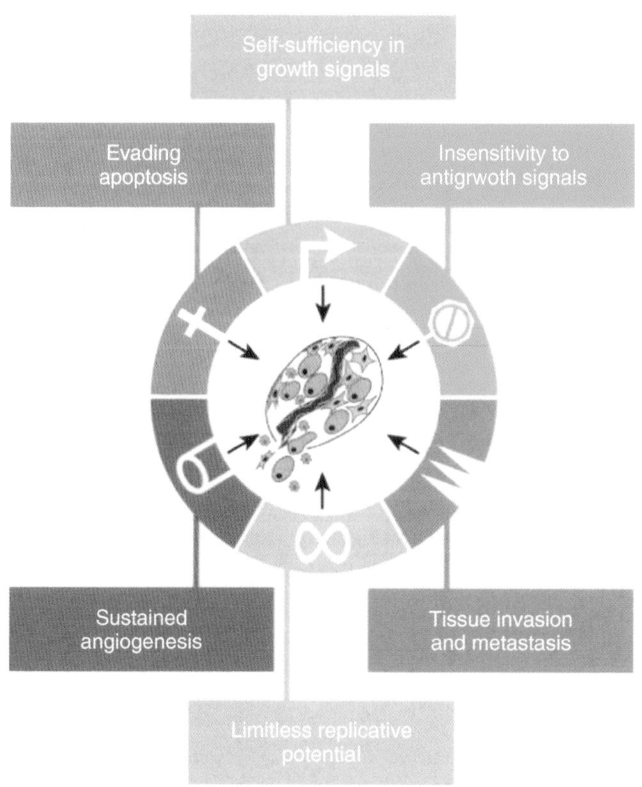

Figure 2.2 See full caption on page 19.

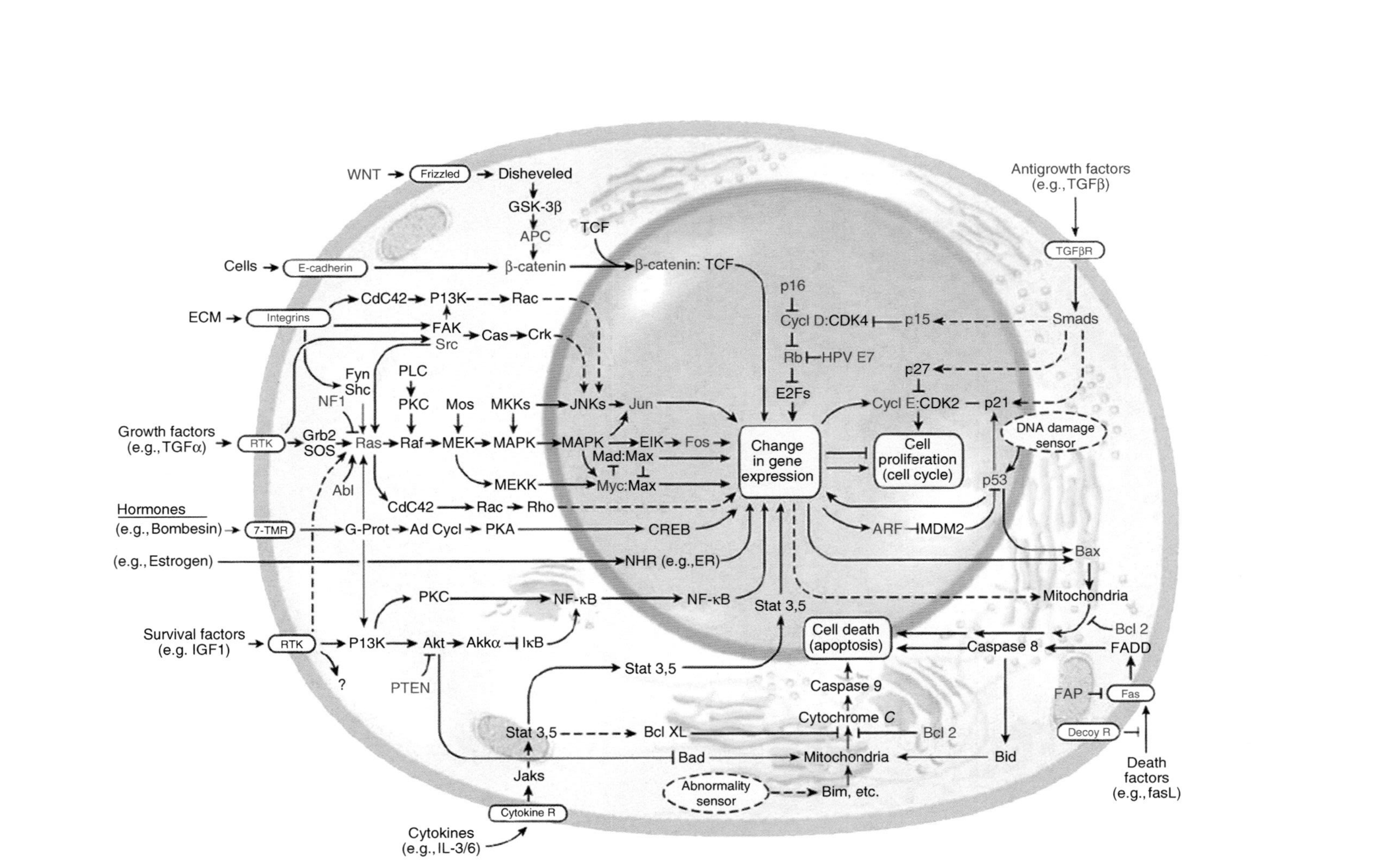

WNT
Frizzled
Disheveled
GSK-3β
APC
TCF
Cells
E-cadherin
β-catenin
β-catenin: TCF
Antigrowth factors
(e.g., TGFβ)
TGFβR
Smads
ECM
Integrins
CdC42
P13K
Rac
FAK
Cas
Crk
Src
p16
Cycl D:CDK4
p15
Rb
HPV E7
E2Fs
p27
Cycl E:CDK2
p21
DNA damage sensor
Fyn
Shc
NF1
PLC
PKC
Mos
MKKs
JNKs
Jun
Growth factors
(e.g., TGFα)
RTK
Grb2
SOS
Ras
Raf
MEK
MAPK
MAPK
ElK
Fos
Mad:Max
MEKK
Myc:Max
Change in gene expression
Cell proliferation (cell cycle)
p53
Abl
CdC42
Rac
Rho
Hormones
(e.g., Bombesin)
7-TMR
G-Prot
Ad Cycl
PKA
CREB
ARF
MDM2
(e.g., Estrogen)
NHR (e.g., ER)
Bax
PKC
NF-κB
NF-κB
Stat 3,5
Mitochondria
Survival factors
(e.g. IGF1)
RTK
P13K
Akt
Akkα
IκB
Cell death (apoptosis)
Caspase 8
Bcl 2
FADD
?
PTEN
Stat 3,5
Caspase 9
FAP
Fas
Cytochrome C
Stat 3,5
Bcl XL
Bcl 2
Decoy R
Bad
Mitochondria
Bid
Death factors
(e.g., fasL)
Jaks
Abnormality sensor
Bim, etc.
Cytokine R
Cytokines
(e.g., IL-3/6)

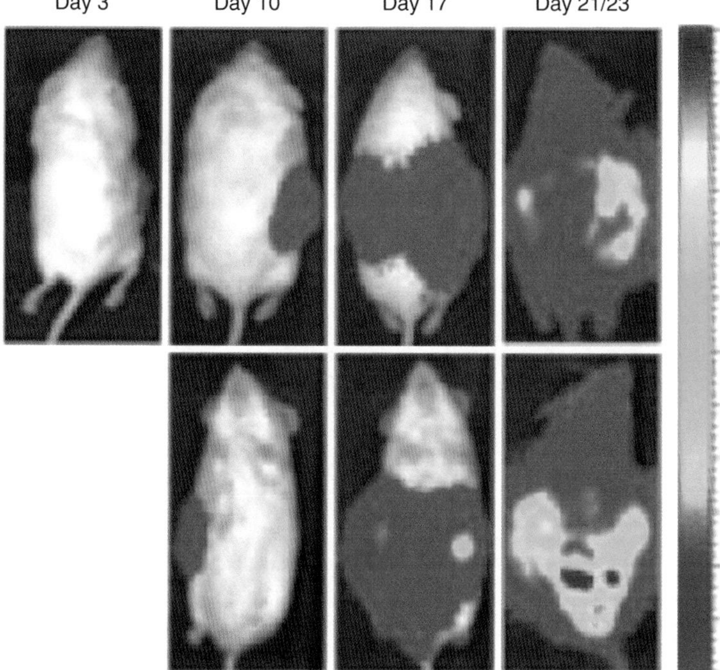

Figure 3.1 See full caption on page 36.

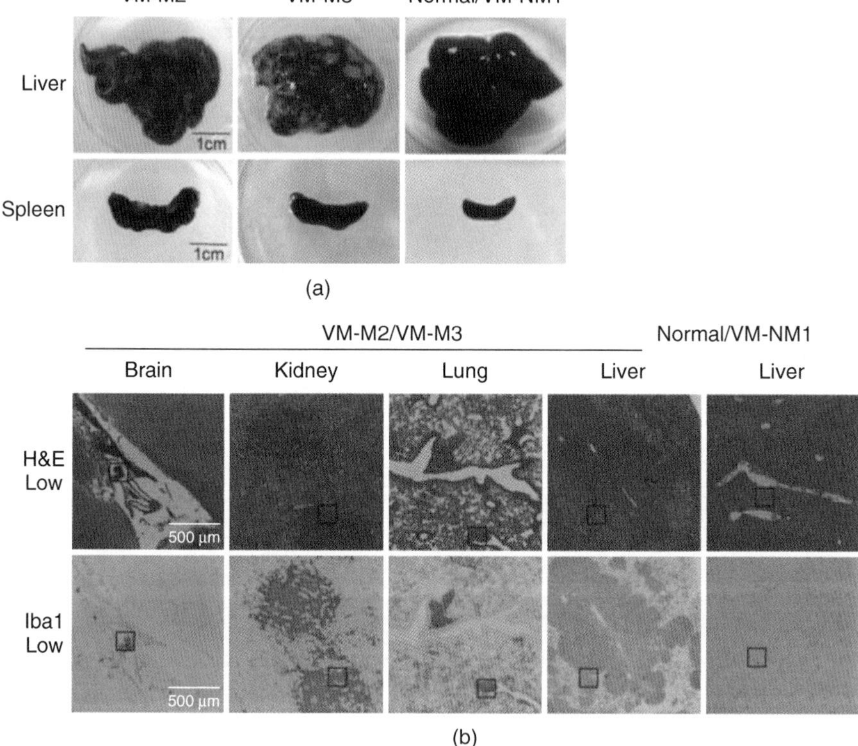

Figure 3.2 See full caption on page 37.

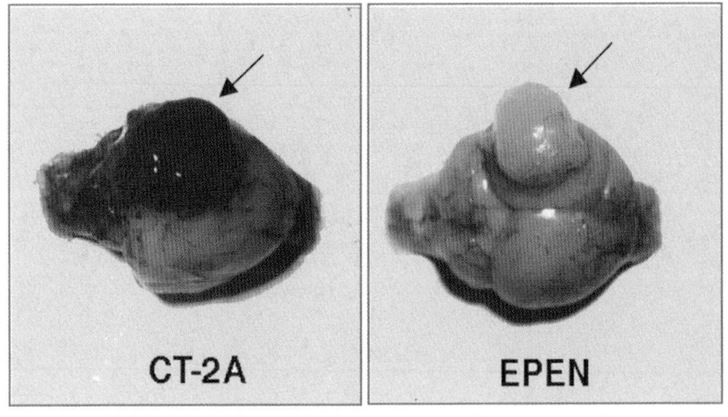

Figure 3.3 See full caption on page 40.

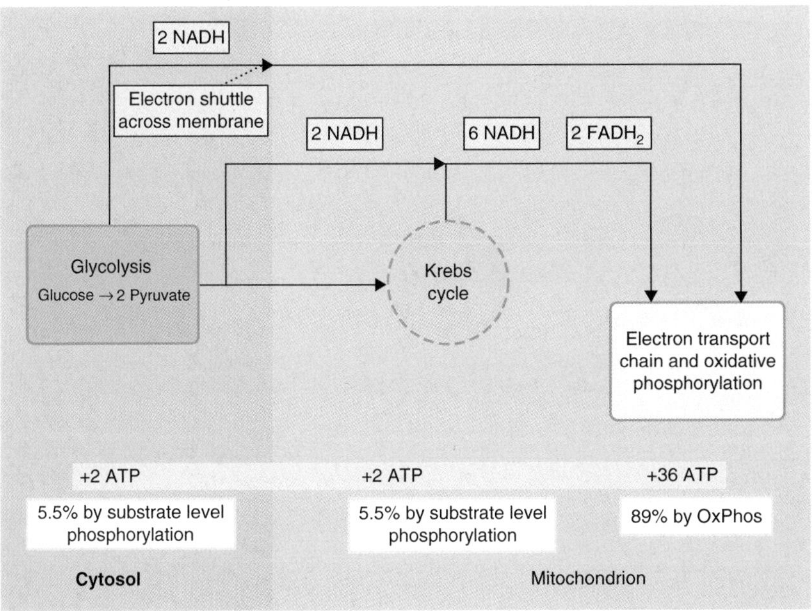

Figure 4.3 See full caption on page 50.

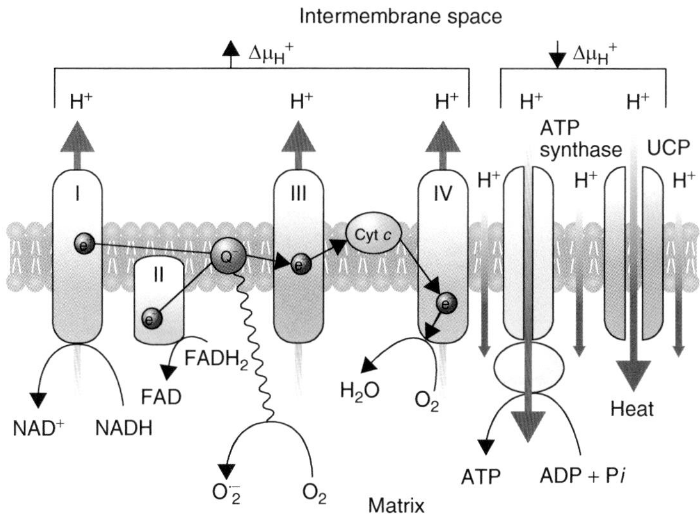

Figure 4.4 See full caption on page 51.

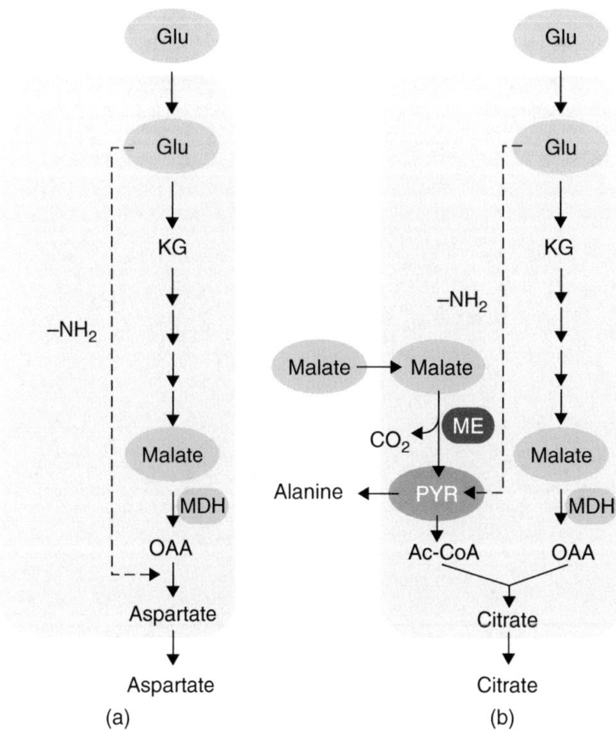

Figure 4.9 See full caption on page 63.

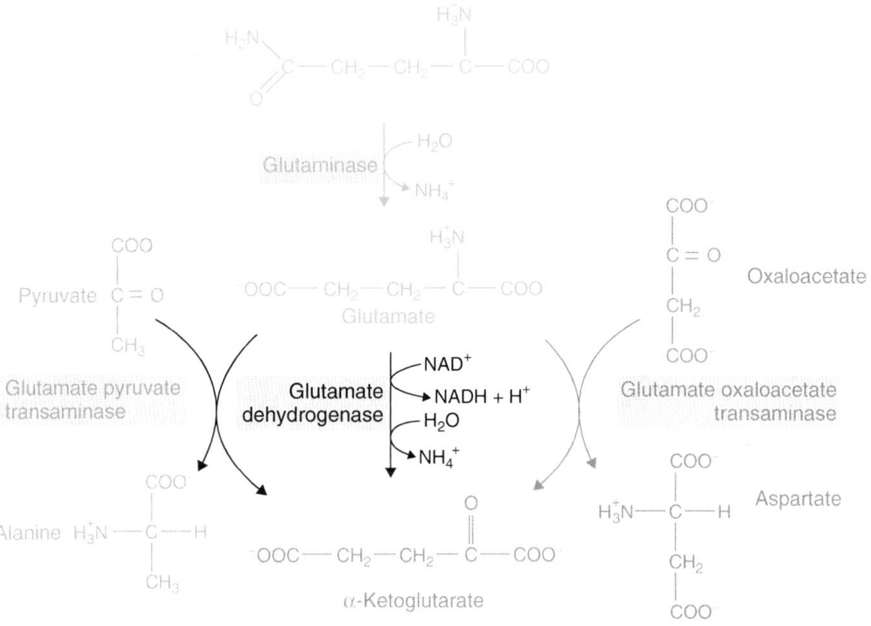

Figure 4.11 See full caption on page 65.

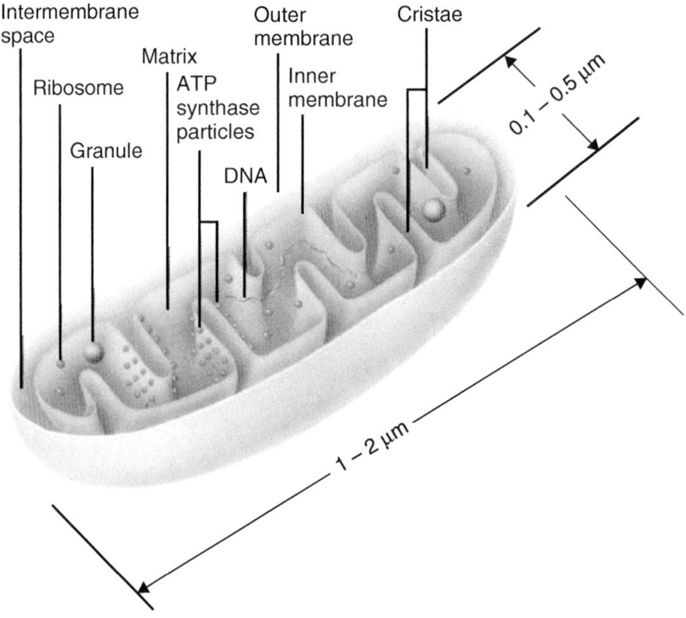

Figure 5.1 See full caption on page 74.

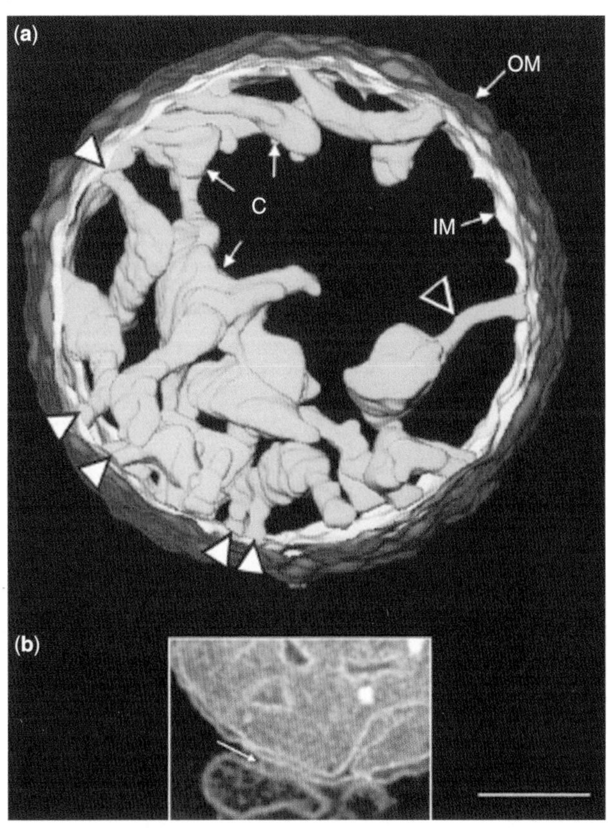

Figure 5.2 See full caption on page 75.

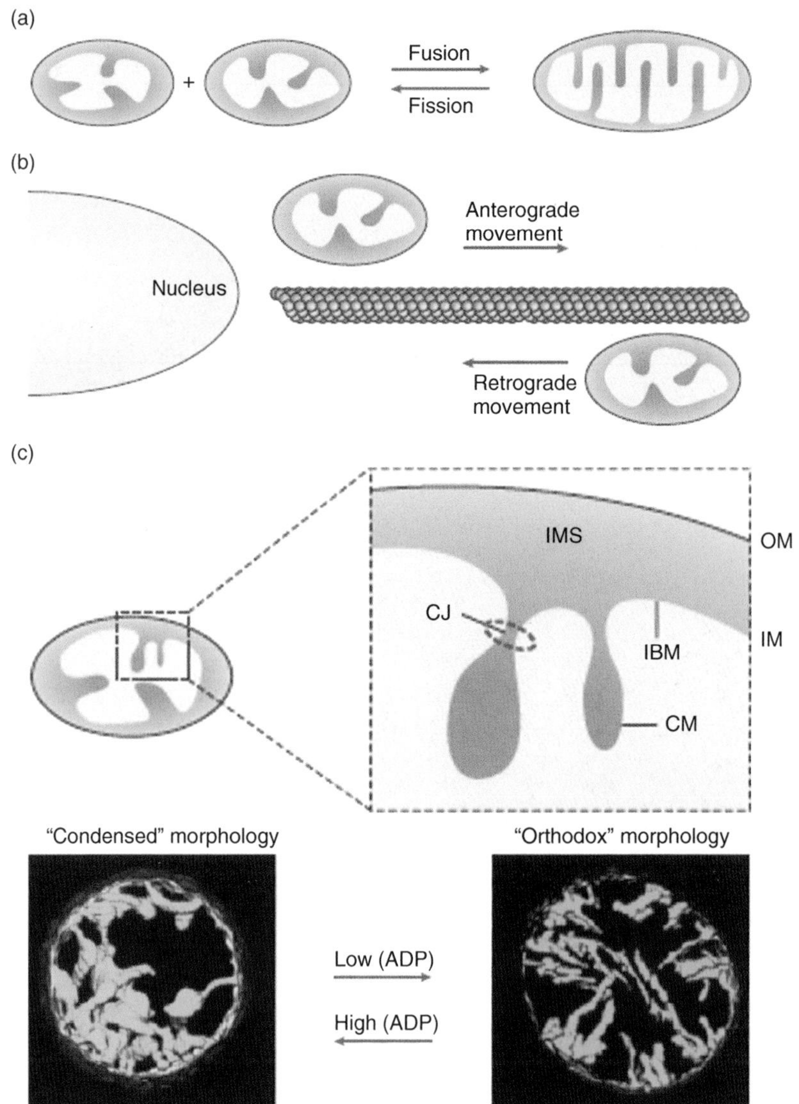

Figure 5.3 See full caption on page 76.

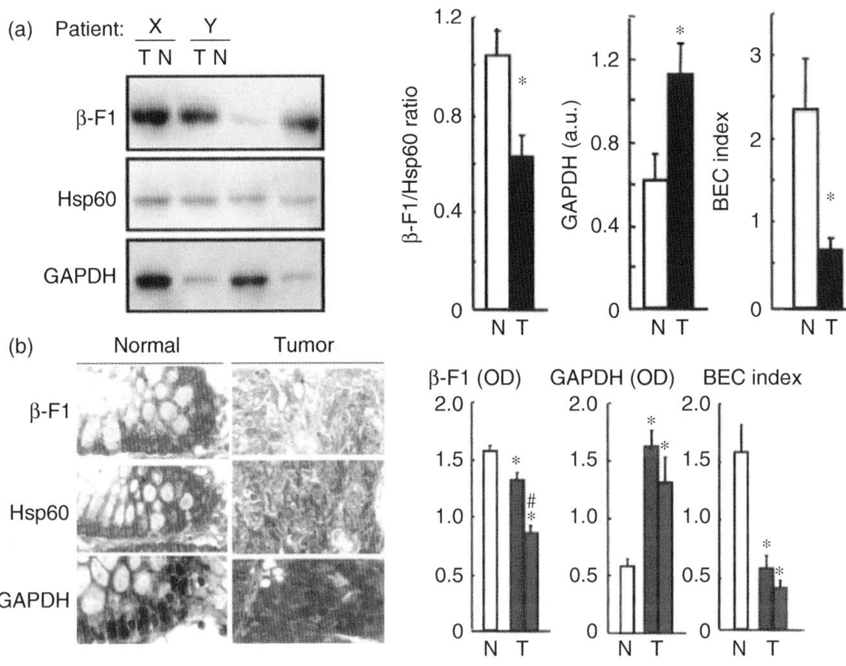

Figure 5.5 See full caption on page 80.

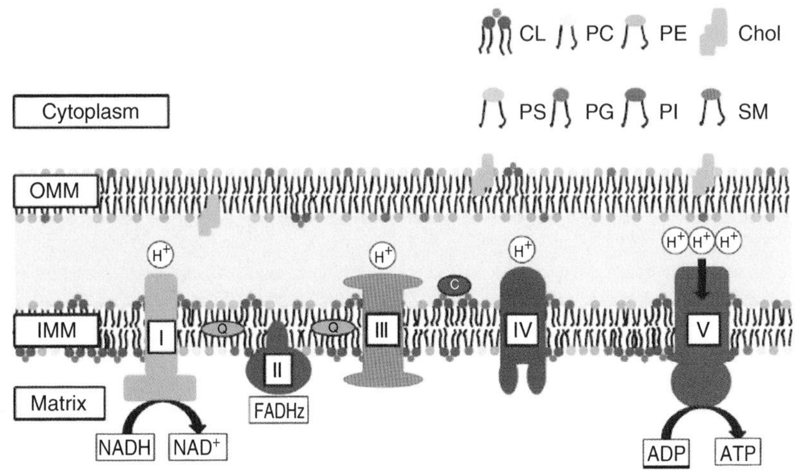

Figure 5.6 See full caption on page 82.

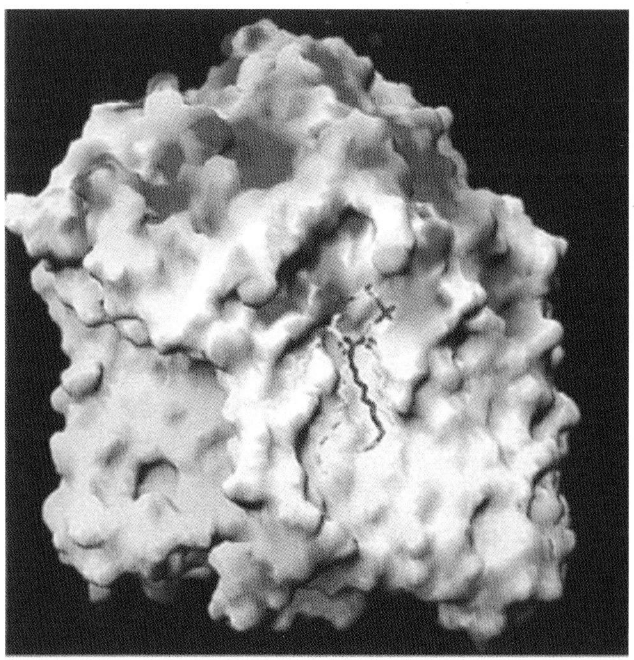

Figure 5.8 See full caption on page 85.

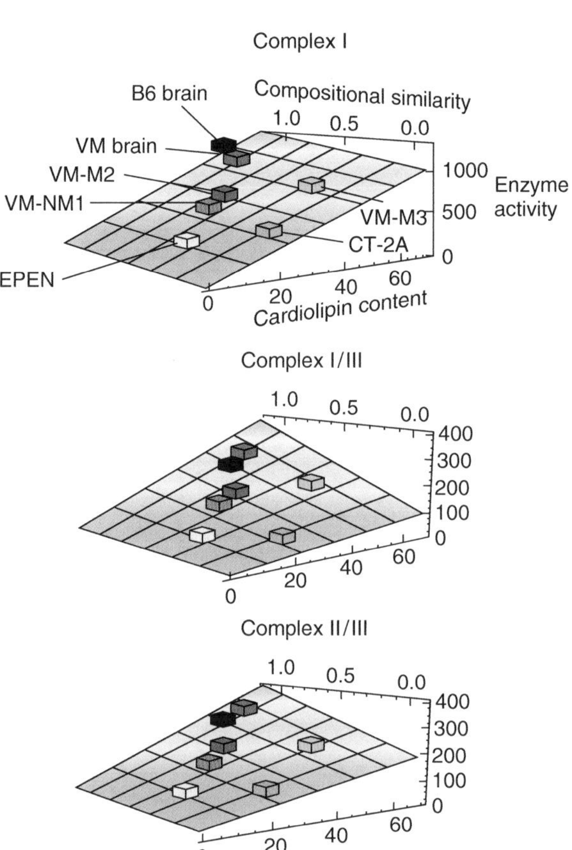

Figure 5.12 See full caption on page 90.

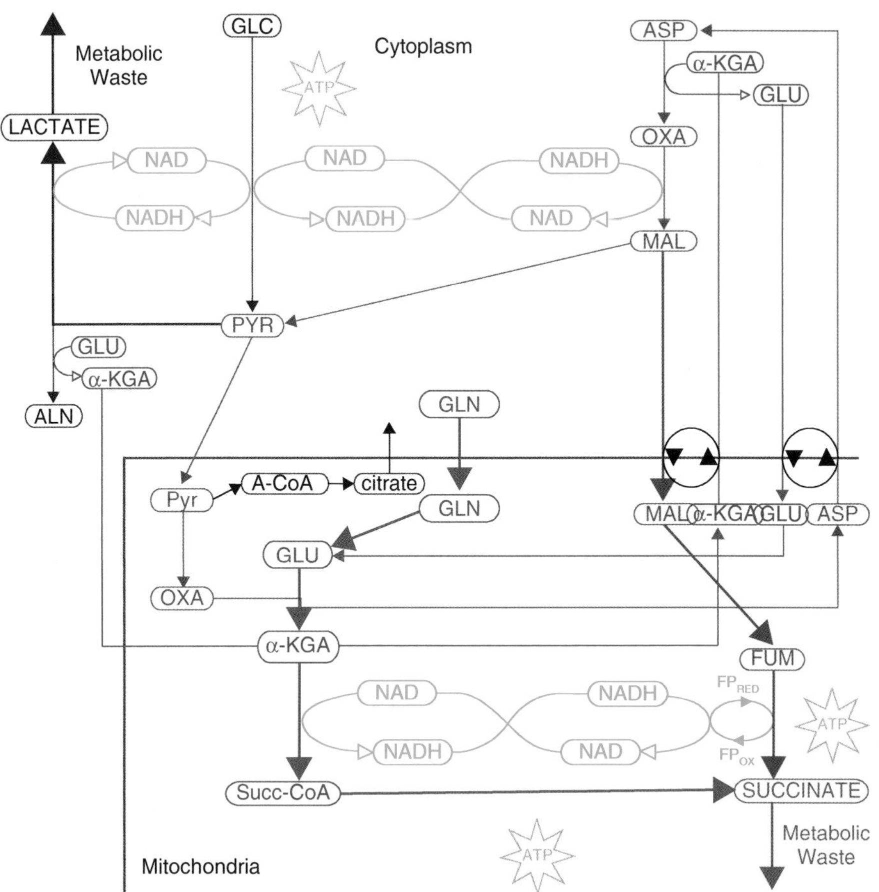

Figure 8.6 See full caption on page 140.

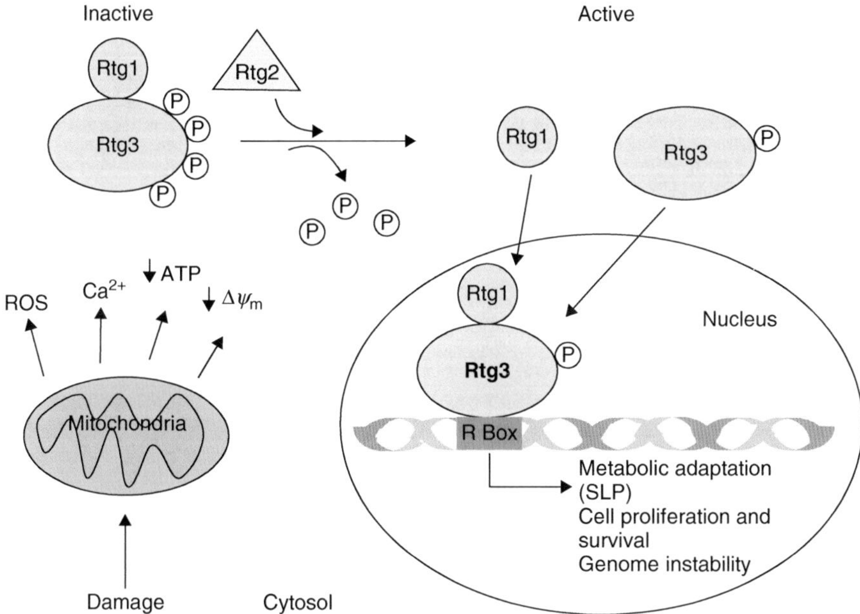

Figure 10.1 See full caption on page 179.

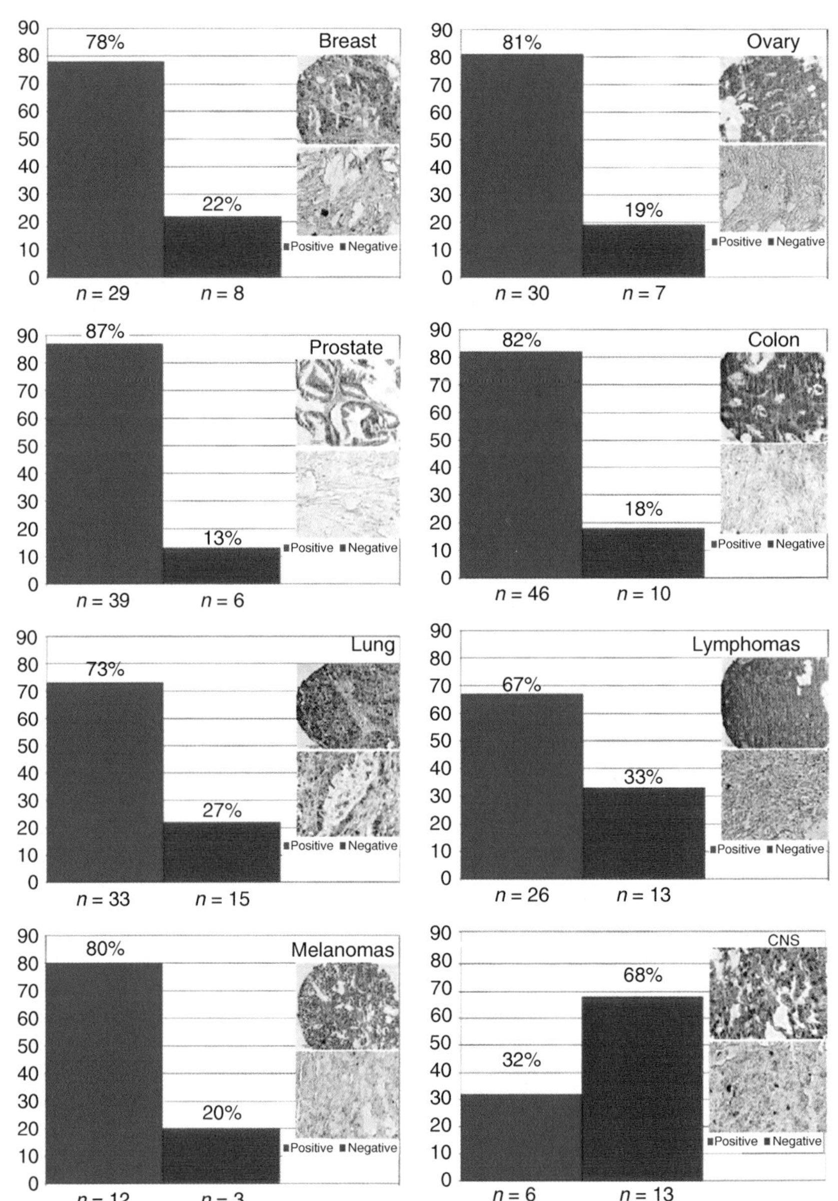

Figure 10.2 See full caption on page 185.

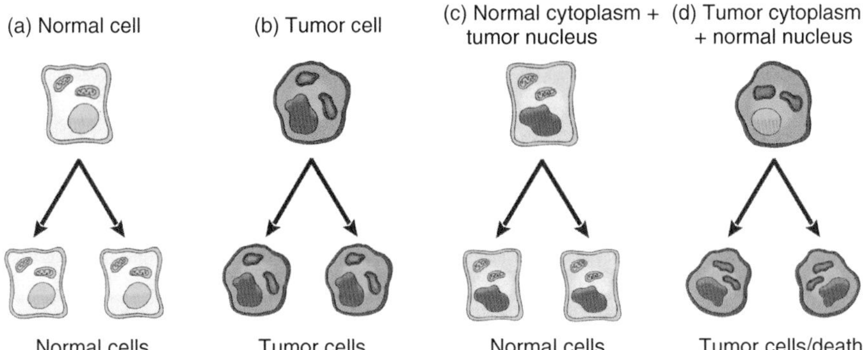

Figure 11.3 See full caption on page 203.

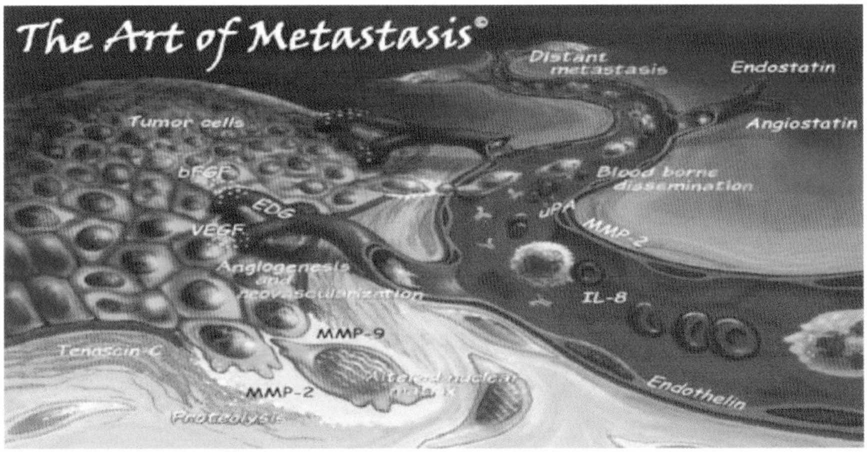

Figure 13.1 See full caption on page 216.

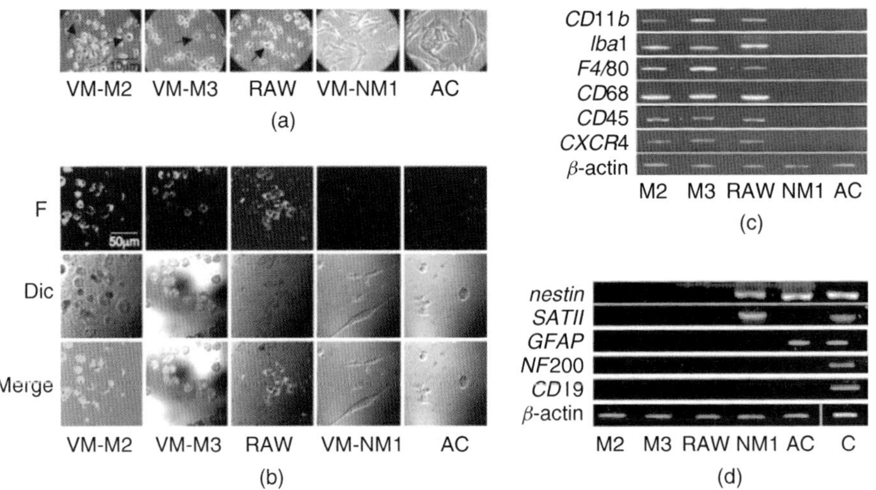

Figure 13.4 See full caption on page 227.

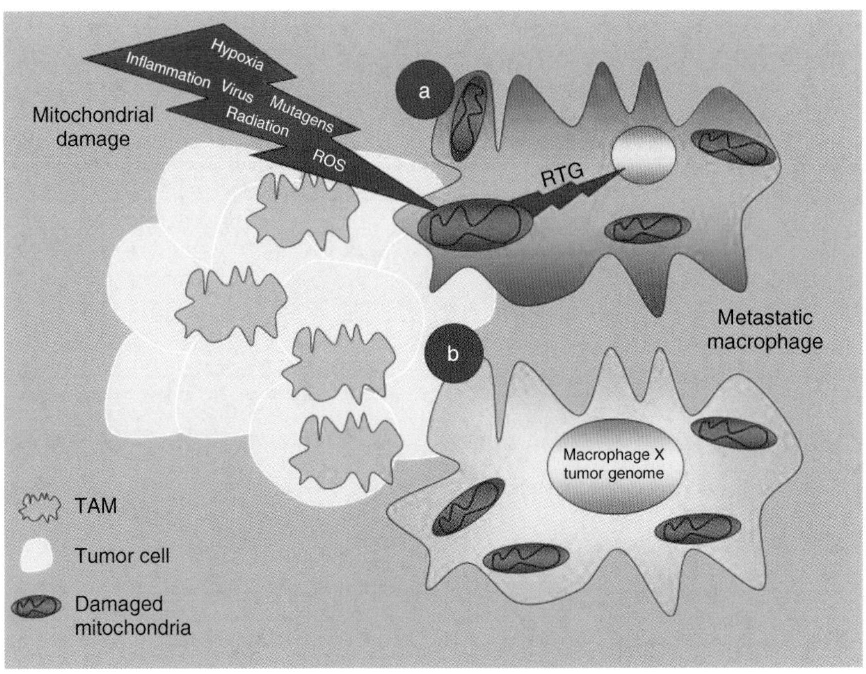

Figure 13.5 See full caption on page 234.

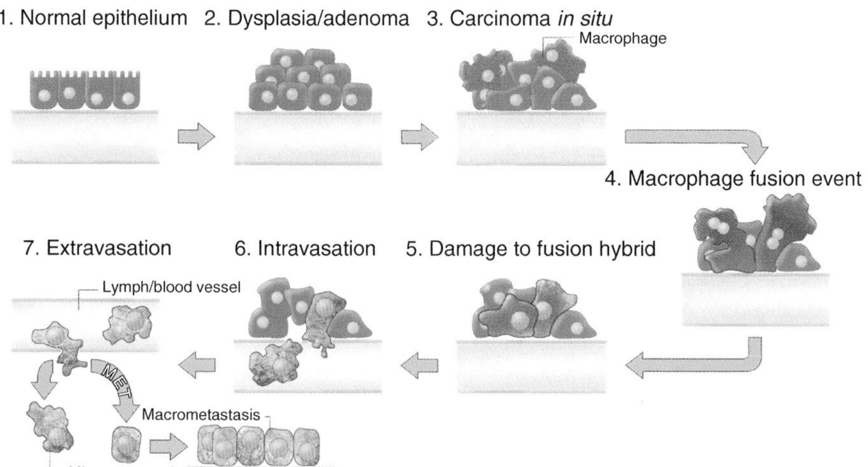

Figure 13.6 See full caption on page 238.

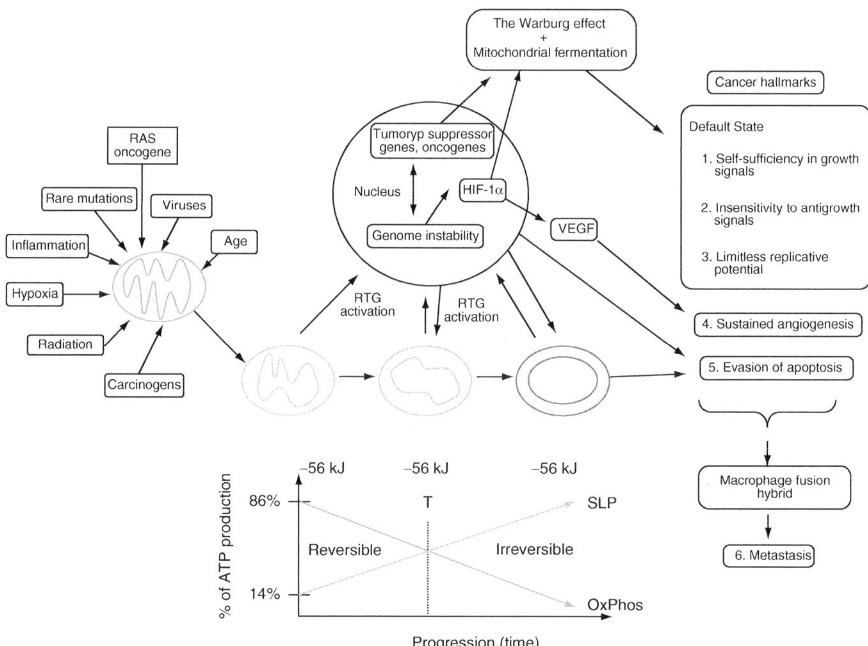

Figure 14.1 See full caption on page 254.

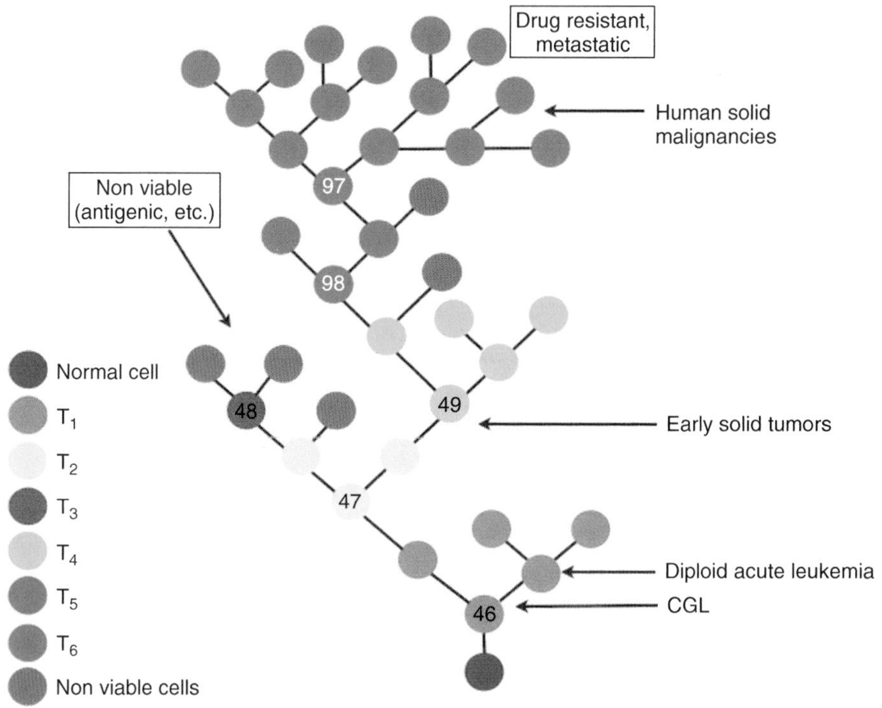

Figure 15.1 See full caption on page 265.

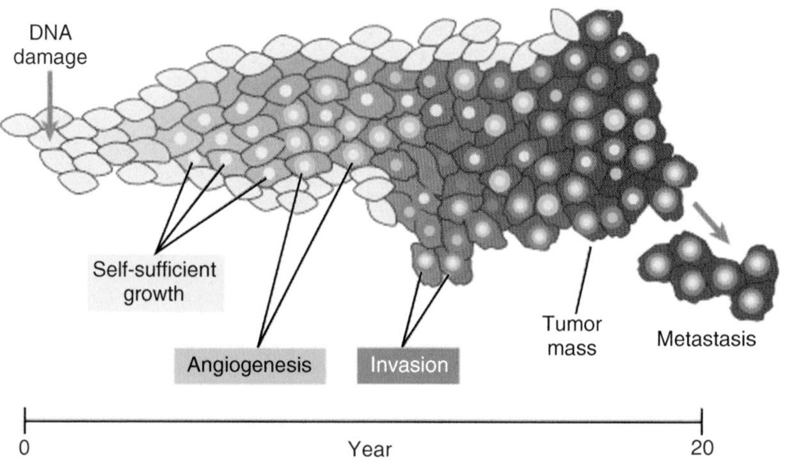

Figure 15.2 See full caption on page 267.

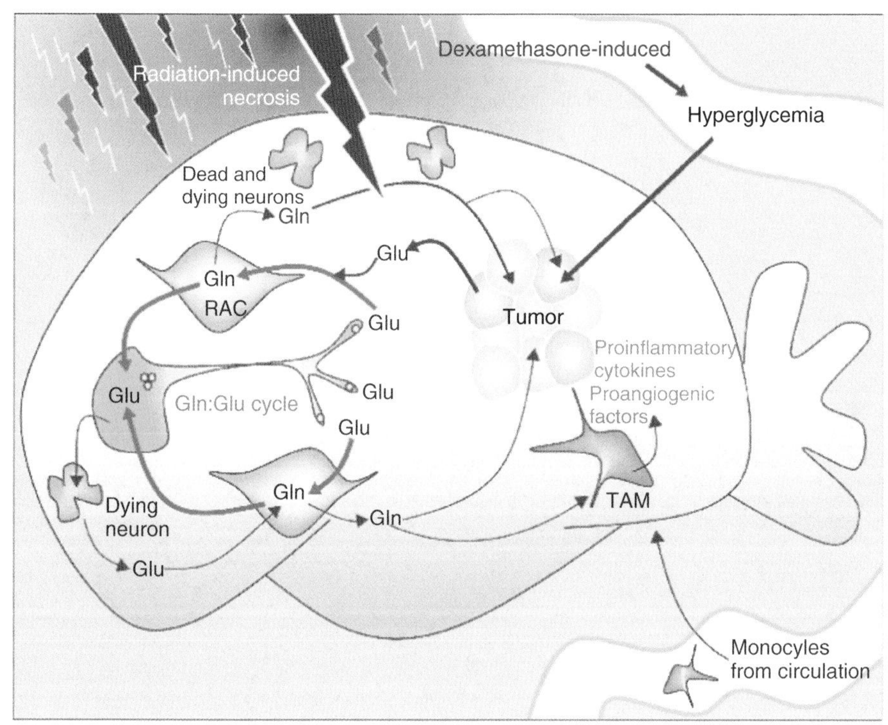

Figure 16.1 See full caption on page 282.

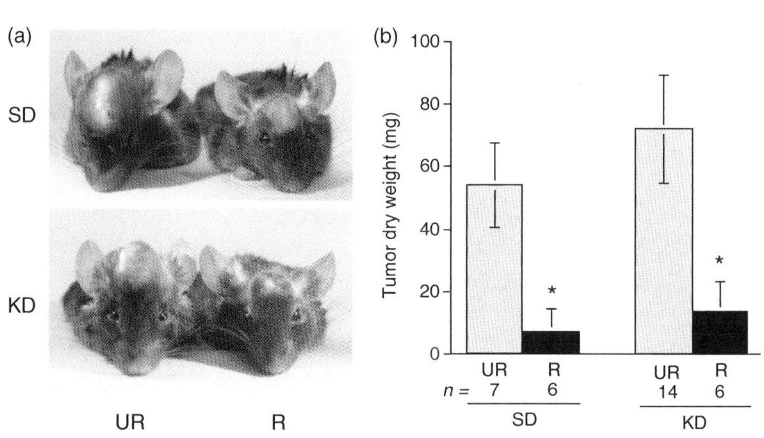

Figure 17.1 See full caption on page 293.

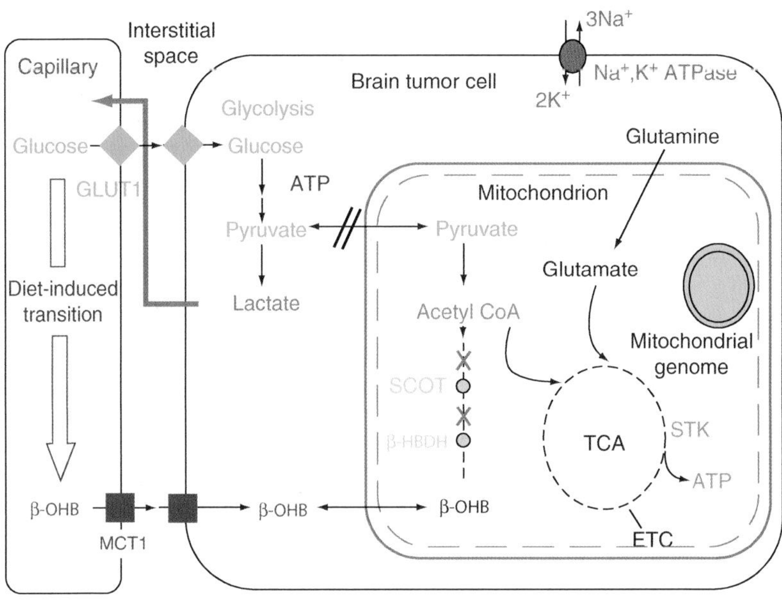

Figure 17.4 See full caption on page 299.

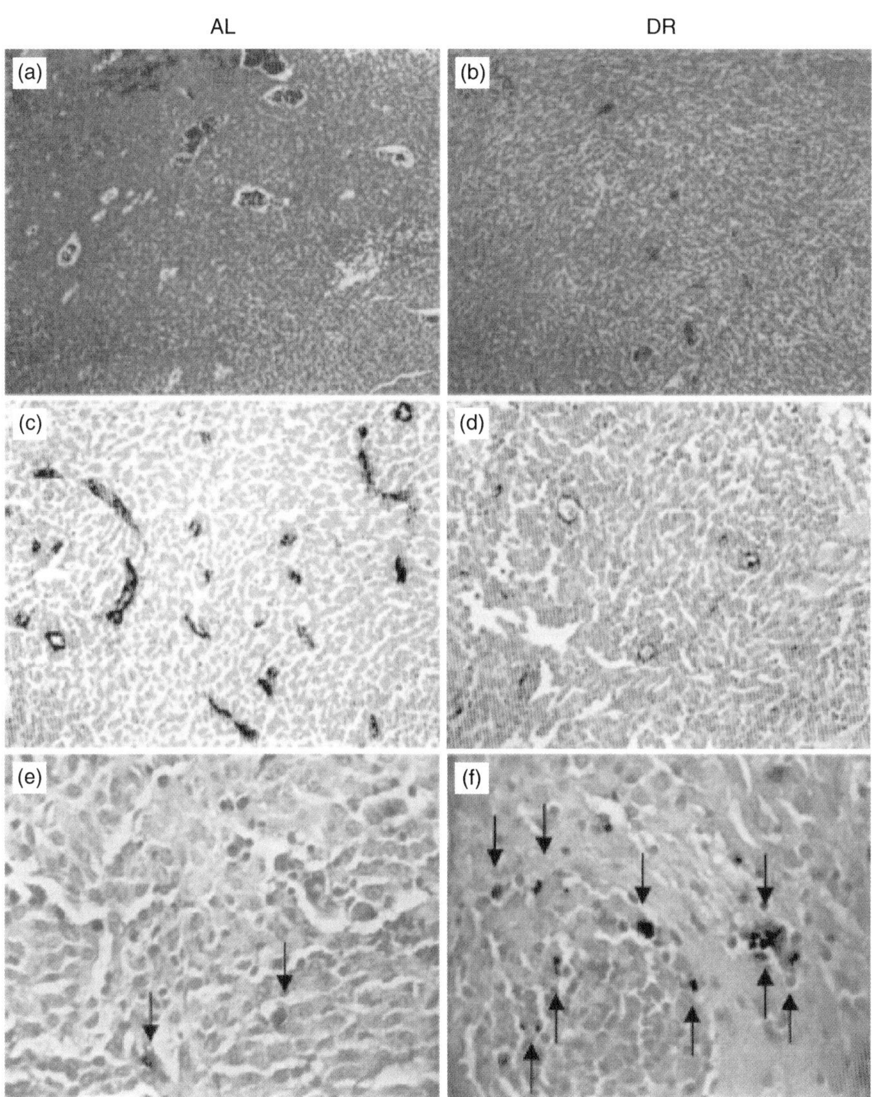

Figure 17.9 See full caption on page 306.

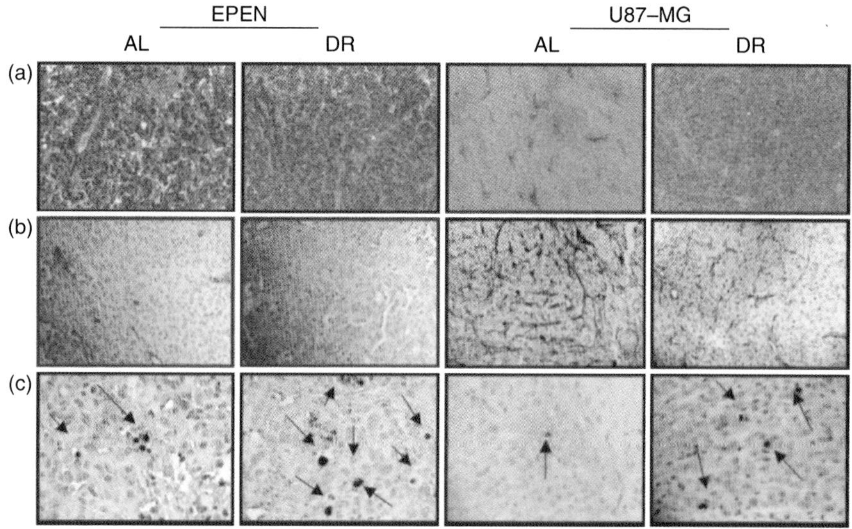

Figure 17.10 See full caption on page 307.

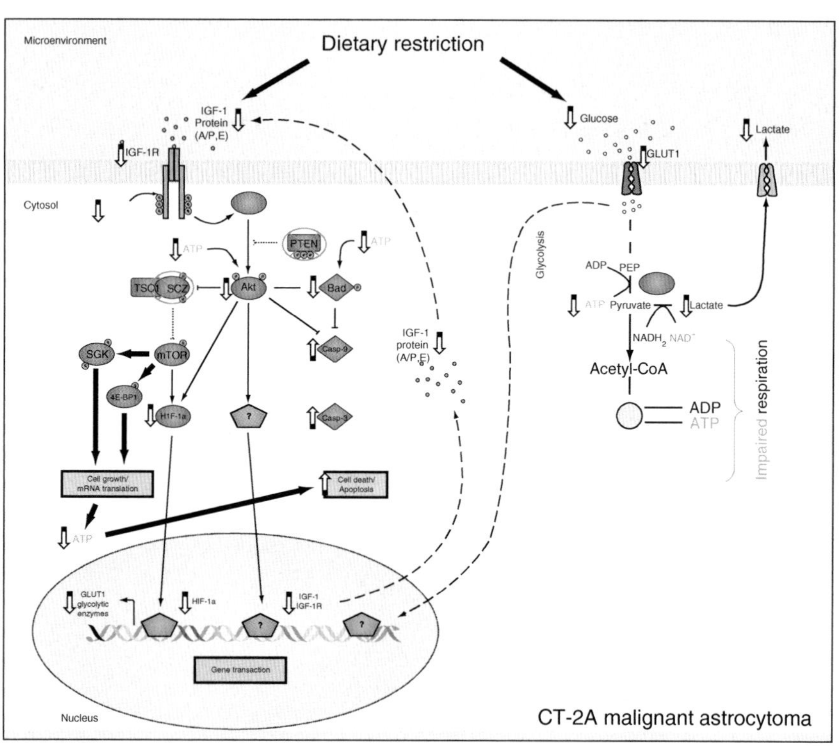

Figure 17.11 See full caption on page 310.

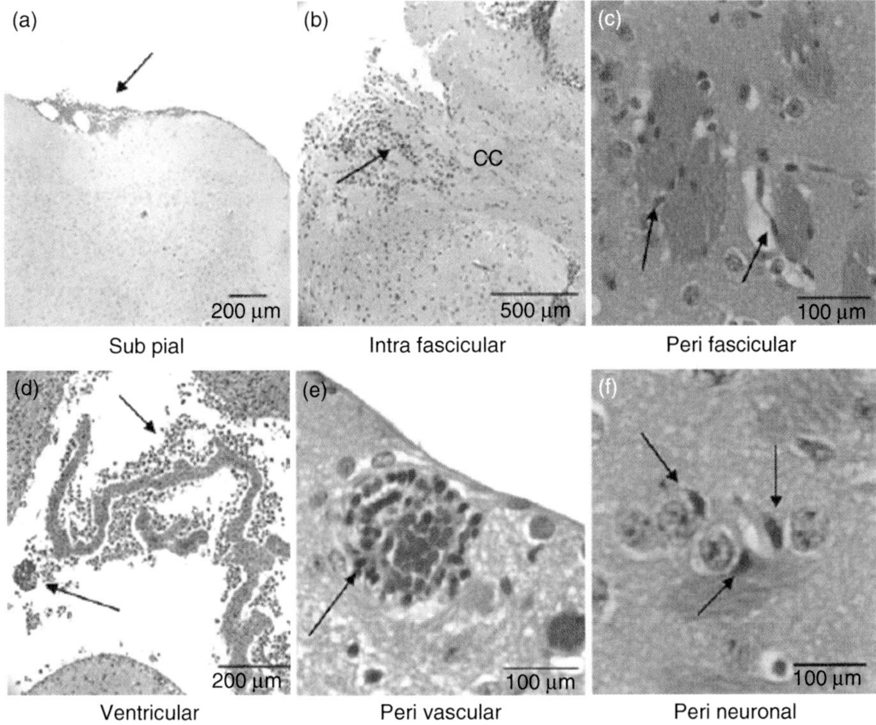

Figure 17.17 See full caption on page 319.

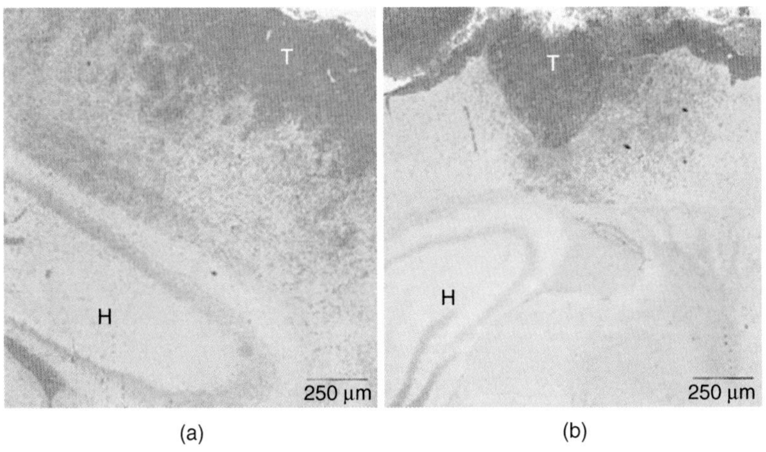

(a) (b)

Figure 17.18 See full caption on page 320.

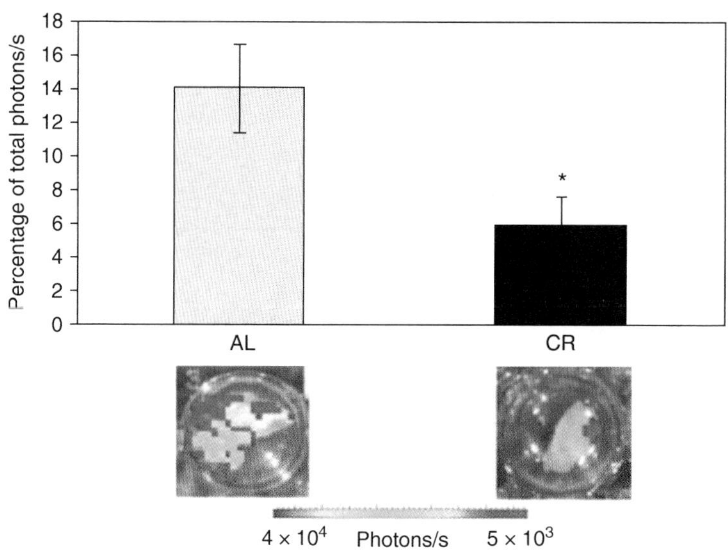

Figure 17.19 See full caption on page 321.

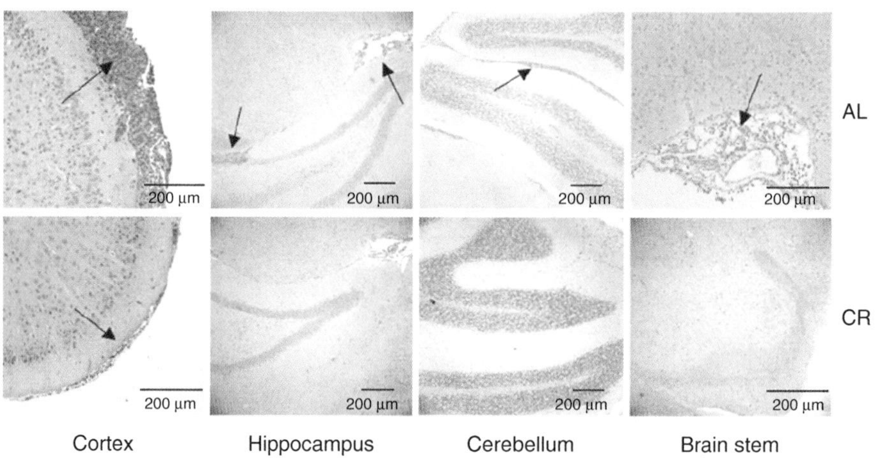

Figure 17.20 See full caption on page 322.

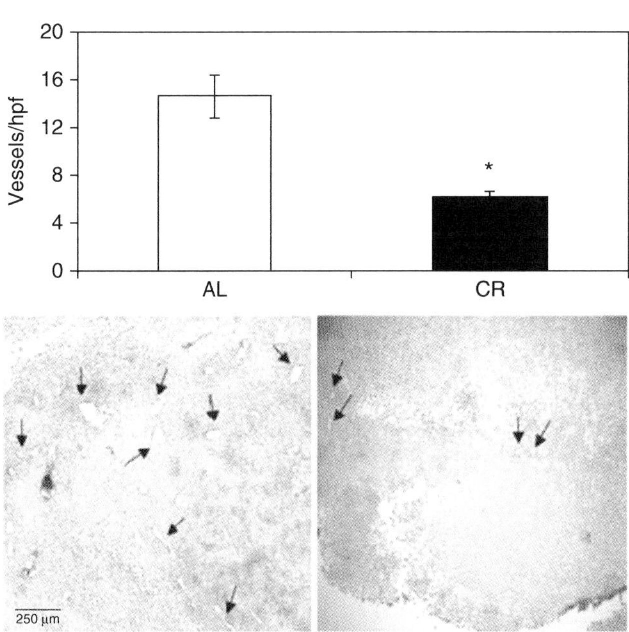

Figure 17.22 See full caption on page 324.

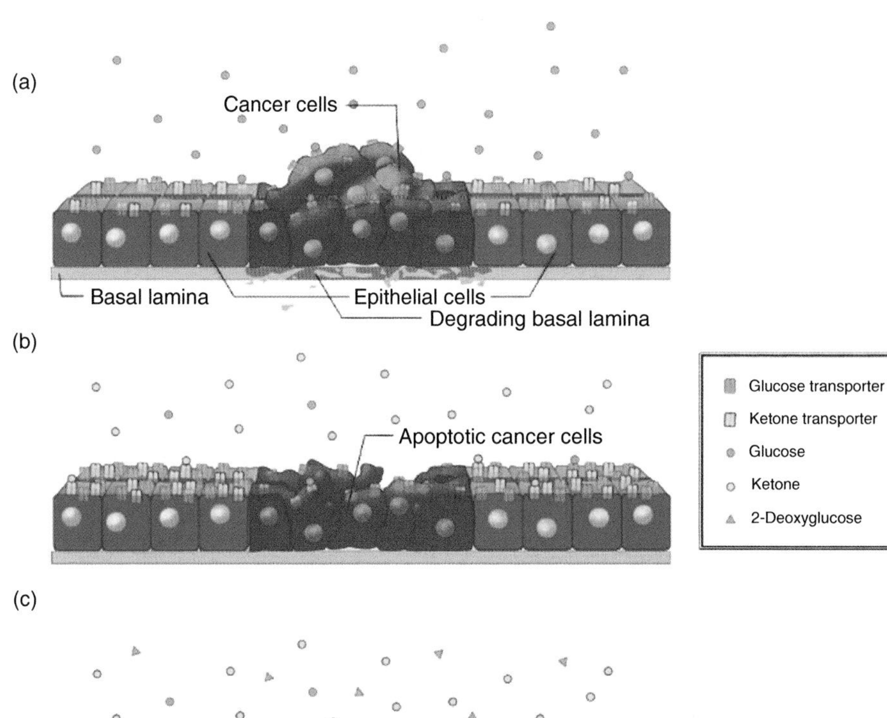

Figure 17.26 See full caption on page 330.

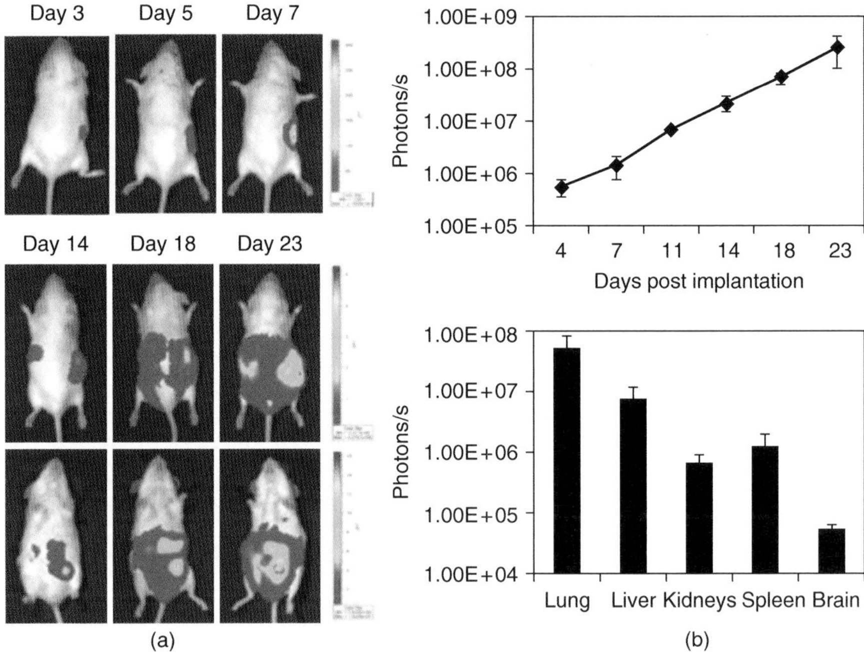

Figure 17.28 See full caption on page 334.

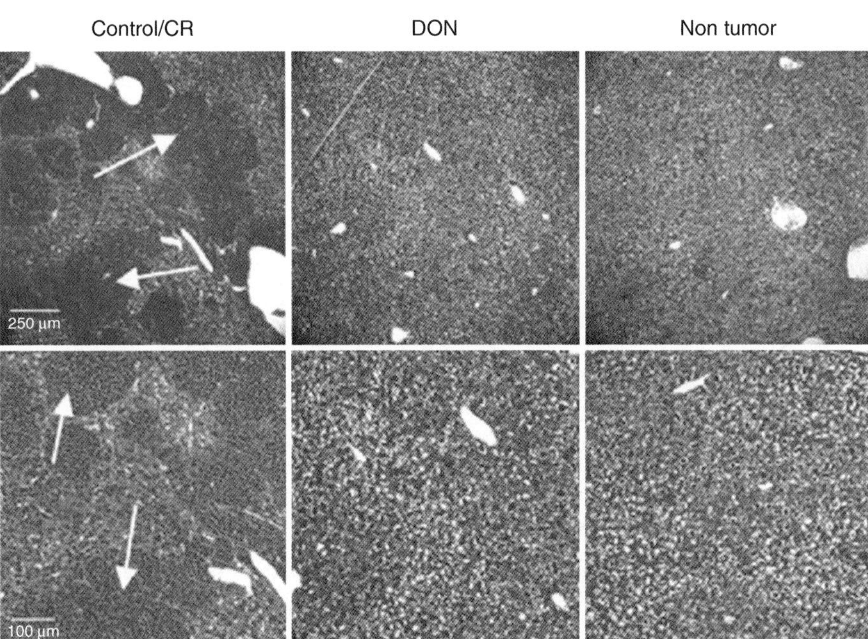

Figure 17.31 See full caption on page 337.

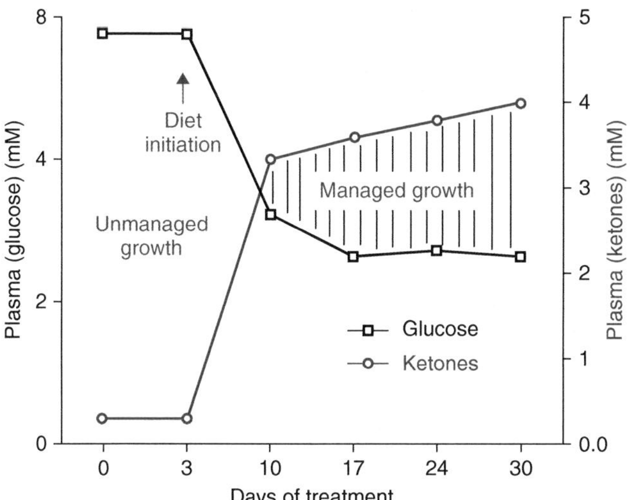

Figure 18.1 See full caption on page 357.

common with benign tumors than with metastatic carcinomas. If the models used to understand the nature of metastases do not accurately model the phenomenon, then the lack of progress in managing metastases should not be surprising.

The in vitro models have shortcomings in not replicating all the steps required for systemic metastasis in vivo. Although the major steps of metastasis are well documented, the process by which metastatic cells arise from within populations of nonmetastatic cells of the primary tumor is largely unknown (3, 7, 14, 15). It would therefore be helpful to highlight current views on the cellular origin of metastasis. Several ideas have been advanced to account for the origin of metastasis.

CELLULAR ORIGIN OF METASTASIS

Epithelial to Mesenchymal Transition (EMT)

The epithelial to mesenchymal transition (EMT) posits that metastatic cells arise from either epithelial stem cells or differentiated epithelial cells through a stepwise accumulation of gene mutations that eventually transform the epithelial cell into a tumor cell with mesenchymal features (8, 9, 16–20). This idea comes from findings that many cancers arise in epithelial tissues where abnormalities in cell–cell and cell–matrix interactions occur during tumor progression. Eventually, neoplastic cells emerge that appear as mesenchymal cells, which lack cell–cell adhesion, are dysmorphic in shape, and eventually spread to distant organs (7, 16, 17). How does this extremely complicated phenomenon actually happen?

Jean Paul Thiery provided a comprehensive overview of how EMT might contribute to metastasis (Fig. 13.2). Recent studies also suggest that misplaced (ectopic) coexpression of only two genes might be all that is necessary to facilitate EMT in some gliomas, though the process is highly complex (21). However, considerable controversy surrounds the EMT hypothesis of metastasis, as EMT is not often detected in tumor pathological preparations (1, 22, 23). The EMT is primarily considered a phenomenon of the in vitro environment (7). It remains debatable whether this in vitro model of metastasis has an in vivo counterpart.

The idea for the EMT arose from attempts to draw parallels between the behavior of normal cells during metazoan morphogenesis and the behavior of cancer cells during tumor progression (9, 16). Adaptation of the EMT into the gene theory of cancer suggested that metastasis is the endpoint of a series of genomic alterations and clonal selection. This then provided the neoplastic cells with a growth advantage over normal cells (17, 20, 24, 25). It is difficult to understand how a collection of gene mutations, many of which are random, could produce cells with the capacity to detach from the primary tumor, intravasate into the circulation and lymphatic systems, evade immune attack, extravasate at distant capillary beds, and recapitulate epithelial characteristics following invasion and proliferation in distant organs. This would be quite a feat for a cell with a disorganized genome.

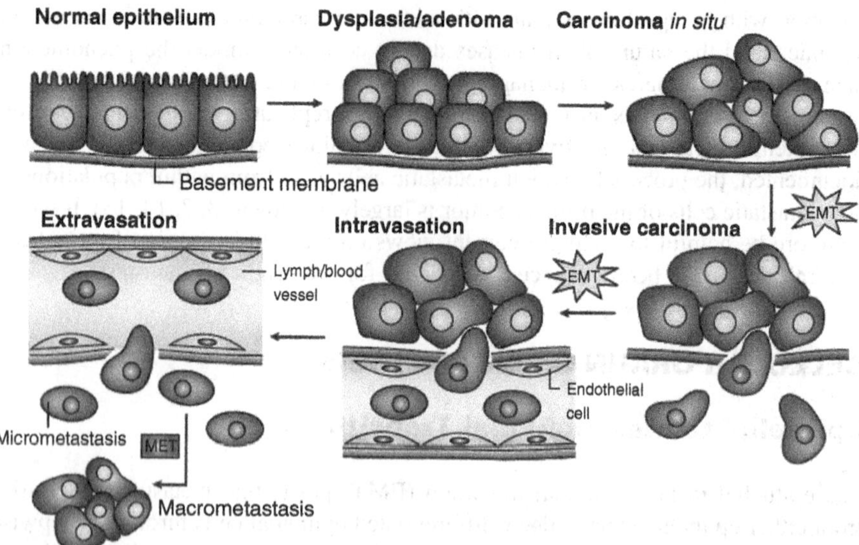

Figure 13.2 The epithelial–mesenchymal transition (EMT) model of tumor metastasis. Sites of EMT and mesenchymal–epithelial transition (MET) in the emergence and progression of carcinoma. According to Jean Paul Thiery, normal epithelia lined by a basement membrane can proliferate locally to give rise to an adenoma. Further transformation by epigenetic changes and genetic alterations leads to a carcinoma *in situ*, still outlined by an intact basement membrane. Further alterations can induce local dissemination of carcinoma cells, possibly through an EMT, and the basement membrane becomes fragmented. The cells can intravasate into lymph or blood vessels, allowing their passive transport to distant organs. At secondary sites, solitary carcinoma cells can extravasate and either remain solitary (micrometastasis) or they can form a new carcinoma through an MET. *Source:* Reprinted with permission from Ref. 16. To see this figure in color please go to ftp://ftp.wiley.com/public/sci_tech_med/cancer_metabolic_disease.

The recapitulation of epithelial characteristics at distant secondary sites is referred to as the *mesenchymal–epithelial transition* (MET) and is thought to involve a reversal of the changes responsible for the EMT (9, 16, 17). No explanation has appeared on how the genomic instability and multiple point mutations and chromosomal rearrangements responsible for the neoplastic mesenchymal phenotype could be reversed or suppressed when the tumor cells recapitulate the epithelial phenotype at distant sites. If many of the nuclear genomic mutations are not reversed, how is it possible that they could be responsible for EMT in the first place? I think the imagination must be stretched to the limits in order to accept the EMT/MET as a credible explanation for metastasis. The changes in cell behavior and morphology linked to this explanation of metastasis and their dramatic reversibility are similar in some ways to those of the *werewolf*.

Our recent studies suggest that random mutations and EMT are not required for the origin of metastasis (26). The massive complexity associated with the EMT hypothesis is largely man-made, especially in attempting to describe the phenomenon as a gene-driven process (9, 16, 17, 21, 27). If one looks closely, many of the gene expression profiles observed in metastatic cancers are similar to those associated with the function of macrophages or other fusogenic cells of the immune system (7, 28, 29). Many gene changes associated with EMT can also be found in most benign tumors (5, 27). As we now know, cancer is not a genetic disease, but is a metabolic disease involving mitochondrial dysfunction and respiratory insufficiency. Credible mechanisms of metastasis must, therefore, be framed in light of the underlying origin of cancer as a mitochondrial respiratory disease. The EMT/MET theory has yet to do this.

Stem Cell Origin of Metastasis

Several investigators hold that metastatic cancer cells arise from populations of tissue stem cells (30–32). Most tissues contain cells in semidifferentiated states that can replace dead or damaged cells due to natural wear and tear. These undifferentiated or semidifferentiated cells are often referred to as *tissue stem cells* and are considered by many to be the origin of metastatic cancers (21, 30, 33, 34). Similarities in gene expression and biological characteristics are often seen in stem cells and cancer cells (35). Observations that tumor cells express characteristics of undifferentiated stem cells come from the fact that embryonic stem cells and tumor cells largely use anaerobic energy (fermentation) for metabolism. As mentioned in Chapter 12, telomerase activity, which is high in tumor cells, is also linked to fermentation energy. It is therefore not surprising that numerous genetic and biochemical phenotypes will be shared between tumor cells and stem cells, as most tumor cells also use energy from fermentation.

As stem cells are known for their ability to proliferate and migrate during tissue morphogenesis and differentiation, it was reasonable to assume that genetic damage to stem cells could give rise to metastatic cancers in various tissues (18, 30, 35). However, many tumor cells with stem cell properties do not express systemic metastasis. Indeed, many of the chemically induced brain tumors I developed in mice over the years express stem cell properties but do not display extensive invasion or metastasis (36, 37). Most of these tumors also express several of the Hanahan–Weinberg cancer hallmarks. However, only those tumor cells that expressed characteristics of macrophages showed *systemic* metastasis (26, 36).

The origin of glioblastoma from stem cells alone is now questioned (38). While metastatic cancers can express properties of stem cells, expression of stem cell properties alone is not synonymous with expression of distant invasion and metastasis. Tumors derived from hematopoietic stem cells, however, may be an exception (31). Hematopoietic stem cells can give rise to myeloid cells, which we consider the cellular origin of most metastatic cancers (26, 36). According to my hypothesis, metastatic cancers arise from respiratory insufficiency in hematopoietic stem cells or their lineage descendants, for example, macrophages or lymphocytes. Chronic hypoxia in the inflamed microenvironment can permanently damage mitochondrial respiration in activated macrophages (39, 40). What is the evidence that metastatic cancer arises from myeloid cells?

Myeloid Cells as the Origin of Metastasis

My former graduate student, Dr. Leanne Huysentruyt, and I recently reviewed evidence showing that many characteristics of myeloid cells are also seen in most human metastatic cancers (26). Indeed, neoplastic brain macrophages (microglia) could also represent the most highly invasive cells in glioblastoma (41). The myeloid cell origin of metastasis proposes that metastatic cancer cells arise from myeloid cells regardless of tissue origin (26).

Myeloid cells are already mesenchymal cells and would therefore not require the complicated genetic mechanisms proposed for the EMT in order to metastasize. Macrophages arise from the myeloid lineage and have long been considered the origin of human metastatic cancer (15, 26, 42–45). Macrophages can fuse with epithelial cells within the inflamed microenvironment, thus manifesting properties of both the epithelial cell and macrophage in the fusion hybrids (29, 46). The origin of metastatic cancer from hematopoietic stem cells, derived from bone marrow cells, is also consistent with the myeloid hypothesis. In his recent excellent review on metastasis, David Tarin states: "*Hence, it would appear that tumor metastasis first appears in the lower chordates in parallel with the origin of lymphocytes and this may indicate that metastasis cannot occur until an organism has evolved the genes for lymphocyte trafficking*." According to our hypothesis, it is hematopoietic stem cells themselves or their lineage descendants that become the metastatic cells either through direct transformation in the inflamed microenvironment or through their fusion with neoplastic tumor cells.

The idea that transformed myeloid cells can give rise to invasive and metastatic cells within tumors is not widely recognized. Rather than being recognized as part of the neoplastic cell population, many investigators consider macrophages and other myeloid cells as part of the tumor stroma (35, 47–54). The macrophages present in tumors are generally referred to as *tumor-associated macrophages (TAMs)* and often comprise significant numbers of the inflammatory cell infiltrate in tumors. TAMs also establish the premetastatic niche, while enhancing tumor inflammation and angiogenesis (55). The properties of TAM facilitate tumor development and progression (35, 47, 49, 56, 57). While some TAMs are certainly part of the stroma, we recently reviewed evidence showing that human tumors also contain neoplastic cells with macrophage properties (26, 41).

It is important to mention that metastatic cells of macrophage origin are generally not found in rodent tumor transplant models. Most macrophages seen in chemically induced tumors are derived from TAM, as we previously showed in experimental mouse brain tumors grown either orthotopically or subcutaneously in the flank (58, 59). I suggest that rodent tissues respond to tumor implants as if they were an *acute* infection or wound. This would involve invasion of TAM and activation of local macrophages. It is also possible that fusion hybrids would form between tumor cells and host macrophages, but this might not cause damage to the macrophage mitochondria.

In contrast to the acute situation in mice, neoplastic transformation is a protracted process in humans. In other words, murine myeloid cells respond acutely to the tumor implant, whereas human myeloid cells in the inflamed microenvironment respond chronically to the tumor initiating insult. It is not really clear why highly metastatic carcinomas, similar to those seen in humans, are rarely found in experimental rodent tumors. There are, however, some exceptions (1, 36).

MACROPHAGES AND METASTASIS

What are the properties of macrophages that would make them prime suspects for the origin of metastasis? Macrophages are among the most versatile cells of the body with respect to their ability to migrate, to change shape, and to secrete growth factors and cytokines (36, 60–62). These macrophage behaviors are also the recognized behaviors of metastatic cells. Macrophages manifest two distinct polarization phenotypes: the classically activated (M1 phenotype) and the alternatively activated (M2 phenotype). Macrophages acquire the M1 phenotype in response to proinflammatory molecules and release inflammatory cytokines, reactive oxygen species, and nitric oxide (28, 51, 53, 63–65). In contrast, macrophages acquire the M2 phenotype in response to anti-inflammatory molecules such as IL-4, IL-13, and IL-10 and to apoptotic cells (51, 66). M2 macrophages promote tissue remodeling and repair but are immunosuppressive and poor antigen presenters (53). Although the M1 and M2 macrophages play distinct roles during tumor initiation and malignant progression, macrophage–epithelial cell fusions can involve either activation state (29, 67).

M1 macrophages facilitate the early stages of tumorigenesis through the creation of an inflammatory microenvironment that can produce nuclear and mitochondrial damage (67). However, TAM can also undergo a phenotypic switch to the M2 phenotype during tumor progression (51). The TAM population comprising M2 macrophages scavenge cellular debris, promote tumor growth, and enhance angiogenesis. M2 macrophages also fuse with tumor cells and are considered facilitators of metastasis (51, 53, 68, 69). It has always been difficult to know for certain, however, whether TAMs are part of the normal stroma or are part of the malignant cell population (26). This is especially the case in human cancers.

Increasing evidence suggests that many of the myeloid/macrophage cells seen within human tumors are also part of the malignant cell population. Aichel proposed over a century ago that tumor progression involved fusion between leukocytes and somatic cells (reviewed in Ref. 44). Several human metastatic cancers express multiple molecular and behavioral characteristics of macrophages, including phagocytosis, cell–cell fusion, and antigen expression (Table 13.1). Tarin (1) also considers the expression of osteopontin (OPN) and CD44 as important in the regulatory gene group/network associated with metastasis. This is interesting as there is strong evidence that both OPN and CD44 are expressed in monocytes and macrophages under various physiological and pathological states (70–72). We argued that an origin of metastatic cancer from myeloid cells could account for many mesenchymal properties of metastatic cancers (26). It is not, therefore, necessary to invoke an EMT to account for metastasis.

Table 13.1 Tumors Expressing Macrophage Characteristics

Tumor	Phagocytosis	Fusogenicity	Gene expression
Bladder	(73)		
Brain	(36, 74–82)	(83–85)	(36, 82)
Breast	(86–93)	(94–98)	(99–101)
Carcinoma of unknown primary	(102)	(103)	
Endometrial	(104)		
Fibrosarcoma	(93)		
Gall bladder		(105)	
Liver		(106)	
Lung	(88, 107–110)	(95)	(111–113)
Lymphoma/leukemia	(114–116)	(117–119)	
Melanoma/skin	(120–124)	(45, 125)	(124, 126–128)
Meth A sarcoma	(129)	(129)	(129)
Multiple myeloma	(130)	(131)	
Ovarian	(93, 132)		(133)
Pancreatic	(134, 135)	(136)	(135)
Rectal/colon		(29, 137)	(138)
Renal	(139)	(140, 141)	(139)
Rhabdomyosarcoma	(142, 143)		
Reviews	(144–147)	(15, 42–44, 46, 98 118, 148–151)	

Source: This table is updated from that previously published (26).

Interestingly, macrophages express most hallmarks of metastatic tumor cells when responding to tissue injury or disease. For example, monocytes (derived from hematopoietic bone marrow cells) extravasate from the vasculature and are recruited to the wound via cytokines released from the damaged tissue (26, 28). Within the wound, monocytes differentiate into alternatively activated macrophages and dendritic cells where they release a variety of proangiogenic molecules, including vascular endothelial growth factor, fibroblast growth factor, and platelet-derived growth factor (28, 152, 153). M2 macrophages also actively phagocytize dead cells and cellular debris (28, 62). Occasionally, macrophages undergo homotypic fusion resulting in multinucleated giant cells with increased phagocytic capacity (43, 69, 154). Following these wound healing activities, macrophages intravasate back into the circulation where they travel to the lymph nodes to participate in the immune response (62, 155, 156).

Some phagocytic macrophages also migrate to lymph nodes and differentiate into dendritic cells (157). These findings indicate that normal macrophages are capable of expressing all hallmarks of metastatic cancer cells, including tissue invasion, release of proangiogenic molecules/cytokines, survival in hypoxic and necrotic environments, intravasation into the circulatory/lymphatic systems, and extravasation from these systems at distant locations. An EMT is not necessary to explain these behaviors, as they are already the evolutionary programmed behaviors of macrophages and cells of myeloid origin.

Phagocytosis: A Shared Behavior of Macrophages and Metastatic Cells

Phagocytosis involves the engulfment and ingestion of extracellular material and is a specialized behavior of M2 macrophages and other professional phagocytes (62). This process is essential for maintaining tissue homeostasis by clearing apoptotic cells, cellular debris, and invading pathogens. Like M2 macrophages, many malignant tumor cells are phagocytic both in vitro and in vivo (Table 13.1).

Tumor cell phagocytosis was first described over a century ago from histopathological observations of foreign cell bodies within in the cytoplasm of cancer cells, which displayed crescent-shaped nuclei (44). This cellular phenotype resulted from the ingested material pushing the nucleus to the periphery of the phagocytic cell. These cells were commonly referred to as either *bird's-eye* or *signet-ring* cells (144, 158). While this phagocytic/cannibalistic phenomenon is commonly seen in feeding microorganisms, cell cannibalism is also seen in malignant human tumor cells (120, 144, 158, 159). Fais and colleagues provided dramatic evidence of tumor cell phagocytosis in showing how malignant melanoma cells eat T-cells (Fig. 13.3). This is remarkable as T-cells are thought to target and kill tumor cells.

There is also evidence that some tumor cells can eat NK cells (159). If macrophage-derived metastatic cells can eat T-cells and possibly NK cells, then it is possible that immune therapies involving these cells might not be effective for long-term management of some metastatic cancers. Indeed, cancer immunotherapies have had little impact in reducing the yearly death rate from advanced metastatic cancers. It is not often mentioned how potential immunotherapies for melanoma and other cancers might deal with the issue of phagocytic metastatic macrophages (160).

Melanocytes are the resident macrophages of the skin. Expression of cathepsins B and D are elevated in the phagocytic melanoma cells just as they are in malignant melanomas (120). These tumor cell phagocytic/cannibalistic behaviors are not to be confused with autophagy, a cellular self-digestion process often associated with starvation conditions (158, 161, 162). Many human cancers and some murine cancers can phagocytize other tumor cells, erythrocytes, leukocytes, platelets, dead cells, as well as extracellular particles (Table 13.1). Hence, the characteristics of phagocytosis appear similar in resident skin macrophages and in malignant melanoma.

Phagocytic Cancers

Numerous reports have described the phagocytic behaviors seen in aggressive human cancers (Table 13.1). We previously identified two spontaneous

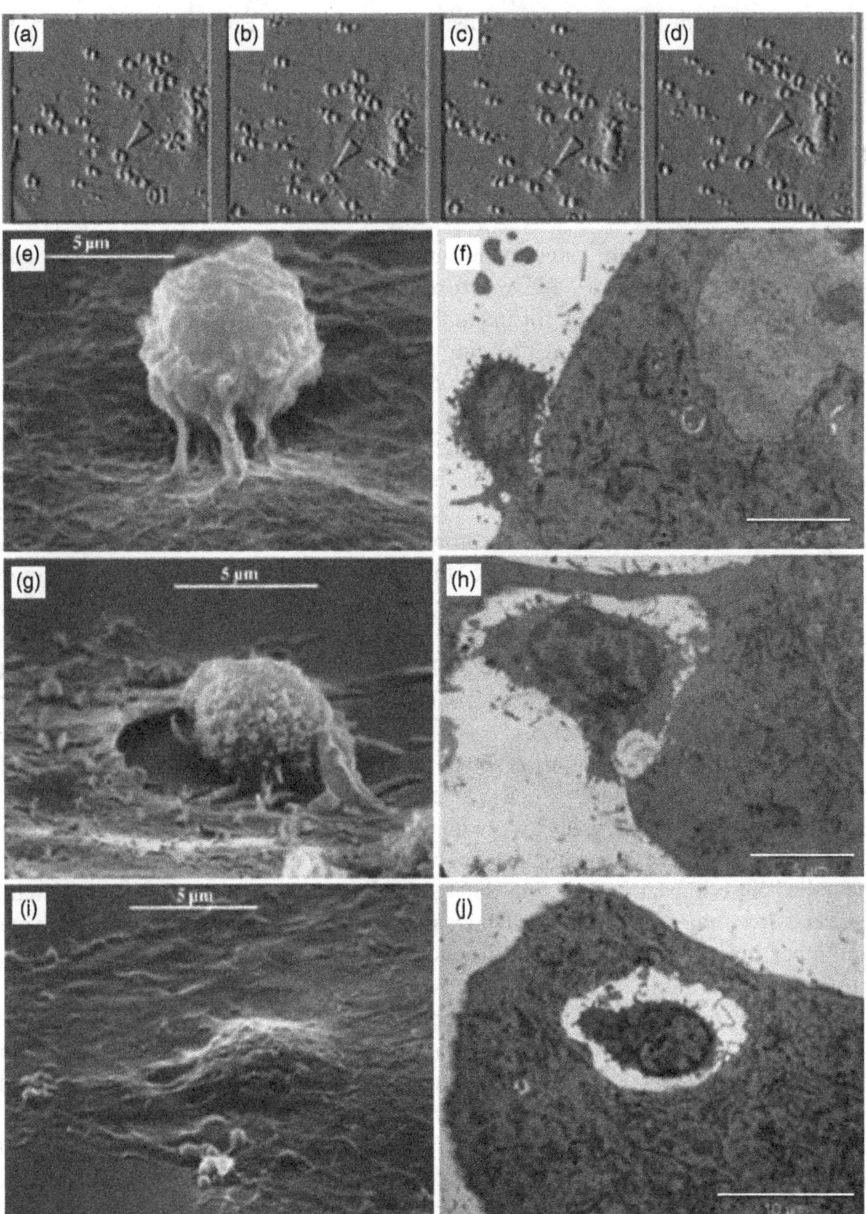

Figure 13.3 Lymphocyte–melanoma cell interaction in cannibalism. (a–d) Time-lapse cinematography analysis of a coculture of live autologous CD8+ T lymphocytes and metastatic melanoma cells (melanoma cell to lymphocyte ratio of 1:2.5). Scanning electron microscopy (e, g, and i) and transmission electron microscopy (f, h, and j) of a coculture of live autologous CD8+ T lymphocytes and metastatic melanoma cell monolayer (melanoma cell to lymphocyte ratio of 1:2.5). (e and f) Intimate contact between a lymphocyte and a melanoma cell. (g and h) Lymphocyte embraced by melanoma cells. (i and j) Internalized lymphocyte. *Source:* Reprinted with permission from Ref. 120.

invasive/metastatic murine brain tumors (VM-M2 and VM-M3) that express many macrophage characteristics, including phagocytosis (36). These metastatic tumor cells can engulf fluorescent beads. One of the more interesting features of these natural mouse brain tumors was their metastatic behavior when grown outside the central nervous system. The cells spread to multiple organ systems following implantation into most extracranial sites. While extracranial metastasis of central nervous system tumors is not common, many gliomas, especially glioblastoma multiforme, are highly metastatic if the tumor cells gain access to extraneural tissue (41). Indeed, several investigators have documented the metastatic behavior of malignant brain cancers, especially GBM (36, 163–168). Hence, the VM-M2 and VM-M3 tumors replicate this feature of GBM behavior.

One report showed that recipients of organs from a donor with GBM developed metastatic cancer (169). This indicates that neoplastic cells from this GBM metastasized from the brain and infiltrated extraneural tissues without detection. As extraneural tissues are not often examined in patients dying from GBM, it is not clear if this was a rare event or was part of a more general phenomenon. Dr. Brent Reynolds, a leader in the stem cell field, mentioned to me that circulating metastatic cells are not uncommon in GBM patients (personal communication). Moreover, extracranial metastasis of brain tumors portends an extremely poor survival, with the vast majority of patients surviving <6 months from the diagnosis of metastatic GBM disease (170). The widely held view that metastasis does not occur for GBM should be reevaluated (168). Many GBM patients often die before detection of systemic metastasis. A take-home message is that GBM patients should not donate their organs!

While it might be difficult to prove a myeloid origin of invasive GBM cells, substantial evidence shows that subpopulations of neoplastic GBM cells display the phagocytic behavior of macrophages/microglia. As microglia are the resident macrophages of the brain, we considered that some of the cells in these tumors could arise from neoplastic microglia/macrophages (35, 41, 50). Human GBM contain mixtures of numerous neoplastic cell types, many of which have mesenchymal properties, do not express glial fibrillary acidic protein (GFAP), and are of unknown origin (171–173). Indeed, the original nineteenth-century observations of Virchow (1863/1865) described glioblastomas as *gliosarcomas* of mesenchymal origin (174, 175). While numerous mesenchymal cells are frequently seen in GBM, the specific classification of all tumor cell types within human GBM remains ambiguous at best (41, 171, 176–178).

According to my hypothesis, many of the neoplastic mesenchymal cells seen in GBM arise from transformed macrophages or microglia that fuse with neoplastic stem cells (41). Such hybrid cells would represent the most invasive neoplastic cells within the tumor. It also appears that bevacizumab (Avastin) selects for the most invasive cells within GBM (179). This could account for why tumor recurrence following bevacizumab therapy is universally fatal (180). Complicated EMT explanations for the mesenchymal properties of GBM are unnecessary once it becomes recognized that the same properties can arise from neoplastic transformation of microglia/macrophages.

Phagocytic behaviors have been reported for many human cancers including skin, breast, lymphoma, lung, brain, ovarian, pancreatic, renal, endometrial, rhabdomyosarcoma, myeloma, fibrosarcoma, and bladder (Table 13.1). For most of these tumors, the phagocytic phenotype was restricted primarily to those cells that were also highly invasive and metastatic (36, 74, 75, 86–88, 107, 108, 120, 121, 134). Hence, the most potentially deadly cells within tumors are those with macrophage properties.

Lugini and coworkers measured the phagocytic behavior of cell lines derived from primary human melanomas ($n = 8$) and metastatic lesions ($n = 11$) (121). Interestingly, the phagocytic behavior in all the cell lines derived from metastatic lesions was similar to that of the macrophage controls, whereas phagocytic behavior was not found in any of the cell lines derived from primary melanomas (121). Histological examination of in vivo metastatic melanoma lesions confirmed the presence of phagocytic tumor cells (120).

Numerous phagocytic tumor cells were identified within metastatic breast cancer lesions and were not observed within the primary tumor of the same patient (87). This is similar to the appearance of phagocytic signet-ring cells observed in secondary metastatic lesions in other breast cancer patients (181). Additionally, the number of phagocytic tumor cells present within the tumor stroma correlates with breast cancer malignancy and grade (89). Hence, phagocytosis is a common macrophage phenotype seen in many metastatic human cancers.

Metastatic Behavior of the RAW 264.7 Mouse Macrophages

RAW 264.7 cells are considered a normal mouse macrophage cell line and are widely used to study a broad range of macrophage properties. RAW 264.7 cells were transformed with Abelson leukemia virus and derived from BALB/c mice. It is known that viruses damage mitochondrial function (Chapter 9). We used the RAW 264.7 cells as a control cell line for our metastatic VM-M2 and VM-M3 metastatic cancer cells (36). Using fluorescent microspheres, we found that the phagocytic activity of the metastatic VM-M2 and VM-M3 tumor cells was similar to that of the RAW 264.7 macrophage cell line (36). Not only were there similarities between the RAW cells and the metastatic VM tumor cells for phagocytic behavior, but these cells also were similar in their morphology, gene expression, and lipid composition (Fig. 13.4).

How was it possible that the RAW macrophages could be so similar to the metastatic VM cells, yet be considered normal cells? To confirm the normality of these cells, I asked my students to implant the RAW 264.7 cells subcutaneously into the flanks of immunodeficient BALB/SCID mice to confirm that they did not form tumors. The results were shocking. The RAW 264.7 cells not only formed tumors, but the cells from these tumors also metastasized throughout the mice!

We discovered that the RAW 264.7 macrophage cell line is highly metastatic following intracerebral and subcutaneous transplantation into SCID mice. The

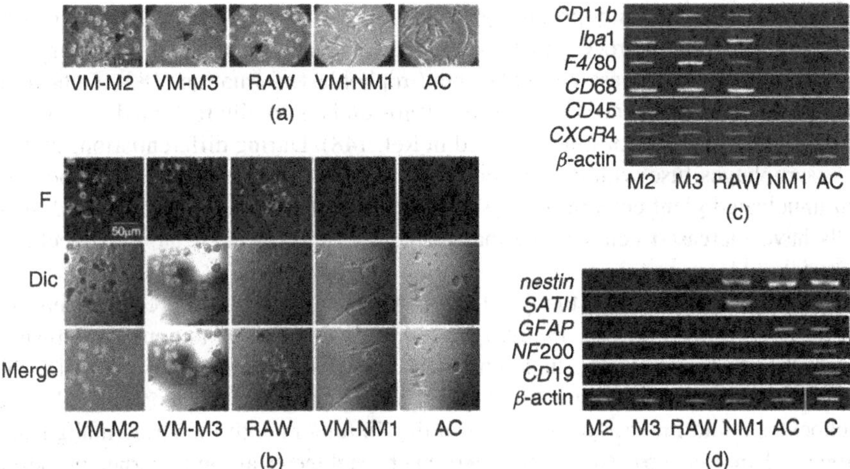

Figure 13.4 Characteristics of mouse metastatic cancer cells. In vitro morphology (a), phagocytic behavior (b), and gene expression (c and d) in VM tumors and control cell lines. Phagocytic behavior was assessed from merging (Merge) the fluorescence (F) images and the differential interference contrast (DIC) images (b). C designates control tissue. Embryonic brain was used for nestin and SAT II, adult brain was used for GFAP and NF200, and spleen was used for CD19. β-actin was used the control. Total lipid composition was also similar in the metastatic cancer cells and macrophages, suggesting a common origin (36). The scale bars shown in (a) and (b) are the same for all images in that panel. At least three independent samples were analyzed for each experiment. *Source:* Reprinted with permission from Ref. 36. See color insert.

metastatic properties of the RAW 264.7 cells are similar to those of our VM-M2 and VM-M3 cell lines in forming metastases in lung, liver, spleen, and kidney.

The metastatic behaviors of the RAW cells also appear similar to the metastatic behaviors of the tumors described by Kerbel et al. (117, 182). Like the VM-M2 and VM-M3 tumor cell lines, the RAW 264.7 cells express little ganglioside GM3 and metastasize to multiple organ systems (liver, spleen, kidney, lung, and brain) when grown subcutaneously outside the brain (36). Ganglioside GM3 inhibits angiogenesis and blocks tumor cell invasion (36).

These findings provided direct evidence showing that cells with macrophage properties can give rise to metastatic cancer regardless of how the cells are classified. Moreover, our findings suggest that a single mechanism was responsible for the metastatic phenotype regardless of cell origin. We think that this mechanism involves fusion hybridization followed by or commensurate with mitochondrial damage and respiratory insufficiency.

Fusogenicity: A Shared Behavior of Macrophages and Metastatic Cells

Fusogenicity is the ability of a cell to fuse with another cell through the merging of their plasma membranes (29, 154). This process can arise in vitro as is seen with the formation of antibody-producing hybridomas. However, fusion in human cells is a highly regulated process that is essential for fertilization (sperm and egg) and skeletal muscle (myoblasts) and placenta (trophoblast) formation (183). Outside of these developmental processes, cell-to-cell fusion is normally restricted to differentiated cells of myeloid origin (reviewed in Ref. 148). During differentiation, subsets of macrophages fuse with each other to form multinucleated osteoclasts in bone or multinucleated giant cells in response to foreign bodies (43). Osteoclasts and giant cells have increased cell volume that facilitates engulfment of large extracellular materials (43).

Macrophages are also thought to fuse with damaged somatic cells during the process of tissue repair (29, 43, 69, 154, 184, 185). In addition to homotypic fusion, macrophages are known to undergo heterotypic fusion with tumor cells (15, 29, 43, 68, 148, 186). Aichel suggested in 1911 that fusion between somatic cells and leukocytes could induce aneuploidy resulting in tumors with increased malignancy (reviewed in Ref. 45). Nearly 60 years later, Mekler et al. and Warner proposed that fusion of committed tumor cells with host myeloid cells would produce tumor hybrids capable of migrating throughout the body and invading distant organs (149, 187). Recent studies from Wong and coworkers described how macrophages fuse with tumor epithelial cells (29, 188). Besides inflammation, radiation also increases the fusion hybrid process (188). Is it possible that decreased long-term survival in some irradiated cancer patients results from enhanced production of macrophage–epithelial fusion hybrids? We have stated that the human brain should rarely if ever be irradiated (189). It is my opinion that radiation will contribute to brain tumor recurrence.

John Pawelek and colleagues strongly favor the fusion hypothesis for the origin of metastatic cancers (44, 45, 99, 125, 140, 141, 190, 191). They provide compelling evidence showing that fusion hybrids could account for the diversity of cell phenotypes observed within tumors. Fusion between neoplastic tumor cells and myeloid cells, with subsequent nuclear fusion, could produce new phenotypes in the absence of new mutations, as the hybrids would express genetic and functional traits of both parental cells (45). These neoplastic hybrids would have the ability of macrophages to intravasate, extravasate, and migrate to distant organs, while also possessing the unlimited proliferative potential of the cancer cells. Since myeloid cells are part of the immune system (29), it would be easy to see how tumor hybrids would also be able to evade immune surveillance.

Fusogenic Cancers

Many tumor cells are fusogenic (148). Fusogenic tumor cells are found in a wide variety of cancer types including, melanoma, breast, renal, liver, gall bladder, lymphoma, and brain (Table 13.1). Tumor cell hybrids can form either in vitro or in vivo from fusions between two tumor cells or between a tumor cell and a normal somatic cell. One of the first reports of tumor cell fusion hybrids showed that human glioma cells, when implanted within the cheeks of hamsters, spontaneously fused with nontumorigenic host cells, resulting in metastatic hybrid human–hamster tumor cells (83). Many of the early reports for fusogenic cancers described fusions between lymphomas and myeloid cells. For example, spontaneous in vivo fusion between the nonmetastatic murine MDW4 lymphoma and host bone marrow cells resulted in aneuploid metastatic tumor cells (26, 117).

Munzarova et al. recognized that numerous traits expressed in macrophages were also expressed in metastatic melanoma cells and suggested that the tumor metastasis could result from fusions between tumor cells and macrophages (42, 150). Pawelek and coworkers tested this hypothesis by inducing fusions between cultured nonmetastatic Cloudman S91 melanoma cells and murine peritoneal macrophages. The majority of the resulting macrophage–melanoma hybrids displayed increased metastatic potential when grown in vivo (45). Further studies revealed that the Cloudman S91 melanoma cells underwent spontaneous fusion with the murine host cells in vivo resulting in secondary lesions that were comprised mostly of tumor–host cell hybrids. The fusion of tumor cells with host myeloid cells was a compelling explanation (125).

Artificial fusions of human monocytes and mouse melanoma cells revealed that the resulting hybrids expressed both human and mouse genes (45). Other investigators also showed that the macrophage-specific antigens F4/80 and Mac-1 were expressed in murine Meth A sarcoma cells after spontaneous in vivo fusion with host cells. Interestingly, latex bead phagocytosis was also expressed in the Meth A sarcoma–host cell fusion hybrids (129). Since these fusion hybrids expressed genotypes and phenotypes of both parental cells, it appears that the nonmetastatic tumor cells could acquire an invasive/metastatic phenotype without new mutations. Such findings are at odds with the EMT hypothesis of metastasis.

It is well documented that TAMs promote tumor progression through the release of cytokines and proangiogenic and prometastatic molecules (reviewed in Ref. 35 and 47). However, the fusion of cancer cells with tissue macrophages could also accelerate tumor progression. Fusion among tumor cells in human solid tumors is difficult to detect. Several reports provide evidence for fusions between tumor cells and myeloid cells in human bone marrow transplant (BMT) recipients (140, 141). Such fusions would accelerate tumor progression.

Wong and colleagues recently conducted parabiosis experiments, where one mouse is surgically attached to another mouse, to show how bone marrow-derived cells of one mouse fuse with intestinal tumor cells of the other mouse (29). Moreover, they identified the macrophage as the driver for this process. They also showed that the fused hybrid cells retained a transcriptome identity characteristic of both parental derivatives, while also expressing unique transcripts (29). These findings show how fusions between macrophages and tumor cells, within the inflamed wound environment, could give rise to the metastatic phenotype of cancer cells, thus enhancing tumor progression.

It is important to recognize that both radiation therapy and immunosuppression can increase the incidence of metastatic cancers (192). DNA analysis of microdissected metastatic cells from a child diagnosed with renal cell carcinoma after a BMT revealed DNA from both the BMT donor and the recipient in the metastatic cells (141). Bone marrow and tumor cell hybrids were also identified in a female, who developed renal carcinoma after receiving a BMT from a male donor (140). These reports provided genetic evidence that spontaneous fusions can occur between human myeloid cells and tumor cells. It should not be surprising that the radiation therapy given to many cancer patients might actually exacerbate formation of metastases and disease progression as I have mentioned above (138, 193, 194).

Macrophage–macrophage fusions could also induce aneuploidy in the fused hybrids (191, 195). The numerous in vitro studies and in vivo reports suggest that myeloid hybrids could be responsible for the metastatic progression of numerous cancers. Multinucleated giant cells, a signature of hybrid formation, are frequently seen in human cancers, suggesting that cell fusions are not rare events (Table 13.1). Regardless of the mechanism, metastatic cells express numerous behaviors of mesenchymal/myeloid cells and, if exploited, could generate novel therapeutic strategies for managing metastatic cancers.

Myeloid Biomarkers Expressed in Tumor Cells

Myeloid cells express a wide variety of biomarkers that are unique to their ontogeny and function (196). Routine histological and immunohistochemical analyses are often preformed to assess tumor type and grade. Since TAMs are often correlated with a poor patient prognosis, tumor biopsies are frequently evaluated for macrophage markers. The macrophage antigen-expressing cells within the tumor stroma are usually classified as TAMs. However, several reports show that macrophage-specific antigens and biomarkers are also expressed on a wide variety of human cancer cells (Table 13.1).

One of the most interesting studies was that of Ruff and Pert (111), who demonstrated that several macrophage antigens (CD26, C3bi, and CD11b) were expressed on tumor cells from small cell lung carcinoma (SCLC). Levels of expression were comparable to that seen in the monocyte controls. It is important to note that the macrophage antigens were also expressed in the cultured tumor cells themselves. This tumor cell expression was confirmed from in vivo tissue preparations. This eliminated the possibility that the antigen expression was derived from TAMs. These investigators concluded that the SCLC tumor cells in their specimens were not of lung epithelial origin but rather were of "myeloid origin." A malignant transformation of recruited myeloid cells, from smoking-related tissue damage, was offered as an explanation for the origin of tumor cells with myeloid/macrophage properties (111). Although this interpretation was controversial (100, 197), the authors showed additional myeloid properties of these tumor cells to support their hypothesis of a macrophage origin (112). The findings of Ruff and Pert that myeloid cell phenotypes were expressed in SCLC were confirmed in other independent studies of this tumor type (113, 197). In light of the above discussion, it is also possible that the myeloid properties of the SCLC were derived from fusions of macrophages and neoplastic lung epithelial cells.

Besides SCLC, myeloid-associated antigens (CD14 and CD11b) were also expressed in five metastatic breast cancer cell lines (100). None of the breast cancer cell lines, however, expressed markers for B- or T-cells (100). The authors suggested that common antigen sharing between different cell types could be related to common cellular interactions (100). Further evidence for a mesenchymal origin of metastatic cancer comes from tissue microarray analysis of 127 breast cancer patients (138). The CD163 macrophage scavenger receptor was expressed on the tumor cells of 48% of the patients, while MAC387 macrophage marker was expressed on the tumor cells of 14% of the patients (138). Pathology confirmed that the staining was localized to the tumor cells and not solely to the tumor-infiltrating macrophages. Interestingly, cancers that contained CD163-expressing tumor cells had a more advanced histological grade, a higher occurrence of distant metastasis, and reduced patient survival (138). This report demonstrated, for the first time, that tumor cells expressing macrophage antigens could be identified in more than half of breast cancer patients.

Similar studies were conducted on patients with rectal cancer (138). As in breast cancer patients, CD163 was expressed on tumor cells in many patients with rectal cancer (138). Moreover, CD163 expression was found in 31% of the rectal tumors from patients in the preoperative irradiation group but was expressed in only 17% in the nonirradiation group. Prognosis was also worse for those patients with CD163-positive cancer cells than in those patients with CD163-negative cancer cells. Inflammation and radiation is known to enhance formation of macrophage–epithelial cell fusion hybrids (137). In addition to these studies on human cancers, Maniecki et al. (198) presented findings at the 2011 meeting of the Amer. Assoc. Cancer Res. (AACR) showing that expression of CD163 could be a common phenotype of many metastatic cancers arising from heterotypic cell fusions between tumor cells and macrophages. The findings in these metastatic cancers are consistent with the origin of metastatic cells from macrophage fusion hybrid cells, which are increased by radiation and inflammation.

These findings provide additional evidence that radiation therapy can be counterproductive to long-term survival of patients (194). Although radiation therapy can help some cancer patients, radiation therapy will also enhance mitochondrial damage and fusion hybridization, thus potentially making the disease much worse. These findings are consistent with role of radiation in inducing tumor cell–macrophage fusions and in exacerbating the metastatic properties of some cancers (137, 138, 194). It also appears that some antiangiogenic drugs such as bevacizumab and cediranib actually increase the number of invasive cells with macrophage properties in brain tumors (54, 199). In light of the findings presented here, I suggest that these drugs select for invasive tumor cells with macrophage properties. This would not be beneficial to patients. Viewed together, these studies demonstrate that macrophage antigens, which are associated with enhanced metastasis and poor prognosis, are expressed on the tumor cells of patients with breast, bladder, rectal, and brain cancers.

Cathepsins, Ezrin, and E-Cadherin

Macrophages express high levels of lysosomal-enriched cathepsins, which facilitate the digestion of proteins ingested following phagocytosis or pinocytosis (200, 201). This is interesting since lysosomal cathepsins D and B are viewed as prognostic factors in cancer patients (120, 201). Indeed, a high content of these enzymes in tumors of the head and neck, breast, brain, colon, or endometrium was considered a sign for high malignancy, high metastasis, and overall poor prognosis (201). Besides the cathepsins, activated macrophages also express ezrin as part of a protein complex with radixin and moesin (202). The ezrin–radixin–moesin is a family of molecules that play essential roles in tissue remodeling by linking the cell surface with the actin cytoskeleton and facilitating signal-transduction pathways (203). There is increasing awareness that ezrin is also expressed in metastatic cancer cells, suggesting an important role in metastatic phenotype of cancer cells (122, 144, 204–206). The transition from the EMT is associated with downregulation of the cell adhesion molecule, E-cadherin (16). It is important to recognize that E-cadherin is either unexpressed or expressed in low levels in macrophages (207, 208). Viewed collectively, these findings provide further evidence linking macrophage phenotypes with the properties of metastatic cancers.

Anemia and Increased Hepcidin in Metastatic Cancer

Iron-deficient anemia is a comorbid trait in many patients with metastatic cancers (209, 210). Hepcidin is a key regulator of iron metabolism and plasma iron levels by controlling the efflux of iron from enterocytes, hepatocytes, and macrophages and by internalizing and degrading the iron exporter, ferroportin (211). Chris Tselepis, Douglas Ward, and colleagues suggested that hepcidin contributes to the systemic anemia in colorectal cancer patients by acting at the level of the macrophage (209). Activated macrophages express IL-6, which induces expression of hepcidin. Macrophages are the major cell type responsible for systemic iron recycling (209, 212). The findings of Ward et al. are therefore consistent with our hypothesis that metastatic cancer is a disease of myeloid cells, especially macrophages. Many characteristics of metastatic cancers can be explained once it becomes recognized that metastatic cancer is a macrophage metabolic disease. Hence, iron-deficient anemia should not be unexpected for metastatic cancers derived from transformed macrophages or macrophage fusion hybrids.

CARCINOMA OF UNKNOWN PRIMARY ORIGIN

Carcinoma of unknown primary (CUP) origin is a systemic metastatic disease without an identifiable primary tumor and is often associated with poor prognosis. Approximately 5% of all newly diagnosed cancers are classified as CUP (213, 214). These cancers are often classified as adenocarcinomas, squamous cell carcinomas, poorly differentiated carcinoma, and neuro-endocrine carcinomas (213, 214). It is thought that these cancers metastasize before the primary tumor has had time to develop into a macroscopic lesion (214). Signet-ring cells were found in some CUP, indicating that subsets of these cancers exhibit phagocytic behavior like other metastatic cancers (102). Interestingly, aneuploidy was identified in 70% of CUP adenocarcinomas but was not found in about 30% of the tumors (103). Aneuploidy can arise in part from cell fusion events (117). Survival was better in patients with aneuploid tumors than with diploid tumors, showing that patients with diploid tumors do not have a more favorable prognosis. This is interesting and is consistent with findings that aneuploidy actually slows cell growth (215, 216). Owing to their high aggressiveness, we suggested that some CUPs could arise from macrophage fusion hybrids (26).

MANY METASTATIC CANCERS EXPRESS MULTIPLE MACROPHAGE PROPERTIES

The evidence presented here and in our recent review shows that many metastatic cancers express multiple myeloid characteristics (Table 13.1). For instance, many phagocytic or fusogenic tumors also express myeloid antigens, further supporting a myeloid origin of these metastatic cancers. It is important to mention that the myeloid properties are expressed in the tumor cells themselves and should not be confused with myeloid properties expressed in TAM, which are also present in the tumors but are not part of the neoplastic cell population. The Pawelek, Lazebnik, and Wong groups have amassed compelling evidence that cell fusion events involving macrophages can give rise to cells that metastasize (29, 148, 191, 217, 218). In contrast to the EMT/MET explanation of metastasis, the macrophage cell fusion explanation of metastasis does not require the induction and reversion of extremely complicated gene regulatory systems. It is only a matter of time before the myeloid fusion hypothesis becomes the dominant explanation of cancer metastasis.

LINKING METASTASIS TO MITOCHONDRIAL DYSFUNCTION

As I showed in previous chapters, substantial evidence indicates that cancer is a mitochondrial disease arising from respiratory insufficiency. When permanent respiratory damage occurs in cells of myeloid origin, including hematopoietic stem cells and their fusion hybrids, metastasis would be a potential outcome. It is not necessary to blame mutations or to invent complicated genetic regulatory systems to explain the phenomenon of metastasis.

Numerous studies indicate that mitochondria from a broad range of metastatic cancers are abnormal and incapable of generating energy through normal respiration

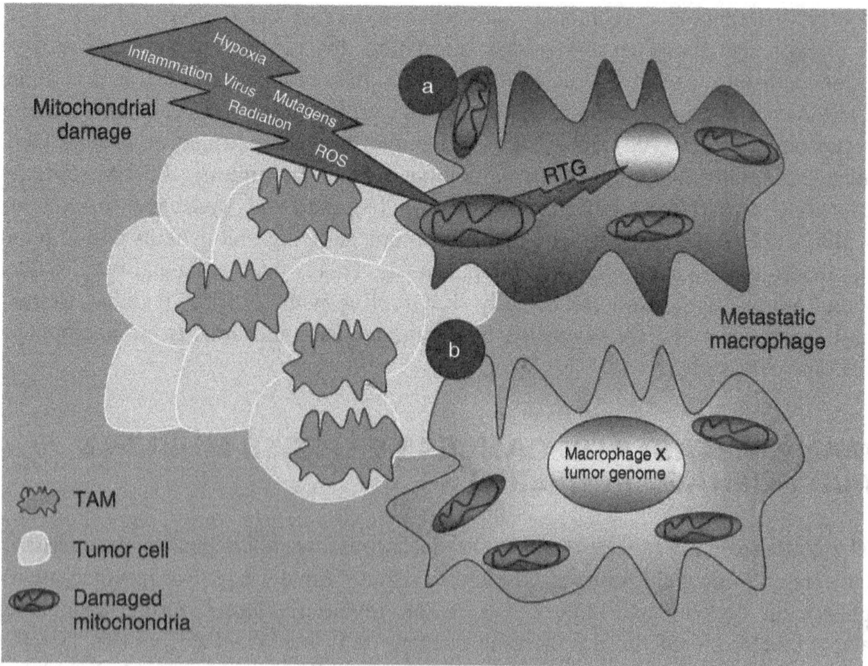

Figure 13.5 Proposed mechanisms of macrophage transformation and metastasis. The tumor microenvironment consists of numerous mitochondria damaging elements, which could impair mitochondria energy production in TAM and tissue macrophages. This would eventually produce genetic instability through the mitochondrial stress or RTG response (a). Fusions between macrophages or between macrophages and cancer stem cells could result in cells expressing both the tumor and macrophage genomes (b). The end result would be cells that can survive in hypoxic environments, can proliferate, and can spread to multiple sites through the circulation. *Source*: Reprinted with permission from Ref. 26. See color insert.

(219, 220). Energy through fermentation is the single most common hallmark of all cancer cells, including those with metastatic potential. This phenotype arises from mitochondrial dysfunction as I have discussed before. Mitochondrial damage can arise in any cell within the inflammatory microenvironment of the incipient tumor, including TAM, homotypic fusion hybrids of hematopoietic cells, or heterotypic fusion hybrids of macrophages and neoplastic epithelial cells. The end result would be cells with metastatic potential. Although metastatic cells will differ in their morphology from one organ system to the next, they all suffer from insufficient respiration. The origin of metastatic cancer from myeloid cells and fusion hybrids can explain the substantial morphological and genetic diversity seen among different tumor types (148). It is clear to me that metastasis can arise in macrophage fusion hybrids that sustain irreversible mitochondrial damage (Fig. 13.5).

What About the Tumor Suppressive Effects of Mitochondria?

I have reviewed substantial evidence in Chapter 11 showing that normal mitochondrial function suppresses tumorigenesis. How do these findings relate to the origin of metastatic cancer cells following macrophage fusions with other cells? If normal macrophages fuse with neoplastic stem cells, would not the normal function of the macrophages suppress tumorigenicity in the fused hybrid? Although normal mitochondria would initially suppress tumorigenicity in fused hybrids, persistent inflammation in the microenvironment will eventually damage the majority of mitochondria in the fused hybrids, thus initiating the path to metastasis. As macrophages also evolved to survive in hypoxic and inflammatory environments, considerable time would likely be required to damage mitochondria in the fusion hybrids from these environments. However, progressive damage to mitochondria in fusion hybrids would be unlikely in nonhypoxic or noninflammatory microenvironments. It is also noteworthy that radiation exposure would not only enhance fusion hybrid formation but would also damage respiration. It should not be surprising why long-term survival is reduced or why more aggressive tumors recur in many patients that receive radiation to treat their cancers.

As respiration is responsible for maintaining genomic stability and the differentiated state, respiratory insufficiency will eventually induce the default state of unbridled proliferation. If this occurs in cells of myeloid origin such as macrophages, then emergence of cells with enhanced metastatic potential would be a predicted outcome. Macrophages are genetically programmed to exist in the circulation and to enter and exit tissues (221). While cells of myeloid origin can serve as the body's best friend during wound healing and in killing pathogenic bacteria, these same cells can become the body's worst enemy if they become transformed during tumorigenesis.

REVISITING THE "SEED AND SOIL" HYPOTHESIS OF METASTASIS

It is well documented that metastatic tumor cells do not invade distant organs randomly. Rather, metastatic cancer cells invade in a nonrandom pattern with lung, liver, and bone as primary sites of metastases (2, 3). The English surgeon, Stephen Paget, was the first to record this phenomenon in his "seed and soil" hypothesis of breast cancer metastasis (222). He proposed that certain tumor cells (the seed) have a preferential affinity to invade certain organs (the soil) (222).

Although the nonrandom dissemination of metastatic cancer cells has engaged the attention of numerous investigators for decades, no credible genetic mechanism has been able to account for the phenomenon (2, 3, 223). The seed and soil hypothesis is extremely difficult to explain if cancer is viewed as a genetic disease (181, 223). There are simply no clear connections between the nonrandom invasion of distant organs and the genetic abnormalities found in metastatic cells. On the other hand, a credible explanation of the seed and soil hypothesis emerges if cancer is viewed as a mitochondrial disease involving macrophages.

Basically, respiratory insufficiency in cells of myeloid origin can explain the seed and soil phenomenon. This comes from findings showing that mature cells of monocyte origin (macrophages) enter and engraft tissues in a nonrandom manner (224). Macrophages are genetically programmed to exist in the circulation and to preferentially enter various tissues during wound healing and the replacement of resident myeloid cells (221, 224). Some macrophage populations in liver are regularly replaced with bone marrow-derived monocytic cells, whereas other macrophage populations are more permanent and require fewer turnovers (225). It is reasonable to assume that metastatic cancer cells derived from macrophages or fusions of monocytic cells with epithelial cells will also preferentially home to those tissues that naturally require regular replacement of resident macrophages.

This prediction comes from findings that many metastatic cells express characteristics of macrophages (29). Macrophage turnover should be greater in tissues such as liver and lung where the degree of bacterial exposure and the wear-and-tare on the resident macrophage populations is considerable (226). This could explain why these organs are a preferred soil of many metastatic cancer cells. Bone marrow should also be a common target of metastatic cells because this site is the origin of the hematopoietic stem cells, which give rise to myeloid cells. Liver, lung, and bone are also preferential sites for metastatic spread for the VM mouse tumor cells (36). This is one reason why the natural tumors in the VM mouse, which preferentially home to these tissues, are an excellent model for metastatic cancer (36).

Because the metastatic cells express insufficient respiration with compensatory fermentation, these cells will enter their default state of proliferation, as would any neoplastic cell. In addition to those organs receiving high macrophage turnover, macrophages also target sites of inflammation and injury (226). This is interesting in light of findings showing that metastatic cancer cells from lung and breast can appear in the mouth following recent tooth extraction or along needle tracts following biopsy (227–229). An unhealed wound is an ideal "soil" for macrophage infiltration (226, 230). This phenomenon is referred to as *inflammatory oncotaxis* and can explain in part the seed and soil hypothesis (231). If metastasis were a metabolic disease of myeloid cells, then the appearance of metastatic cells in recent tooth extraction or wounds would not be unexpected. While the mechanistic details of these phenomena will require further examination, the general principle is clear. The nonrandom pattern of metastasis to visceral organs, bone marrow, and wounds (the soil) is consistent with a macrophage (the seed) origin of metastasis.

REVISITING THE MESENCHYMAL EPITHELIAL TRANSITION (MET)

In contrast to the EMT, the MET involves proliferation and reexpression of epithelial characteristics following extravasation, invasion, and proliferation at distant cites (Fig. 13.2). The MET is considered a reversibility of the EMT (17). How is it possible that a series of somewhat random somatic mutations orchestrate the sophisticated series of behaviors associated the EMT and then have most of these behaviors reversed during the MET? *This explanation seems preposterous and highlights the inability of the gene theory of cancer to provide a credible explanation for these phenomena* (16, 17, 223). However, an origin from myeloid cells provides a more credible explanation of metastasis.

Metastatic cells arising from myeloid cell fusions would retain the genetic architecture necessary for entering and exiting the circulation at recognized sites. It is not necessary to construct complicated mutation-based regulatory systems to explain these phenomena. Macrophages naturally enter and exit the circulatory and lymphatic systems. The circulatory system is not a "hostile" environment for cells in the macrophage lineage. These cells also express the cell-surface adhesion molecules (*selectins*) necessary for extravasation at designated organs. They already express the batteries of metalloproteases necessary for degradation of basement membranes and invasion. When these capabilities occur together with impaired respiration, dysregulated proliferation would be an expected outcome. While these properties certainly implicate myeloid cells as the origin of metastatic cells, the fusogenic properties of myeloid cells can also explain how metastatic cells can recapitulate the epithelial characteristics of the primary tumor at secondary growth sites (Fig. 13.6).

Previous studies of fusion hybrids showed that functional hepatocytes could be derived from bone marrow-derived macrophages or myelomonocytic cells following cell fusions (232). Rizvi et al. (217) also showed that expression of epithelial characteristics was found in fusion hybrids between bone marrow-derived cells and either normal epithelium or neoplastic intestinal epithelium. More recently, Wong and colleagues showed how macrophage/epithelial cell hybrids could recapitulate phenotypes of epithelial cells, while retaining the properties of macrophages (29). It is clear that phenotypes of epithelial cells and macrophages can be maintained in fusion hybrids of macrophages and intestinal epithelial tumor cells. Moreover, these characteristics are passed on to daughter cells through somatic inheritance.

Fusions of activated macrophages with epithelial cells in the primary tumor microenvironment will bestow the capability of the fused cells to degrade basement membranes, to enter and exit the circulatory and lymphatic systems, and to recapitulate the epithelial characteristics of the primary tumor at distant secondary sites. The dysregulated growth at secondary sites is the consequence of damaged respiration in theses cells (Chapter 10). Hence, the origin of metastatic cells from macrophage fusion hybrids with dysfunctional mitochondria can explain the phenomenon of metastasis (Fig. 13.6).

GENETIC HETEROGENEITY IN CANCER METASTASES

Considerable genetic heterogeneity is observed on comparing tumor tissue from primary growth sites with tissue from distant metastases (14, 181, 223, 233). Genetic heterogeneity is seen not only between patients with similar tumor histopathology

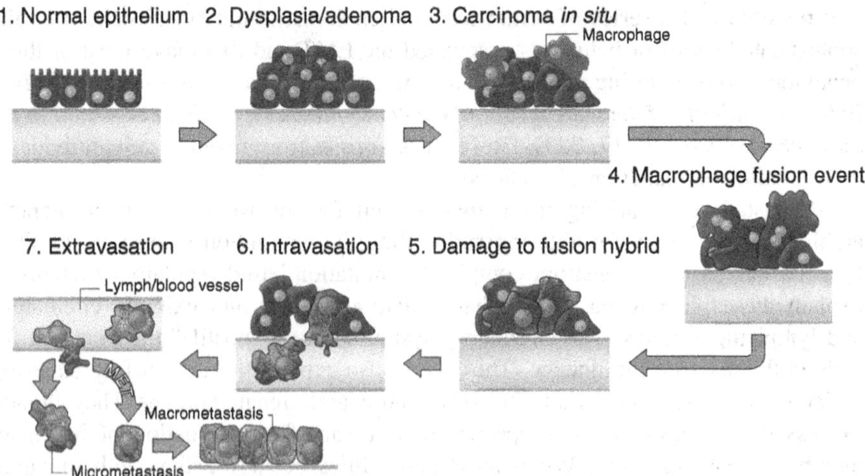

Figure 13.6 Fusion hybrid hypothesis of cancer cell metastasis. According to our hypothesis, metastatic cancer cells arise following fusion hybridization between neoplastic epithelial cells and myeloid cells (macrophages). Macrophages are known to invade *in situ* carcinoma as if it were an unhealed wound. This creates a protracted inflammatory microenvironment leading to fusion hybridization between the neoplastic epithelial cell and the macrophage. Fusion hybridization can explain the phenomenon of EMT without invoking new mutations. Inflammation damages mitochondria leading to enhanced fermentation and acidification of the microenvironment. Mitochondrial damage becomes the driver for the neoplastic transformation of the epithelial cell and of the fusion hybrids (Figure 13.5). As macrophages are already mesenchymal cells that naturally possess the capability to enter (intravasate) and exit (extravasate) the circulation, the neoplastic fusion hybrid will behave as a *rogue* macrophage. The fusogenic properties of macrophage cells can also explain how metastatic cells can recapitulate the epithelial characteristics of the primary tumor at secondary micrometastatic growth sites. This process can explain the phenomenon of MET without invoking a mutation suppression mechanism. See color insert.

but also for the tumors growing at different sites within the same patient (Fig. 13.7). Almost every type of genetic heterogeneity imaginable from point mutations to major genomic rearrangements can be found in metastatic and highly invasive cancers, including those from breast, brain, and pancreas (181, 223, 233–235).

The mostly nonuniform distribution of mutations in these tumors is consistent with findings that each neoplastic cell within a given tumor can have a profile of changes uniquely different from any other cell within the tumor (236). Moreover, if the spread of metastatic cells to some organs (such as liver and lung) occurs earlier than spread to other organs, it is possible that genetic heterogeneity would be greater in these organs than in organs that receive metastatic cells later in the disease progression. This is expected if the number of divisions is greater for tumor cells that arrive earlier in these organs than for tumor cells that arrive later in other organs. This could explain why genomic heterogeneity is more diverse in some organs than in other organs or in the primary tumor (223). These complications can obscure attempts to accurately define the clonal origin of tumor cells.

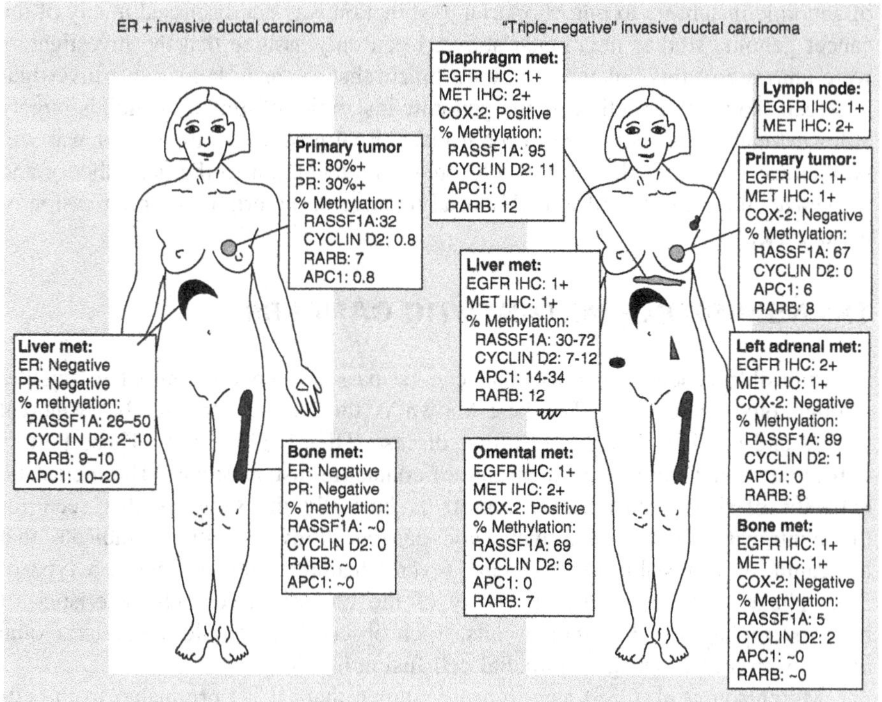

Figure 13.7 Representative cases from a rapid autopsy program. These images and data were compiled from the study of Wu et al. (181). Metastases (Met) were harvested within 4 h of death from 10 patients and compared with the matched primary tumor. The results from two cases are diagrammed. Immunohistochemical (IHC) staining for estrogen receptor (ER) and progesterone receptor (PR) is based on the percentage of positive cells, which for EGFR (epidermal growth factor receptor) and c-Met (MET) is on a 0 to 3+ intensity scale, and cyclooxygenase-2 (COX-2) is positive/negative. Percentage DNA methylation of gene promoters was determined by quantitative multiplex methylation-specific PCR assay. Methylation of the HIN1, Twist, and ERa genes was relatively uniform in all samples. *Source*: Reprinted with permission from Ref. 14. To see this figure in color please go to ftp://ftp.wiley.com/public/sci_tech_med/cancer_metabolic_disease.

In their analysis of the genomic heterogeneity observed in pancreatic cancer, Campbell et al. (223) conclude that, "the biological pathways underlying these forms of genomic instability remain unclear." As genomic stability is dependent on normal mitochondrial function, as I have described in Chapter 10, it should not be surprising that there is a "richness of genetic variation in cancer" as Campbell et al. (223) describe. The richness is the likely consequence of damaged respiration in populations of fusion hybrids that differ from each other in genetic architecture. A nonuniform or random distribution of mutations arises from the migration of these hybrid cells to other organs.

The driver of the metastatic phenomenon is not a gene but is respiratory insufficiency in macrophages or their fusion hybrids. The gene mutations arise as downstream epiphenomena, as I have described in Chapter 10. As the linkage of genomic instability to mitochondrial dysfunction was not discussed in any of the cancer genome studies mentioned above, I can only assume that the investigators were unaware of this linkage. It is unfortunate that so many industrious investigators focus so much attention on the genomic instability of tumors, which is largely a downstream epiphenomenon of the disease. Real progress in the cancer war will be realized only after the cancer field breaks its addiction to the gene theory and recognizes the centrality of mitochondrial damage in the origin and progression of the disease.

TRANSMISSIBLE METASTATIC CANCERS

Transmissible cancers are those that can be passed from one animal to another through physical contact. The best known is the canine transmissible venereal tumors and the Tasmanian devil tumor disease (DFTD) (237, 238). These tumors will often spread from the primary site of contact to distant organs. The metastatic behavior of these transmissible tumors is basically the same as that seen for the nontransmissible human metastatic cancers. Previous studies indicate that the canine transmissible tumors share several features with histocytes, a type of macrophage (239, 240). Indeed, many of the tumors express characteristics of both macrophages and epithelial cells. Such observations would suggest a clonal origin from a macrophage–epithelial cell fusion hybrid.

Murchison et al. (238) have recently shown that DFTD originated from cells expressing Schwann cell and epithelial characteristics. It is important to mention that hematopoietic bone marrow cells can elaborate Schwann cell-like phenotypes in injury conditions (241). Is it possible that these transmissible metastatic cancers arise as fusion hybrids involving myeloid cells and epithelial cells? Further studies will be necessary to determine if these metastatic cancers arise from similar mechanisms responsible for the origin of metastasis from macrophage–epithelial fusion hybrids.

Like all cancers, mitochondrial dysfunction and respiratory insufficiency would be the expected driver phenotype of these transmissible cancers. However, Tasmanian devils living in the western part of the island are resistant to the disease. It appears that the resistance results from a unique DNA polymorphism in the mitochondria of these animals (238). This is interesting in light of findings showing that transmissible cancers will occasionally acquire mitochondria from the host (242). Is it possible that properties of mitochondria determine the origin of transmissible cancers? Further studies are needed to evaluate the linkage between fusion hybridization and role of mitochondria in the origin of transmissible cancers.

THE ABSENCE OF METASTASES IN CROWN-GALL PLANT TUMORS

The crown-gall disease in plants shares many features with tumors in animals (243–245). Crown-gall tumors arise from bacterial infections that enter damaged areas of the plant leading to plant cell proliferation. The mechanisms by which bacteria induce crown-gall disease in plants are similar to those by which viruses induce tumors in animals (245). Robinson (243) first suggested that Warburg's cancer theory might account for the abnormal cell proliferation in crown-gall tumors following bacterial damage to respiration in the affected plant cells. Indeed, defects in mitochondrial morphology and energy metabolism were later described in crown-gall tumors (246–248).

It is interesting that the crown-gall tumors express four of the Hanahan and Weinberg hallmarks of cancer, that is, self-sufficiency in growth signals, insensitivity to growth inhibitory (antigrowth) signals, evasion of programmed cell death (apoptosis), and limitless replicative potential. However, these tumors do not express invasion or metastasis (245). With the exception of metastasis and invasion, the abnormalities in growth and physiology are similar in crown-gall disease and in animal tumors. If metastasis arises from damaged respiration in macrophages or in their fusion hybrids as I have discussed above, then it becomes clear why the grown-gall tumors do not display invasion or metastasis despite expressing other hallmarks of tumors. The crown-gall tumors do not metastasize because they do not have macrophages or myeloid cells as part of their immune system (249). The findings in crown-gall are also consistent with Tarin's (1) hypothesis, "that metastasis cannot occur until an organism has evolved the genes for lymphocyte trafficking." Plants have not evolved these genes as far as I know. According to our hypothesis, metastasis occurs predominantly in cells that express properties of macrophages.

CHAPTER SUMMARY

A transition from an epithelial cell to a mesenchymal cell is considered an underlying characteristic of metastasis. However, it is improbable that random mutations acquired through a Darwinian selection process could account for all the myeloid cell behaviors necessary for the completion of the metastatic cascade. As an alternative to a series of gain-of-function mutations and clonal selection, I propose that the metastatic mesenchymal phenotype arises primarily from respiratory damage in macrophages or in epithelial–macrophage fusion hybrids. Inflammation and radiation damage enhances hybridization, while also damaging mitochondrial function over time. It is my opinion that the myeloid origin of metastasis is the most compelling explanation for the origin of metastasis and tumor progression. I anticipate major advances in management of metastatic cancer once this explanation becomes more widely recognized.

REFERENCES

1. Tarin D. Cell and tissue interactions in carcinogenesis and metastasis and their clinical significance. Semin Cancer Biol. 2011;21:72–82.
2. Chambers AF, Groom AC, MacDonald IC. Dissemination and growth of cancer cells in metastatic sites. Nat Rev. 2002;2:563–72.

3. Fidler IJ. The pathogenesis of cancer metastasis: the 'seed and soil' hypothesis revisited. Nat Rev. 2003;3:453–8.
4. Welch DR. Defining a Cancer Metastasis. Philadelphia (PA): AACR Education Book; 2006. p.111–5.
5. Lazebnik Y. What are the hallmarks of cancer? Nat Rev. 2010;10:232–3.
6. Tarin D. Comparisons of metastases in different organs: biological and clinical implications. Clin Cancer Res. 2008;14:1923–5.
7. Bacac M, Stamenkovic I. Metastatic cancer cell. Annu Rev Pathol. 2008;3:221–47.
8. Hanahan D, Weinberg RA. The hallmarks of cancer. Cell. 2000;100:57–70.
9. Chaffer CL, Weinberg RA. A perspective on cancer cell metastasis. Science (New York). 2011;331:1559–64.
10. Sporn MB. The war on cancer. Lancet. 1996;347:1377–81.
11. Faguet G. The War on Cancer: An Anatomy of a Failure, A Blueprint for the Future. Dordrecht, The Netherlands: Springer; 2008.
12. Duffy MJ, McGowan PM, Gallagher WM. Cancer invasion and metastasis: changing views. J Pathol. 2008;214:283–93.
13. Steeg PS. Tumor metastasis: mechanistic insights and clinical challenges. Nat Med. 2006;12: 895–904.
14. Steeg PS. Heterogeneity of drug target expression among metastatic lesions: lessons from a breast cancer autopsy program. Clin Cancer Res. 2008;14:3643–5.
15. Pawelek JM. Cancer-cell fusion with migratory bone-marrow-derived cells as an explanation for metastasis: new therapeutic paradigms. Future Oncol (London, England). 2008;4:449–52.
16. Thiery JP. Epithelial-mesenchymal transitions in tumour progression. Nat Rev. 2002;2:442–54.
17. Weinberg RA. The Biology of Cancer. New York: Garland Science; 2007.
18. Kalluri R. EMT: when epithelial cells decide to become mesenchymal-like cells. J Clin Invest. 2009;119:1417–9.
19. Yokota J. Tumor progression and metastasis. Carcinogenesis. 2000;21:497–503.
20. Nowell PC. The clonal evolution of tumor cell populations. Science (New York). 1976;194:23–8.
21. Carro MS, Lim WK, Alvarez MJ, Bollo RJ, Zhao X, Snyder EY, et al. The transcriptional network for mesenchymal transformation of brain tumours. Nature. 2010;463:318–25.
22. Hart IR. New evidence for tumour embolism as a mode of metastasis. J Pathol. 2009;219:275–6.
23. Garber K. Epithelial-to-mesenchymal transition is important to metastasis, but questions remain. J Natl Cancer Inst. 2008;100:232–3.
24. Nowell PC. Tumor progression: a brief historical perspective. Semin Cancer Biol. 2002;12: 261–6.
25. Fearon ER, Vogelstein B. A genetic model for colorectal tumorigenesis. Cell. 1990;61:759–67.
26. Huysentruyt LC, Seyfried TN. Perspectives on the mesenchymal origin of metastatic cancer. Cancer Metastasis Rev. 2010;29:695–707.
27. Larue L, Bellacosa A. Epithelial-mesenchymal transition in development and cancer: role of phosphatidylinositol 3′ kinase/AKT pathways. Oncogene. 2005;24:7443–54.
28. Martin P, Leibovich SJ. Inflammatory cells during wound repair: the good, the bad and the ugly. Trends Cell Biol. 2005;15:599–607.
29. Powell AE, Anderson EC, Davies PS, Silk AD, Pelz C, Impey S, et al. Fusion between intestinal epithelial cells and macrophages in a cancer context results in nuclear reprogramming. Cancer Res. 2011;71:1497–505.
30. Trosko JE. Review paper: cancer stem cells and cancer nonstem cells: from adult stem cells or from reprogramming of differentiated somatic cells. Vet Pathol. 2009;46:176–93.
31. Reya T, Morrison SJ, Clarke MF, Weissman IL. Stem cells, cancer, and cancer stem cells. Nature. 2001;414:105–11.
32. Shackleton M, Quintana E, Fearon ER, Morrison SJ. Heterogeneity in cancer: cancer stem cells versus clonal evolution. Cell. 2009;138:822–9.
33. Sell S, Pierce GB. Maturation arrest of stem cell differentiation is a common pathway for the cellular origin of teratocarcinomas and epithelial cancers. Lab Invest. 1994;70:6–22.

34. Yuan X, Curtin J, Xiong Y, Liu G, Waschsmann-Hogiu S, Farkas DL, et al. Isolation of cancer stem cells from adult glioblastoma multiforme. Oncogene. 2004;23:9392–400.
35. Seyfried TN. Perspectives on brain tumor formation involving macrophages, glia, and neural stem cells. Perspect Biol Med. 2001;44:263–82.
36. Huysentruyt LC, Mukherjee P, Banerjee D, Shelton LM, Seyfried TN. Metastatic cancer cells with macrophage properties: evidence from a new murine tumor model. Int J Cancer. 2008;123:73–84.
37. Seyfried NT, Huysentruyt LC, Atwood JA 3rd, Xia Q, Seyfried TN, Orlando R. Up-regulation of NG2 proteoglycan and interferon-induced transmembrane proteins 1 and 3 in mouse astrocytoma: a membrane proteomics approach. Cancer Lett. 2008;263:243–52.
38. Siebzehnrubl FA, Reynolds BA, Vescovi A, Steindler DA, Deleyrolle LP. The origins of glioma: E Pluribus Unum? Glia. 2011;59:1135–47.
39. Navarro A, Boveris A. Hypoxia exacerbates macrophage mitochondrial damage in endotoxic shock. Am J Physiol Regul Integr Comp Physiol. 2005;288:R354–5.
40. Frost MT, Wang Q, Moncada S, Singer M. Hypoxia accelerates nitric oxide-dependent inhibition of mitochondrial complex I in activated macrophages. Am J Physiol Regul Integr Comp Physiol. 2005;288:R394–400.
41. Huysentruyt LC, Akgoc Z, Seyfried TN. Hypothesis: are neoplastic macrophages/microglia present in glioblastoma multiforme? ASN Neuro. 2011;3.
42. Munzarova M, Kovarik J. Is cancer a macrophage-mediated autoaggressive disease? Lancet. 1987;1:952–4.
43. Vignery A. Macrophage fusion: are somatic and cancer cells possible partners? Trends Cell Biol. 2005;15:188–93.
44. Pawelek JM. Tumour cell hybridization and metastasis revisited. Melanoma Res. 2000;10: 507–14.
45. Chakraborty AK, de Freitas Sousa J, Espreafico EM, Pawelek JM. Human monocyte x mouse melanoma fusion hybrids express human gene. Gene. 2001;275:103–6.
46. Pawelek JM. Tumour-cell fusion as a source of myeloid traits in cancer. Lancet Oncol. 2005;6:988–93.
47. Lewis CE, Pollard JW. Distinct role of macrophages in different tumor microenvironments. Cancer Res. 2006;66:605–12.
48. Joyce JA, Pollard JW. Microenvironmental regulation of metastasis. Nat Rev. 2009;9:239–52.
49. Pollard JW. Macrophages define the invasive microenvironment in breast cancer. J Leukoc Biol. 2008;84:623–30.
50. Morantz RA, Wood GW, Foster M, Clark M, Gollahon K. Macrophages in experimental and human brain tumors. Part 2: studies of the macrophage content of human brain tumors. J Neurosurg. 1979;50:305–11.
51. Biswas SK, Sica A, Lewis CE. Plasticity of macrophage function during tumor progression: regulation by distinct molecular mechanisms. J Immunol. 2008;180:2011–7.
52. Leek RD, Lewis CE, Whitehouse R, Greenall M, Clarke J, Harris AL. Association of macrophage infiltration with angiogenesis and prognosis in invasive breast carcinoma. Cancer Res. 1996;56:4625–9.
53. Mantovani A, Sozzani S, Locati M, Allavena P, Sica A. Macrophage polarization: tumor-associated macrophages as a paradigm for polarized M2 mononuclear phagocytes. Trends Immunol. 2002;23:549–55.
54. di Tomaso E, Snuderl M, Kamoun WS, Duda DG, Auluck PK, Fazlollahi L, et al. Glioblastoma recurrence after cediranib therapy in patients: lack of "rebound" revascularization as mode of escape. Cancer Res. 2011;71:19–28.
55. Peinado H, Rafii S, Lyden D. Inflammation joins the "niche". Cancer Cell. 2008;14:347–9.
56. Talmadge JE, Donkor M, Scholar E. Inflammatory cell infiltration of tumors: Jekyll or Hyde. Cancer Metastasis Rev. 2007;26:373–400.
57. Bingle L, Brown NJ, Lewis CE. The role of tumour-associated macrophages in tumour progression: implications for new anticancer therapies. J Pathol. 2002;196:254–65.

58. BRIGANDE JV, PLATT FM, SEYFRIED TN. Inhibition of glycosphingolipid biosynthesis in the cultured mouse embryo [abstract]. Glycobiology. 1996;6:722.
59. ECSEDY JA, YOHE HC, BERGERON AJ, SEYFRIED TN. Tumor-infiltrating macrophages influence the glycosphingolipid composition of murine brain tumors. J Lipid Res. 1998;39:2218–27.
60. STOSSEL TP. Mechanical Responses of White Blood Cells. In: GALLIN JI, SNYDERMAN R, editors. Inflammation: Basic Principles and Clinical Correlates. New York: Lippincott Williams & Wilkins; 1999. p.661–79.
61. GORDON S. Development and Distribution of Mononuclear Phagocytes: Relevance to Inflammation. In: GALLIN JI, SNYDERMAN R, editors. Inflammation: Basic Principles and Clinical Correlates. New York: Lippincott Williams & Wilkins; 1999. p.35–48.
62. BURKE B, LEWIS CE. The Macrophage. 2nd ed. New York: Oxford University Press; 2002.
63. SICA A, SACCANI A, MANTOVANI A. Tumor-associated macrophages: a molecular perspective. Int Immunopharmacol. 2002;2:1045–54.
64. SICA A, SCHIOPPA T, MANTOVANI A, ALLAVENA P. Tumour-associated macrophages are a distinct M2 polarised population promoting tumour progression: potential targets of anti-cancer therapy. Eur J Cancer. 2006;42:717–27.
65. BISWAS SK, MANTOVANI A. Macrophage plasticity and interaction with lymphocyte subsets: cancer as a paradigm. Nat Immunol. 2010;11:889–96.
66. GORDON S, MARTINEZ FO. Alternative activation of macrophages: mechanism and functions. Immunity. 2003;32:593–604.
67. QIAN BZ, POLLARD JW. Macrophage diversity enhances tumor progression and metastasis. Cell. 2010;141:39–51.
68. VAN DEN BOSSCHE J, BOGAERT P, VAN HENGEL J, GUERIN CJ, BERX G, MOVAHEDI K, et al. Alternatively activated macrophages engage in homotypic and heterotypic interactions through IL-4 and polyamine-induced E-cadherin/catenin complexes. Blood. 2009;114:4664–74.
69. SICA A, MANTOVANI A. Macrophage fusion cuisine. Blood. 2009;114:4609–10.
70. UNDERHILL CB, NGUYEN HA, SHIZARI M, CULTY M. CD44 positive macrophages take up hyaluronan during lung development. Dev Biol. 1993;155:324–36.
71. GIACHELLI CM, LOMBARDI D, JOHNSON RJ, MURRY CE, ALMEIDA M. Evidence for a role of osteopontin in macrophage infiltration in response to pathological stimuli in vivo. Am J Pathol. 1998;152:353–8.
72. CULTY M, O'MARA TE, UNDERHILL CB, YEAGER H Jr., SWARTZ RP. Hyaluronan receptor (CD44) expression and function in human peripheral blood monocytes and alveolar macrophages. J Leukoc Biol. 1994;56:605–11.
73. KOJIMA S, SEKINE H, FUKUI I, OHSHIMA H. Clinical significance of "cannibalism" in urinary cytology of bladder cancer. Acta Cytol. 1998;42:1365–9.
74. BJERKNES R, BJERKVIG R, LAERUM OD. Phagocytic capacity of normal and malignant rat glial cells in culture. J Natl Cancer Inst. 1987;78:279–88.
75. YOUNESS E, BARLOGIE B, AHEARN M, TRUJILLO JM. Tumor cell phagocytosis. Its occurrence in a patient with medulloblastoma. Arch Pathol Lab Med. 1980;104:651–3.
76. ZIMMER C, WEISSLEDER R, POSS K, BOGDANOVA A, WRIGHT SC Jr., ENOCHS WS. MR imaging of phagocytosis in experimental gliomas. Radiology. 1995;197:533–8.
77. VAN LANDEGHEM FK, MAIER-HAUFF K, JORDAN A, HOFFMANN KT, GNEVECKOW U, SCHOLZ R, et al. Post-mortem studies in glioblastoma patients treated with thermotherapy using magnetic nanoparticles. Biomaterials. 2009;30:52–7.
78. PERSSON A, ENGLUND E. The glioma cell edge—winning by engulfing the enemy? Med Hypotheses. 2009;73:336–7.
79. NITTA T, OKUMURA K, SATO K. Lysosomal enzymic activity of astroglial cells. Pathobiology. 1992;60:42–4.
80. CHANG GH, BARBARO NM, PIEPER RO. Phosphatidylserine-dependent phagocytosis of apoptotic glioma cells by normal human microglia, astrocytes, and glioma cells. Neuro-Oncology. 2000;2:174–83.
81. JEYAKUMAR M, NORFLUS F, TIFFT CJ, CORTINA-BORJA M, BUTTERS TD, PROIA RL, et al. Enhanced survival in Sandhoff disease mice receiving a combination of substrate deprivation

therapy and bone marrow transplantation. Blood. 2001;97:327–9.

82. Leenstra S, Das PK, Troost D, de Boer OJ, Bosch DA. Human malignant astrocytes express macrophage phenotype. J Neuroimmunol. 1995;56:17–25.
83. Goldenberg DM, Pavia RA, Tsao MC. In vivo hybridisation of human tumour and normal hamster cells. Nature. 1974;250:649–51.
84. Takeuchi H, Kubota T, Kitai R, Nakagawa T, Hashimoto N. CD98 immunoreactivity in multinucleated giant cells of glioblastomas: an immunohistochemical double labeling study. Neuropathology. 2008;28:127–31.
85. Deininger MH, Meyermann R, Trautmann K, Morgalla M, Duffner F, Grote EH, et al. Cyclooxygenase (COX)-1 expressing macrophages/microglial cells and COX-2 expressing astrocytes accumulate during oligodendroglioma progression. Brain Res. 2000;885:111–6.
86. Ghoneum M, Gollapudi S. Phagocytosis of Candida albicans by metastatic and non metastatic human breast cancer cell lines in vitro. Cancer Detect Prev. 2004;28:17–26.
87. Marin-Padilla M. Erythrophagocytosis by epithelial cells of a breast carcinoma. Cancer. 1977;39:1085–9.
88. Spivak JL. Phagocytic tumour cells. Scand J Haematol. 1973;11:253–6.
89. Abodief WT, Dey P, Al-Hattab O. Cell cannibalism in ductal carcinoma of breast. Cytopathology. 2006;17:304–5.
90. Ghoneum M, Matsuura M, Braga M, Gollapudi S. S. cerevisiae induces apoptosis in human metastatic breast cancer cells by altering intracellular Ca2+ and the ratio of Bax and Bcl-2. Int J Oncol. 2008;33:533–9.
91. Coopman PJ, Do MT, Thompson EW, Mueller SC. Phagocytosis of cross-linked gelatin matrix by human breast carcinoma cells correlates with their invasive capacity. Clin Cancer Res. 1998;4:507–15.
92. Ghoneum M, Wang L, Agrawal S, Gollapudi S. Yeast therapy for the treatment of breast cancer: a nude mice model study. In Vivo. 2007;21:251–8.
93. Berx G, Raspe E, Christofori G, Thiery JP, Sleeman JP. Pre-EMTing metastasis? Recapitulation of morphogenetic processes in cancer. Clin Exp Metastasis. 2007;24:587–97.
94. Bjerregaard B, Holck S, Christensen IJ, Larsson LI. Syncytin is involved in breast cancer-endothelial cell fusions. Cell Mol Life Sci. 2006;63:1906–11.
95. Atladottir HO, Pedersen MG, Thorsen P, Mortensen PB, Deleuran B, Eaton WW, et al. Association of family history of autoimmune diseases and autism spectrum disorders. Pediatrics. 2009;124:687–94.
96. Hochberg FH, Miller DC. Primary central nervous system lymphoma. J Neurosurg. 1988;68: 835–53.
97. Athanasou NA, Wells CA, Quinn J, Ferguson DP, Heryet A, McGee JO. The origin and nature of stromal osteoclast-like multinucleated giant cells in breast carcinoma: implications for tumour osteolysis and macrophage biology. Brit J Cancer. 1989;59:491–8.
98. Acebo P, Giner D, Calvo P, Blanco-Rivero A, Ortega AD, Fernandez PL, et al. Cancer abolishes the tissue type-specific differences in the phenotype of energetic metabolism. Trans Oncol. 2009;2:138–45.
99. Handerson T, Camp R, Harigopal M, Rimm D, Pawelek J. Beta1,6-branched oligosaccharides are increased in lymph node metastases and predict poor outcome in breast carcinoma. Clin Cancer Res. 2005;11:2969–73.
100. Calvo F, Martin PM, Jabrane N, De Cremoux P, Magdelenat H. Human breast cancer cells share antigens with the myeloid monocyte lineage. Brit J Cancer. 1987;56:15–9.
101. Shabo I, Stal O, Olsson H, Dore S, Svanvik J. Breast cancer expression of CD163, a macrophage scavenger receptor, is related to early distant recurrence and reduced patient survival. Int J Cancer. 2008;123:780–6.
102. Heidemann J, Gockel HR, Winde G, Herbst H, Domschke W, Lugering N. Signet-ring cell carcinoma of unknown primary location. Metastatic to lower back musculature—remission following FU/FA chemotherapy. Z Gastroenterol. 2002;40:33–6.
103. Hedley DW, Leary JA, Kirsten F. Metastatic adenocarcinoma of unknown primary site: abnormalities of cellular DNA content and survival. Eur J Cancer Clin Oncol. 1985;21:185–9.

104. Chandrasoma P. Polymorph phagocytosis by cancer cells in an endometrial adenoacanthoma. Cancer. 1980;45:2348–51.
105. Caruso RA, Muda AO, Bersiga A, Rigoli L, Inferrera C. Morphological evidence of neutrophil-tumor cell phagocytosis (cannibalism) in human gastric adenocarcinomas. Ultrastruct Pathol. 2002;26:315–21.
106. Hu J, La Vecchia C, Negri E, Chatenoud L, Bosetti C, Jia X, et al. Diet and brain cancer in adults: a case-control study in northeast China. Int J Cancer. 1999;81:20–3.
107. Molad Y, Stark P, Prokocimer M, Joshua H, Pinkhas J, Sidi Y. Hemophagocytosis by small cell lung carcinoma. Am J Hematol. 1991;36:154–6.
108. Falini B, Bucciarelli E, Grignani F, Martelli MF. Erythrophagocytosis by undifferentiated lung carcinoma cells. Cancer. 1980;46:1140–5.
109. DeSimone PA, East R, Powell RD Jr. Phagocytic tumor cell activity in oat cell carcinoma of the lung. Hum Pathol. 1980;11:535–9.
110. Richters A, Sherwin RP, Richters V. The lymphocyte and human lung cancers. Cancer Res. 1971;31:214–22.
111. Ruff MR, Pert CB. Small cell carcinoma of the lung: macrophage-specific antigens suggest hemopoietic stem cell origin. Science (New York). 1984;225:1034–6.
112. Ruff MR, Pert CB. Origin of human small cell lung cancer. Science (New York). 1985;229:680.
113. Bunn PA Jr., Linnoila I, Minna JD, Carney D, Gazdar AF. Small cell lung cancer, endocrine cells of the fetal bronchus, and other neuroendocrine cells express the Leu-7 antigenic determinant present on natural killer cells. Blood. 1985;65:764–8.
114. Amaravadi RK, Yu D, Lum JJ, Bui T, Christophorou MA, Evan GI, et al. Autophagy inhibition enhances therapy-induced apoptosis in a Myc-induced model of lymphoma. J Clin Invest. 2007;117:326–36.
115. Radosevic K, van Leeuwen AM, Segers-Nolten IM, Figdor CG, de Grooth BG, Greve J. Occurrence and a possible mechanism of penetration of natural killer cells into K562 target cells during the cytotoxic interaction. Cytometry. 1995;20:273–80.
116. Koren HS, Handwerger BS, Wunderlich JR. Identification of macrophage-like characteristics in a cultured murine tumor line. J Immunol. 1975;114:894–7.
117. Kerbel RS, Lagarde AE, Dennis JW, Donaghue TP. Spontaneous fusion in vivo between normal host and tumor cells: possible contribution to tumor progression and metastasis studied with a lectin-resistant mutant tumor. Mol Cell Biol. 1983;3:523–38.
118. Larizza L, Schirrmacher V, Pfluger E. Acquisition of high metastatic capacity after in vitro fusion of a nonmetastatic tumor line with a bone marrow-derived macrophage. J Exp Med. 1984;160:1579–84.
119. De Baetselier P, Roos E, Brys L, Remels L, Feldman M. Generation of invasive and metastatic variants of a non-metastatic T-cell lymphoma by in vivo fusion with normal host cells. Int J Cancer. 1984;34:731–8.
120. Lugini L, Matarrese P, Tinari A, Lozupone F, Federici C, Iessi E, et al. Cannibalism of live lymphocytes by human metastatic but not primary melanoma cells. Cancer Res. 2006;66:3629–38.
121. Lugini L, Lozupone F, Matarrese P, Funaro C, Luciani F, Malorni W, et al. Potent phagocytic activity discriminates metastatic and primary human malignant melanomas: a key role of ezrin. Lab Invest. 2003;83:1555–67.
122. Fais S. A role for ezrin in a neglected metastatic tumor function. Trends Mol Med. 2004;10: 249–50.
123. Breier F, Feldmann R, Fellenz C, Neuhold N, Gschnait F. Primary invasive signet-ring cell melanoma. J Cutan Pathol. 1999;26:533–6.
124. Monteagudo C, Jorda E, Carda C, Illueca C, Peydro A, Llombart-Bosch A. Erythrophagocytic tumour cells in melanoma and squamous cell carcinoma of the skin. Histopathology. 1997;31:367–73.
125. Chakraborty AK, Sodi S, Rachkovsky M, Kolesnikova N, Platt JT, Bolognia JL, et al. A spontaneous murine melanoma lung metastasis comprised of host x tumor hybrids. Cancer Res. 2000;60:2512–9.

126. Brocker EB, Suter L, Sorg C. HLA-DR antigen expression in primary melanomas of the skin. J Invest Dermatol. 1984;82:244–7.
127. Facchetti F, Bertalot G, Grigolato PG. KP1 (CD 68) staining of malignant melanomas. Histopathology. 1991;19:141–5.
128. Munzarova M, Rejthar A, Mechl Z. Do some malignant melanoma cells share antigens with the myeloid monocyte lineage? Neoplasma. 1991;38:401–5.
129. Busund LT, Killie MK, Bartnes K, Seljelid R. Spontaneously formed tumorigenic hybrids of Meth A sarcoma cells and macrophages in vivo. Int J Cancer. 2003;106:153–9.
130. Savage DG, Zipin D, Bhagat G, Alobeid B. Hemophagocytic, non-secretory multiple myeloma. Leuk Lymphoma. 2004;45:1061–4.
131. Andersen TL, Soe K, Sondergaard TE, Plesner T, Delaisse JM. Myeloma cell-induced disruption of bone remodelling compartments leads to osteolytic lesions and generation of osteoclast-myeloma hybrid cells. Br J Haematol.148:551–61.
132. Yasunaga M, Ohishi Y, Nishimura I, Tamiya S, Iwasa A, Takagi E, et al. Ovarian undifferentiated carcinoma resembling giant cell carcinoma of the lung. Pathol Int. 2008;58:244–8.
133. Talmadge JE, Key ME, Hart IR. Characterization of a murine ovarian reticulum cell sarcoma of histiocytic origin. Cancer Res. 1981;41:1271–80.
134. Khayyata S, Basturk O, Adsay NV. Invasive micropapillary carcinomas of the ampullo-pancreatobiliary region and their association with tumor-infiltrating neutrophils. Mod Pathol. 2005;18:1504–11.
135. Kern HF, Bosslet K, Sedlacek HH, Schorlemmer HU. Monocyte-related functions expressed in cell lines established from human pancreatic adenocarcinoma. II. Inhibition of stimulated activity by monoclonal antibodies reacting with surface antigens on tumor cells. Pancreas. 1988;3:2–10.
136. Imai S, Sekigawa S, Ohno Y, Yamamoto H, Tsubura Y. Giant cell carcinoma of the pancreas. Acta Pathol Jpn. 1981;31:129–33.
137. Davies PS, Powell AE, Swain JR, Wong MH. Inflammation and proliferation act together to mediate intestinal cell fusion. PloS One. 2009;4:e6530.
138. Shabo I, Olsson H, Sun XF, Svanvik J. Expression of the macrophage antigen CD163 in rectal cancer cells is associated with early local recurrence and reduced survival time. Int J Cancer. 2009;125:1826–31.
139. Chetty R, Cvijan D. Giant (bizarre) cell variant of renal carcinoma. Histopathology. 1997;30:585–7.
140. Yilmaz Y, Lazova R, Qumsiyeh M, Cooper D, Pawelek J. Donor Y chromosome in renal carcinoma cells of a female BMT recipient: visualization of putative BMT-tumor hybrids by FISH. Bone Marrow Transplant. 2005;35:1021–4.
141. Chakraborty A, Lazova R, Davies S, Backvall H, Ponten F, Brash D, et al. Donor DNA in a renal cell carcinoma metastasis from a bone marrow transplant recipient. Bone Marrow Transplant. 2004;34:183–6.
142. Etcubanas E, Peiper S, Stass S, Green A. Rhabdomyosarcoma, presenting as disseminated malignancy from an unknown primary site: a retrospective study of ten pediatric cases. Med Pediatr Oncol. 1989;17:39–44.
143. Tsoi WC, Feng CS. Hemophagocytosis by rhabdomyosarcoma cells in bone marrow. Am J Hematol. 1997;54:340–2.
144. Fais S. Cannibalism: a way to feed on metastatic tumors. Cancer Lett. 2007;258:155–64.
145. Matarrese P, Ciarlo L, Tinari A, Piacentini M, Malorni W. Xeno-cannibalism as an exacerbation of self-cannibalism: a possible fruitful survival strategy for cancer cells. Curr Pharm Des. 2008;14:245–52.
146. Overholtzer M, Brugge JS. The cell biology of cell-in-cell structures. Nat Rev Mol Cell Biol. 2008;9:796–809.
147. Gupta K, Dey P. Cell cannibalism: diagnostic marker of malignancy. Diagn Cytopathol. 2003;28:86–7.
148. Duelli D, Lazebnik Y. Cell fusion: a hidden enemy? Cancer Cell. 2003;3:445–8.
149. Warner TF. Cell hybridization: an explanation for the phenotypic diversity of certain tumours. Med Hypotheses. 1975;1:51–7.

150. MUNZAROVA M, LAUEROVA L, CAPKOVA J. Are advanced malignant melanoma cells hybrids between melanocytes and macrophages? Melanoma Res. 1992;2:127–9.
151. DUELLI D, LAZEBNIK Y. Cell-to-cell fusion as a link between viruses and cancer. Nat Rev. 2007;7:968–76.
152. SUNDERKOTTER C, STEINBRINK K, GOEBELER M, BHARDWAJ R, SORG C. Macrophages and angiogenesis. J Leukoc Biol. 1994;55:410–22.
153. CHETTIBI S, FERGUSON MWJ. Wound Repair: An Overview. In: GALLIN JI, SNYDERMAN R, editors. Inflammation: Basic Principles and Clinical Correlates. New York: Lippincott Williams & Wilkins; 1999. p.865–81.
154. VIGNERY A. Osteoclasts and giant cells: macrophage-macrophage fusion mechanism. Int J Exp Pathol. 2000;81:291–304.
155. SERHAN CN, SAVILL J. Resolution of inflammation: the beginning programs the end. Nat Immunol. 2005;6:1191–7.
156. BELLINGAN GJ, CALDWELL H, HOWIE SE, DRANSFIELD I, HASLETT C. In vivo fate of the inflammatory macrophage during the resolution of inflammation: inflammatory macrophages do not die locally, but emigrate to the draining lymph nodes. J Immunol. 1996;157:2577–85.
157. RANDOLPH GJ, INABA K, ROBBIANI DF, STEINMAN RM, MULLER WA. Differentiation of phagocytic monocytes into lymph node dendritic cells in vivo. Immunity. 1999;11:753–61.
158. AKST J. It's a cell-eat-cell world. Scientist. 2011;25:44–9.
159. WANG S, GUO Z, XIA P, LIU T, WANG J, LI S, et al. Internalization of NK cells into tumor cells requires ezrin and leads to programmed cell-in-cell death. Cell Res. 2009;19:1350–62.
160. COUZIN-FRANKEL J. Immune therapy steps up the attack. Science (New York). 2010;330:440–3.
161. KLIONSKY DJ. Cell biology: regulated self-cannibalism. Nature. 2004;431:31–2.
162. MIZUSHIMA N, LEVINE B, CUERVO AM, KLIONSKY DJ. Autophagy fights disease through cellular self-digestion. Nature. 2008;451:1069–75.
163. HOFFMAN HJ, DUFFNER PK. Extraneural metastases of central nervous system tumors. Cancer. 1985;56:1778–82.
164. TAHA M, AHMAD A, WHARTON S, JELLINEK D. Extra-cranial metastasis of glioblastoma multiforme presenting as acute parotitis. Br J Neurosurg. 2005;19:348–51.
165. LAERUM OD, BJERKVIG R, STEINSVAG SK, DE RIDDER L. Invasiveness of primary brain tumors. Cancer Metastasis Rev. 1984;3:223–36.
166. RUBINSTEIN LJ. Tumors of the Central Nervous System. Washington (DC): Armed Forces Institute of Pathology; 1972.
167. GOTWAY MB, CONOMOS PJ, BREMNER RM. Pleural metastatic disease from glioblastoma multiforme. J Thorac Imaging. 2011;26:W54–8.
168. KALOKHE G, GRIMM SA, CHANDLER JP, HELENOWSKI I, RADEMAKER A, RAIZER JJ. Metastatic glioblastoma: case presentations and a review of the literature. J Neuro-oncol. 2012;107:21–7.
169. ARMANIOS MY, GROSSMAN SA, YANG SC, WHITE B, PERRY A, BURGER PC, et al. Transmission of glioblastoma multiforme following bilateral lung transplantation from an affected donor: case study and review of the literature. Neuro-Oncology. 2004;6:259–63.
170. NG WH, YEO TT, KAYE AH. Spinal and extracranial metastatic dissemination of malignant glioma. J Clin Neurosci. 2005;12:379–82.
171. STROJNIK T, KAVALAR R, ZAJC I, DIAMANDIS EP, OIKONOMOPOULOU K, LAH TT. Prognostic impact of CD68 and kallikrein 6 in human glioma. Anticancer Res. 2009;29:3269–79.
172. DUFFY PE. Astrocytes: Normal, Reactive, and Neoplastic. New York: Raven Press; 1983.
173. HAN SJ, YANG I, OTERO JJ, AHN BJ, TIHAN T, MCDERMOTT MW, et al. Secondary gliosarcoma after diagnosis of glioblastoma: clinical experience with 30 consecutive patients. J Neurosurg. 2011;112:990–6.
174. ZAGZAG D, ESENCAY M, MENDEZ O, YEE H, SMIRNOVA I, HUANG Y, et al. Hypoxia- and vascular endothelial growth factor-induced stromal cell-derived factor-1alpha/CXCR4 expression in glioblastomas: one plausible explanation of Scherer's structures. Am J Pathol. 2008;173:545–60.
175. SCHERER HJ. A critical review: the pathology of cerebral gliomas. J Neurol Neuropsychiatr. 1940;3:147–77.

176. Tso CL, Shintaku P, Chen J, Liu Q, Liu J, Chen Z, et al. Primary glioblastomas express mesenchymal stem-like properties. Mol Cancer Res. 2006;4:607–19.
177. Fan X, Salford LG, Widegren B. Glioma stem cells: evidence and limitation. Semin Cancer Biol. 2007;17:214–8.
178. Yates AJ. An overview of principles for classifying brain tumors. Mol Chem Neuropathol. 1992;17:103–20.
179. de Groot JF, Fuller G, Kumar AJ, Piao Y, Eterovic K, Ji Y, et al. Tumor invasion after treatment of glioblastoma with bevacizumab: radiographic and pathologic correlation in humans and mice. Neuro-Oncology. 2010;12:233–42.
180. Iwamoto FM, Abrey LE, Beal K, Gutin PH, Rosenblum MK, Reuter VE, et al. Patterns of relapse and prognosis after bevacizumab failure in recurrent glioblastoma. Neurology. 2009;73:1200–6.
181. Wu JM, Fackler MJ, Halushka MK, Molavi DW, Taylor ME, Teo WW, et al. Heterogeneity of breast cancer metastases: comparison of therapeutic target expression and promoter methylation between primary tumors and their multifocal metastases. Clin Cancer Res. 2008;14:1938–46.
182. Kerbel RS, Twiddy RR, Robertson DM. Induction of a tumor with greatly increased metastatic growth potential by injection of cells from a low-metastatic H-2 heterozygous tumor cell line into an H-2 incompatible parental strain. Int J Cancer. 1978;22:583–94.
183. Chen EH, Grote E, Mohler W, Vignery A. Cell-cell fusion. FEBS Lett. 2007;581:2181–93.
184. Camargo FD, Chambers SM, Goodell MA. Stem cell plasticity: from transdifferentiation to macrophage fusion. Cell Prolif. 2004;37:55–65.
185. Camargo FD, Finegold M, Goodell MA. Hematopoietic myelomonocytic cells are the major source of hepatocyte fusion partners. J Clin Invest. 2004;113:1266–70.
186. Paris S, Sesboue R. Metastasis models: the green fluorescent revolution? Carcinogenesis. 2004;25:2285–92.
187. Mekler LB, Drize OB, Osechinskii IV, Shliankevich MA. Transformation of a normal differentiated cell of an adult organism, induced by the fusion of this cell with another normal cell of the same organism but with different organ or tissue specificity. Vestn Akad Med Nauk SSSR. 1971;26:75–80.
188. Levin TG, Powell AE, Davies PS, Silk AD, Dismuke AD, Anderson EC, et al. Characterization of the intestinal cancer stem cell marker CD166 in the human and mouse gastrointestinal tract. Gastroenterology. 2010;139:2072–82.e5.
189. Seyfried TN, Kiebish MA, Marsh J, Shelton LM, Huysentruyt LC, Mukherjee P. Metabolic management of brain cancer. Biochim Biophys Acta. 2010;1807:577–94.
190. Rachkovsky M, Pawelek J. Acquired melanocyte stimulating hormone-inducible chemotaxis following macrophage fusion with Cloudman S91 melanoma cells. Cell Growth Differ. 1999;10:517–24.
191. Pawelek JM, Chakraborty AK. The cancer cell–leukocyte fusion theory of metastasis. Adv Cancer Res. 2008;101:397–444.
192. Ades L, Guardiola P, Socie G. Second malignancies after allogeneic hematopoietic stem cell transplantation: new insight and current problems. Blood Rev. 2002;16:135–46.
193. Tebeu P-M, Verkooijen HM, Bouchardy C, Ludicke F, Usel M, Major AL. Impact of external radiotherapyon survival after stage I endometrial cancer: results from a population based study. J Cancer Sci Ther. 2011;3:041–6.
194. Seyfried TN, Shelton LM, Mukherjee P. Does the existing standard of care increase glioblastoma energy metabolism? Lancet Oncol. 2010;11:811–3.
195. Pawelek JM, Chakraborty AK. Fusion of tumour cells with bone marrow-derived cells: a unifying explanation for metastasis. Nat Rev. 2008;8:377–86.
196. Guillemin GJ, Brew BJ. Microglia, macrophages, perivascular macrophages, and pericytes: a review of function and identification. J Leukoc Biol. 2004;75:388–97.
197. Gazdar AF, Bunn PA Jr., Minna JD, Baylin SB. Origin of human small cell lung cancer. Science (New York). 1985;229:679–80.
198. Maniecki MB, Damsky WE, Ulhoi BP, Steiniche T, Orntoft TE, Dyrskjot L, et al. The expression of monocyte/macrophage-restricted scavenger receptor CD163 by malignant cells may

be a consequence of cell fusion with tumor-associated macrophages: a novel target for cancer therapy. Am Assoc Cancer Res. 2011.

199. Keunen O, Johansson M, Oudin A, Sanzey M, Rahim SA, Fack F, et al. Anti-VEGF treatment reduces blood supply and increases tumor cell invasion in glioblastoma. Proc Natl Acad Sci USA. 2011;108:3749–54.
200. Diment S, Leech MS, Stahl PD. Cathepsin D is membrane-associated in macrophage endosomes. J Biol Chem. 1988;263:6901–7.
201. Stehle G, Sinn H, Wunder A, Schrenk HH, Stewart JC, Hartung G, et al. Plasma protein (albumin) catabolism by the tumor itself—implications for tumor metabolism and the genesis of cachexia. Crit Rev Oncol Hematol. 1997;26:77–100.
202. Moon Y, Kim JY, Choi SY, Kim K, Kim H, Sun W. Induction of ezrin-radixin-moesin molecules after cryogenic traumatic brain injury of the mouse cortex. Neuroreport. 2011;22:304–8.
203. Fehon RG, McClatchey AI, Bretscher A. Organizing the cell cortex: the role of ERM proteins. Nat Rev Mol Cell Biol. 2010;11:276–87.
204. Krishnan K, Bruce B, Hewitt S, Thomas D, Khanna C, Helman LJ. Ezrin mediates growth and survival in Ewing's sarcoma through the AKT/mTOR, but not the MAPK, signaling pathway. Clin Exp Metastasis. 2006;23:227–36.
205. Park HR, Cabrini RL, Araujo ES, Paparella ML, Brandizzi D, Park YK. Expression of ezrin and metastatic tumor antigen in osteosarcomas of the jaw. Tumori. 2009;95: 81–6.
206. Hunter KW. Ezrin, a key component in tumor metastasis. Trends Mol Med. 2004;10:201–4.
207. Bobryshev YV, Lord RS, Watanabe T, Ikezawa T. The cell adhesion molecule E-cadherin is widely expressed in human atherosclerotic lesions. Cardiovasc Res. 1998;40:191–205.
208. Armeanu S, Buhring HJ, Reuss-Borst M, Muller CA, Klein G. E-cadherin is functionally involved in the maturation of the erythroid lineage. J Cell Biol. 1995;131:243–9.
209. Ward DG, Roberts K, Brookes MJ, Joy H, Martin A, Ismail T, et al. Increased hepcidin expression in colorectal carcinogenesis. World J Gastroenterol. 2008;14:1339–45.
210. Leonard RC, Untch M, Von Koch F. Management of anaemia in patients with breast cancer: role of epoetin. Ann Oncol. 2005;16:817–24.
211. Kamai T, Tomosugi N, Abe H, Arai K, Yoshida K. Increased serum hepcidin-25 level and increased tumor expression of hepcidin mRNA are associated with metastasis of renal cell carcinoma. BMC Cancer. 2009;9:270.
212. Ganz T. Hepcidin—a regulator of intestinal iron absorption and iron recycling by macrophages. Best Pract Res Clin Haematol. 2005;18:171–82.
213. Pavlidis N, Fizazi K. Carcinoma of unknown primary (CUP). Crit Rev Oncol Hematol. 2009;69:271–8.
214. Carlson HR. Carcinoma of unknown primary: searching for the origin of metastases. JAAPA. 2009;22:18–21.
215. Torres EM, Williams BR, Tang YC, Amon A. Thoughts on aneuploidy. Cold Spring Harbor Symposia on Quantitative Biology. 2010;75:445–51.
216. Torres EM, Williams BR, Amon A. Aneuploidy: cells losing their balance. Genetics. 2008;179:737–46.
217. Rizvi AZ, Swain JR, Davies PS, Bailey AS, Decker AD, Willenbring H, et al. Bone marrow-derived cells fuse with normal and transformed intestinal stem cells. Proc Natl Acad Sci USA. 2006;103:6321–5.
218. Lazova R, Chakraborty A, Pawelek JM. Leukocyte-cancer cell fusion: initiator of the warburg effect in malignancy? Adv Exp Med Biol. 2011;714:151–72.
219. Carew JS, Huang P. Mitochondrial defects in cancer. Mol Cancer. 2002;1:9.
220. Pedersen PL. Tumor mitochondria and the bioenergetics of cancer cells. Prog Exp Tumor Res. 1978;22:190–274.
221. Auffray C, Sieweke MH, Geissmann F. Blood monocytes: development, heterogeneity, and relationship with dendritic cells. Annu Rev Immunol. 2009;27:669–92.

222. Paget S. The distribution of secondary growths in cancer of the breast. Lancet. 1889;1:571–3.
223. Campbell PJ, Yachida S, Mudie LJ, Stephens PJ, Pleasance ED, Stebbings LA, et al. The patterns and dynamics of genomic instability in metastatic pancreatic cancer. Nature. 2010;467:1109–13.
224. Kennedy DW, Abkowitz JL. Mature monocytic cells enter tissues and engraft. Proc Natl Acad Sci USA. 1998;95:14944–9.
225. Klein I, Cornejo JC, Polakos NK, John B, Wuensch SA, Topham DJ, et al. Kupffer cell heterogeneity: functional properties of bone marrow derived and sessile hepatic macrophages. Blood. 2007;110:4077–85.
226. Hofer SO, Molema G, Hermens RA, Wanebo HJ, Reichner JS, Hoekstra HJ. The effect of surgical wounding on tumour development. Eur J Surg Oncol. 1999;25:231–43.
227. Hirshberg A, Leibovich P, Horowitz I, Buchner A. Metastatic tumors to postextraction sites. J Oral Maxillofac Surg. 1993;51:1334–7.
228. Cho E, Kim MH, Cha SH, Cho SH, Oh SJ, Lee JD. Breast cancer cutaneous metastasis at core needle biopsy site. Ann Dermatol. 2010;22:238–40.
229. Hirshberg A, Leibovich P, Buchner A. Metastases to the oral mucosa: analysis of 157 cases. J Oral Pathol Med. 1993;22:385–90.
230. Jones FS, Rous P. On the cause of the localization of secondary tumors at points of injury. J Exp Med. 1914;20:404–12.
231. Walter ND, Rice PL, Redente EF, Kauvar EF, Lemond L, Aly T, et al. Wound healing after trauma may predispose to lung cancer metastasis: review of potential mechanisms. Am J Respir Cell Mol Biol. 2011;44:591–6.
232. Willenbring H, Bailey AS, Foster M, Akkari Y, Dorrell C, Olson S, et al. Myelomonocytic cells are sufficient for therapeutic cell fusion in liver. Nat Med. 2004;10:744–8.
233. Jones S, Zhang X, Parsons DW, Lin JC, Leary RJ, Angenendt P, et al. Core signaling pathways in human pancreatic cancers revealed by global genomic analyses. Science (New York). 2008;321:1801–6.
234. Ohgaki H, Kleihues P. Genetic alterations and signaling pathways in the evolution of gliomas. Cancer Sci. 2009;100:2235–41.
235. Parsons DW, Jones S, Zhang X, Lin JC, Leary RJ, Angenendt P, et al. An integrated genomic analysis of human glioblastoma multiforme. Science (New York). 2008;321:1807–12.
236. Salk JJ, Fox EJ, Loeb LA. Mutational heterogeneity in human cancers: origin and consequences. Annu Rev Pathol. 2010;5:51–75.
237. Park MS, Kim Y, Kang MS, Oh SY, Cho DY, Shin NS, et al. Disseminated transmissible venereal tumor in a dog. J Vet Diagn Invest. 2006;18:130–3.
238. Murchison EP, Tovar C, Hsu A, Bender HS, Kheradpour P, Rebbeck CA, et al. The Tasmanian devil transcriptome reveals Schwann cell origins of a clonally transmissible cancer. Science (New York). 2010;327:84–7.
239. Mozos E, Mendez A, Gomez-Villamandos JC, Martin De Las Mulas J, Perez J. Immunohistochemical characterization of canine transmissible venereal tumor. Vet Pathol. 1996;33:257–63.
240. Marchal T, Chabanne L, Kaplanski C, Rigal D, Magnol JP. Immunophenotype of the canine transmissible venereal tumour. Vet Immunol Immunopathol. 1997;57:1–11.
241. Zhao FQ, Zhang PX, He XJ, Du C, Fu ZG, Zhang DY, et al. Study on the adoption of Schwann cell phenotype by bone marrow stromal cells in vitro and in vivo. Biomed Environ Sci. 2005;18:326–33.
242. Rebbeck CA, Leroi AM, Burt A. Mitochondrial capture by a transmissible cancer. Science (New York). 2011;331:303.
243. Robinson W. Some features of crown-gall in plants in reference to comparisons with cancer. Proc Roy Soc Med 1927;20:1507–9.
244. Nester EW, Gordon MP, Amasino RM. Crown gall: a molecular and physiological analysis. Ann Rev Plant Physiol. 1984;35:387–413.
245. Levine M. Plant tumors and their relationship to cancer. Botanical Rev. 1936;2:439–55.

246. Tamaoki T, Hildebrandt AC, Burris RH, Riker AJ, Hagihara B. Respiration and phosphorylation of mitochondria from normal and crown-gall tissue cultures of tomato. Plant Physiol. 1960;35:942–7.
247. Klein RM. Nitrogen and phosphorus fractions, respiration, and structure of normal and crown gall tissues of tomato. Plant Physiol. 1952;27:335–54.
248. Fogelberg SO, Struckmeyer E, Roberts RH. Morphological variations of mitochondria in the presence of plant tumors. Am J Botany. 1957;44:454–9.
249. Jones JD, Dangl JL. The plant immune system. Nature. 2006;444:323–9.

Chapter 14

Mitochondrial Respiratory Dysfunction and the Extrachromosomal Origin of Cancer

The credibility of any theory to explain a complicated phenomenon is dependent on the extent to which it can explain all or most observations associated with the phenomenon (1, 2). As I mentioned in previous chapters, there are serious inconsistencies with the somatic mutation theory of cancer. These inconsistencies have undermined the credibility of this theory to explain the origin of the disease. The gene theory has now reached a critical point of disbelief. The current acceptance of the gene theory as an explanation of cancer must be based more on ideology than on reason (1, 3).

Unlike Darwin, who incorporated most observations on the origin of species into his theory of natural selection, Warburg failed to explain how his theory of mitochondrial injury could explain metastasis or why some cancer cells might appear to respire. These omissions contributed in part for the failure of Warburg's theory to become the dominant explanation for the origin of cancer. However, no data have disproved Warburg's central hypothesis that damaged or insufficient respiration is the origin of cancer. As I discussed in Chapters 7 and 8, amino acid fermentation and anaerobic respiration in tumor mitochondria can give the appearance that aerobic respiration is normal when, in fact, it is not.

In Chapter 13, I have discussed how mitochondrial dysfunction can account for the phenomenon of metastasis in macrophage fusion hybrids and, in Chapters 7 and 8, how amino acid fermentation might simulate OxPhos. This evidence more strongly supports cancer as a metabolic disease than as a genetic disease. That the mitochondrion, rather than the nucleus, dictates the origin of tumorigenesis is now

Cancer as a Metabolic Disease: On the Origin, Management and Prevention of Cancer, First Edition. Thomas Seyfried.

incontrovertible. The Warburg effect (aerobic glycolysis) is seen in most cancers. It is becoming clear how respiratory insufficiency arising from mitochondrial damage underlies the Warburg effect and all other phenomena associated with the disease. The evidence supporting cancer as a disease of mitochondrial respiratory insufficiency is overwhelming. As mitochondria constitute a classic extrachromosomal epigenetic system, cancer can be considered an epigenetic metabolic disease.

CONNECTING THE LINKS

The path from normal cell physiology to malignant behavior, where all major cancer hallmarks are expressed, is depicted in Figure 14.1, and is based on the evidence reviewed in previous chapters. The diagram has been modified slightly from our original diagram that was first published in *Nutrition & Metabolism* (4). Any unspecific condition that damages a cell's respiratory capacity, but is not

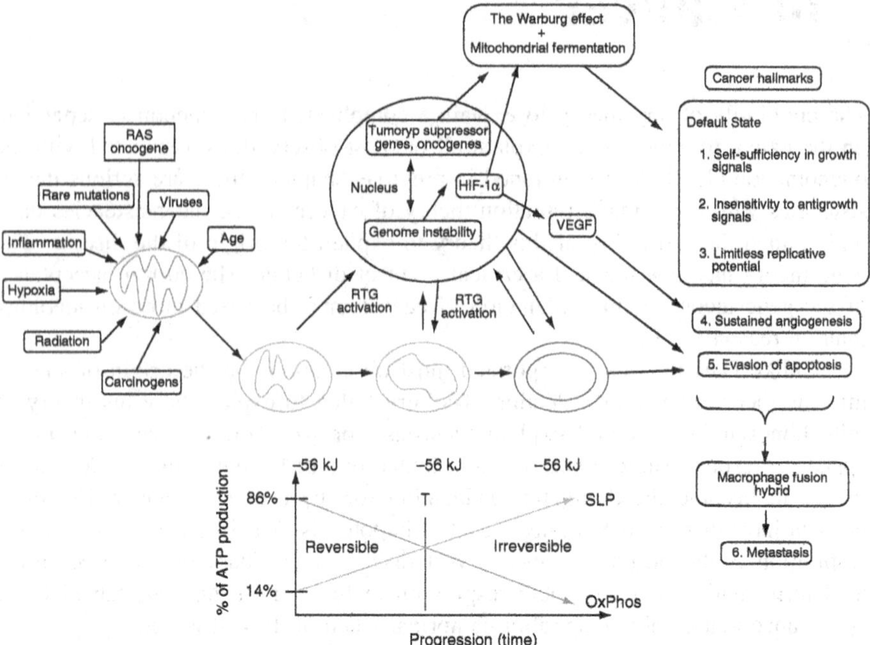

Figure 14.1 Mitochondrial respiratory dysfunction as the origin of cancer. Cancer can arise from any number of nonspecific events that damage the respiratory capacity of cells over time. The path to carcinogenesis will occur only in those cells that are capable of enhancing energy production through SLP (fermentation). Despite the shift from respiration to fermentation the $\Delta G'$ of ATP hydrolysis remains fairly constant at approximately −56 kJ. Oncogene upregulation and tumor-suppressor gene inactivation are necessary to maintain viability of incipient cancer cells when respiration becomes damaged. Metastasis arises from respiratory damage in cells of myeloid/macrophage origin. This scenario links all major cancer hallmarks to respiratory dysfunction. *Source*: Reprinted with modifications from Ref. 4. See color insert.

severe enough to kill the cell, can potentially initiate the path to a malignant cancer. Reduced respiratory capacity could arise from damage to any mitochondrial protein, lipid, or mtDNA. Some of the many unspecific conditions that can damage a cell's respiratory capacity thus initiating carcinogenesis include inflammation, carcinogens, radiation (ionizing or ultraviolet), intermittent hypoxia, rare germline mutations, viral infections, and age.

Inflammation has long been recognized in the initiation and promotion of cancer. Inflammation produces ROS and elevates TGF-β, which damage mitochondria while disrupting tissue morphogenetic fields (Chapters 10 and 12). Besides producing mutations, *carcinogens* also produce ROS (Chapter 9). Carcinogens, in addition to causing mutations, disrupt OxPhos and cause permanent injury to mitochondria. It is this effect of the carcinogen on mitochondrial energy production rather than its mutagenic effect that primarily initiates cancer. It is unfortunate that the Ames tests focused only on the mutagenic effects of carcinogens rather than on the mitochondrial damaging effects of these compounds (5). *Radiation* not only causes mutations, but also injures mitochondria (Chapters 7 and 9). Radiation causes necrotic cell death and inflammation (6). It is the production of ROS and the injurious effect of radiation on OxPhos that causes cancer (7, 8). While radiation can certainly kill cancer cells, radiation can also initiate cancer through its effect on the mitochondrial energy production. Similar to inflammation, *hypoxia* produces high levels of ROS in the microenvironment, which will damage mitochondrial respiratory capacity thus facilitating cancer initiation and progression. Although we did not include *age* in our original discussion of cancer inducing agents (4), it is certainly a cancer risk factor. The accumulation of ROS with age damages mitochondrial respiratory energy production. If mitochondrial damage underlies the origin of cancer according to my central hypothesis, then it is predictable that cancer risk should increase with age. Finally, rare *germline mutations* increase cancer risk through a direct effect on mitochondrial function (Chapter 9). Hence, the plethora of nonspecific factors known to increase the risk of cancer can all be linked to the disease through their protracted and deleterious effects on mitochondrial function, which leads to respiratory insufficiency.

ADDRESSING THE ONCOGENIC PARADOX

Szent-Gyorgyi stated,

> *The malignant transformation of tissues involves a paradox which, to my knowledge, has not been pointed out before. This transformation is a very specific process, which must involve very specific changes in a very specific chemical machinery. Accordingly, one would expect that such transformation can be brought about only by a very specific process, as locks can be opened only by their own keys. Contrary to this, a malignant transformation can be brought about by an infinite number of unspecific influences, such as pieces of asbestos, high-energy radiation, irritation, chemicals, viruses, etc. It is getting more and more difficult to find something that is not carcinogenic. That a very specific process should be elicited in such an unspecific way is very unexpected* (9).

According to the evidence presented in my treatise, protracted damage to the respiratory capacity of cells that are capable of upregulating fermentation can explain, in large part, Szent-Gyorgyi's paradox.

Chronic injury to the structure and function of mitochondria, which impairs respiration, will activate the mitochondrial RTG response within the damaged cell (Chapter 10). The RTG response is an epigenetic system that upregulates those genes needed to derive energy through fermentation. Fermentation involves SLP through glycolysis in the cytoplasm and through amino acid fermentation in the mitochondria (Chapter 8). Uncorrected mitochondrial damage will require continuous compensatory energy through fermentation involving SLP in order to maintain the $\Delta G'_{\mathrm{ATP}}$ of approximately −56 kJ/mol. This standard energy of ATP hydrolysis is essential for cell viability. This ATP hydrolysis remains mostly constant regardless of whether the ATP is synthesized through respiration or fermentation (Chapter 4).

Although fermentation energy can temporally compensate for disruptions to respiration in order to maintain cell viability, persistent energy production through fermentation can compromise cellular differentiation. Tumor cells require energy production through fermentation because their mitochondrial respiration is insufficient to maintain energy homeostasis. If their respiration was sufficient, fermentation would not persist. Confusion arises from amino acid fermentation, which can simulate properties of normal respiration. Cancer cells appear to respire while also fermenting glucose (aerobic glycolysis). Hence, tumor cells differ from normal cells because they generate significant amount of energy through fermentation (Chapter 8).

Tumor progression is linked to irreversible respiratory damage with fermentation becoming the permanent compensatory energy source for the tumor cells. The change in shape of mitochondrial cristae from convoluted- to smooth illustrates the transition from respiration to fermentation as shown in Figure 14.1. Persistent and cumulative mitochondrial damage underlies the initiation and progression of cancer. To illustrate this point further, I have also inserted a threshold on the progression (time) line in the figure. The threshold (T) passes through the intersection of the OxPhos and SLP lines. This concept is based on Warburg's findings that fermentation gradually displaces respiration after a long time period (8). It is only when fermentation compensates for the greater part of total cellular energy production that tumor progression becomes irreversible according to our model. Tumor progression can be reversed, however, as long as some functional mitochondrial respiration remains. Mitochondrial enhancement therapies can help restore impaired respiration (Chapters 17–19). The failure to restore respiratory energy coupled with a greater dependence on fermentation energy underlies all hallmarks of cancer, including the Warburg effect. In addition to the fermentation of glucose to lactate, which is needed to drive glycolysis, many cancer cells might also ferment glutamine in the mitochondria. It is the fermentation of glucose and glutamine that primarily drives tumor progression and makes tumor cells unresponsive to most conventional therapies.

Most of the gene changes associated with tumor progression arise as direct or indirect consequences of insufficient respiration and elevated fermentation. Oncogene upregulation coupled with tumor-suppressor-gene downregulation becomes necessary in order to increase those metabolic pathways needed for fermentation. If the oncogenes needed to drive cellular fermentation are not expressed then the cell will die from energy failure. Oncogene expression is essential to maintain cell viability following respiratory damage. This perspective addresses the NCI provocative question #22 (provocativequestions.nci.nih.gov). Succinate produced through mitochondrial glutamine fermentation could be responsible, in part, for stabilization of Hif-1α (Chapter 8). Hif-1α is required for maintaining elevated glucose uptake and glycolysis. Respiratory injury becomes the driver for the gene regulatory changes needed for increasing compensatory energy production through fermentation. Insufficient respiration drives oncogene expression, not the reverse.

As DNA repair mechanisms are dependent on the efficiency of respiratory energy production, the continued impairment of respiration will gradually undermine nuclear genome integrity leading to a mutator phenotype and the plethora of somatic mutations identified in tumor cells. Specifically, the integrity of the nuclear genome is dependent on normal cellular respiration. When cellular respiration becomes compromised genomic instability increases. Activation of oncogenes, inactivation of tumor-suppressor genes, and aneuploidy will be the natural consequences of protracted mitochondrial dysfunction (Chapter 9). These gene abnormalities will contribute to accumulative mitochondrial dysfunction while also enhancing those energy pathways needed to upregulate and sustain fermentation energy. The greater the dependency on fermentation and SLP over time, the greater will be the degree of malignancy.

As respiration is necessary for maintaining cellular differentiation, loss of respiration leads to dedifferentiation and a return to the default state of proliferation. Szent-Gyorgyi considered this cellular state as that which existed in the α-period in the history of life on the planet.

> To make life perennial, the living systems, in this period, had to proliferate as fast as conditions permitted. Energy for this proliferation had to be produced by fermentation so that the α-period could also be called the fermentative period of unbridled proliferation (9).

The first three cancer hallmarks are consequences of the cell's return to its mode of existence during the α-period (Chapter 2). This would naturally involve increased aerobic glycolysis and resistance to apoptosis. A large number of fermenting cells will also produce excess of lactate and succinate. This would naturally produce an acidic microenvironment. Angiogenesis is a natural response to wound healing and to the metabolic state in the tumor microenvironment. All of these cancer hallmarks arise as a consequence of insufficient respiration and tumor cell fermentation.

According to the recent commentary of Lazebnik, all hallmarks of cancer with the exception of invasion and metastasis can be found in benign tumors or nonmetastatic cancers (10). I have also mentioned in Chapter 13 that four of the five hallmarks of cancer are also found in the crown-gall tumors of plants. In contrast to animal cancers, crown-gall tumors do not invade or develop metastases. Hence, it is the hallmark of invasion and metastasis that primarily makes cancer the deadly disease that it is.

Although EMT is currently viewed as a credible explanation for the cancer cell invasion and metastasis, this hypothesis does not link metastasis to defects in mitochondria, but rather to changes in developmental regulatory programs (11). As an alternative to the EMT explanation of metastasis, I have showed in Chapter 13 how macrophage fusion hybridization with neoplastic epithelial cells can logically account for all characteristics of the metastatic cascade. Many of the gene expression profiles observed in metastatic cancers are similar to those associated with the function of macrophages or other fusogenic cells of the immune system. Damaged respiration in these fusion hybrids can account for the invasive and metastatic properties found in cancer cells.

IS CANCER MANY DISEASES OR A SINGULAR DISEASE OF ENERGY METABOLISM?

If all cancer cells suffer from respiratory insufficiency, then respiratory insufficiency becomes the central hallmark of the disease. The current view of cancer as a hodgepodge of many diseases is fundamentally inaccurate in view of the central defect of the disease. Cancer appears as many diseases only if viewed from its histological appearance and from its genomic changes. I consider the histological appearance and gene expression profiles of cancer cells as "red herrings." When viewed in the light of energy metabolism, cancer is a singular disease of respiratory insufficiency.

The most convincing evidence supporting my view comes from the reduced growth response of all tumors when their ability to ferment glucose and glutamine becomes interrupted (Chapters 17–19). How long will it take before the cancer field comes to recognize that all cancer cells suffer from some degree of disabled respiration? It is my opinion that there will be no real progress in managing cancer until this fact becomes widely recognized and accepted.

REFERENCES

1. Dobzhansky T. Nothing in biology makes sense except in the light of evolution. Am Biol Teach. 1973;35:125–9.
2. Lands B. A critique of paradoxes in current advice on dietary lipids. Prog Lipid Res. 2008;47:77–106.
3. Mayr E. The Growth of Biological Thought: Diversity, Evolution, and Inheritance. Cambridge, MA: Harvard University Press; 1982.
4. Seyfried TN, Shelton LM. Cancer as a metabolic disease. Nutr Metab. 2010;7:7.
5. Gold LS, Slone TH, Manley NB, Garfinkel GB, Hudes ES, Rohrbach L, et al. The carcinogenic potency database: analyses of 4000 chronic animal cancer experiments published in the general literature and by the U.S. National Cancer Institute/National Toxicology Program. Environ Health Perspect. 1991;96:11–5.
6. Lawrence T, Gilroy DW. Chronic inflammation: a failure of resolution? Int J Exp Pathol. 2007;88:85–94.

7. Smith AE, Kenyon DH. A unifying concept of carcinogenesis and its therapeutic implications. Oncology. 1973;27:459–79.
8. Warburg O. On the origin of cancer cells. Science. 1956;123:309–14.
9. Szent-Gyorgyi A. The living state and cancer. Proc Natl Acad Sci USA. 1977;74:2844–7.
10. Lazebnik Y. What are the hallmarks of cancer? Nat Rev. 2010;10:232–3.
11. Hanahan D, Weinberg RA. Hallmarks of cancer: the next generation. Cell. 2011;144:646–74.

Chapter 15

Nothing in Cancer Biology Makes Sense Except in the Light of Evolution

The title of this chapter is a paraphrase from that of Theodosius Dobzhansky's famous article describing how Darwinian concepts of evolution are not incompatible with religious faith, but are incompatible with creationist's views of evolution (1). The article focuses on the biological diversity among organisms that could arise *only* through the process of natural selection as Darwin had described in his original theory (2). Many investigators in the cancer field have attempted to force the concepts of the Darwinian theory to the phenomenon of tumor progression (3–10). If cancer were a genetic disease, then cancer progression might follow rules of Darwinian evolution. On the other hand, if cancer were not a genetic disease, but a metabolic disease, then major inconsistencies should become apparent in attempting to link Darwinian concepts to cancer development. In addition to Darwin's theory, Jean-Baptiste Pierre Antoine de Monet, Chevalier de la Marck's (Lamarck) had also entertained evolutionary concepts in his "Philosophie zoologique ou exposition des considérations relatives à l'histoire naturelle des animaux" (11, 12). Although Lamarck's theory was largely discredited as an explanation for the origin of species, his theory of acquired characteristics is relevant to the origin and progression of cancer (12, 13). I will illustrate the connection of Lamarck's ideas to the cancer problem after first revealing the inconsistencies with Darwin's theory.

According to the view of cancer as a Darwinian process, collections of mutations accumulate in various cells of the expanding tumor. Some of the mutations are assumed to bestow a growth advantage on certain cells. In other words, some tumor cells grow faster than other tumor cells owing to unique types of genomic damage and rearrangements (drivers). The descendants of these mutant cells then contribute greater numbers of cells to the tumor population as the tumor progresses, thus expanding the genetic heterogeneity and adaptability of the tumor cells. Some

Cancer as a Metabolic Disease: On the Origin, Management and Prevention of Cancer, First Edition. Thomas Seyfried.

investigators also consider that the mutation-riddled tumor cells will have a growth advantage over the naturally quiescent or differentiated normal cells of the host. The view of cancer progression as a type of Darwinian evolutionary process is founded on the premise that cancer is a genetic disease.

REVISITING GROWTH ADVANTAGE OF TUMOR CELLS, MUTATIONS, AND EVOLUTION

The concept that cancer progression simulates Darwinian evolution can be best appreciated from the comments of those who espouse this view. John Cairns has proposed that carcinogens induce mutations in genes that further enhance mutagenesis (6). These mutations were considered to produce cells with a greater fitness (better able to survive) and adaptability than normal cells. According to Cairns' view, natural selection becomes a liability during cancer progression by selecting dangerous mutations that confer an increased survival advantage on a cell (6). This view is essentially opposite to that of established evolutionary concepts where Darwin states, "*. . . we may feel sure that any variation in the least degree injurious would be rigidly destroyed. This preservation of favorable variations and the rejection of injurious variations, I call Natural Selection*" (p. 81) (2). In contrast to Darwin's view that natural selection will purge injurious variations from populations while preserving favorable variations, Cairns considers that natural selection will select for injurious mutations that enhance fitness and survival. How is it possible that natural selection would purge unfit organisms from the natural world according to Darwin's theory, but would select positively for the most genetically damaged cells in tumors according to Cairns' hypothesis? Cairns' view of evolution is clearly at odds with Darwin's central theory.

It is hard for me to see how any cells with multiple types and kinds of injurious mutations could possibly be considered fitter and better able to survive than cells that do not contain these mutations. Just because genetically damaged cells are enriched in tumors does not mean that these cells are more fit than cells that do not contain genetic damage. Most mutations by nature are injurious. Mckusick's catalogue of human inborn errors makes this fact quite clear (14). Fitness and survival is generally less for humans that inherit deleterious mutations than for humans that do not inherit these mutations. Are we to consider mutations bad for survival when they occur in the germ line and are expressed in all cells of the organism, but consider them advantageous for survival when they occur through somatic inheritance in cancer cells? This makes no sense in the light of Darwin's theory. While a subtle genomic change might enhance the survival of cells under certain stressful conditions, large numbers of deletions, duplications, broken chromosomes, and aneuploidy cannot be considered favorable traits with respect to cell strength and viability.

According to my view, cancer cells proliferate and survive not because of their genomic instability, but because of their respiratory insufficiency. Respiratory insufficiency enhances fermentation, destabilizes the genome, and causes entry into the default state of unbridled proliferation (15). Unlike mammalian embryos, which would abort if they expressed the types of mutations found in cancer cells, cancer cells survive and grow with these mutations. The mutations are not lethal and are tolerated because the cancer cells depend more on fermentation than on respiration for energy. Glycolysis-derived pyruvate also enhances the p-glycoprotein activity (16). The p-glycoprotein is responsible for pumping toxic drugs out of cells, and, when activated, makes tumor cells resistant to most chemotherapy (17). This is often referred to as the *multidrug resistance* (MDR)*phenotype*, which is glycolysis dependent (18). Hence, this aspect of cancer cell drug resistance arises as a consequence of the glycolytic phenotype.

Moreover, the accumulated mutations arise as downstream effects of inadequate respiration coupled with compensatory fermentation. As nuclear genomic repair mechanisms are dependent on the integrity of mitochondrial respiration, mutations and genomic instability are the expected consequences of protracted energy transition from respiration to fermentation. The mutations arise following mitochondrial damage and are largely irrelevant to the origin of the disease despite reports to the contrary (3, 19). Mutations could, however, contribute to cancer progression and irreversibility of the disease (Chapter 14).

If mutations cause cancer, then how is it possible that normal embryos can be generated from tumor nuclei, which contain genomic instability (Fig. 11.5, Chapter 11)? Aborted development, rather than neoplasia, arises from the epigenetic reprograming of tumor nuclei. That normal tissues can be derived from the nuclei of cancer cells provides compelling evidence against the notion that somatic mutations are the origin of cancer or drive the disease. There is no conceivable explanation in Darwin's doctrine that could account for enhanced survival of organisms that express multiple types and kinds of deleterious mutations. If this statement were true, then how do cancer cells survive and proliferate when expressing profound genomic instability?

The survival and proliferation of tumor cells in hypoxic environments is not due to the numbers or types of mutations they express, but it is due to the replacement of their respiration with fermentation. Cells adapted to ferment glucose and glutamine can survive better in hypoxic environments than cells that require respiration for survival. Adaptation to fermentation is the consequence of protracted damage to respiration during the initiation and progression of cancers. *Fermentation energy is primal energy*. Fermentation is linked to unbridled proliferation (15, 20). Somatic mutations do not drive this process, but arise as the result of the process. I consider all the mutations found in somatic cancers as passengers with none as drivers. This is probably not the type of answer expected from the NCI provocative question #10 (provocativequestions.nci.nih.gov). Unfortunately, it might take some time before the cancer field comes to appreciate my perspective on this question.

Moreover, somatic mutations and aneuploidy reduce rather than enhance the rate of tumor growth. Support of the inhibitory effects of mutations on tumor growth comes from the well-documented findings that glioma progression is generally slower in patients with chromosome 1p/19q codeletions, promoter hypermethylation of the *O*6-methylguanine methyltransferase (*MGMT*) gene, or mutations in the gene for isocitrate dehydrogenase 1 (*IDH1*) (21–24). Mouse lung tumors expressing mutations in the Ki-ras oncogene grow slower than lung tumors not expressing Ki-ras mutations (25). If mutations are supposed to confer a growth advantage, why do tumors containing these mutations grow slower than tumors not containing these mutations? Are we to consider these as good mutations or as bad mutations? The work from the Amon and the Compton groups shows that aneuploidy retards cell growth (26–29). These findings are also inconsistent with the view that mutations enhance the fitness and growth advantage of tumor cells. The paradox is resolved once cancer becomes recognized as a mitochondrial metabolic disease rather than as a genetic disease.

Similar to the concepts of Cairns, Nowell has also outlined an evolutionary process by which mutations and chromosomal rearrangements could give rise to cancer cells with a growth advantage over normal cells (9, 30). Nowell writes in his 1976 paper in *Science*, "*The nature of the alteration from a normal cell to the first neoplastic cell... must still be defined arbitrarily. For the purposes of this model, 'neoplasia' is considered as some degree of escape from normal growth control (whether these controls are intracellular, local 'chalones,' or hormonal) that provides the cell with a selective growth advantage over the normal cell from which it was derived. In some instances, the process may include a latent period, until the altered cell is triggered from a resting state (G_0) into active proliferation (G_1); in other circumstances, the initial event may involve a stem cell that is already dividing, and simply increases the proportion of progeny remaining in the mitotic cycle instead of proceeding to terminal differentiation. The fundamental nature of this initial step, and degree to which it is specific for each neoplasm, remains a basic problem in cancer research.*" These concepts are illustrated in Figure 15.1 similar to what Nowell specified in Figure 1 from his paper (9).

According to Nowell's model, tumor cells with extra chromosomes would have a selective growth advantage over nonneoplastic cells with normal chromosomes. Eventually tumor cells emerge that express the highest degree of malignancy. Nowell indicates that "*the key point is not the specificity or lack of specificity of the chromosomal rearrangements, but the fact that major genetic errors do occur in tumor cell populations with sufficient frequency to permit sequential selection of mutant subpopulations over time*" (9).

According to Nowell's argument, it makes no difference what type of genomic defects are present in tumor cells to provide a selective advantage, but only that genomic defects are present. What does this mean in light of the massive cancer genome projects that attempt to define the minutia of genomic alterations in tumor cells? Nowell's argument would suggest that there would be no specific "driver" genes in cancer progression and that any genomic abnormality would be effective in driving progression. According to Nowell's hypothesis, the efforts of the cancer genome projects would be unproductive, as the specificity of the genomic defects in cancer cells is not considered important for cancer progression. This is interesting since Nowell's findings together with the early studies of Boveri were considered to be justification for detailed studies of the cancer genome (3). Nowell also mentions that the mechanism for the increased mutability in the neoplastic

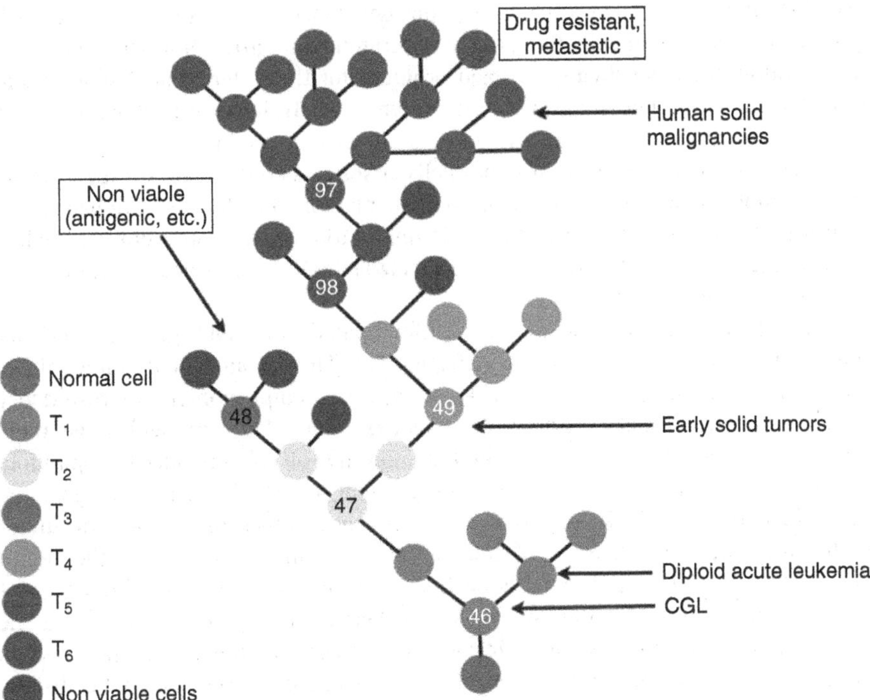

Figure 15.1 Model of clonal evolution in neoplasia (from Nowell (9). Carcinogen-induced change in progenitor normal cell (N) produces a diploid tumor cell (T_1, 46 chromosomes) with growth advantage permitting clonal expansion to begin. Genetic instability of T_1 cells leads to production of variants (illustrated by changes in chromosome number, T_2 to T_6). Most variants die due to metabolic or immunologic disadvantage (dark circles); occasionally one has an additional selective advantage (e.g., T_2, 47 chromosomes), and its progeny becomes the predominant subpopulation until an even more favorable variant appears (e.g., T_4). The stepwise sequence in each tumor differs (being partially determined by environmental pressures on selection), and results in a different, aneuploid karyotype in each fully developed malignancy (T_6). Biological characteristics of tumor progression (e.g., morphological and metabolic loss of differentiation, invasion and metastasis, resistance to therapy) parallel the stages of genetic evolution. Human tumors with minimal chromosome change (diploid acute leukemia, chronic granulocytic leukemia) are considered to be early in clonal evolution; human solid cancers, typically highly aneuploid, are viewed as late in the developmental process. *Source*: Modified from Reference 9. See color insert.

cells remains unknown. On the basis of the information amassed in my treatise, a rational explanation for Nowell's observations is now possible.

The apparent growth advantage of tumor cells over normal cells is not really an advantage but is an abnormal phenotype. If normal cells needed to expand their population in response to a wound or physiological stress, they could grow as fast or faster than any tumor cell. Support for my position comes from the findings of Dean Burk, who had first shown that the growth rate was similar for normal liver cells during regeneration and for hepatoma cells during tumor progression

(31, 32). It is therefore erroneous to consider cancer cells as having a growth advantage over normal cells that are not programmed to grow. Normal cells of any tissue follow their genetically scripted program for the differentiated state. Rapid growth is generally not part of this scripted program. If, however, growth becomes necessary for tissue repair, then the normal cells can grow fast.

The dysregulated growth of tumor cells arises as a consequence of their abnormal metabolic state involving enhanced fermentation. As I have mentioned above and in previous chapters, fermentation is linked to unbridled proliferation. Unbridled proliferation is the default state of metazoan cells when released from active negative control (33–35).

Active respiration maintains the differentiated state and genome integrity through the RTG signaling system (Chapter 10). The chromosomal abnormalities and gene mutations in cancer arise as downstream epiphenomena of respiratory damage and the mitochondrial stress response. The fidelity of nuclear genomic repair mechanisms is dependent on the integrity of mitochondrial respiration. Carcinogens induce mitochondrial damage that eventually leads to genomic instability. Indeed, Burk and Warburg have shown that the degree of tumor malignancy was correlated with the degree of fermentation within the tumor cells (32, 36–39). Although Nowell's scenario described in Figure 15.1 does not accurately parallel the stages of Darwinian genetic evolution as he suggests, his view that genomic instability increases during cancer progression is compatible with the evidence of cancer as a mitochondrial metabolic disease as I have described in earlier chapters. The greater the fermentation, the greater will be the genomic instability, inflammation, and disruption of the microenvironment (40).

The views of Nowell and Cairns on the role of mutations in the evolution of cancer are essentially mirrored in the views of Loeb and colleagues (8, 41). I have included their Figure 3 as it appears from their review illustrating cancer as a somatic evolutionary process (Fig. 15.2). The Salk et al. model of cancer evolution is also in "*lock step*" with the view of the Fearon and Vogelstein (42) that cancer progression is essentially a linear process. Although not mentioned in their review, the model is also consistent with the epithelial–mesenchymal transition (EMT) for the origin of metastasis (43). I have reviewed the fallacy of the EMT as a logical explanation of metastasis in Chapter 13. Previous studies from Petrelli and coworkers indicated that an average of 11,000 genomic events may occur in any given colon carcinoma cell (19). The Loeb group also inferred that cancer cells in any given tumor are unlikely to contain the same complement of mutations (8, 41). This information can account for why most gene-based targeted therapies have done little to arrest the disease and are unlikely to have a major impact on cancer management (44). When will people come to know this?

In their recent review, Stratton and colleagues (3) mentioned that approximately 100,000 somatic mutations from cancer genomes have been described in the quarter of a century since the first somatic mutation was found in *HRAS*. They predict that large-scale, complete sequencing of cancer genomes will identify *several hundred million more mutations* over the next few years. Moreover, the data collected are

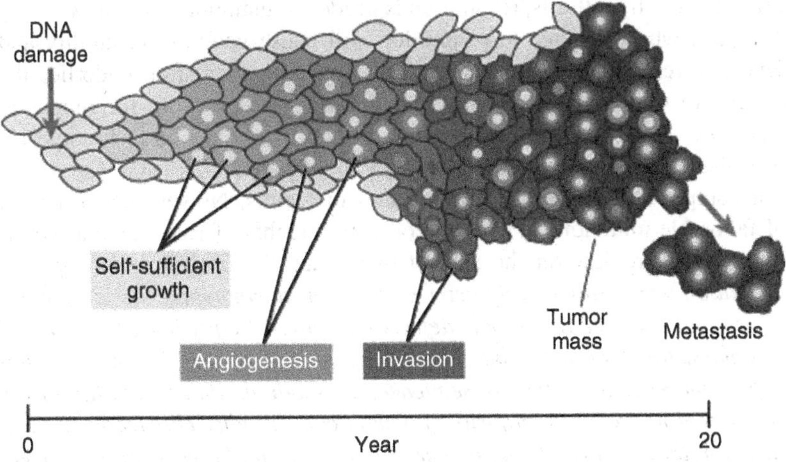

Figure 15.2 Model showing how cancer recapitulates evolution Within a developing tumor, mutations accumulate over time as a result of unrepaired DNA damage. Most of these mutations are either neutral or detrimental; only a small number bestow growth and survival benefits upon a cell. Cells with these beneficial variants preferentially multiply and additional mutations occur that may undergo further selection and expansion. Advantageous phenotypes for tumor growth include, among others, the ability to divide independently of extracellular signals (light shade), the ability to recruit a blood supply (gray shade), and the ability to invade adjacent and distant tissues (dark shade). *Source*: Reprinted with permission from Salk et al. (8). See color insert.

predicted to provide what they consider "*a fine-grained picture of the evolutionary processes that underlie our commonest genetic disease, providing new insights into the origins and new directions for the treatment of cancer.*" (3). Nothing could be further from the truth in my opinion.

It is now recognized that *Ras* oncogenes damage the mitochondria to induce cell senescence (45, 46). The transition from respiration to fermentation allows tumor cells to circumvent the *Ras*-induced senescence checkpoint. This also allows the tumor cells to survive despite high production of oxidative stress. The origin of Ras-induced oncogenic transformation is damaged mitochondria. Nuclear genome integrity unravels as a result of protracted respiratory dysfunction. The plethora of mutations collected in cancer cells arise as a consequence, rather than as a cause, of respiratory damage. A focus on cancer mutations is unlikely to impact cancer management (47). Although the large-scale cancer genome projects have had little impact on the development of effective cancer drugs, these projects have significantly advanced our ability to sequence DNA, as Dr. Linda Chen had mentioned in her plenary lecture at the 2011 meeting of the AACR. It is unclear, however, if DNA sequencing efficiency was the intended purpose of the cancer genome projects. Certainly the large cancer genome projects have provided a wealth of new information for mathematicians and for the field of bioinformatics, but have yet to provide much benefit to those suffering from the disease.

According to my hypothesis, respiration is needed to maintain genomic stability and DNA repair, while damage to respiration will cause genomic instability and a dependence on fermentation for energy. The shift to fermentation underlies the ability of cells to divide independently and is largely responsible for increased angiogenesis. Metastasis is initiated following respiratory damage in fusion hybrids between neoplastic cells and myeloid cells (macrophages).

In near-complete agreement with the views of Cairns, Nowell, and Loeb on the role of favorable mutations in cancer progression is that of Hanahan and Weinberg. In their recent review on the subject they state: "*Although the specifics of genome alteration vary dramatically between different tumor types, the large number of genome maintenance and repair defects that have already been documented in human tumors, together with abundant evidence of widespread destabilization of gene copy number and nucleotide sequence, persuade us that instability of the genome is inherent to the great majority of human cancer cells. This leads, in turn, to the conclusion that the defects in genome maintenance and repair are selectively advantageous and therefore instrumental for tumor progression, if only because they accelerate the rate at which evolving premalignant cells can accumulate favorable genotypes. As such, genome instability is clearly an enabling characteristic that is causally associated with the acquisition of hallmark capabilities.*" (7).

It is clear that this view differs little from previous views of cancer as a Darwinian evolutionary process. Consequently, this view also suffers from inconsistencies with Darwin's original theory. It is apparent from this statement that the specifics of genomic instability in cancer are less important than the fact that genomic instability exists, which enables progression. This is essentially a recapitulation of Nowell's argument. How important can specific genomic defects be if cancer progression occurs despite dramatic genomic differences between different tumor types? These inconstancies will persist in discussions of cancer progression as long as cancer is viewed as a genetic disease.

One of the more comprehensive views of cancer as an evolutionary process is that of Maley and colleagues (5). These investigators have done a good job in linking aspects of the evolution theory to the problem of cancer. Although their treatise also favors the idea that cancer is a genetic disease, they raise a number of important issues that go beyond this fallacy. This is especially the case for their discussion on the "fitness landscape" in tumors and the role of the microenvironment. They state: "*The fitness of genotypes, and therefore the topography of the fitness landscape, depends on the local microenvironment, including the ecology of other cells present.*" (5). According to my view, the appearance of fitness arises from the replacement of respiration with fermentation. This could be viewed as a type of somatic evolution, but one that would be driven more by epigenetics (mitochondrial dysfunction) than by nuclear gene mutations (48, 49).

More recently, Davies and Lineweaver provided insightful views on the evolutionary origin of cancer (10). They consider cancer as an atavistic state of multicellular life where long-suppressed ancestral cellular functions become reactivated or switched on. According to their view, cancer genetic or epigenetic mutations unlock an ancient "toolkit" of preexisting adaptations that allow cancer cells to survive in hypoxic environments. The Davies and Lineweaver evolutionary view of cancer is consistent in some ways with my hypothesis and with the views of Sonnenschein and Soto (33) and Szent-Gyorgyi (20). Unbridled proliferation is the default state of metazoan cells. Unbridled proliferation existed during the oxygen-sparse α period of species evolution. This was also a highly reduced state where the ancient pathways of fermentation predominated in driving cell physiology. The appearance of oxygen gave rise to the oxidized state and the emergence of respiration. The emergence of respiration facilitated greater complexity in biological systems.

Respiration largely maintains the differentiated state of metazoan cells. Irreversible damage to respiration, coincident with a rise in fermentation, would unlock the toolkit of preexisting adaptations needed to survive in low oxygen environments. According to my view, protracted respiratory injury gives rise to compensatory fermentation or the atavistic condition in order to maintain cell viability. Mutations and genomic instability are not the cause of the process, but are the effects of the process. It is not necessary to force these phenomena into a Darwinian model of evolution.

TUMOR CELL FITNESS IN LIGHT OF THE EVOLUTIONARY THEORY OF RICK POTTS

Potts (50–52), a paleoanthropologist at the Smithsonian Institution, has suggested that the evolutionary success of our species has been due largely to the germ line inheritance of traits that have bestowed adaptive versatility. Adaptability was defined in terms of (i) the ability of an organism to persist through major environmental shifts, (ii) to spread to new habitats, and (iii) to respond in novel ways to its surroundings (52). These characteristics were honed over millions of years and have enabled humans to adapt rapidly to abrupt changes in the physical environment including changes in moisture, temperature, food resources, and so on. The adaptability to abrupt environmental change is a property of the genome, which was selected for in order to ensure survival under environmental extremes (53).

This hypothesis is an extension of Darwin's original theory (Chapter IV, Natural Selection) and can be applied to the individual cells of the organism, which exist as an integrated society of cells. The success in dealing with environmental stress and disease is therefore dependent on the integrated action of all cells in the organism. Further, this integrated action depends on the flexibility of each cell's genome, which responds to both internal and external signals according to the needs of the organism. More specifically, only those cells that possess flexibility in nutrient utilization will be able to survive under nutrient stress. Environmental forcing has therefore selected those genomes that are most capable of adapting to change in order to maintain metabolic homeostasis (2, 51–53).

The widely held notion that tumor cells have a growth advantage and are more fit than normal cells is not only inconsistent with Darwin's theory, but is also inconsistent with Potts' concept of adaptive versatility (52, 53). It is difficult to conceive how the nonuniform genomic instability seen in cancer cells could enhance their adaptability. As long as the tumor cells have access to the metabolic fuels needed for fermentation, they will give the appearance of having a growth advantage over normal cells. According to Darwin and Potts, mutations that bestow a selective advantage are those that will enhance survival under environmental stress. If the multiple pathogenic mutations, chromosomal rearrangements, and mitochondrial abnormalities confer a fitness or survival advantage to tumor cells, then survival under environmental stress should be better in tumor cells than in normal cells. This is not what actually happens when the hypothesis is tested.

For example, when mice or people with tumors are placed under energy stress using dietary energy reduction (DER), many tumor cells die while normal cells survive. Indeed, the health and vitality of the normal cells improves with time under DER while the tumor cells die from apoptotic death. Support for my contention comes from our many studies of treating brain-tumor-bearing mice with dietary energy stress (presented in Chapter 17) (54–60). It is clear that the adaptability to environmental stress is greater in normal cells than in tumor cells. This also explains why cell death is generally greater in tumor cells than in normal cells following exposure to toxic radiation and cancer drugs. In contrast to DER, radiation and toxic drugs run the risk of creating tumor cells that are highly resistant to drugs and radiation. This comes in large part from the damage to respiration in bystander precancerous cells. These cells are often those that eventually become heavily dependent on fermentation for energy. I will address this more in Chapter 17 when dealing with tumor management.

The greater adaptability of normal cells than tumor cells to energy stress is predicted based on the theories of Darwin and Potts. Adaptive versatility is a complex phenotype regulated by multiple regulatory systems derived through germ line inheritance. Metabolic flexibility allows the organism to respond in a coordinated way to environmental stress. Energy stress will force all normal cells to work together for the survival of the society. The genomic instability in tumor cells reduces their flexibility when they are placed under metabolic stress. Any type of genomic instability should reduce metabolic flexibility. In other words, the specificity of the genomic defects is less important than the fact that the genomic defects exist. Genomic defects will disrupt metabolic flexibility under energy stress, thus inhibiting adaptability. My concept is the same as that of Nowell's except in viewing genomic instability as a liability rather than as an advantage to progression. Because tumor cells ferment rather than respire, they are dependent on the availability of fermentable fuels (glucose and glutamine). Normal cells shift metabolism from glucose to ketone bodies and fats when placed under energy stress. This is dependent on genomic stability.

Ketone bodies and fats are nonfermentable fuels in mammalian cells. Tumor cells have difficulty in using ketone bodies and fats for fuel when glucose is reduced. Because tumor cells lack genomic stability, they are less able than normal cells to adapt to changes in the metabolic environment. The survival of such cells will be counterproductive for the survival of the society and will be eliminated for the good of the society. This might be considered communism in the ideal sense.

Our studies in mice with brain tumors are proof of principle that tumor cells are less adaptable than normal cells when placed under energy stress. Apoptosis under energy stress is greater in tumor cells than in normal cells.

Cancer cells survive and multiply only in physiological environments that provide fuels necessary for fermentation through substrate-level phosphorylation (61). If these fuels become restricted, tumor cells will have difficulty in surviving and growing regardless of their complement of genomic changes. Multiple genetic defects will reduce genomic flexibility, thus increasing the likelihood of cell death under environmental stress. Regardless of when or how genomic defects become involved in the initiation or progression of tumors, these defects can be exploited for the destruction or management of the tumor. The view of cancer progression as a Darwinian process is inconsistent with the facts.

CANCER DEVELOPMENT AND LAMARCKIAN INHERITANCE

The growth of biological thought is dependent on the evolution of ideas. When biological facts support the idea, a new generalization arises. If the facts supporting the origin and progression of cancer are incompatible with the concepts of Darwinian evolution, then what evolutionary concepts might be compatible with the facts?

If cancer were viewed as a mitochondrial metabolic disease, as I have described in this treatise, then the general concepts of Lamarck might be better suited than those of Darwin to explain the phenomenon. It is my opinion that Lamarck's theory of evolution, involving the use and disuse of organs and the inheritance of acquired characters, can better explain the origin and progression of cancer than can Darwin's views (13, 62). According to Lamarck, the environment produces changes in biological structures. Through adaptation and differential use, these changes lead to modifications in the structures. The modifications would then be passed on to successive generations as acquired traits.

According to Lamarck, any animal organ that receives more sustained and frequent use will be strengthened in proportion to the length of time that it has been used for. On the other hand, constant disuse will weaken the function of an organ until it disappears [62, p. 355]. Lamarck's evolutionary synthesis was based on his belief in the innate tendency toward increasing organizational complexity or progress (13). The degree of use or disuse has shaped biological evolution along with the inheritance of acquired adaptability. Lamarck's ideas could also accommodate a dominant role for epigenetics and horizontal gene transfer, as factors that could facilitate progression (12, 13). Epigenetic mechanisms in the form of cell fusion and horizontal gene transfer also contribute to cancer progression and metastasis (63–65).

How could Lamarck's evolutionary concepts be linked to the phenomenon of cancer progression? This linkage is seen if we replace organs with organelles and consider cytoplasmic inheritance as somatic inheritance of acquired characters.

Considering the dynamic behavior of mitochondria involving regular fusions and fissions, abnormalities in mitochondrial structure and function can be rapidly disseminated throughout the cellular mitochondrial network and passed along to daughter cells somatically through cytoplasmic inheritance (61, 66). The degree of mitochondrial respiratory function becomes progressively less with each cell division as adaptability to fermentation increases.

The somatic progression of cancer would therefore embody the concept of acquired inheritance. The most malignant cancer cells would sustain the near-complete replacement of their respiration with fermentation. This process could be viewed as Lamarckian inheritance. Although Lamarck considered the inheritance of acquired characteristics as enhancing biological complexity and perfection, the opposite effect would occur in adapting his evolutionary concepts to cancer progression. More specifically, a Lamarckian view can account for the escalation situation of biological chaos and the nonuniform accumulation of mutations seen during cancer progression. Hence, the evolutionary concepts of Lamarck can better explain the phenomena of tumor progression than can the evolutionary concepts of Darwin.

CAN TELEOLOGY EXPLAIN CANCER?

I find it remarkable that teleological considerations or explanations are occasionally mentioned in scientific publications dealing with cancer (7, 67–72). Teleology involves design with a purpose and is the cornerstone of arguments involving *intelligent design* or *creationism*. A teleological explanation for complicated phenomena assumes that the system examined has an intended purpose and was designed to accomplish that purpose (73, 74). Evolution operates without purpose, but rather by genetic chance and environmental necessity (75).

The Reverend William Paley (1743–1805) was better than anyone else at elaborating the teleological argument for creation by design (73). Cybernetic-type diagrams, which are sometimes used to describe the complexity of cancer signaling systems, can also be considered as teleological explanations (76). Although teleological explanations might appear appealing on the surface, they muddle mechanistic explanations for the phenomena under investigation. Intentions and purpose are not assumed in mechanistic explanations (73). Descriptions of cancer cells as being motivated, choosing to do something, or having an agenda are examples of teleology. It is unlikely in my opinion that teleological explanations will provide much insight into the origin or progression of cancer.

REFERENCES

1. Dobzhansky T. Nothing in biology makes sense except in the light of evolution. Am Biol Teach. 1973;35:125–9.
2. Darwin C. On the Origin of Species by Means of Natural Selection, or on the Preservation of Favored Races in the Struggle for Life. London: John Murry; 1859.
3. Stratton MR, Campbell PJ, Futreal PA. The cancer genome. Nature. 2009;458:719–24.

4. Crespi B, Summers K. Evolutionary biology of cancer. Trends Ecol Evol. 2005;20:545–52.
5. Merlo LM, Pepper JW, Reid BJ, Maley CC. Cancer as an evolutionary and ecological process. Nat Rev. 2006;6:924–35.
6. Cairns J. Mutation selection and the natural history of cancer. Nature. 1975;255:197–200.
7. Hanahan D, Weinberg RA. The hallmarks of cancer. Cell. 2000;100:57–70.
8. Salk JJ, Fox EJ, Loeb LA. Mutational heterogeneity in human cancers: origin and consequences. Annu Rev Pathol. 2010;5:51–75.
9. Nowell PC. The clonal evolution of tumor cell populations. Science. 1976;194:23–8.
10. Davies PC, Lineweaver CH. Cancer tumors as Metazoa 1.0: tapping genes of ancient ancestors. Phys Biol. 2011;8:015001.
11. Mayr E. The Growth of Biological Thought: Diversity, Evolution, and Inheritance. Cambridge: Belknap Harvard; 1982.
12. Handel AE, Ramagopalan SV. Is Lamarckian evolution relevant to medicine? BMC Med Genet. 2010;11:73.
13. Koonin EV, Wolf YI. Is evolution Darwinian or/and Lamarckian? Biol Direct. 2009;4:42.
14. McKusick VA. Mendelian Inheritance in Man: A Catalog of Human Genes and Genetic Disorders. 12th ed. Baltimore, MD: The Johns Hopkins University Press; 1998.
15. Moase CE, Trasler DG. Delayed neural crest cell emigration from Sp and Spd mouse neural tube explants. Teratology. 1990;42:171–82.
16. Wartenberg M, Richter M, Datchev A, Gunther S, Milosevic N, Bekhite MM, et al. Glycolytic pyruvate regulates P-Glycoprotein expression in multicellular tumor spheroids via modulation of the intracellular redox state. J Cell Biochem. 2010;109:434–46.
17. Aller SG, Yu J, Ward A, Weng Y, Chittaboina S, Zhuo R, et al. Structure of P-glycoprotein reveals a molecular basis for poly-specific drug binding. Science. 2009;323:1718–22.
18. Xu RH, Pelicano H, Zhou Y, Carew JS, Feng L, Bhalla KN, et al. Inhibition of glycolysis in cancer cells: a novel strategy to overcome drug resistance associated with mitochondrial respiratory defect and hypoxia. Cancer Res. 2005;65:613–21.
19. Stoler DL, Chen N, Basik M, Kahlenberg MS, Rodriguez-Bigas MA, Petrelli NJ, et al. The onset and extent of genomic instability in sporadic colorectal tumor progression. Proc Natl Acad Sci USA. 1999;96:15121–6.
20. Szent-Gyorgyi A. The living state and cancer. Proc Natl Acad Sci USA. 1977;74:2844–7.
21. Denny CA, Desplats PA, Thomas EA, Seyfried TN. Cerebellar lipid differences between R6/1 transgenic mice and humans with huntington's disease. J Neurochem. 2010;115:748–58.
22. Krex D, Klink B, Hartmann C, von Deimling A, Pietsch T, Simon M, et al. Long-term survival with glioblastoma multiforme. Brain. 2007;130:2596–606.
23. Stupp R, Mason WP, van den Bent MJ, Weller M, Fisher B, Taphoorn MJ, et al. Radiotherapy plus concomitant and adjuvant temozolomide for glioblastoma. N Engl J Med. 2005;352:987–96.
24. van den Bent MJ, Dubbink HJ, Marie Y, Brandes AA, Taphoorn MJ, Wesseling P, et al. IDH1 and IDH2 mutations are prognostic but not predictive for outcome in anaplastic oligodendroglial tumors: a report of the European organization for research and treatment of cancer brain tumor group. Clin Cancer Res. 2010;16:1597–604.
25. Ramakrishna G, Bialkowska A, Perella C, Birely L, Fornwald LW, Diwan BA, et al. Ki-ras and the characteristics of mouse lung tumors. Mol Carcinog. 2000;28:156–67.
26. Torres EM, Williams BR, Tang YC, Amon A. Thoughts on aneuploidy. Cold Spring Harb Symp Quant Biol. 2011;75:445–51.
27. Torres EM, Williams BR, Amon A. Aneuploidy: cells losing their balance. Genetics. 2008;179: 737–46.
28. Compton DA. Mechanisms of aneuploidy. Curr Opin Cell Biol. 2011;23:109–13.
29. Thompson SL, Compton DA. Chromosomes and cancer cells. Chromosome Res. 2010;19: 433–44.
30. Nowell PC. Tumor progression: a brief historical perspective. Semin Cancer Biol. 2002;12:261–6.
31. Burk D, Behrens OK, Sugiura K. Metabolism of butter yellow rat liver cancers. Cancer Res. 1941;1:733–4.
32. Warburg O. On the origin of cancer cells. Science. 1956;123:309–14.

33. Sonnenschein C, Soto AM. The Society of Cells: Cancer and the Control of Cell Proliferation. New York: Springer-Verlag; 1999.
34. Sonnenschein C, Soto AM. Somatic mutation theory of carcinogenesis: why it should be dropped and replaced. Mol Carcinog. 2000;29:205–11.
35. Soto AM, Sonnenschein C. The somatic mutation theory of cancer: growing problems with the paradigm?. Bioessays. 2004;26:1097–107.
36. Warburg O. The Metabolism of Tumours. New York: Richard R. Smith; 1931.
37. Warburg O. On the respiratory impairment in cancer cells. Science. 1956;124:269–70.
38. Warburg O. Revidsed Lindau Lectures: The prime cause of cancer and prevention - Parts 1 & 2. In: Burk D, editor. Meeting of the Nobel-Laureates Lindau, Lake Constance, Germany: K.Triltsch; 1969. http://www.hopeforcancer.com/OxyPlus.htm.
39. Burk D, Schade AL. On respiratory impairment in cancer cells. Science. 1956;124:270–272.
40. Bissell MJ, Hines WC. Why don't we get more cancer? A proposed role of the microenvironment in restraining cancer progression. Nat Med. 2011;17:320–9.
41. Loeb LA. A mutator phenotype in cancer. Cancer Res. 2001;61:3230–9.
42. Fearon ER, Vogelstein B. A genetic model for colorectal tumorigenesis. Cell. 1990;61:759–67.
43. Weinberg RA. The Biology of Cancer. New York: Garland Science; 2007.
44. Gabor Miklos GL. The human cancer genome project–one more misstep in the war on cancer. Nat Biotechnol. 2005;23:535–7.
45. Moiseeva O, Bourdeau V, Roux A, Deschenes-Simard X, Ferbeyre G. Mitochondrial dysfunction contributes to oncogene-induced senescence. Mol Cell Biol. 2009;29:4495–507.
46. Hu Y, Lu W, Chen G, Wang P, Chen Z, Zhou Y, et al. K-ras(G12V) transformation leads to mitochondrial dysfunction and a metabolic switch from oxidative phosphorylation to glycolysis. Cell Res. 2012;22:399–412.
47. Hambley TW, Hait WN. Is anticancer drug development heading in the right direction? Cancer Res. 2009;69:1259–62.
48. Nanney DL. Epigenetic control systems. Proc Natl Acad Sci USA. 1958;44:712–7.
49. Smiraglia DJ, Kulawiec M, Bistulfi GL, Gupta SG, Singh KK. A novel role for mitochondria in regulating epigenetic modification in the nucleus. Cancer Biol Ther. 2008;7:1182–90.
50. Potts R. Environmental hypotheses of hominin evolution. Am J Phys Anthropol. 1998;27:93–136.
51. Potts R. Humanity's Descent: The Consequences of Ecological Instability. New York: William Morrow & Co., Inc; 1996.
52. Potts R. Complexity of Adaptibility in Human Evolution. In: Goodman M, Moffat AS, editors. Probing Human Origins. Cambridge (MA): American Academy of Arts & Sciences; 2002. p.33–57.
53. Seyfried TN, Mukherjee P. Targeting energy metabolism in brain cancer: review and hypothesis. Nutr Metab. 2005;2:30.
54. Zuccoli G, Marcello N, Pisanello A, Servadei F, Vaccaro S, Mukherjee P, et al. Metabolic management of glioblastoma multiforme using standard therapy together with a restricted ketogenic diet: Case Report. Nutr Metab. 2010;7:33.
55. Marsh J, Mukherjee P, Seyfried TN. Akt-dependent proapoptotic effects of dietary restriction on late-stage management of a phosphatase and tensin homologue/tuberous sclerosis complex 2-deficient mouse astrocytoma. Clin Cancer Res. 2008;14:7751–62.
56. Mukherjee P, El-Abbadi MM, Kasperzyk JL, Ranes MK, Seyfried TN. Dietary restriction reduces angiogenesis and growth in an orthotopic mouse brain tumour model. Br J Cancer. 2002;86:1615–21.
57. Mukherjee P, Abate LE, Seyfried TN. Antiangiogenic and proapoptotic effects of dietary restriction on experimental mouse and human brain tumors. Clin Cancer Res. 2004;10:5622–9.
58. Seyfried TN, Sanderson TM, El-Abbadi MM, McGowan R, Mukherjee P. Role of glucose and ketone bodies in the metabolic control of experimental brain cancer. Br J Cancer. 2003;89:1375–82.
59. Seyfried TN, Mukherjee P. Anti-angiogenic and pro-apoptotic effects of dietary restriction in experimental brain cancer: role of glucose and ketone bodies. In: Meadows GG, editor. Integration/Interaction of Oncologic Growth. 2nd ed. New York: Kluwer Academic; 2005. p. 259–70.

60. ZHOU W, MUKHERJEE P, KIEBISH MA, MARKIS WT, MANTIS JG, SEYFRIED TN. The calorically restricted ketogenic diet, an effective alternative therapy for malignant brain cancer. Nutr Metab. 2007;4:5.
61. SEYFRIED TN, SHELTON LM. Cancer as a metabolic disease. Nutr Metab. 2010;7:7.
62. MAYR E. The Growth of biological thought. 1982.
63. PAWELEK JM. Tumour cell hybridization and metastasis revisited. Melanoma Res. 2000;10:507–14.
64. HUYSENTRUYT LC, SEYFRIED TN. Perspectives on the mesenchymal origin of metastatic cancer. Cancer Metastasis Rev. 2010;29:695–707.
65. HOLMGREN L, SZELES A, RAJNAVOLGYI E, FOLKMAN J, KLEIN G, ERNBERG I, et al. Horizontal transfer of DNA by the uptake of apoptotic bodies. Blood. 1999;93:3956–63.
66. DETMER SA, CHAN DC. Functions and dysfunctions of mitochondrial dynamics. Nat Rev Mol Cell Biol. 2007;8:870–9.
67. BODE BP, FUCHS BC, HURLEY BP, CONROY JL, SUETTERLIN JE, TANABE KK, et al. Molecular and functional analysis of glutamine uptake in human hepatoma and liver-derived cells. Am J Phys. 2002;283:G1062–73.
68. GILLIES RJ, ROBEY I, GATENBY RA. Causes and consequences of increased glucose metabolism of cancers. J Nucl Med. 2008;49 Suppl 2:24S–42S.
69. ALEDO JC, JIMENEZ-RIVEREZ S, CUESTA-MUNOZ A, ROMERO JM. The role of metabolic memory in the ATP paradox and energy homeostasis. FEBS J. 2008;275:5332–42.
70. BONNET S, ARCHER SL, ALLALUNIS-TURNER J, HAROMY A, BEAULIEU C, THOMPSON R, et al. A mitochondria-K+ channel axis is suppressed in cancer and its normalization promotes apoptosis and inhibits cancer growth. Cancer Cell. 2007;11:37–51.
71. JAHNKE VE, SABIDO O, DEFOUR A, CASTELLS J, LEFAI E, ROUSSEL D, et al. Evidence for mitochondrial respiratory deficiency in rat rhabdomyosarcoma cells. PloS One. 2010;5:e8637.
72. GILLIES RJ, GATENBY RA. Ádaptive landscapes and emergent phenotypes: why do cancers have high glycolysis? J Bioenerg Biomembr. 2007;39:251–7.
73. SCHLESINGER AB. Explaining Life. New York: McGraw-Hill, Inc; 1994.
74. ROSENBLEUTH A, WIENER N, BIGELOW J. Behavior, purpose and teleology. Phil Sci. 1943;10: 18–24.
75. MONOD J. Chance & Necessity: An essay on the natural philosophy of modern biology. New York: Random House; 1971.
76. WIENER N. Cybernetics, Second Edition: or the Control and Communication in the Animal and the Machine. MIT Press: Cambridge, MA; 1965.

Chapter 16

Cancer Treatment Strategies

CURRENT STATUS OF CANCER TREATMENT

At present, surgery, chemo-, and radiation therapy are the standard procedures used for treating most malignant cancers. While these therapies can certainly provide long-term management of benign or nonmetastatic tumors, they have been less effective in providing long-term control of many advanced metastatic cancers. The data presented in Table 1.1 indicate that the current strategies for treating malignant cancer are not effective in reducing the overall number of deaths per year. It is hard to "sugar coat" these numbers. If claims are accurate that many new and effective cancer therapies are available, then why have the yearly death rates remained unchanged for so long? Not only is the cure rate poor for most malignant metastatic cancers, but many current therapies can actually exacerbate the disease. Is it necessary to surgically disfigure, poison, and "nuke" people in attempting to cure them? Chemo- and radiation therapy sicken and weaken patients, thus increasing their susceptibility to infections and diseases. Although these procedures could resolve the disease over a short term (months to years), they can enhance systemic physiological disorder (entropy) over the long term. Enhanced entropy accelerates the aging process thus reducing longevity. It is not clear how many cancer patients die from their disease or die from the toxic effects of the therapies used to treat their disease.

The US Food and Drug Administration (FDA) has recently approved the immunotherapy drug ipilimumab, "ipi", for treatment of malignant melanoma (1). The adverse effects of ipilimumab can include severe diarrhea, colitis (colon inflammation), and endocrine disruption. These adverse effects are generally treated with steroids (2). Indeed, the steroid drug, dexamethasone, is widely used to suppress nausea and vomiting in many cancer patients who receive toxic chemotherapy. Steroids significantly elevate blood glucose levels thus enhancing tumor cell survival and drug resistance (Chapter 17). While only 3 out of 540 persons who

Cancer as a Metabolic Disease: On the Origin, Management and Prevention of Cancer, First Edition. Thomas Seyfried.

received ipilimumab became cancer free, 14 persons died from the drug treatment (3). These findings indicate that the probability of dying from ipilimumab treatment was five times higher than the probability of receiving a cure. Patients treated with ipilimumab can be expected to live on an average for about 4 months longer than those patients treated with other drugs. Ipilimumab is administered in four infusions over a 3-month period with an estimated per patient cost of $120,000 (1). Hence, if ipilimumab does not kill a patient outright, the patient can be expected to pay about $30,000/month to exist on the planet for about 4 months longer. Do most patients with advanced cancer really think this drug is a promising breakthrough for managing their disease?

Besides ipilimumab, the BRAF kinase inhibitor vemurafenib received approval from the FDA for treatment of those melanoma patients who's tumors contain the V600E mutation in the V-RAF murine sarcoma viral homolog B1 (*BRAF*) oncogene (4). About 50% of the patients with this mutation in their tumor responded well to the drug, but about 50% of the patients with the mutation did not respond well. While short-term survival was better in patients who received vemurafenib than in those who received the control drug (dacarbazine), overall survival after 12 months was similar for both drugs (4). Common adverse events associated with vemurafenib were joint pain, rash, fatigue, hair loss, squamous cell carcinoma, photosensitivity, nausea, and diarrhea. If vemurafenib is supposed to manage melanoma by specifically targeting the damaged oncogene, it is not clear why some patients suffer from so many debilitating adverse events. It appears that vemurafenib might do more than simply target the BRAF V600E mutation. Nevertheless, tolerance of the adverse events might be acceptable to those patients who prefer quantity to quality of life.

It is hard for me to get excited about new immunotherapy drugs, especially since we showed that it is the immune system itself (myeloid cells) that gives rise to many metastatic cancers (Chapter 13). According to a recent article in the *Wall Street Journal*, there are currently about 23 cancer immunotherapies in the pipeline development (5). One company executive, Ira Mellman, said, "*We don't have to convince people that this is a good idea*" (5). I am not sure what people are convinced that this *is* a good idea. I remain skeptical. Knowing what I know about the origin and progression of cancer, I can predict that few of these immunotherapy drugs will provide real advances in long-term cancer management unless they can also target tumor energy metabolism. In some cases, however, immunotherapies could target energy metabolism.

For example, any cancer immunotherapy that also produces chills and fever in patients could be effective in causing tumor regression. Fever places global stress on the body, which will indirectly target energy metabolism. I address in Chapter 17 how energy stress specifically targets cancer cells. Cancer cells are less adaptable to hyperthermia energy stress than are normal cells (6). William Coley long ago showed that cancer regression could follow vaccine-induced fever (7). It will therefore be important to determine if the therapeutic action of expensive new immune therapies arises through a gene-mediated mechanism or through the simple induction of fever. I predict that progression-free survival will be better in those patients experiencing chills and fever than in patients not experiencing chills and fever.

Marketing hype will certainly make many immunotherapy drugs profitable, until they are eventually shown to be ineffective and are displaced by other costly and toxic drugs. There will be no real progress in new therapies until cancer becomes recognized as a metabolic disease. I do not, however, rule out the possibility that some low-dose immunotherapies could be effective in targeting surviving tumor cells once the bulk of the tumor is reduced using metabolic therapies. The present situation of treating patients with costly, toxic, and largely ineffective drugs is unacceptable.

A recent study showed that the adverse effects of rash and diarrhea were correlated with very modest increase in survival of brain cancer patients treated with Gefitinib, a small molecule inhibitor of the epidermal growth factor receptor (EGFR) tyrosine kinase (8). Without appropriate control groups for rash and diarrhea, it is difficult to interpret such findings. In other words, was the modest increase in survival due to the effects of Gefitinib or was it due to the effects of rash and diarrhea? As many cancer therapies are toxic to cells and tissues, toxicity has become the norm rather than the exception for new cancer therapies. Unfortunately, many patients will need to endure these failed toxic therapies until the oncology field comes to recognize that cancer is a metabolic disease requiring metabolic solutions.

The cancer field continues with expensive clinical trials using new combinations of radiation and/or toxic drug therapies in the hope of finding a therapeutic approach with improved efficacy (8–11). The BATTLE (Biomarker-integrated Approaches of Targeted Therapy for Lung Cancer Elimination) trail for lung cancer treatment has produced much hype, with little success so far. According to Dr. Edward Kim's presentation at the 2010 AACR meeting in Washington, DC, about 46% of the treated patients had disease management at 8 weeks and lived on an average about 2.7 months longer than patients not receiving the therapy. Viewed alternatively, the disease remained unmanageable in 54% of those treated. It should not therefore be surprising why the number of cancer patients who reject the advice of oncologists is increasing (12).

The BATTLE therapy is considered to be personalized because it uses a cocktail of drugs that target various tumor cell growth factor receptors such as the EGFR and other tumor cell molecular defects. In other words, the BATTLE therapeutic strategy is structured on the view that cancer is a genetic disease. However, we know from the work of Loeb, Stratton, and others that the molecular abnormalities found in tumors are likely to differ from one tumor cell to the next within most malignant neoplasms (13–15). On the basis of my treatise that cancer is a metabolic disease rather than a genetic disease, I predict that the BATTLE therapy will not likely play a major role in the eventual management of metastatic lung cancer or in managing any advanced metastatic cancers for that matter. Unfortunately, many more cancer patients will die before the basis for my prediction becomes recognized.

I find it remarkable that so many cancer patients are recruited for therapies that are toxic, potentially lethal, and offer little hope for improved long-term clinical outcome. Patients often hear from their oncology team that survival can be better with a particular new drug combination than with previous combinations. The data supporting these claims are often exaggerated (16). Why are cancer patients offered less than accurate information?

More than 40 years of clinical research indicates that such approaches are largely ineffective in extending survival or improving the quality of life for those patients with advanced cancers. The resurrection of immunotherapies is an example, despite evidence of past failures (5). Therapeutic approaches to cancer management, which produce adverse effects and reduce quality of life, should not be pursued, especially when more effective and less toxic alternative metabolic therapies are available. As metastatic melanoma and most metastatic cancers arise from macrophages with defective energy metabolism (Chapter 13), it is my opinion that targeting glucose and glutamine under energy restriction will be a more effective long-term therapy than any of the current drugs used to treat these cancers (Chapters 17 and 18).

The current epoch of cancer therapy, extending from the 1960s to the present, can be best described as the unenlightened barbaric period. While some might consider my assessment harsh, it is hard to ignore the unacceptable toxic effects of many current therapies or the death statistics in Table 1.1. Substantial mental and physical suffering accompanies many new treatment strategies. The current epoch will persist as long as cancer is viewed as something other than a metabolic disease. Once cancer becomes recognized as a metabolic disease, more effective and less toxic therapies will emerge.

THE "STANDARD OF CARE" FOR GLIOBLASTOMA MANAGEMENT

The state of cancer management can be best appreciated when considering the current standard of care for glioblastoma multiforme (GBM). I will use GBM as an example primarily because I am more familiar with this disease than with other malignant cancers (17). Nevertheless, the poor success rate in managing GBM parallels the poor success rate in managing other invasive and metastatic cancers such as those involving lung, pancreas, and liver, to name a few (18). It is also recognized that the metastatic behavior of GBM is as bad as that of many other metastatic cancers once the GBM cells leave the central nervous system and spread systemically (19). In contrast to other metastatic cancers, GBM usually kills patients before the disease shows systemic metastasis. As most metastatic cancers, including GBM, arise from respiratory damage in cells of myeloid origin (Chapter 13), the problems encountered in treatment strategies will be common to all or most metastatic cancers.

GBM is the most common primary adult human brain tumor with a median survival time of about 12–14 months from diagnosis (17, 20–22). Secondary GBM can also arise from lower grade astrocytomas, following therapeutic interventions (23). Many of the provocative procedures used to treat low grade gliomas (surgery, radiation, and chemotherapy) contribute to the eventual development of GBM. As we mentioned previously, the human brain should rarely, if ever, be exposed to high dose radiation (24). The survival time of those afflicted with GBM has changed a little in more than 50 years. As for many malignant cancers, the current standard of care for GBM includes maximum surgical resection, radiation therapy, and chemotherapy (22, 25–28). Almost 99% of GBM patients receive perioperative corticosteroids (dexamethasone) as part of the standard of care, which is sometimes extended throughout the course of the disease (28, 29).

GBM is notoriously heterogeneous in cellular composition consisting of tumor stem cells, mesenchymal cells, and host stromal cells (23, 30–32). The number of cells with characteristics of macrophages/monocytes in GBM can sometimes equal the number of tumor cells (33, 34). These cells are sometimes referred to as *tumor-associated macrophages* (*TAMs*), but their origin from either stromal cells or neoplastic cells remains ambiguous. We recently suggested that many cells that appear as TAM are actually part of the neoplastic cell population (35). TAMs contribute to tumor progression through the release of proinflammatory and proangiogenic factors (34, 36, 37). Many of the neoplastic cells in GBM are highly migratory and invade the brain well beyond the main tumor mass making complete surgical resections exceedingly rare (38, 39). Despite the best available treatments, only about 5–10% of GBM patients become long-term survivors (36 months) (17, 20–22, 25, 40).

Although GBM is biologically complex, glucose and glutamine are the primary energy metabolites necessary for driving rapid growth (26, 41–46). As will be shown in Chapter 17, glucose and glutamine are the prime fuels for driving the growth of most malignant cancers. Glucose is necessary for nearly all brain functions under normal physiological conditions (47). Tumor cells metabolize glutamine to glutamate, which is then metabolized to α-ketoglutarate for further metabolism within the mitochondria (Chapter 4). In contrast to extracranial tissues, where glutamine is the most available amino acid, glutamine is tightly regulated in the brain through its involvement in the glutamate–glutamine cycle of neurotransmission (47, 48).

Glutamate is a major excitatory neurotransmitter that must be cleared rapidly following synaptic release in order to prevent excitotoxic damage to neurons (49, 50). Glial cells possess transporters for the clearance of extracellular glutamate, which is then metabolized to glutamine for delivery back to neurons. Neurons metabolize the glutamine to glutamate, which is then repackaged into synaptic vesicles for future release (Fig. 16.1). The glutamine–glutamate cycle maintains low extracellular levels of both glutamate and glutamine in normal neural parenchyma. Disruption of the cycle can provide neoplastic GBM cells access to glutamine. Besides serving as a metabolic fuel for the neoplastic tumor cells, glutamine is also an important fuel for cells of myeloid origin that is, macrophages, monocytes, and microglia (37, 51). As long as GBM cells have access to glucose and glutamine, the tumor will grow making long-term management difficult or impossible.

In contrast to normal glia, some neoplastic glioma cells secrete glutamate. Glioma glutamate secretion is thought to contribute, in part, to neuronal excitotoxicity and tumor expansion (50). Neurotoxicity from mechanical trauma (surgery), radiotherapy, and chemotherapy will increase extracellular glutamate levels thus contributing to tumor progression (50). How might information on glioblastoma

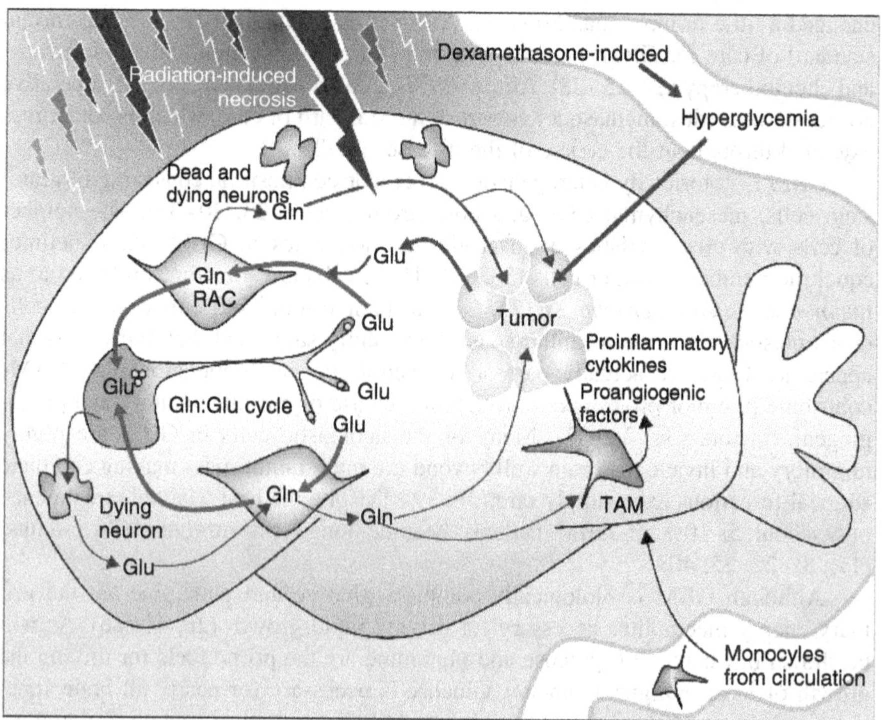

Figure 16.1 How the standard of care can provoke aggressive growth of GBM (17). GBM consists of multiple neoplastic cell types as well as tumor-associated macrophages (TAMs), which release proinflammatory and proangiogenic factors. All of these cells will use glucose and glutamine (Gln) as major metabolic fuels for their growth and survival. These fuels can be fermented for energy in hypoxic environments despite the absence of a supporting vasculature. Elevated glutamate (Glu) levels will arise following radiation/drug-induced necrosis. Reactive astrocytes (RACs) take up and metabolize glutamate to glutamine, whereas hyperglycemia will arise following corticosteroid (dexamethasone) therapy. Together, these standard therapies will provide an environment that facilitates survival and growth of GBM. We predict a scenario similar to this for the current treatment of other cancers. See text for details. *Source*: Reprinted with permission from *Lancet Oncology*. See color insert

energy metabolism relate to disease progression and to the standard of care for this cerebral neoplasm?

It is well documented that radiation and chemotherapies induce necrosis and inflammation, both of which will increase tissue glutamate levels (26, 38, 52–54). Local astrocytes rapidly clear extracellular glutamate metabolizing it to glutamine for release to neurons. In the presence of dead or dying neurons, surviving tumor cells and TAM will use astrocyte-derived glutamine for their energy and growth (Fig. 16.1).

Radiation damage to tumor cell mitochondria will hasten a dependence on glucose and glutamine for growth and survival of neoplastic cells (42, 55, 56). Radiation therapy is known to upregulate the PI3K/Akt pathway, which drives glioma glycolysis, vascularization, and chemotherapeutic drug resistance (42, 57–59). Moreover, radiation therapy will increase horizontal gene transfer from tumor cells to normal cells thus enhancing the formation of fusion hybrids, which will invade the entire brain (60, 61). Besides GBM, radiation therapy can also provoke the invasive and metastatic behaviors of other cancers, including breast, rectum, and endometrium, to name a few (62–64). While radiation can be therapeutic over a short term, it can enhance the risk of recurrent tumor growth over the long term.

Radiation is known to cause cancer. Dentists place lead aprons on patients that receive teeth X rays. Many people fear nuclear reactors because of cancer risk from exposure to radioactive materials leaked into the environment. Why is radiation exposure considered unhealthy for most people, but acceptable for many people with cancer? Thousands of cancer patients regularly receive high dose radiation to treat their disease. While radiation therapy will enhance 5-year survival for some low grade nonmetastatic cancers, it can increase the risk of recurrent disease. Better alternatives to radiation therapy are needed.

High dose glucocorticoids (dexamethasone) are generally prescribed to reduce radiation-associated brain swelling and tumor edema. It is well documented, however, that dexamethasone significantly elevates blood glucose levels (65, 66). These elevations become similar to those seen in patients with type-2 diabetes. Glucose is the major fuel for normal brain metabolism, but it also drives glycolysis-dependent tumor cells and the synthesis of glutamate (46, 47). Glycolysis-derived pyruvate also enhances *p*-glycoprotein activity (67). The *p*-glycoprotein is responsible for pumping toxic drugs out of cells, and, when activated, makes tumor cells resistant to most chemotherapy (68). By elevating blood glucose and providing fuel for glycolysis, steroids can contribute to the drug resistance of tumor cells. More specifically, the steroid drugs used to reduce tissue swelling and edema protect the tumor cells from the chemotherapy used to kill them. Oncologists should consider this information when prescribing steroids to their patients.

It is well documented that the growth of brain tumor is more accelerated, and prognosis is generally worse both in animals and in patients with higher than lower circulating glucose levels (69–71). On the basis of this evidence, I was surprised to read Linda M. Liau's comments at the end of the McGirt et al.'s paper regarding the role of glucose in tumor progression (70). Dr. Liau mentions:

> *Hyperglycemia cannot be directly linked to mortality from tumor recurrence and may reflect overall medical condition. The correlation here was only between hyperglycemia and shorter overall survival, and there are no longitudinal data regarding tumor status (i.e., time to tumor progression or progression-free survival). Patients with severe, chronic, uncontrolled hyperglycemia may be less able to tolerate systemic chemotherapy, require prolonged steroid use, or might be dying from causes not directly resulting from tumor recurrence. Therefore, the direct impact of hyperglycemia on tumor control is unknown.*

These comments suggest to me that Dr. Liau is unfamiliar with Warburg's theory of cancer. I suspect that many investigators working in the cancer field lack knowledge about Warburg's theory. It should therefore be no surprise that so little progress has been made in managing GBM or other highly invasive/metastatic cancers when the underlying mechanism responsible for the disease goes unrecognized.

It is also interesting to note that dexamethasone is given to cancer patients receiving CTLA-4 (ipilimumab). As mentioned earlier, this is an immunotherapy used for treating malignant melanomas and renal cell cancers (2, 3). As in the case of GBM patients, dexamethasone is given to patients with melanoma and renal cancer to reduce the adverse toxic effects of the cancer therapy. While dexamethasone can quickly reduce tissue edema giving an impression of therapeutic efficacy, it will ultimately protect tumor cells from dying, thus increasing the risk of unmanageable growth and advanced metastasis. It is not clear to me which therapy is more hazardous for cancer patients; the toxic one, or the one used to reduce toxicity! If oncologists already know about this, I wonder what they think about when prescribing high dose steroids to their patients.

TAMs will respond to the local tumor environment as if it were an unhealed wound thus releasing proangiogenic growth factors. What develops then is an escalating situation of biological chaos, where the intrinsic properties of TAM to heal wounds enhance the capacity of brain tumor cells to proliferate, invade, and self-renew (34, 72). High glucose levels together with unrestricted glutamine availability will provide the energy necessary for driving the escalating situation within the tumor. It should not therefore be surprising why 90% of GBM patients rarely survive beyond 36 months following treatment using the standard of care. That 10% of GBM patients actually survive the standard of care is quite remarkable and attests to the physiological resilience of the human body.

The current standard of care for GBM creates the "perfect storm" of adverse effects that guarantees the demise of most patients. Radiation therapy provides abundance of glutamine while dexamethasone provides abundance of glucose. To make matters worse, some patients receiving the standard of care are also given the drug bevacizumab (Avastin). Recurrent GBM following bevacizumab therapy is nearly 100% fatal (73, 74). Although bevacizumab targets the tumor vasculature, it exacerbates radiation-induced necrosis and also enhances the invasive properties of the tumor cells (75–78). This also appears to be the case of patients treated with cediranib, a pan-VEGFR tyrosine kinase inhibitor that increases the number of cells with characteristics of macrophages (79).

Dr. Leanne Huysentruyt and I recently presented evidence showing that many cells with properties of macrophages are part of the neoplastic cell population (35). As cells with properties of macrophages are naturally programmed to ferment glucose and glutamine, it is possible that bevacizumab, cediranib, and other antiangiogenic drug therapies select for those cells that are invasive and can survive in hypoxic microenvironments. Such cells would become less dependent on a supporting vasculature for survival. These tumor cell capabilities would render bevacizumab and other similar antiangiogenic drugs fruitless as effective treatments for GBM management. Recent findings with cediranib support my predictions (79).

Owing to a plethora of adverse effects, bevacizumab was discontinued as a therapy for brain cancer in Germany and was revoked by the US FDA as a therapy for breast cancer in the United States. In light of the adverse effects and the potential for driving invasive cancer cells in patients, it is unclear why this drug is still in use for treating GBM or other cancers. The situation appears to be worse in the case of children treated for cancer (80). According to a recent report in the *Wall Street Journal*, Roche Holding, maker of bevacizumab (Avastin), plans to lobby members of the US Congress to insure that Avastin remains in use as a cancer drug (81).

Cancer is a big business. In reading the Review & Outlook section of the *Wall Street Journal* (Wednesday, June 29, 2011), "*Race Against the Cure*", it became clear to me that factors other than patient health are responsible for the resistance to discontinue bevacizumab and other toxic drugs as cancer therapies. I consider those who recommend bevacizumab or who use bevacizumab for cancer treatment as lacking knowledge, ethically challenged, or some combination of these. I applaud the action of FDA commissioner Margaret Hamburg and cancer drug chief Richard Pazdur for their protection of cancer patients who are especially vulnerable to the deceptive insincerity of the cancer drug industry. If patients with brain cancer knew what I know about the origin of cancer, they would avoid bevacizumab like the *plague*.

While the current standard of care for GBM can increase patient survival over a shorter term (months), this therapeutic program will ultimately accelerate GBM energy metabolism and progression. Any therapy that enhances tumor cell energy metabolism runs the risk of reducing long-term patient survival. Figure 16.1 summarizes the influence of radiation therapy and dexamethasone on GBM. The long-term prognosis for GBM is not likely to change anytime soon as those who apply the standard of care seem unaware of the underlying metabolic mechanisms that drive the disease (22). It is not clear to me how long the high-grade glioma field will persist in entertaining therapies that provoke rather than manage the disease.

If the current standard of care for GBM and other advanced metastatic cancers is toxic and might accelerate tumor growth over a long term, what alternative therapeutic approaches might retard malignant tumor growth? We suggest that therapies targeting tumor cell energy metabolism will be more effective than the current standard of care in reducing tumor growth and in extending long-term patient survival (46, 82–85). I will highlight the evidence for using nontoxic metabolic therapies for cancer management in Chapter 17.

REFERENCES

1. Rockoff JD. Cancer therapy approved. Wall St J. 2011. http://online.wsj.com/article/SB10001424052748704517404576222710528145314.html.
2. Beck KE, Blansfield JA, Tran KQ, Feldman AL, Hughes MS, Royal RE, et al. Enterocolitis in patients with cancer after antibody blockade of cytotoxic T-lymphocyte-associated antigen 4. J Clin Oncol. 2006;24:2283–9.
3. Couzin-Frankel J. Immune therapy steps up the attack. Science. 2010; 330: 440–3. http://online.wsj.com/article/SB10001424052702304778304576377892911572686.html.
4. Chapman PB, Hauschild A, Robert C, Haanen JB, Ascierto P, Larkin J, et al. Improved survival with vemurafenib in melanoma with BRAF V600E mutation. N Engl J Med. 2011;364:2507–16.
5. Gryta T. Enlisting the body to fight cancer. Wall St J. 2010;D1–D2.

6. MARTINO F. Alternative inductothermia in cancer. A confirmation with therapeutic applications of Warburg's theory. Cancro. 1962;15:358–85.
7. HOPTION CANN SA, VAN NETTEN JP, VAN NETTEN C. Dr William Coley and tumour regression: a place in history or in the future. Postgrad Med J. 2003;79:672–80.
8. UHM JH, BALLMAN KV, WU W, GIANNINI C, KRAUSS JC, BUCKNER JC, et al. Phase II evaluation of gefitinib in patients with newly diagnosed grade 4 astrocytoma: Mayo/North central cancer treatment group study N0074. Int J Radiat Oncol Biol Phys. 2010;80:347–53.
9. AHLUWALIA MS, PATTON C, STEVENS G, TEKAUTZ T, ANGELOV L, VOGELBAUM MA, et al. Phase II trial of ritonavir/lopinavir in patients with progressive or recurrent high-grade gliomas. J Neuro Oncol. 2010;102:317–21.
10. SEYFRIED NT, KIEBISH M, MUKHERJEE P. Targeting Energy Metabolism in Brain Cancer with Restricted Diets. In: RAY S, editor. Glioblastoma: Molecular Mechanisms of Pathogenesis and Current Therapeutic Strategies. New York: Springer; 2010. p.341–63.
11. SMITH TJ, HILLNER BE. Bending the cost curve in cancer care. N Engl J Med. 2011;364:2060–5.
12. KONIGSBERG RD. The refuseniks: why some cancer patients reject their doctor's advise. TIME. 2011;177:72–7.
13. LOEB LA. A mutator phenotype in cancer. Cancer Res. 2001;61:3230–9.
14. SALK JJ, FOX EJ, LOEB LA. Mutational heterogeneity in human cancers: origin and consequences. Annu Rev Pathol. 2010;5:51–75.
15. STRATTON MR, CAMPBELL PJ, FUTREAL PA. The cancer genome. Nature. 2009;458:719–24.
16. FISHMAN J, TEN HAVE T, CASARETT D. Cancer and the media: how does the news report on treatment and outcomes? Arch Intern Med. 2010;170:515–8.
17. SEYFRIED TN, SHELTON LM, MUKHERJEE P. Does the existing standard of care increase glioblastoma energy metabolism? Lancet Oncol. 2010;11:811–3.
18. MARSHALL E. Cancer research and the $90 billion metaphor. Science. 2011;331:1540–1.
19. LUN M, LOK E, GAUTAM S, WU E, WONG ET. The natural history of extracranial metastasis from glioblastoma multiforme. J Neuro Oncol. 2011;105:261–73.
20. STUPP R, HEGI ME, MASON WP, VAN DEN BENT MJ, TAPHOORN MJ, JANZER RC, et al. Effects of radiotherapy with concomitant and adjuvant temozolomide versus radiotherapy alone on survival in glioblastoma in a randomised phase III study: 5-year analysis of the EORTC-NCIC trial. Lancet Oncol. 2009;10:459–66.
21. YOVINO S, GROSSMAN SA. Treatment of glioblastoma in "Elderly" patients. Curr Treat Options Oncol. 2011;12:253–62.
22. PREUSSER M, DE RIBAUPIERRE S, WOHRER A, ERRIDGE SC, HEGI M, WELLER M, et al. Current concepts and management of glioblastoma. Ann Neurol. 2011;70:9–21.
23. OHGAKI H, KLEIHUES P. Genetic alterations and signaling pathways in the evolution of gliomas. Cancer Sci. 2009;100:2235–41.
24. SEYFRIED TN, KIEBISH MA, MARSH J, SHELTON LM, HUYSENTRUYT LC, MUKHERJEE P. Metabolic management of brain cancer. Biochim Biophys Acta. 2010;1807:577–94.
25. SOUHAMI L, SEIFERHELD W, BRACHMAN D, PODGORSAK EB, WERNER-WASIK M, LUSTIG R, et al. Randomized comparison of stereotactic radiosurgery followed by conventional radiotherapy with carmustine to conventional radiotherapy with carmustine for patients with glioblastoma multiforme: report of radiation therapy oncology group 93-05 protocol. Int J Radiat Oncol Biol Phys. 2004;60:853–60.
26. YANG I, AGHI MK. New advances that enable identification of glioblastoma recurrence. Nat Rev Clin Oncol. 2009;6:648–57.
27. MASON WP, MAESTRO RD, EISENSTAT D, FORSYTH P, FULTON D, LAPERRIERE N, et al. Canadian recommendations for the treatment of glioblastoma multiforme. Curr Oncol. 2007;14:110–17.
28. CHANG SM, PARNEY IF, HUANG W, ANDERSON FA, Jr., ASHER AL, BERNSTEIN M, et al. Patterns of care for adults with newly diagnosed malignant glioma. JAMA. 2005;293:557–64.
29. KOEHLER PJ. Use of corticosteroids in neuro-oncology. Anticancer Drugs. 1995;6:19–33.
30. CHEN R, NISHIMURA MC, BUMBACA SM, KHARBANDA S, FORREST WF, KASMAN IM, et al. A hierarchy of self-renewing tumor-initiating cell types in glioblastoma. Cancer Cell. 2010;17:362–75.

31. Prestegarden L, Svendsen A, Wang J, Sleire L, Skaftnesmo KO, Bjerkvig R, et al. Glioma cell populations grouped by different cell type markers drive brain tumor growth. Cancer Res. 2010;70:4274–9.
32. Tso CL, Shintaku P, Chen J, Liu Q, Liu J, Chen Z, et al. Primary glioblastomas express mesenchymal stem-like properties. Mol Cancer Res. 2006;4:607–19.
33. Morantz RA, Wood GW, Foster M, Clark M, Gollahon K. Macrophages in experimental and human brain tumors. Part 2: studies of the macrophage content of human brain tumors. J Neurosurg. 1979;50:305–11.
34. Seyfried TN. Perspectives on brain tumor formation involving macrophages, glia, and neural stem cells. Perspect Biol Med. 2001;44:263–82.
35. Huysentruyt LC, Seyfried TN. Perspectives on the mesenchymal origin of metastatic cancer. Cancer Metastasis Rev. 2010;29:695–707.
36. Qian BZ, Pollard JW. Macrophage diversity enhances tumor progression and metastasis. Cell. 2010;141:39–51.
37. Lewis C, Murdoch C. Macrophage responses to hypoxia: implications for tumor progression and anti-cancer therapies. Am J Pathol. 2005;167:627–35.
38. Kallenberg K, Bock HC, Helms G, Jung K, Wrede A, Buhk JH, et al. Untreated glioblastoma multiforme: increased myo-inositol and glutamine levels in the contralateral cerebral hemisphere at proton MR spectroscopy. Radiology. 2009;253:805–12.
39. Talacchi A, Turazzi S, Locatelli F, Sala F, Beltramello A, Alessandrini F, et al. Surgical treatment of high-grade gliomas in motor areas. The impact of different supportive technologies: a 171-patient series. J Neuro Oncol. 2010;100:417–26.
40. Krex D, Klink B, Hartmann C, von Deimling A, Pietsch T, Simon M, et al. Long-term survival with glioblastoma multiforme. Brain. 2007;130:2596–606.
41. DeBerardinis RJ, Cheng T. Q's next: the diverse functions of glutamine in metabolism, cell biology and cancer. Oncogene. 2010;29:313–24.
42. Seyfried TN, Shelton LM. Cancer as a metabolic disease. Nutr Metab. 2010;7:7.
43. Wise DR, DeBerardinis RJ, Mancuso A, Sayed N, Zhang XY, Pfeiffer HK, et al. Myc regulates a transcriptional program that stimulates mitochondrial glutaminolysis and leads to glutamine addiction. Proc Natl Acad Sci USA. 2008;105:18782–7.
44. Spence AM, Muzi M, Graham MM, O'Sullivan F, Krohn KA, Link JM, et al. Glucose metabolism in human malignant gliomas measured quantitatively with PET, 1-[C-11]glucose and FDG: analysis of the FDG lumped constant. J Nucl Med. 1998;39:440–8.
45. DeBerardinis RJ, Mancuso A, Daikhin E, Nissim I, Yudkoff M, Wehrli S, et al. Beyond aerobic glycolysis: transformed cells can engage in glutamine metabolism that exceeds the requirement for protein and nucleotide synthesis. Proc Natl Acad Sci USA. 2007;104:19345–50.
46. Seyfried TN, Mukherjee P. Targeting energy metabolism in brain cancer: review and hypothesis. Nutr Metab. 2005;2:30.
47. McKenna MC, Gruetter R, Sonnewald U, Waagepetersen HS, Schousboe A. Energy Metabolism of the Brain. In: Siegel GJ, Albers RW, Bradey ST, Price DP, editors. Basic Neurochemistry: Molecular, Cellular, and Medical Aspects. New York: Elsevier Academic Press; 2006. p.531–57.
48. Hawkins RA. The blood-brain barrier and glutamate. Am J Clin Nutr. 2009;90:867S–74S.
49. Allen NJ, Karadottir R, Attwell D. A preferential role for glycolysis in preventing the anoxic depolarization of rat hippocampal area CA1 pyramidal cells. J Neurosci. 2005;25:848–59.
50. Takano T, Lin JH, Arcuino G, Gao Q, Yang J, Nedergaard M. Glutamate release promotes growth of malignant gliomas. Nat Med. 2001;7:1010–5.
51. Newsholme P. Why is L-glutamine metabolism important to cells of the immune system in health, postinjury, surgery or infection? J Nutr. 2001;131:2515S–22S. discussion 23S–4S.
52. Monje ML, Vogel H, Masek M, Ligon KL, Fisher PG, Palmer TD. Impaired human hippocampal neurogenesis after treatment for central nervous system malignancies. Ann Neurol. 2007;62:515–20.
53. Lee WH, Sonntag WE, Mitschelen M, Yan H, Lee YW. Irradiation induces regionally specific alterations in pro-inflammatory environments in rat brain. Int J Radiat Biol. 2010;86:132–44.

54. Di Chiro G, Oldfield E, Wright DC, De Michele D, Katz DA, Patronas NJ, et al. Cerebral necrosis after radiotherapy and/or intraarterial chemotherapy for brain tumors: PET and neuropathologic studies. AJR Am J Roentgenol. 1988;150:189–97.
55. Warburg O. On the origin of cancer cells. Science. 1956;123:309–14.
56. Smith AE, Kenyon DH. A unifying concept of carcinogenesis and its therapeutic implications. Oncology. 1973;27:459–79.
57. Elstrom RL, Bauer DE, Buzzai M, Karnauskas R, Harris MH, Plas DR, et al. Akt stimulates aerobic glycolysis in cancer cells. Cancer Res. 2004;64:3892–9.
58. Marsh J, Mukherjee P, Seyfried TN. Akt-dependent proapoptotic effects of dietary restriction on late-stage management of a phosphatase and tensin homologue/tuberous sclerosis complex 2-deficient mouse astrocytoma. Clin Cancer Res. 2008;14:7751–62.
59. Zhuang W, Qin Z, Liang Z. The role of autophagy in sensitizing malignant glioma cells to radiation therapy. Acta Biochim Biophys Sin (Shanghai). 2009;41:341–51.
60. Espejel S, Romero R, Alvarez-Buylla A. Radiation damage increases Purkinje neuron heterokaryons in neonatal cerebellum. Ann Neurol. 2009;66:100–9.
61. Holmgren L, Szeles A, Rajnavolgyi E, Folkman J, Klein G, Ernberg I, et al. Horizontal transfer of DNA by the uptake of apoptotic bodies. Blood. 1999;93:3956–63.
62. Shabo I, Olsson H, Sun XF, Svanvik J. Expression of the macrophage antigen CD163 in rectal cancer cells is associated with early local recurrence and reduced survival time. Int J Cancer. 2009;125:1826–31.
63. Shabo I, Stal O, Olsson H, Dore S, Svanvik J. Breast cancer expression of CD163, a macrophage scavenger receptor, is related to early distant recurrence and reduced patient survival. Int J Cancer. 2008;123:780–6.
64. Tebeu PM, Verkooijen HM, Bouchardy C, Ludicke F, Usel M, Major AL. Impact of external radiotherapy on survival after stage I endometrial cancer: results from a population based study. J Cancer Sci Ther. 2011;3:041–6.
65. Lukins MB, Manninen PH. Hyperglycemia in patients administered dexamethasone for craniotomy. Anesth Analg. 2005;100:1129–33.
66. Hans P, Vanthuyne A, Dewandre PY, Brichant JF, Bonhomme V. Blood glucose concentration profile after 10mg dexamethasone in non-diabetic and type 2 diabetic patients undergoing abdominal surgery. Br J Anaesth. 2006;97:164–70.
67. Wartenberg M, Richter M, Datchev A, Gunther S, Milosevic N, Bekhite MM, et al. Glycolytic pyruvate regulates P-Glycoprotein expression in multicellular tumor spheroids via modulation of the intracellular redox state. J Cell Biochem. 2010;109:434–46.
68. Aller SG, Yu J, Ward A, Weng Y, Chittaboina S, Zhuo R, et al. Structure of P-glycoprotein reveals a molecular basis for poly-specific drug binding. Science. 2009;323:1718–22.
69. Derr RL, Ye X, Islas MU, Desideri S, Saudek CD, Grossman SA. Association between hyperglycemia and survival in patients with newly diagnosed glioblastoma. J Clin Oncol. 2009;27: 1082–6.
70. McGirt MJ, Chaichana KL, Gathinji M, Attenello F, Than K, Ruiz AJ, et al. Persistent outpatient hyperglycemia is independently associated with decreased survival after primary resection of malignant brain astrocytomas. Neurosurgery. 2008;63:286–91. discussion 91.
71. Seyfried TN, Sanderson TM, El-Abbadi MM, McGowan R, Mukherjee P. Role of glucose and ketone bodies in the metabolic control of experimental brain cancer. Br J Cancer. 2003;89: 1375–82.
72. Staw BM, Ross J. Understanding behavior in escalation situations. Science. 1989;246:216–20.
73. Zhang W, Lin Y, Chen B, Song SW, Jiang T. Recurrent glioblastoma of childhood treated with bevacizumab: case report and molecular features. Childs Nerv Syst. 2010;26:137–43.
74. Iwamoto FM, Abrey LE, Beal K, Gutin PH, Rosenblum MK, Reuter VE, et al. Patterns of relapse and prognosis after bevacizumab failure in recurrent glioblastoma. Neurology. 2009;73:1200–6.
75. Jeyaretna DS, Curry WT Jr, Batchelor TT, Stemmer-Rachamimov A, Plotkin SR. Exacerbation of cerebral radiation necrosis by bevacizumab. J Clin Oncol. 2011;29:e159–62.

76. Verhoeff JJ, van Tellingen O, Claes A, Stalpers LJ, van Linde ME, Richel DJ, et al. Concerns about anti-angiogenic treatment in patients with glioblastoma multiforme. BMC Cancer. 2009;9:444.
77. de Groot JF, Fuller G, Kumar AJ, Piao Y, Eterovic K, Ji Y, et al. Tumor invasion after treatment of glioblastoma with bevacizumab: radiographic and pathologic correlation in humans and mice. Neuro Oncol. 2010;12:233–42.
78. Keunen O, Johansson M, Oudin A, Sanzey M, Rahim SA, Fack F, et al. Anti-VEGF treatment reduces blood supply and increases tumor cell invasion in glioblastoma. Proc Natl Acad Sci USA. 2011;108:3749–54.
79. di Tomaso E, Snuderl M, Kamoun WS, Duda DG, Auluck PK, Fazlollahi L, et al. Glioblastoma recurrence after cediranib therapy in patients: lack of "rebound" revascularization as mode of escape. Cancer Res. 2011;71:19–28.
80. Parekh C, Jubran R, Erdreich-Epstein A, Panigrahy A, Bluml S, Finlay J, et al. Treatment of children with recurrent high grade gliomas with a bevacizumab containing regimen. J Neuro Oncol. 2011;103:673–80.
81. Mundy A. Showdown over cancer drug. Wall St J. 2011; 27. http://online.wsj.com/article/SB10001424052702303627104576409981132510402.html.
82. Nebeling LC, Miraldi F, Shurin SB, Lerner E. Effects of a ketogenic diet on tumor metabolism and nutritional status in pediatric oncology patients: two case reports. J Am Coll Nutr. 1995;14:202–8.
83. Michelakis ED, Sutendra G, Dromparis P, Webster L, Haromy A, Niven E, et al. Metabolic modulation of glioblastoma with dichloroacetate. Sci Transl Med. 2010;2:31ra4.
84. Zuccoli G, Marcello N, Pisanello A, Servadei F, Vaccaro S, Mukherjee P, et al. Metabolic management of glioblastoma multiforme using standard therapy together with a restricted ketogenic diet: Case Report. Nutr Metab. 2010;7:33.
85. Seyfried TN, Mukherjee P, Kalamian M, Zuccoli G. The Restricted Ketogenic Diet: An Alternative Treatment Strategy for Glioblastoma Multiforme. In: Holcroft R, editor. Treatment Strategies Oncology. London: Cambridge Research Center; 2011. p. 24–35.

Chapter 17

Metabolic Management of Cancer

If cancer is primarily a disease of energy metabolism, then rational strategies for cancer management should be found in those therapies that specifically target tumor cell energy metabolism. These therapeutic strategies should be applicable for the majority of cancers regardless of tissue origin, as nearly all cancers suffer from the same underlying disorder, that is, damaged respiration with compensatory fermentation. In this chapter, I review information showing how changes in availability of glucose and glutamine target both tumor cells and the tumor microenvironment. Numerous studies show that dietary energy reduction (DER) is a general metabolic therapy that significantly reduces growth and progression of numerous tumor types, including cancers of the mammary, brain, colon, pancreas, lung, and prostate (1–11). DER naturally lowers circulating glucose levels, which many tumors depend on for growth and survival. David Kritchevsky, as well as Stephen Hursting and Frank Kari, provide historical overviews and comprehensive evidence showing how dietary calorie reduction reduces growth and progression of many tumor types (1, 12–14). *All oncologists should know that dietary energy reduction is the nemesis of many cancers*. This therapeutic approach will be most effective soon after cancer is first diagnosed when most individuals are in good health.

In this chapter, I use the term *dietary energy reduction* to refer to either *calorie restriction* (CR) or *dietary restriction*. The term *calorie restriction* is often used interchangeably with the term *dietary restriction* since their therapeutic effects against tumors are due largely to the reduction in foods that provide energy to the body (11, 15). In the absence of energy molecules from foods, the body will generate energy from internal stores, largely involving fats and proteins. Carbohydrates will be synthesized from these molecules through gluconeogenesis. Humans evolved to function efficiently for extended periods in the absence of food (16). Therapeutic fasting enhances systemic energy conservation to achieve

Cancer as a Metabolic Disease: On the Origin, Management and Prevention of Cancer, First Edition. Thomas Seyfried.

a new homeostatic state. Respiratory insufficiency and genomic instability will prevent tumor cells from entering this new energy state.

DER is produced from a total reduction in dietary nutrients and differs from starvation in that DER reduces total calorie energy intake without causing anorexia or malnutrition (4, 11, 17–20). As a natural therapy, DER improves health, prevents tumor formation, and reduces inflammation (17, 19, 21–25). Reduced calorie intake is ideally suited as a therapy for reducing tumor growth without the adverse effects associated with conventional cancer therapies. Indeed, fasting can reduce the toxic effects of some chemotherapies (26). Research from Gary Meadows' laboratory also shows that tumor metabolism and growth can be affected from restriction of certain amino acids (27, 28). I have addressed in Chapter 18 the issue of cachexia and how DER can also be used to target this energy state.

IS IT DIETARY CONTENT OR DIETARY COMPOSITION THAT PRIMARILY REDUCES TUMOR GROWTH?

Albert Tannenbaum first showed that the anticancer action of DER mostly involved CR itself rather than restriction of any specific dietary component (11, 20). On the basis of the data from his 1953 study with Herbert Silverstone, Tannenbaum stated that "*Underfeeding or caloric restriction of mice bearing mammary carcinoma of spontaneous origin increased their life span, decreased the rate of growth of the tumors, hindered the formation of additional neoplasms of the mammae, and decreased the frequency of lung metastases*" (29). It is clear from these and numerous other studies that the simple process of DER inhibits tumor growth and metastasis. We also showed that the therapeutic efficacy of DER against brain cancer could be significantly enhanced when combined with drugs that also target glycolysis (30).

We confirmed and extended the findings of Tannenbaum and coworkers in a series of orthotopic mouse brain tumor models treated with the reduced intake of the ketogenic diet (KD). We refer to this as a *restricted* or *reduced intake ketogenic diet* (KD-R). The KD-R produces antitumor effects similar to that of DER (15, 31–34). We showed that reduced intake of either a high carbohydrate diet or a high fat, low carbohydrate KD could reduce aggressive brain tumor growth to a similar degree (Fig. 17.1). The energy composition of a typical KD compared to a normal high carbohydrate, low fat diet is shown in Table 17.1.

It is important to mention that mouse body weight was similar in the two unrestricted groups and was reduced to a similar level (about 20% reduction) in the two restricted diet groups. This is very important as some mouse strains gain weight on the KD, while other mouse strains might loose weight on the KD. It is not possible to accurately compare the effects of different diets on tumor growth if the diets differentially influence body weight. Use of isocaloric diets is relevant only if the diets maintain similar body weights in the test subjects. If body weights differ in mice fed isocaloric diets, the diets are not metabolically equivalent.

We showed that it is better to use body weight rather than degree of CR as the independent variable for assessing the influence of DER on tumor growth in mice. For example, a 20% reduction in a KD causes less body weight reduction than does

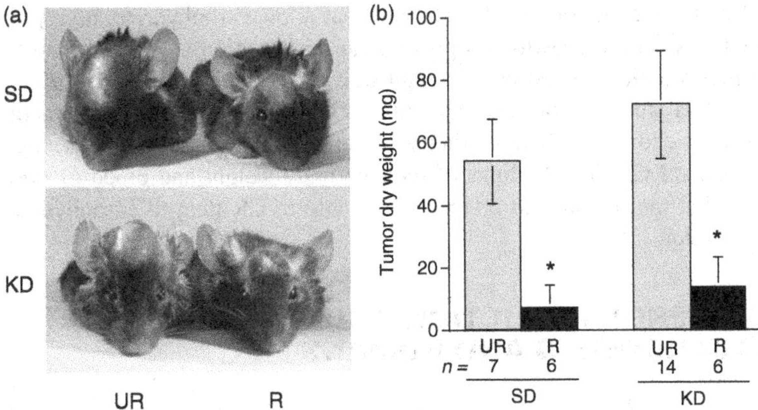

Figure 17.1 Influence of diet on the intracerebral growth of the CT-2A brain tumor. The visual representation (a) and quantitative assessment (b) of the tumor growth in C57BL/6J mice receiving the standard high carbohydrate diet (SD) or the ketogenic diet (KD) under either unrestricted (UR) or restricted (R) feeding as we described (15). Unrestricted feeding is the same as AL feeding. Values in (b) are expressed as means with 95% confidence intervals, and n = the number of mice examined in each group. The dry weights of the tumors in the R-fed mice were significantly lower than those in the unrestricted UR-fed mice at $P < 0.01$. The results show that DER significantly reduces tumor growth whether the mice are fed a standard high carbohydrate diet (SD) or a high fat, low carbohydrate KD. No adverse effects were seen in the mice maintained on the 30–40% DER. Body weights were similar for mice in both UR groups and were reduced similarly in both R groups (15). Despite a reduction in total body weight, the R-fed mice were more healthy and active than the UR-fed mice as assessed by ambulatory and grooming behavior. No signs of vitamin or mineral deficiency were observed in the DER-fed mice according to standard criteria for mice. These findings are consistent with the well-recognized health benefits of mild to moderate diet restriction in rodents and in humans (32, 33). *Source:* Modified from data presented in Reference 33. See color insert.

Table 17.1 Composition (%) of a Standard Diet High Carbohydrate Diet and a Typical Ketogenic Diet

Components	Standard Diet (SD)	Ketogenic Diet (KC)
Carbohydrate	62	3
Fat	6	72
Protein	27	15
Energy (kcal/g)	4.4	7.2
$F/(P+C)^a$	0.07	4

$^a F/(P + C)$ = ratio of fats to protein + carbohydrates.

a 20% reduction in a high carbohydrate diet. It is difficult to compare the influence of one diet with another diet on tumor growth if the components of the two diets are metabolized differently. When body weight is used as the independent variable, data interpretation becomes more accurate (15, 34). Our results show that brain tumor growth is influenced more by diet energy content than by diet nutrient composition.

Gerald Krystal and coworkers also showed that a low carbohydrate, high protein diet could slow tumor growth and prevent cancer initiation (35). Their results showed that reduced glucose and body weight were associated with reduced tumor growth. While this study is conceptually important, it was not clear if the therapeutic effects of the low carbohydrate diet were due to reduced carbohydrate, high protein, or a general CR. The findings of reduced body weight and blood glucose suggest to me that the therapeutic effect is due more to CR than to carbohydrate or protein restriction (36).

DIETARY ENERGY REDUCTION AND THERAPEUTIC FASTING IN RODENTS AND HUMANS

The DER-induced inhibition of brain tumor growth in mice is directly correlated with reduced levels of glucose and elevated levels of ketone bodies (15). Ketone bodies [β-hydroxybutyrate (β-OHB) and acetoacetate] become an alternative fuel for tissue energy metabolism when glucose levels are reduced as would occur during consumption of very low calorie diets or water-only fasting (37–39) (Fig. 17.2). Acetone is also a by-product of ketone synthesis, but acetone is not used for energy and is released in the breath or urine. β-OHB is the major circulating ketone body and is used primarily for energy metabolism when glucose levels become reduced. Although β-OHB is metabolized to acetoacetate in most tissues except liver, tissue uptake from the circulation is faster for β-OHB than for acetoacetate (40). A greater number of surface receptors for β-OHB than for acetoacetate might account for the more rapid uptake of β-OHB (17, 41).

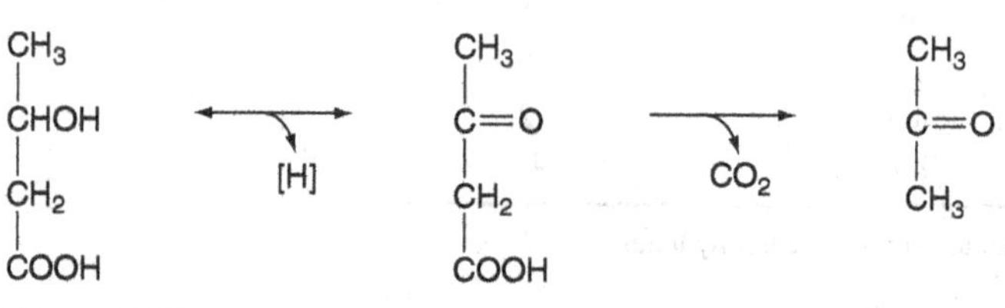

Figure 17.2 Glucose and ketone bodies. Glucose is the major metabolic fuel for most tissues and cells and is the sole fuel for brain during normal physiological conditions. When DER lowers glucose levels, ketone bodies, produced in the liver, will substitute for glucose as a major energy metabolite (38, 42). A [H] is removed from b-OHB (left structure) to form aetoacetate (center structure). A carbon and two oxygen molecules are removed from acetoacetate to form acetone (right structure). Acetone is a nonenzymatic metabolite of ketone synthesis and is eliminated through the lungs.

KETOGENIC DIETS

KDs were developed originally to manage epileptic seizures in children, but they are also effective in managing brain cancers, especially when administered in reduced amounts to reduce glucose levels (15, 43–46). KD consumption can also lower blood glucose levels in some persons. This is usually due to self-restriction because of diet unpalatability. Additionally, the high fat composition of the KD can reduce overall consumption through an effect on expression the cholecystokinin peptide. Intestinal cells release cholecystokinin in response to high fat diets. Cholecystokinin activates vagal sensory neurons to inhibit feeding behavior (47).We also showed that simply feeding mice less total food in the form of DER reduces circulating glucose levels, while elevating circulating levels of ketone bodies (15, 36, 37, 48). The efficacy of the KetoCal KD is optimal for brain cancer management when the ratio of dietary fats to combined carbohydrate/protein is 4:1 (49) (Table 17.1). I believe that this 4:1 ratio will be effective in targeting energy metabolism in most tumors that rely heavily on glucose for survival and growth. Reduced glucose availability will target aerobic glycolysis and the pentose phosphate shunt, critical metabolic pathways required for the survival and proliferation of many tumor types (50).

We found that the KD reduced growth and vascularization in mouse astrocytoma, but only when the diet was administered in reduced amounts that also lowered body weights (15, 34, 51). The KD had no therapeutic efficacy against tumor growth when consumed *ad libitum* (AL) or in unrestricted amounts. The data in Figure 17.1 show that unrestricted consumption of the KD has no inhibitory influence on mouse astrocytoma growth. The data in Figure 17.3 show that blood glucose levels remain high in mice that consume the KD in unreduced amounts. If glucose levels remain high, body weights remain stable or increase (36). When the KD is fed to mice in unrestricted amounts, blood glucose levels remain high and ketones are largely excreted in the urine. We clearly showed, however, that blood ketones were higher in tumor-bearing mice under DER than under AL feeding (34). Under DER, ketones are retained in the body for use in metabolism rather than excreted in the urine. This information is critical when designing metabolic therapies for tumor management.

It is also important to mention that Dr. Adrienne Scheck and colleagues reported growth inhibition of the mouse GL261 glioma cells from a KD fed to mice in unrestricted amounts. These findings suggest that some tumors might be susceptible to KD growth inhibition without CR or glucose reduction (52). Drs Scheck and Mohammed Abdelwahab also reported at the 2011 meeting of the American Association of Cancer Research (AACR) that the KD can improve survival in mice receiving radiation therapy for brain cancer. It is my belief, however, that the anticarcinogenic effects of the KD will be best when the diet is consumed in lower amounts rather than in higher amounts, as the unrestricted consumption of the KD can produce adverse events due to the high fat content of the diet (49, 53). Adverse events are reduced when the KD is consumed in restricted amounts.

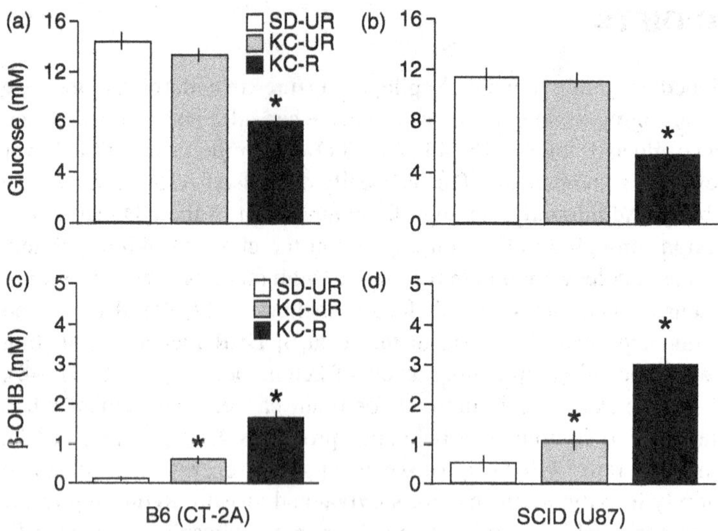

Figure 17.3 Influence of diet on plasma glucose and ketone (β-OHB) levels in mice bearing orthotopic experimental brain tumors. The mice were fed either a standard high carbohydrate diet (SD) or the KetoCal ketogenic diet (KC) (34). (a and c) Show the diet influence on C57BL/6J (B6) mice bearing the CT-2A astrocytoma. (b and d) Show the diet influence on SCID mice bearing the human U87 glioma. Values are expressed as mean ± SEM ($n = 12-14$ mice per group) and the asterisks indicate that the values differ from the control SD-UR group at the $P < 0.01$ level. *Source:* Reprinted with permission from Reference 34.

The results in Figure 17.3 show that lowered intake of the KetoCal KD (KC-R) lowers circulating glucose levels. However, blood glucose levels are not lowered in the tumor-bearing mice if they consume the KD in unrestricted amounts. This means that it is the amount of the diet consumed rather than the composition of the diet that determines blood glucose levels. Many people have difficulty appreciating this fact because they often think that low carbohydrate diets will produce low blood glucose levels. This is clearly not the case here. We reported similar findings in our previous investigation of glucose and ketones in epileptic mice (36, 48). Our data show that blood glucose levels are influenced more by the amount of calories consumed than by the composition of the calories consumed. Nutritional oncologists and cancer patients also need to know this information.

Although ketone (β-OHB) levels are higher in the mice consuming the KD than in mice consuming the standard diet (SD), the β-OHB levels are even higher in mice that consume the KD in restricted amounts (KC-R). Why would blood ketone levels be higher in mice that eat less KD than in mice that eat more KD? The answer is simple. Ketones are retained in the body when glucose levels are low. Ketones serve as an energy substitute for glucose. If glucose is not reduced as in the KC-UR groups, then most ketones will be excreted in the urine. This is why it is better to measure blood ketone levels than to measure urine ketones as an indicator of ketosis. Cancer cells are placed under metabolic stress when glucose levels are reduced and ketone levels are elevated (also see Fig. 18.1). The therapeutic action of ketones is best when blood glucose levels are low.

DER, which lowers glucose and elevates ketone bodies (36, 48), improves mitochondrial respiratory function and glutathione redox state in normal cells (54–56). Glutathione plays an important role in protecting cells and tissues from oxygen radical damage. Reactive oxygen species (ROS), which are elevated in many cancer cells, can damage DNA, lipids, and proteins. Ketone bodies can also protect normal cells from damage associated with aggressive tumor growth through a variety of neuroprotective mechanisms, including elevated glutathione levels (38, 57–65). Hence, the natural elevation of blood ketones accompanying reduced food intake or consumption of KDs can reduce the pathological effects of a broad range of diseases, especially cancer. Ketone body metabolism also reduces inflammation as I have described later in this chapter.

GLUCAGON AND INSULIN

The hormone glucagon is responsible for the elevation of ketone bodies in the blood. Glucagon becomes elevated during food restriction. Besides stimulating fat breakdown, glucagon also stimulates the synthesis of glucose from stored proteins and fats in order to maintain a basal level of glucose in blood (66). The glycerol part of the triglycerides is used for glucose synthesis. A basal glucose level is also needed to maintain brain function. However, ketone bodies will gradually replace glucose as the major fuel for the brain and other tissues (67). Ketone bodies allow the brain to maintain normal function under hypoglycemia. In addition to ketone bodies, fatty acids released from stored triglycerides (body fat) will become a predominant fuel for most tissues except brain, which primarily burns ketone bodies for energy under low glucose conditions (68).

Interestingly, most organs reduce glucose metabolism under DER in order to spare what little glucose is available for the brain metabolism. Evidence suggests that rat brain can metabolize small amounts of fatty acids for energy (69). However, fatty acid metabolism can generate heat, which can interfere with normal brain function. The brain does not dissipate heat as well as other organs due to skull confinement. More heat is generated from metabolism of fatty acids than from ketones, as fatty acid metabolism can increase expression of uncoupling proteins (54, 69). Uncoupling of the proton motive gradient produces heat as shown in Figure 4.4. Ketone metabolism is therefore more energetically efficient than fatty acid metabolism since less heat is released from ketone body metabolism than from fatty acid metabolism. The use of fatty acids and ketone bodies as metabolic fuel arises in large part from the inability of the liver and kidneys to synthesize enough glucose (gluconeogenesis) to maintain metabolic homeostasis under periods of prolonged food restriction.

Glucagon also works in an opposite way to insulin. In contrast to elevating glucagon levels, reduced food intake reduces blood insulin levels. Glucose derived from consumed foods will elevate insulin, which enhances glucose uptake and glycolysis in cells and tissues. As insulin stimulates glycolysis, insulin can stimulate the growth of those tumors that depend on glucose and glycolysis. Blood insulin levels become low in the absence of food intake because blood glucose levels become low. Hence, insulin and glucagon regulate metabolic homeostasis in the presence and absence of food, respectively.

BASAL METABOLIC RATE

The basal metabolic rate (BMR) is the energy required for maintaining body temperature, blood circulation, cellular respiration, and glandular activities under conditions of rest (68). It is important to recognize that the physiological response to DER in rodents is not the same as in humans due to differences in BMR. The BMR is about sevenfold greater in the mouse than in the human (70). Consequently, the ability to maintain metabolic homeostasis under food restriction is far greater in humans than in rodents. The health benefits documented in mice under 40% dietary restriction (DR) can be realized in humans under very low calorie intake (400–500 kcal) or with water-only therapeutic fasting (37). Alternatively, these health benefits can also be achieved using the restricted KD, which increases circulating levels of ketone bodies, while maintaining low blood glucose levels (34, 36). The KD-R can replace a draconian therapeutic fast for cancer management. Additionally, recent studies by Kashiwaya, Veech, and coworkers show that diets supplemented with ketone esters could also be effective in reducing blood glucose and glutamine, while elevating ketone levels (69). Dominic D'Augostino from the University of South Florida is also evaluating new formulations of ketone esters that might eventually replace some aspects of the KD-R as a cancer therapy (personal communication). These data suggest that dietary supplements of ketones could enhance the therapeutic action of DER without requiring drastic reductions in total calorie intake.

KETONES AND GLUCOSE

Only those cells with normal mitochondrial respiratory capacity can effectively use ketone bodies for energy, as ketones cannot be effectively metabolized for energy without an intact electron transport chain or with the mitochondrial enzymes needed to metabolize ketones (39, 54, 71). We found that ketone bodies are unable to maintain tumor cell viability in the absence of either glucose or glutamine (72). The inability of ketone bodies alone to maintain tumor cell viability has also been found in human glioma cells (73). Moreover, ketone bodies can actually be toxic to some tumor cells even in the presence of glucose, for example, neuroblastoma (74). Many human and mouse tumors also express deficiencies in the key enzymes needed to process β-OHB to acetyl CoA (34, 75, 76). More specifically, many tumor cells cannot effectively use ketone bodies to fuel their growth or maintain their survival. This is illustrated further for brain cancer in Figure 17.4. The same should hold true for any cancer type with defective respiration. Therapies that reduce glucose

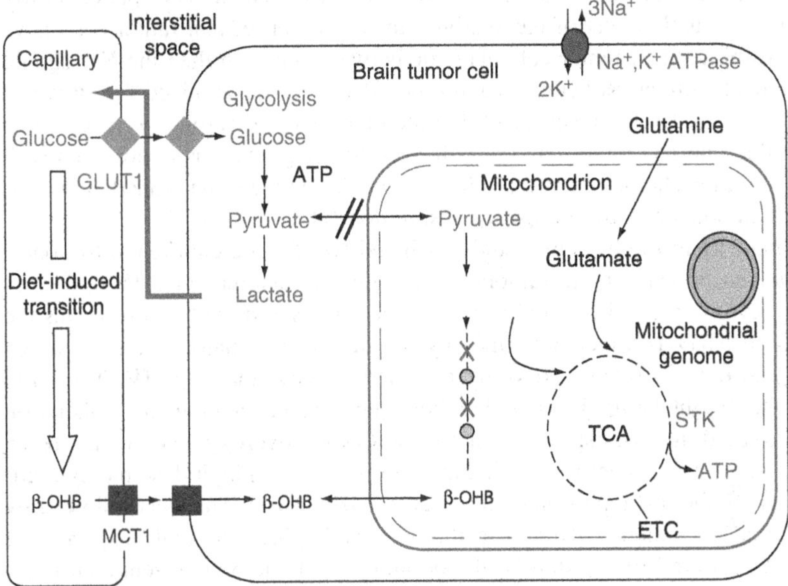

Figure 17.4 Perspectives on the metabolic management of brain cancer through a dietary reduction of glucose and elevation of ketone bodies. Glucose transporters are elevated in cancer cells (77). A dietary reduction in circulating glucose will increase ketone utilization for energy in normal neurons and glia. This will induce an energy transition from glycolysis to respiration. Cancer cells, however, are unable to transition from glucose to ketones due to alterations in mitochondrial structure or function (dashed lines). The "X" indicates defects in the activity of the enzymes needed to metabolize ketones to acetyl CoA. It is not clear in all cases whether it is defective enzymes or defective mitochondria that are responsible for the reduced ketone body metabolism in cancer cells (34, 73–76, 78). The double slash indicates a disconnection between glycolysis and respiration according to the Warburg theory. Glutamine can act synergistically with glucose to drive tumor cell fermentation (79) (Chapter 8). Membrane pumps, for example, Na^+, K^+, and ATPase, consume most ATP synthesized in cells. Abbreviations: GLUT-1, glucose transporter; MCT-1, monocarboxylate transporter; SCOT, succinyl-CoA-acetoacetate-CoA transferase; β-OHB, β-hydroxybutyrate; β-HBDH, β-hydroxybutyrate dehydrogenase; ETC, electron transport chain; SKT, succinate thiokinase. *Source:* Modified from our original figure (8). See color insert.

and elevate blood ketone bodies will therefore starve glucose-dependent tumor cells, while protecting and feeding normal cells. I know of no conventional cancer treatments that can do this.

METABOLIC MANAGEMENT OF BRAIN CANCER USING THE KD

In 1995, Nebeling et al. (46) attempted the first nutritional metabolic therapy for human malignant brain cancer using the KD. The objective of this study was to shift the prime substrate for energy metabolism from glucose to ketone bodies in order to disrupt tumor metabolism, while maintaining the nutritional status of patients (46). This study is actually a test of the Warburg theory of cancer that tumor cells have defective respiration. If tumor cells have defective respiration then the KD should have therapeutic effect, as ketones cannot be effectively metabolized for energy if respiration is defective or insufficient. On the other hand, if respiratory capacity is sufficient then the KD might have no effect on tumor growth, as tumor cells with normal respiration should use ketones for energy. As Warburg mentioned, however, no tumors are known with unimpaired respiration (80).

The patients in the Nebeling study included two female children with nonresectable advanced stage brain tumors (anaplastic astrocytoma stage IV and cerebellar astrocytoma stage III). I refer to this as a *landmark study* because it was the first to use the KD as a metabolic therapy for cancer in humans. The first patient was a 3-year-old girl diagnosed with anaplastic astrocytoma stage IV. The child received the "Eight Drugs in One Day Regimen," which involves administration of highly toxic drugs with steroids [vincristine (VCR), hydroxyurea, procarbazine, CCNU (Lomustine), cisplatin, cytosine arabinoside (Ara-C) high dose methylprednisolone, and either cyclophosphamide or dacarbazine] (81). This was followed by hyperfractionated radiation therapy to the head and spine. The child experienced seizures and suffered from extensive blood and renal toxicity. Conventional treatment was eventually discontinued due to continued tumor progression. The second patient was an 8.5-year-old girl diagnosed with grade III cerebellar astrocytoma that had progressed from an initially low grade tumor diagnosed at age 6. This patient suffered hearing loss from cisplatin toxicity. Measurable tumor remained in both subjects following extensive radiation and chemotherapy. Neither patient was expected to survive for long after the conventional treatments (46).

Dr. Nebeling treated these patients using a KD consisting of medium chain triglycerides (46, 82). Although severe life threatening adverse effects occurred from the radiation and chemotherapy, both children responded remarkably well to the KD and experienced long-term tumor management without further chemo or radiation therapy. Positron emission tomography with fluoro-deoxy-glucose (FDG-PET) also showed a 21.8% reduction in glucose uptake at the tumor site in both subjects on the KD (46). Dr. Nebeling reflects on this study in Chapter 20.

These findings showed that a mildly restricted KD, which lowered glucose and elevated ketone bodies, could reduce glycolytic energy metabolism in these brain tumors. More recently we published a case report showing that a KD and therapeutic fasting could also help manage glioblastoma growth in an older female patient (45). The KD can improve the quality of life of some cancer patients (83). It is my opinion that the therapeutic effects would be even greater if the KD were administered in restricted amounts with drugs that target glucose and glutamine as I have discussed further in Chapter 18. Viewed together, these findings show that the KD is well tolerated and can be an effective nontoxic therapy for malignant cancer in both children and adults. This therapeutic strategy should be effective for any cancer that depends heavily on glucose for survival.

GLUCOSE ACCELERATES TUMOR GROWTH!

Glucose fuels tumor cell glycolysis and provides precursors for the pentose phosphate shunt as well as for glutamate synthesis (8, 15, 50, 84–86). We used linear regression analysis to show that growth of experimental astrocytoma was directly related to the levels of circulating glucose levels (Fig. 17.5). It is clear from the figure that the higher is the glucose level, the faster is the growth. As glucose levels fall, tumor size (weight) and growth rate falls. As mentioned earlier, *hyperglycemia* is directly linked to poor prognosis in humans with malignant brain cancer and is connected to the rapid growth of most malignant cancers (50, 87–90). These findings in human brain cancer patients have corroborated our studies in mice with brain tumors. In light of these findings, it is difficult to understand why some oncologists would encourage cancer patients to consume high calorie foods and drinks during their treatment. *Glucose accelerates tumor growth!*

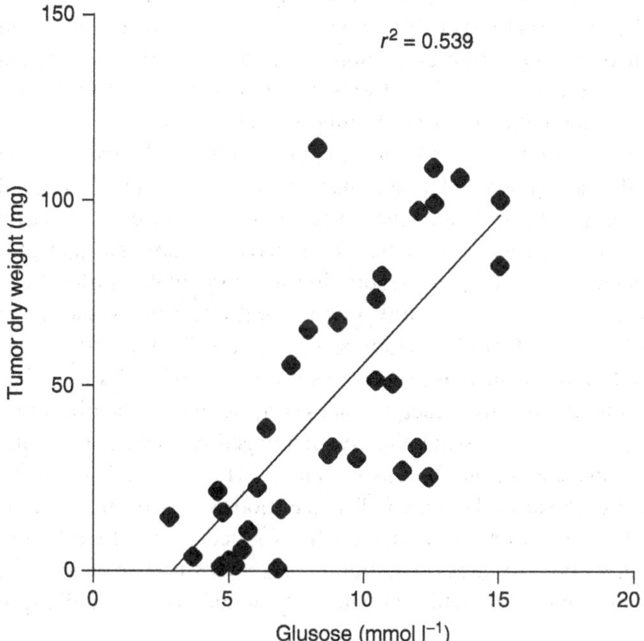

Figure 17.5 Linear regression analysis of plasma glucose and orthotopic CT-2A astrocytoma growth in C57BL/6 mice from both standard diet and KD dietary groups combined ($n = 34$). These analyses included the values for plasma glucose and tumor growth of individual mice from both food-restricted and unrestricted groups. The linear regression was highly significant ($^*P < 0.001$) and indicates that circulating glucose levels are highly predictive of astrocytoma growth. *Source:* Reprinted with permission from Reference 15.

GLUCOSE REGULATES BLOOD LEVELS OF INSULIN AND INSULIN-LIKE GROWTH FACTOR 1

It is well recognized that glucose regulates blood insulin levels. Insulin levels become elevated in response to increasing blood glucose. Indeed, simply smelling or looking at food can elevate blood insulin levels in some subjects (91). Insulin drives glycolysis and cellular energy metabolism (71). In addition to showing that glucose controls tumor growth, we also showed that glucose controls circulating levels of insulin-like growth factor 1 (IGF-1) (15, 77) (Fig. 17.6a). IGF-1 is a cell surface receptor linked to rapid tumor growth (13, 77). The association of elevated plasma IGF-1 levels with rapid tumor growth rate is due primarily to high circulating glucose levels (Fig. 17.6b). Just as DER lowers insulin levels, DER also lowers IGF-1 levels (15). This happens because DER lowers blood glucose levels. Glucose drives tumor cell energy metabolism, while IGF-1 drives tumor cell growth through the IGF-1/PI3K/Akt/hypoxia-inducible factor-1α (HIF-1α) signaling pathway (77). This pathway underlies several cancer hallmarks to include cell proliferation, evasion of apoptosis, and angiogenesis (4, 5, 9, 13, 15, 31, 32, 77, 92–97). Therapies that target this signaling pathway will prevent rapid tumor growth, while also making tumor cells more vulnerable to drugs that target energy metabolism. DER reduces IGF-1 and elevates apoptosis (5, 98, 99). We showed that DER targets this pathway in brain tumors (77).

The transition from glucose to ketone bodies requires multiple changes in gene expression and metabolic adjustments. These adjustments readily occur in normal cells of the body, as the transition from glucose to ketones is an evolutionarily conserved adaptation to food restriction. Genes for glucose metabolism and glycolysis are downregulated, while genes for respiration are upregulated (100–102). Insufficient respiration and genomic instability will prevent adaptive versatility of tumor cells, thus contributing to their elimination as I have described in Chapter 15.

As most tumor cells require increased glycolysis for growth and survival, a transition from glucose energy to ketone energy places considerable metabolic stress on tumor cells (84). Tumor cells are metabolically challenged compared to normal cells and cannot effectively use ketone bodies for energy (Fig. 17.4). Treatments that reduce glucose while elevating ketones will place more stress on tumor cells than on normal cells. Hence, an energy transition from glucose to ketone bodies becomes a rational therapeutic strategy to tumor management that enhances the metabolic efficiency of normal cells, while targeting the metabolically challenged tumor cells.

DIETARY ENERGY REDUCTION IS ANTIANGIOGENIC

A significant literature suggests that vascularity is rate limiting for the formation of solid tumors (103–108). The malignancy and invasiveness of tumors is also correlated with the degree of their vascularity since prognosis is generally better for tumors that are less vascular than for those that are more vascular (105, 109,

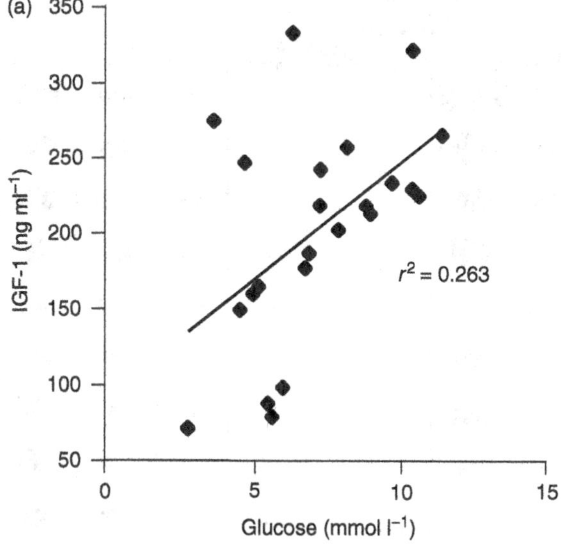

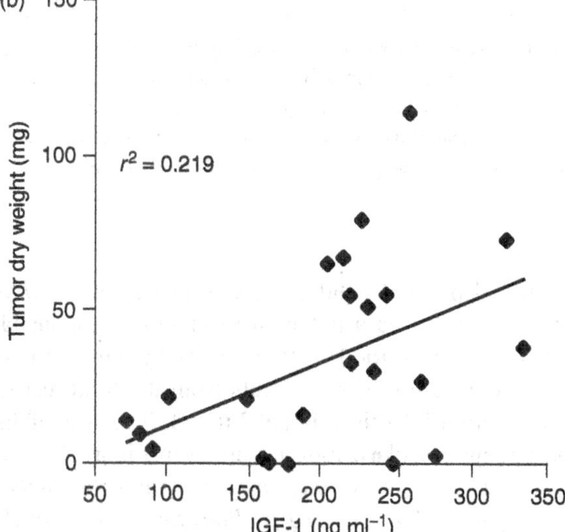

Figure 17.6 Linear regression analysis of plasma glucose and IGF-1 levels (a) and plasma IGF-1 levels and CT-2A astrocytoma growth (b). The mice used for these studies were fed with either a standard diet or a KD ($n = 23$). These analyses included the values for plasma glucose and tumor growth of individual mice from both the UR- and R-fed groups. The linear regressions were highly significant ($P < 0.01$) and indicate that circulating levels of IGF-1 and tumor growth are largely dependent on circulating levels of glucose (15). *Source:* Reprinted with permission from Reference 15.

110). Inhibition of vascularity is therefore considered an important therapeutic strategy for managing tumor growth (103, 111–115) (Fig. 1.3). The challenge is to target tumor angiogenesis without harming patients or reducing the quality of life.

Payton Rous (116) first suggested that the restriction of food intake inhibited tumor growth by delaying tumor vascularity (angiogenesis) from the host. Angiogenesis involves neovascularization or the formation of new capillaries from existing blood vessels and is associated with the processes of tissue inflammation, wound healing, and tumorigenesis (104, 117, 118). Although Rous did not use

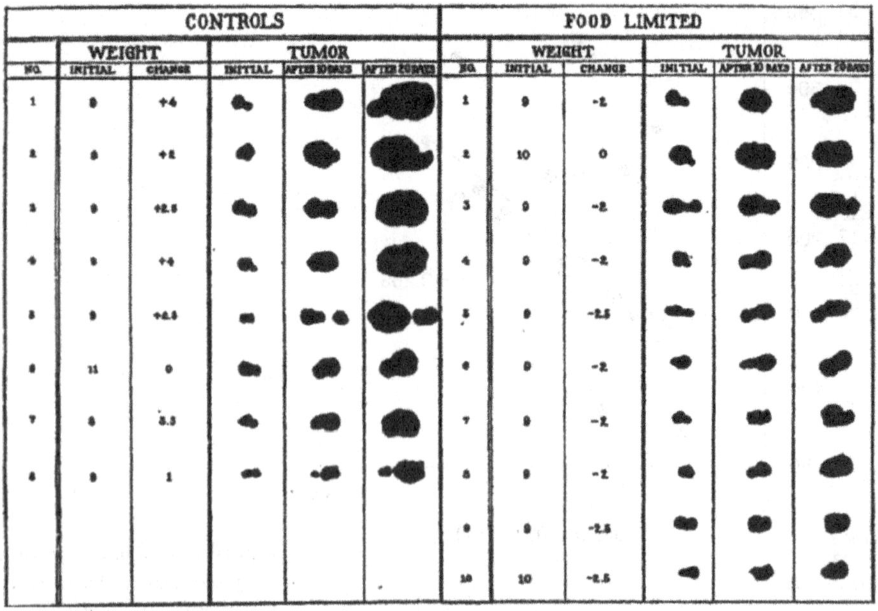

CONTROLS						FOOD LIMITED					
	WEIGHT		TUMOR				WEIGHT		TUMOR		
NO.	INITIAL	CHANGE	INITIAL	AFTER 10 DAYS	AFTER 20 DAYS	NO.	INITIAL	CHANGE	INITIAL	AFTER 10 DAYS	AFTER 20 DAYS
1	9	+4				1	9	-2			
2	9	+2				2	10	0			
3	9	+2.5				3	9	-2			
4	9	+4				4	9	-2			
5	9	+4.5				5	9	-2.5			
6	11	0				6	9	-2			
7	8	3.5				7	9	-2			
8	9	1				8	9	-2			
						9	9	-2.5			
						10	10	-2.5			

Figure 17.7 Influence of limited food intake on tumor growth in mice. In this study, Rous evaluated the influence of underfeeding on the rate of growth of a transplantable adenocarcinoma in mice. Restricted feeding was started after the tumors reached an appreciable size. The control mice were fed a mixed ration unrestricted. This was one of the first studies showing that reduced calorie intake could reduce tumor growth. *Source:* Reprinted with permission from Reference 116.

the term *angiogenesis* in his original paper, he did use the term *vascularization*. He considered that tumor vascularization was a potential mechanism responsible for rapid tumor growth and that calorie or food restriction might inhibit tumor growth by indirectly targeting vascularization. Rous' results from the food restriction experiments are shown in Figure 17.7 (his original Fig. 3). The test of his hypothesis that food restriction targets vascularization is shown in Figure 17.8. In summarizing his findings, Rous stated, "*In these facts may be found the method whereby dieting delays tumor growth. With a lessened proliferative activity of the host tissue, the elaboration of a vascularizing and supporting stroma such as most tumors depend upon for their growth, at least indirectly, is much delayed.*" We know from our extensive studies in mice with gliomas that it is the reduced availability of glucose that reduces tumor cell proliferation, angiogenesis, and inflammation.

In agreement with the early findings of Rous, we found that growth of mouse astrocytoma was about 80% less under moderate DER than under unrestricted or AL feeding (Fig. 17.1). This reduction in tumor growth greatly exceeded the 12% reduction in body weight during the 22-day experiment (32). While the studies that I mentioned earlier in this chapter showed that moderate DER, involving a 25–40% restriction in food content, could reduce the growth of histologically

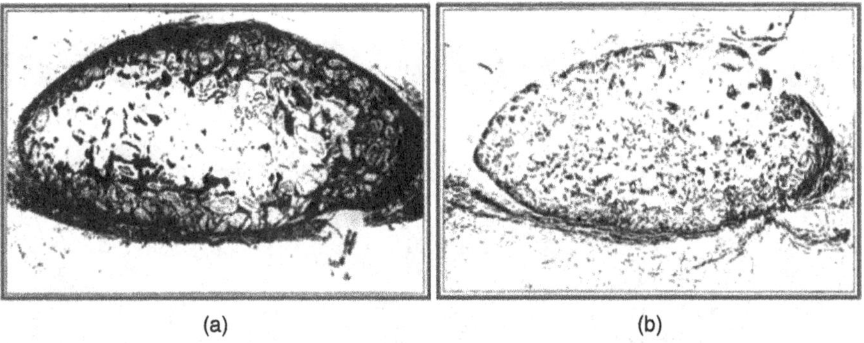

Figure 17.8 Influence of restricted food intake on vascularization of subcutaneous agar implants in rats. Rous found that vascularization (dark shading) was noticeably less in the agar implants of rats given limited food (b) than for the control fed rats given unlimited food (a). *Source:* Reprinted with permission from Reference 116. To see this figure in color please go to ftp://ftp.wiley.com/public/sci_tech_med/cancer_metabolic_disease.

diverse nonneural tumors, we were the first to document this phenomenon in a brain tumor model (32). On the basis of these findings, we suggested that brain tumors should be especially vulnerable to the growth inhibitory effects of DER.

We confirmed further our hypothesis that DER is antiangiogenic in mouse and human experimental brain tumors (Figs. 17.9 and 17.10). Biomarkers for angiogenesis, including IGF-1 and vascular endothelial growth factor (VEGF), were significantly lower in all tumors when grown under DER than when grown under unrestricted or AL conditions (Table 17.2) (31). DER also reduces angiogenesis in prostate and mammary cancer (4, 9, 119). DER of KDs is also antiangiogenic, indicating that the antiangiogenic effects are related to the amount rather than the composition of the diet.

As DER targets brain tumor angiogenesis naturally, while also enhancing the health and vitality of normal brain cells, we suggest that the antiangiogenic effects of DER or calorically restricted KDs will be superior to that of most known antiangiogenic drug therapies, including those involving metronomic applications, where multiple antiangiogenic drugs are given together (120). DER targets the entire tumor microenvironment.

In light of our findings, it is surprising that the cancer field would persist in treating cancer patients with toxic antiangiogenic drugs such as bevacizumab and cediranib, which show marginal efficacy and appear to enhance the invasive behavior of tumor cells (121–123). Compared to bevacizumab (Avastin), which targets angiogenesis, while producing adverse effects and enhancing tumor cell invasion (112, 122, 124–126), DER targets angiogenesis, while improving general health and inhibiting tumor cell invasion (34, 45). *Is it better for oncologists to target tumor angiogenesis using toxic drugs with marginal efficacy or is it better to use nontoxic metabolic strategies such as DER with robust efficacy?* Oncologists should consider this question. Patients with advanced cancers should be presented with therapeutic options.

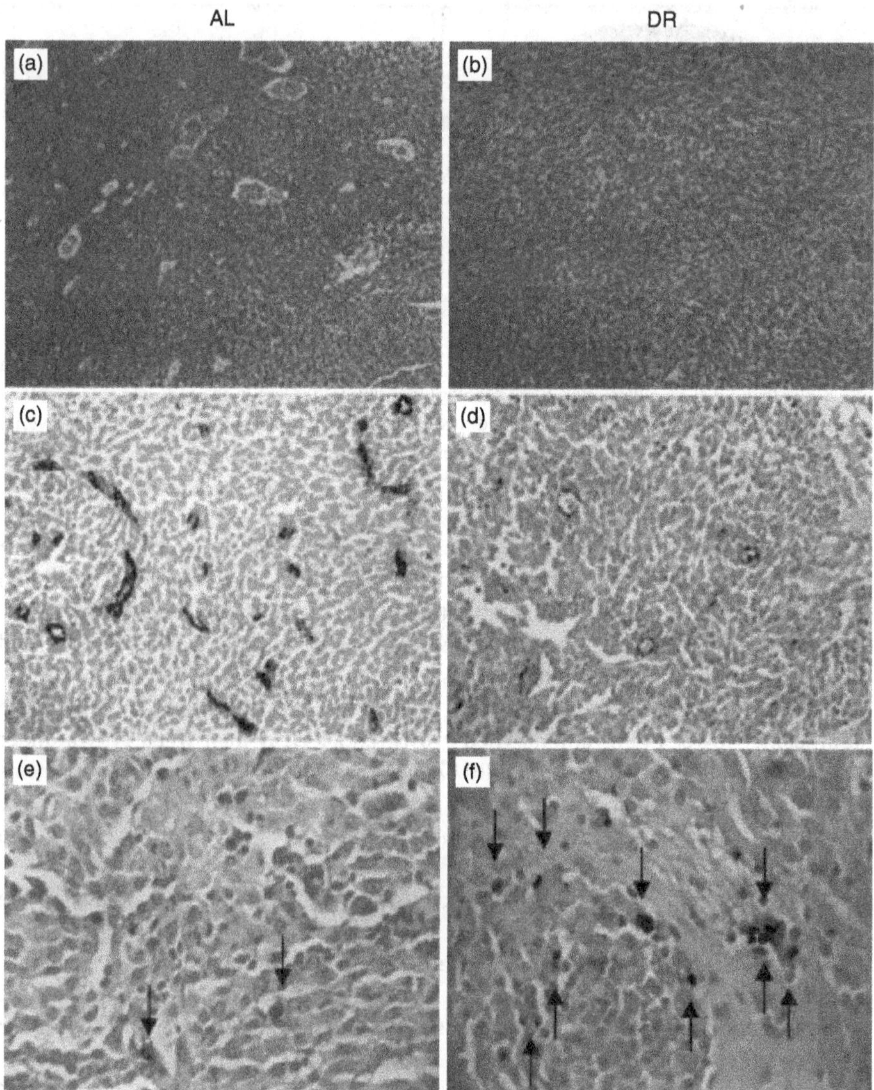

Figure 17.9 (a) Influence of dietary energy restriction on microvessel density and apoptosis in the CT-2A brain tumor. DR was initiated 7 days before intracerebral tumor implantation and was continued for 11 days. H&E-stained tumor sections in an *ad libitum* (AL)-fed mouse and in a DR-fed mouse (b) (100×). Factor VIII immunostaining from the tumor grown in an AL mouse (c) and in a DR mouse (d) (200×). TUNEL-positive apoptotic cells (arrows) from the tumor grown in an AL mouse (e) and in a DR mouse (f) (400×). Each stained section was representative of the entire tumor. All images were produced from digital photography. The results show that DER (left panels) targets tumor blood vessels, while enhancing apoptosis. *Source:* Reprinted with permission from Reference 32. See color insert.

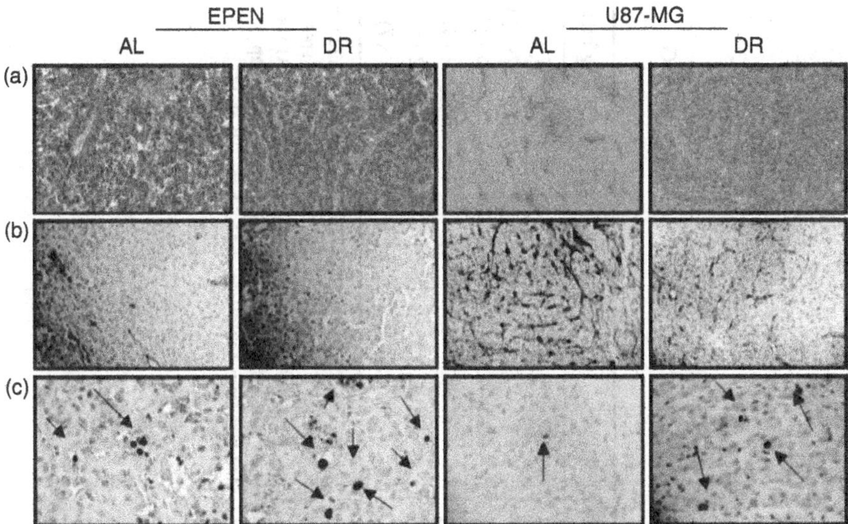

Figure 17.10 Morphology, vascularity, and apoptosis in *ad libitum*-fed and DR-fed mice bearing the intracerebral mouse EPEN and human U87-MG brain tumors. Dietary restriction was initiated 24 h after tumor implantation and continued for 11 days. (a) H&E-stained tumor sections (100×). (b) Factor-VIII-immunostained microvessels (200×). (c) TUNEL-positive apoptotic cells (arrows; 400×). All images were produced from digital photography. As seen for the CT-2A mouse astrocytoma in Figure 17.9, DER targets tumor blood vessels, while enhancing apoptosis. *Source:* Reprinted with permission from Reference 31. See color insert.

DIETARY ENERGY REDUCTION TARGETS ABNORMAL TUMOR VESSELS

It is important to mention that blood vessel structure and function are different in the tumor microenvironment than in normal microenvironment. Puchowicz et al. (127) showed that diet-induced ketosis increases capillary density in normal rat brain. In contrast to normal tissue vasculature, tumor blood vessels express leakiness and immaturity (absence of a pericyte smooth muscle sheath) (128, 129). DER reduces the abnormal vasculature in tumors (Fig. 17.9) (31, 32). My former undergraduate student, Ivan Urits, recently found that DER enhances expression of α-smooth muscle actin (α-SMA) in the tumor vasculature (130). The beautiful images from Ivan's study were not available while I was writing the book, but I encourage readers to examine the images and data in the original publication (130). α-SMA is a marker for vessel maturation and integrity (131). A restoration of blood vessel integrity should reduce local inflammation arising from vessel leakiness. Hence, DER has a marked effect on the structure and function of tumor blood vessels that reduces tumor growth.

Enhancement of vessel maturation in tumors could facilitate delivery of therapeutic drugs and possibly microRNAs to the tumor (132). Support for this

Table 17.2 Effects of DER on biomarkers for vascularity and apoptosis in the CT-2A, and EPEN mouse brain tumors, and in the U87-MG human brain tumor

Tumors	Diet	MVD	Apoptotic index %	Proliferation index %	IGF-I (ng/ml)	VEGF (pg/ml)
CT-2A	AL	24.3 ± 1.4 (5)	3.7 ± 0.4 (5)	71 ± 3 (5)	273 ± 63 (12)	118 ± 17 (5)
	DR	10.3 ± 3.1† (5)	8.1 ± 1.2† (5)	68 ± 2 (5)	170 ± 29† (17)	80 ± 17* (5)
EPEN	AL	7.7 ± 2.4 (6)	3.4 ± 0.9 (6)	48 ± 3 (3)	149 ± 19 (4)	86 ± 19 (4)
	DR	3.6 ± 1.2* (5)	8.1 ± 2.9† (5)	43 ± 2 (3)	77 ± 44† (4)	94 ± 43 (4)
U87-MG	AL	51.0 ± 9.4 (7)	0.9 ± 0.1 (3)	85 ± 5 (3)	370 ± 134 (5)	136 ± 22 (5)
	DR	28.3 ± 3.3† (3)	3.7 ± 1.8* (3)	65 ± 5† (3)	158 ± 25† (6)	100 ± 8* (7)

Animals were fed either ad libitum (AL) or under DER (DR) as we described (31). All values are expressed as means ± 95% confidence intervals. The details for each measurement and statistics are described in (31). To determine microvessel density (MVD), we averaged Factor VIII-positive microvessels in three hot spot areas of each tumor section per high-powered field. The TUNEL assay was used to determine Apoptotic index. The proliferating cell nuclear antigen (PCNA) was used to determine Proliferation index. Numbers in parentheses represent the number of independent tumor tissue samples analyzed. Values from the DR group differed from those of the AL group at $P < 0.05$* or at $P < 0.01$†, as determined by analysis of variance. Modified from Table 1 from Mukherjee et al. (31).

prediction comes from our findings showing that the KD-R enhances brain delivery of *N*B-DNJ (N-butyldeoxynojirimycin), a small imino sugar molecule that inhibits ganglioside biosynthesis (133). Hence, DER or restricted KDs target abnormal tumor blood vessels, while enhancing formation of normal vasculature. I find this intriguing and potentially very important for therapeutic applications.

DIETARY ENERGY REDUCTION IS PROAPOPTOTIC

Besides reducing the tumor vasculature, we also found that DER kills tumor cells through apoptotic mechanisms (4, 31, 32). The DER-induced reduction in brain tumor growth was also associated with significant elevation in TUNEL-positive cells (apoptosis) (Table 17.2 and Figs. 17.7 and 17.8). TUNEL is an acronym for "terminal deoxynucleotidyl transferase-mediated deoxyuridine triphosphate biotin nick end labeling." DNA begins to disintegrate in a specific way in cells that die from programmed cell death or apoptosis.

Apoptotic cell death differs from necrotic cell death, which is usually associated with inflammation (134). Apoptotic tumor cell death would therefore be less provocative to the tumor microenvironment than would necrotic cell death, as tissue inflammation is less during apoptosis than during necrosis. This is important since the current standard of care for many cancers often involve radiation therapy together with toxic chemotherapy that causes inflammation and necrotic tumor cell death as I have shown in Figure 16.1. In contrast to most conventional cancer therapies, which cause tissue necrosis and inflammation, metabolic therapies involving reduced calorie intake primarily kill tumor cells through apoptotic cell death. *Is it better to kill tumor cells using toxic drugs, as is currently done in the oncology field, or is it better to kill tumor cells using a nontoxic metabolic therapy like DER?* I favor the later approach.

Phosphorylation and inactivation of BAD (BCL2 agonist of cell death) and procaspase-9 mediate, in part, the antiapoptotic actions of Akt (protein kinase B) activation (135, 136). BAD transmits proapoptotic signals generated during glucose deprivation. My associates Jeremy Marsh and Purna Mukherjee found that BAD was constitutively phosphorylated in mouse astrocytoma compared with contralateral normal brain and showed that DER suppressed BAD phosphorylation and increased procaspase-9/-3 cleavage (77). BAD stimulates apoptosis by forming heterodimers with and by inactivating the antiapoptotic proteins Bcl-2 and Bcl-xL (135, 136). DER is known to reduce Bcl-2 and Bcl-xL expression and to increase the expression of Bax, Apaf-1, caspase-9, and caspase-3 in experimental carcinomas (96).

Our studies suggest that DER inhibits tumor growth by inducing mitochondrial-dependent apoptosis mediated by the dephosphorylation of BAD. These findings are consistent with evidence that DER is proapoptotic in malignant astrocytomas and support evidence that BAD coordinates glucose/IGF-1 homeostasis and the induction of apoptosis (31, 77, 135–137). Our findings also show that reduced glucose availability and IGF-1 expression play a key role in suppressing Akt and in mediating the proapoptotic effects of DER in the CT-2A mouse astrocytoma that is PTEN (phosphatase and tensin homologue)/TSC2-(tuberous sclerosis complex 2) deficient (Fig. 17.11). On the basis of our extensive studies in mice and the common mechanisms of apoptosis signaling, I believe that systemic energy restriction will also enhance apoptosis in most malignant human tumors.

The proapoptotic effects of DER occur in large part from reduced glycolytic energy that most tumors rely on for growth (77, 138, 139). DER kills tumor cells by depleting available energy and by creating tumor-specific oxidative stress through glucose deprivation (8, 140). In contrast to producing oxidative stress in tumor cells, DER will reduce oxidative stress in normal cells through elevation of ketone bodies (57–60). *Is it better for oncologists to treat patients using only toxic drugs or is it also appropriate for oncologists to consider a nontoxic therapy such as DER?* I think those cancer patients who are mindful of their disease might want to participate in the discussion.

The widely held notion that tumor cells are resistant to apoptosis is inconsistent with our findings that DER enhances tumor cell apoptosis. Tumor cell resistance to apoptosis arises largely from enhanced glucose and glutamine fermentation. The upregulation of metabolic pathways, involving c-Myc, Hif-1α, and so on, will inhibit apoptosis (77, 141–143). If energy from glycolysis is reduced, then many tumor cells will die or growth arrest from catastrophic energy failure. Tumor cells have difficulty growing once their access to glucose and glutamine becomes limited. Indeed, Yuneva (144) considers the dependence of tumor cells on glucose and glutamine for survival as the "Achilles heel" of cancer. I concur with Dr. Yuneva's general assessment. I address this issue more in Chapter 18.

DER is a simple natural process by which tumor glycolysis can be targeted without causing toxicity to normal cells. Restricted KDs can also reduce availability of glutamine to tumors since ketone bodies and the KD enhances glutamine export from the brain (145). Recent studies also show that ketone ester diets, which elevate blood ketones, can also reduce brain glutamate and glutamine, while reducing food intake and blood glucose levels (69). Dominic D'Augostino from the University of South Florida is also evaluating the action of new ketone ester formulations on tumor cell viability. It is my opinion that DER supplemented with ketone esters could be a simple nontoxic therapy for targeting energy metabolism in a broad range of tumor cells.

DIETARY ENERGY REDUCTION IS ANTI-INFLAMMATORY

Inflammation not only initiates tumorigenesis but also drives tumor progression (146–150). Inflammation damages OxPhos, which is the origin of many cancers (Chapter 10). Nuclear factor kappa B (NF-κB) is a transcription factor largely responsible for enhancing tissue inflammation. Phosphorylation and activation of NF-κB results in the transactivation of many genes, including those encoding cyclooxygenase-2 (COX-2) and allograft inflammatory

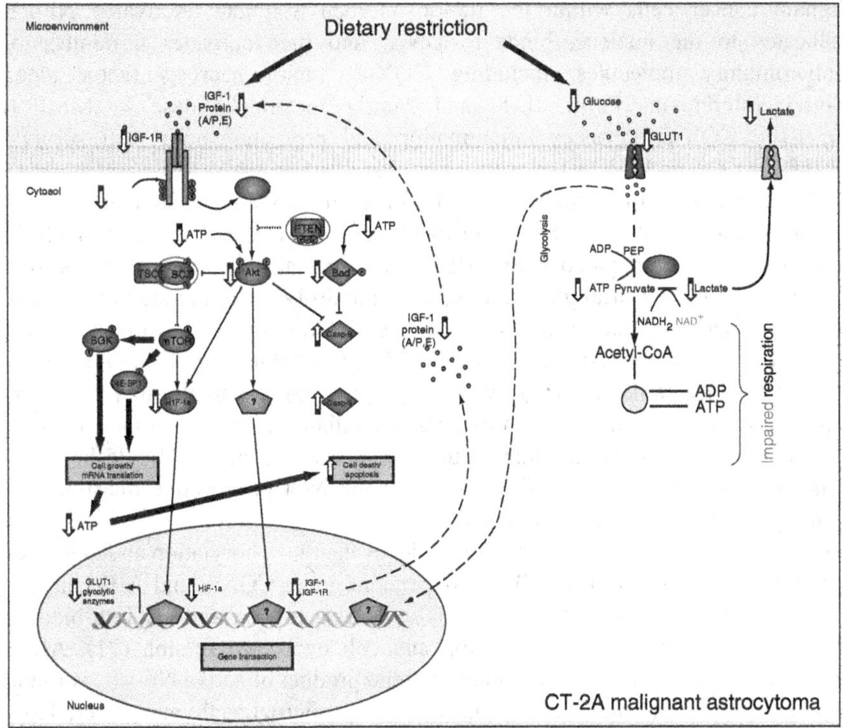

Figure 17.11 Proposed mechanism by which late-onset DER reduces glucose and IGF-I metabolism. Reduced production of IGF-I inhibits signaling through the IGF-IR/Akt pathway and leads to activation of apoptotic pathways induced by the dephosphorylation of BAD (on S-136) and cleavage of procaspase-9/-3. The expression of HIF-1α and GLUT1 are regulated in part by the level of Akt phosphorylation. Consequently, increased expression of HIF-1α and GLUT1 may confer protection against apoptosis. The DR-induced suppression of Akt phosphorylation leads to reduced transcription and translation of HIF-1α as well as to decreased expression of GLUT1. We propose that the disruption of glucose metabolism by DER plays a central role in mediating the antagonistic effects of DER in managing the metabolically inflexible PTEN/TSC2-deficient astrocytomas. More specifically, the loss of PTEN and TSC2 expression in malignant astrocytomas impairs adaptation to DER-induced energy stress. Moreover, the inability of CT-2A to shutdown protein synthesis during DER, owing partially to loss of the PTEN and TSC2 tumor suppressors, may also contribute to DER-induced cell death by accelerating ATP depletion. The shapes with green backgrounds represent signal transduction molecules in the cytosol; the pentagons with red backgrounds represent transcription factors. Upward facing arrows represent increased expression, whereas downward facing arrows represent decreased expression. Question marks represent unknown transcription factors [autocrine (*A*)/paracrine (*P*) and endocrine (*E*)]. *Source*: Reprinted with permission from Reference 77. See color insert.

factor-1 (AIF-1), both of which are primarily expressed by inflammatory and malignant cancer cells within the tumor microenvironment. Activated NF-κB translocates to the nucleus, binds to DNA, and then activates a number of proinflammatory molecules, including COX-2, tumor necrosis factor alpha (TNF-α), interleukin (IL)-6, IL-8, and Matrix metallopeptidase 9 (MMP-9) (150, 151). COX-2 enhances inflammation and promotes tumor cell survival (152, 153).

My associate Purna Mukherjee and former undergraduate student, Tiernan Mulrooney, found that the p65 subunit of NF-κB was expressed constitutively in mouse astrocytoma compared with contralateral normal brain tissue (21). NF-κB also activates mitochondrial glutaminase, which hydrolyzes glutamine to glutamate (154). Glutamate is used as an energy metabolite for tumor growth and, when secreted, can enhance tumor progression (155, 156). Inhibition of NF-κB activation would reduce rapid tumor growth and progression in part through inhibition of glutamine metabolism. As NF-κB-mediated inflammation is common to most malignant cancers, any therapy that reduces expression of NF-κB should be effective in managing cancer (157).This addresses the NCI provocative question # 5 (http://provocativequestions.nci.nih.gov).

Purna and Tiernan showed that DER reduces the phosphorylation and degree of transcriptional activation of the NF-κB-dependent genes COX-2 and AIF-1 in CT-2A tumor. AIF-1, also known as *Iba1*, is a 17-kDa protein and calcium-binding molecule involved in cellular activation and cell cycle progression (21). AIF-1 expression, an established proinflammatory gene product of active NF-κB, is found in human cells and tissues such as macrophages, microglia, thymus, liver, lung, and subtypes of invasive malignant gliomas (21, 158, 159). They also showed that DER reduces expression of proinflammatory markers lying downstream of NF-κB, for example, macrophage inflammatory protein-2 (MIP-2) (21). The evidence supporting these statements is shown in Figures 17.12 and 17.13. On the whole, they showed that the NF-κB inflammatory pathway is constitutively activated in mouse astrocytoma and that calorie restriction targets this pathway and inflammation. This is interesting in light of the findings from Guido Franzoso and colleagues showing that low glucose suppression of NF-κB expression plays an important role in cellular energy homeostasis (157). As the inhibitory effects of calorie restriction on CT-2A growth are similar to those of the restricted KD, I predict that any metabolic therapy, which reduces glucose and elevates ketones, would target tumor inflammation through similar pathways.

There are no oncology drugs known to my knowledge that can simultaneously target inflammation and angiogenesis, while, at the same time, killing tumor cells through an apoptotic mechanism. Mantovani and colleagues (160) have recently reviewed the role played by inflammation in the initiation and progression of cancer. Question 10 raised on page 442 of their review asks "*What is the best way to target cancer-related inflammation in patients with cancer? This is the most difficult question.*" The results from our studies in mice provide an accurate and simple answer to this question. DER, which targets multiple inflammatory biomarkers, is the simplest and best way to target cancer-related inflammation. Weindruch's group

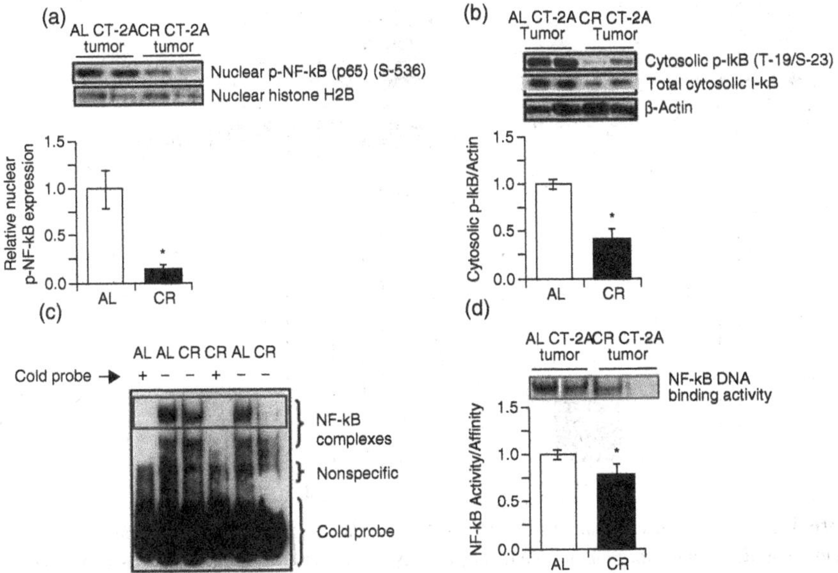

Figure 17.12 Influence of calorie restriction (CR) on NF-κB expression and activation in CT-2A mouse astrocytoma. Nuclear expression of phosphorylated NF-κB (p65) (a); cytosolic expression of phosphorylated IkB and total lkB (b) as assessed by Western blot analysis, DNA promoter-binding activity of activated NF-κB in the CT-2A astrocytoma (c and d) as assessed by electrophoretic mobility shift assay (EMSA). The histograms illustrate the average relative expression of phosphorylated to total protein normalized to the indicated loading control in either nuclear or cytoplasmic extracts of the indicated tissue (a and b). Values are expressed as normalized means ± SEM of 4–5 independent tissue samples/group. The asterisks indicate that the CR value is significantly lower than the AL value at $P < 0.05$ (student's t-test). Two representative samples are shown for each tissue type. (c) Evaluation of the extent of DNA proinflammatory gene promoter-binding activity by activated NF-κB in nuclear extracts of CT-2A under AL and CR condition. (d) DNA promoter-binding activity of activated NF-κB in nuclear extracts of the NF-κB in the CT-2A astrocytoma under AL and CR feeding. Two representative samples are shown for each tissue type. These findings indicate that the CT-2A astrocytoma shows constitutive expression of NF-κB compared to normal brain parenchyma and that CR reduces NF-κB activation and subsequent DNA binding to target promoters. In summary, reduced caloric intake targets NF-κB-mediated inflammation. *Source:* Reprinted with permission from Reference 21. To see this figure in color please go to ftp://ftp.wiley.com/public/sci_tech_med/cancer_metabolic_disease.

also showed that energy restriction is a simple way to target inflammation in the microenvironment (161).

Oncologists and cancer patients should know that water-only therapeutic fasting is one way to reduce inflammation in the tumor microenvironment. This therapeutic strategy can be continued beyond the fast using calorie-restricted KDs or drugs that can simulate the effects of CR (Chapter 18). The therapeutic efficacy of this approach is even more dramatic when the KD-R is administered with drugs that also target energy metabolism (discussed below).

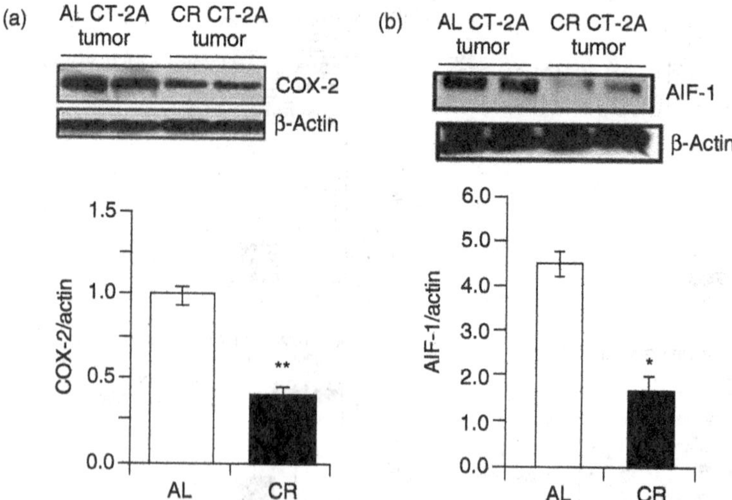

Figure 17.13 CR reduces inflammation in CT-2A mouse astrocytoma. Cyclooxygenase-2 (COX-2) (a) and allograft inflammatory factor 1 (AIF-1) (b) in cytosolic extracts of the CT-2A astrocytoma. COX-2 and AIF-1 have both been reported to be downstream proinflammatory gene-product effectors of the activated NF-κB. The histograms illustrate the average relative expression of the indicated protein normalized to β-actin in CT-2A tumors (a and b). Values are expressed as normalized means $\pm$ SEM of 4–5 independent tissue samples/group. The asterisks indicate that the value is significantly different in the CR tumor than that in the AL tumor at $P < 0.05$ and $P < 0.001$ (student's t-test). Two representative samples are shown for each tissue type. In summary, reduced caloric intake targets NF-κB-mediated inflammation in part through effects on COX-2 and AIF-1. *Source:* Reprinted with permission from Reference 21.

TARGETING ENERGY METABOLISM IN ADVANCED CANCER

Although many studies showed that DER could reduce tumor progression when initiated soon after tumor implantation, fewer studies have evaluated the influence of DER when tumors are already advanced and are heavily inflamed and vascularized. Payton Rous showed that underfeeding could slow advanced subcutaneous adenocarcinoma implants in mice (Fig. 17.7). We also showed that *late*-onset DER (i.e., DER initiated 10 days after tumor implantation rather than only 2–3 days after) could reduce the growth of large tumors (Fig. 17.14). Late-onset DER also delayed malignant progression and significantly extended mouse survival. These effects were associated with changes in metabolic biomarkers, including blood glucose and lactate levels (Fig. 17.15). We showed that expression of these biomarkers was linked to the downregulation of the IGF-I/Akt/Hif-1α signaling pathway (77). Our findings emphasize an important role for activation of the IGF-I/Akt/Hif-1α signaling pathway in potentiating the antiapoptotic phenotype of astrocytomas and indicate that DER targets this signaling pathway. A goal of the cancer drug industry

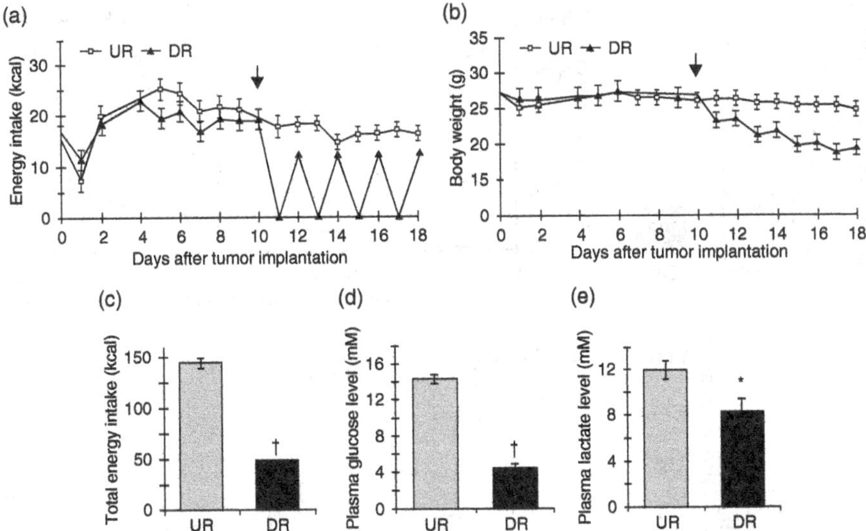

Figure 17.14 Influence of DR on energy biomarkers in advanced brain tumor growth. Energy intake (a and c), body weight (b), plasma glucose (d), and plasma lactate levels (e) in mice bearing the CT-2A astrocytoma. Tumors were implanted intracranially on day 0. All mice were fed *ad libitum* for 10 days and were then separated into unrestricted (UR) and DR groups ($n = 9$ mice per group). DR was initiated on day 10 (arrow, a and b) and all mice were sacrificed 18 days after tumor implantation. (c) Average total energy intake per mouse from day 10 to day 18. Sample size was 8-9 mice per group; SEM. In (c) to (e), the value is significantly lower in the DR group than that in the UR group: $^*P < 0.05$; †, $P < 0.005$, student's t-test. *Source:* Reprinted with permission from Reference 77.

is to target the IGF-I/Akt/Hif-1α pathway for tumor management. Our results show that DER can target this pathway without the use of expensive and toxic cancer drugs. Cancer patients and their oncologists should know about this.

DER reduces the pyruvate kinase isoform M2, which regulates ATP production through glycolysis. Although PKM2 was thought responsible for the Warburg effect (162), recent studies from Ralser and coworkers suggest that this might not be the case (163). The Ralser studies illustrate the importance of comparing energy metabolism in tumor cells with normal cells from the same tissue. On the basis of the extensive information that I have presented in previous chapters, it is now clear that the Warburg effect arises from insufficient mitochondrial respiration, as Warburg originally predicted. It will eventually become recognized by most reasonable investigators that the majority of genetic and metabolic defects seen in tumor cells are linked either directly or indirectly to respiratory insufficiency with compensatory fermentation.

DER-induced reduction in glycolysis is associated with declines in both circulating glucose and lactate levels as well as in the expression of HIF-1α and the type 1 glucose transporter (GLUT1) (77). These reductions were also associated with a reduction in signaling through the IGF-I/Akt pathway. Reduced glycolytic

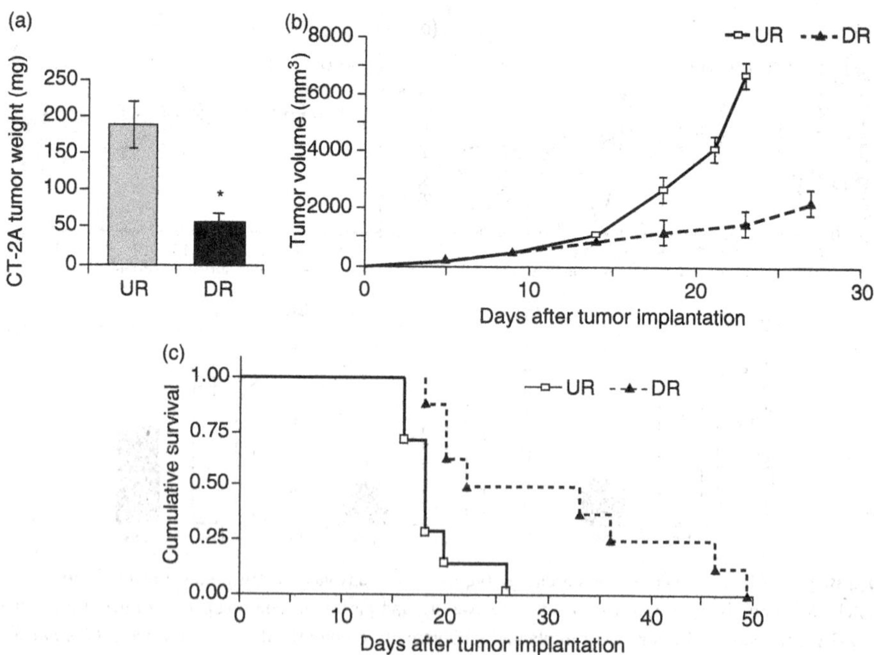

Figure 17.15 Influence of DR on advanced brain tumor growth and survival. Influence of DR on tumor growth and survival in CT-2A-bearing mice. Intracerebral tumor weight (a), subcutaneous tumor volume (b), and Kaplan–Meier survival analysis (c) in mice bearing the CT-2A astrocytoma. Conditions for (a) are as in Figure 17.14. For (b) and (c), the CT-2A tumor tissue was injected subcutaneously on day 0 and DR was initiated on day 14 when tumors were about 1000 mm^3 in volume. Subcutaneous CT-2A tumor volume was significantly lower from day 18 to day 22 ($P < 0.01$, student's t-test) and mouse survival was significantly longer ($P = 0.01$, Kaplan–Meier survival analysis followed by log-rank test) in the DR group than that in the UR-fed control group. The average CT-2A tumor weight was significantly lower in the DR group than that in the UR group: $^*P < 0.005$, student's t-test. *Source:* Reprinted with permission from Reference 77.

energy could increase ROS-related cell death in tumor cells, while reducing ROS levels in normal cells (140). Normal cells switch to ketone bodies for energy under low glucose. Ketone metabolism reduces ROS production in normal cells and is neuroprotective (39, 42, 57, 164, 165). DER reduces availability of a prime fuel (glucose) needed for tumor metabolism, while elevating availability of the prime fuel (ketones) needed to maintain energy homeostasis in normal cells during energy stress.

DIFFERENTIAL RESPONSE OF NORMAL CELLS AND TUMOR CELLS TO ENERGY STRESS

GLUT1 expression is significantly higher in astrocytoma cells than in normal brain cells under AL feeding conditions. Unrestricted food availability maintains high

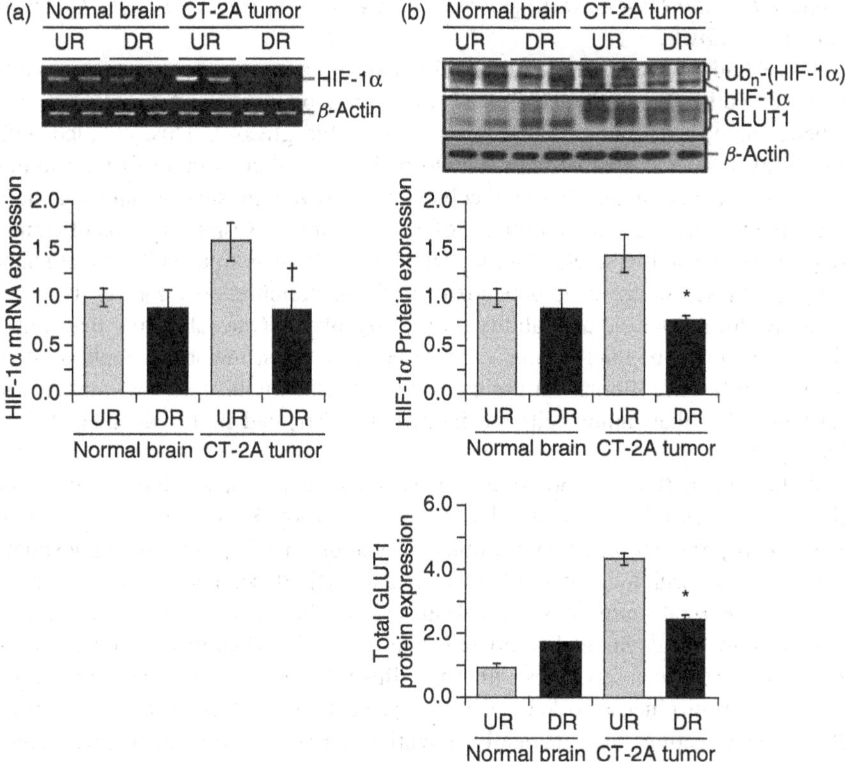

Figure 17.16 Influence of DR on HIF-1α mRNA expression (a) and on HIF-1α and GLUT1 protein expression (b) in contralateral normal brain and in CT-2A astrocytoma. Protein and mRNA expression in contralateral normal brain and in the CT-2A was determined by Western blot analysis and semiquantitative RT-PCR, respectively. The values in the DR tumor are significantly less than those in the UR tumors: $*P < 0.05$; c, $P < 0.01$, student's t-test. The results show that DR downregulates expression of HIF-1α in tumor tissue but not in normal brain tissue. HIF-1α drives tumor glycolysis, including expression of the GLUT1. The results also show that DR increases GLUT1 expression in normal brain tissue, but reduces its expression in the CT-2A tumor tissue. If tumor cells were hardier, tough, and more flexible than normal cells, then expression of HIF-1α GLUT1 should be higher under DR in the tumor than in the normal tissue. This is clearly not the case and indicates that tumor tissue is more vulnerable to energy stress than is normal tissue. *Source*: Reprinted with permission from Reference 77.

blood glucose, which enhances tumor growth (Fig. 17.16). GLUT1 expression is higher in astrocytoma under AL feeding conditions than under DER (77). DER downregulates GLUT1 expression in tumor tissue. In contrast to tumor tissue, DER upregulates GLUT1 expression in normal tissue. These findings show that GLUT1 expression behaves in opposite ways to glucose availability in normal cells and tumor cells. *If tumor cells were more fit or adaptable than normal cells, then GLUT1 expression would be expected to increase more in tumor cells than in normal cells under DER because glucose becomes scarce. The tumor cells, however, do just the opposite.*

These findings indicate a differential response to energy stress in normal cells and tumor cells. By upregulating GLUT1 expression during DER, normal cells are better able than tumor cells to acquire available glucose. This, coupled with their ability to metabolize ketones, would make normal cells more fit than tumor cells under energy stress. Normal cells have evolved to survive and maintain energy homeostasis under conditions of energy stress. Respiratory insufficiency and genomic instability make tumor cells less fit than normal cells. Insufficient respiration makes tumor cells dependent on fermentation energy for survival and growth. Reduced glucose availability targets glycolysis. Our data show that tumor cells are not more fit than normal cells, especially when metabolic fuels used in fermentation become limited in the environment. Our findings *do not* support the persistent belief that tumor cells are hardier, tougher, and more advantaged than normal cells.

Reduction in IGF-1 expression can be lethal to glycolysis-dependent tumor cells but not harmful to normal cells (15, 31, 77, 166). Recent studies show that dietary energy restriction enhances phosphorylation of adenosine monophosphate kinase (AMPK), which induces apoptosis in glycolytic-dependent astrocytoma cells but protects normal brain cells from death (137). This is additional evidence that tumor cells are disadvantaged compared to normal cells when placed under energy stress. Viewed collectively, these findings illustrate further that a shift in energy metabolism from glucose to ketone bodies protects respiratory competent normal cells, while targeting the genetically defective and respiratory challenged tumor cells, which depend more heavily on glycolysis than normal cells for survival (51, 77, 142). Do oncologists and cancer patients know about this?

DIETARY ENERGY REDUCTION IS ANTI-INVASIVE IN EXPERIMENTAL GLIOBLASTOMA

It is the highly invasive and metastatic nature of malignant tumors that makes them difficult to manage using most conventional therapies. The VM-M3 mouse brain tumor is highly invasive when grown in the brain. VM-M3 cells, like human glioblastoma cells, are highly metastatic if the cells gain access to extra neural cites (167, 168). The invasive behavior of VM-M3 in brain is similar to that seen in human glioblastoma multiforme and is considered excellent natural model for this disease (169). Although restricted KDs can be effective in managing invasive brain cancer in children and adults (45, 46), few studies have evaluated the therapeutic effect of calorie or dietary reduction on invasive brain cancer in mice.

The invasive properties of many malignant human brain tumors follow the "secondary structures of Scherer," which include diffuse parenchymal invasion, perivascular growth, subpial surface growth, and growth along white matter tracts (170, 171). Scherer was a German pathologist and one of the first to carefully

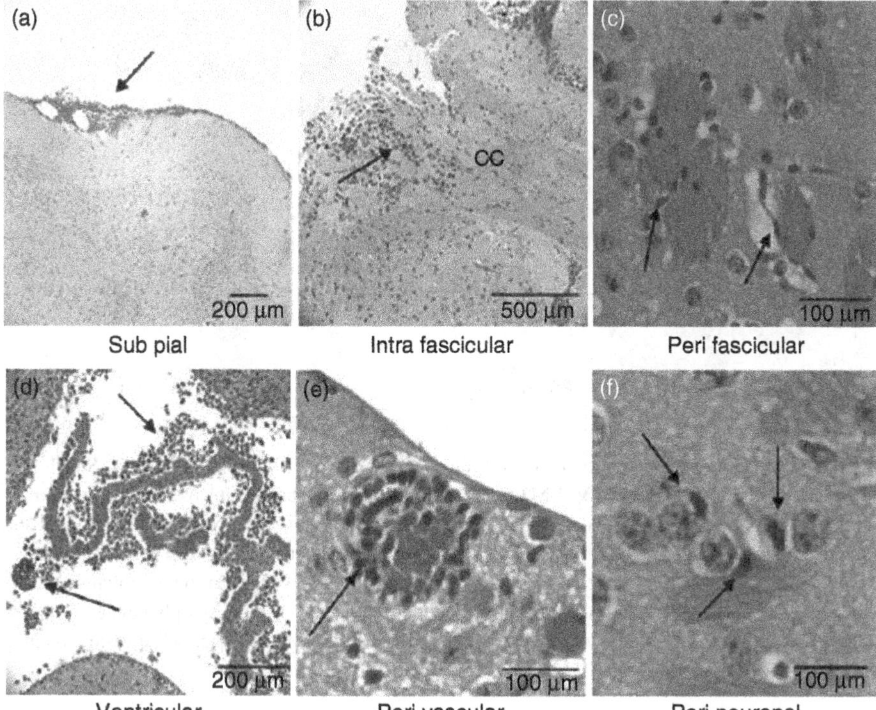

Figure 17.17 VM-M3glioblastoma cells migrate through the brain using Scherer's structures. Histological analysis (H&E) was used to validate the presence of tumor cells (169). The VM-M3 tumor cells are shown invading along the pial surface (arrow, a), within the corpus callosum (CC, arrow, b), along myelinated axons crossing through the striatum (arrow, c), through the ventricular system (arrows, d), around the blood vessels (arrow, e), and around neurons (arrow, f).These invasion routes replicate the *secondary structures of Scherer*, which define the invasion routes seen in invasive malignant brain tumors. VM-M3 tumor fragments or cells were implanted into the cerebral cortex as we described (169). Images are shown at 100×(a), 50×(b), 400×(c), 200×(d), and 400×(e and f). Arrows identify regions containing tumor cells. *Source*: Reprinted with permission from Reference 169. See color insert.

describe the cellular growth patterns of primary malignant brain tumors. We showed that our invasive glioblastoma model is the only syngeneic mouse brain tumor to our knowledge that expresses the full complement of Scherer's secondary structures (169) (Fig. 17.17). As seen in Figure 17.17, DER reduced the invasion of the VM-M3 primary tumor.

We have shown that DER in the form of CR could reduce cerebral invasion of the VM-M3 tumor (Fig. 17.18). Compared to the diffuse, ill-defined borders of the mouse glioblastoma observed in the unrestricted control mice, the tumor grown in the DER mice appears denser with more defined borders. DER also reduced the invasion of tumor cells from the implanted ipsilateral cerebral hemisphere into the contralateral hemisphere (Fig. 17.19). While invading tumor cells were

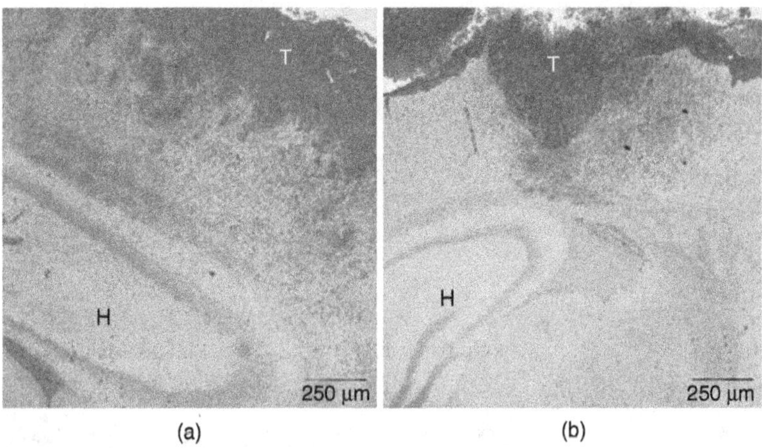

Figure 17.18 Calorie restriction reduces diffuse brain invasion of VM-M3 GBM cells. Small fragments of the highly invasive VM-M3 tumor, containing an established microenvironment, were implanted into the cerebral cortex, were fixed, and then stained with hematoxylin and eosin (H&E) as described (172). Images are shown at 50×(T, tumor; H, hippocampus). Tumor cell invasion through the neural parenchyma (dark blue cells) is less in the CR-fed mice (b) than in the *ad libitum* (AL)-fed mice (a). The boarder between tumor tissue and normal brain tissue is more sharply defined in the CR-fed mice than in the AL-fed mice. The results show that CR reduces tumor cell invasion. *Source*: Reprinted with permission from *ASN Neuro* (172). See color insert.

identified in all regions of the opposite (contralateral) hemisphere of the control AL-fed mice, only subpial invasion was found in the contralateral hemisphere of the food-restricted group (Fig. 17.20).

The total percentage of proliferating tumor cells (Ki-67-positive cells) within the primary tumor and the total number of blood vessels was also significantly lower in the DER-treated mice than in the mice fed AL, indicating that reduced caloric intake is also antiproliferative and antiangiogenic in this tumor (172) (Figs. 17.21 and 17.22). These findings indicate that DER can inhibit proliferation and invasion of tumor cells throughout much of the brain. Considering that invasion is primarily responsible for patient death, I suggest that survival will be increased in those patients who would use DER therapies to treat their brain tumors. I have addressed this concept more in Chapter 18.

Our findings with DER therapy in the invasive glioblastoma model contrast with those in patients treated with bevacizumab. Bevacizumab appears to enhance glioma invasion without reducing Ki-67-positive tumor cells (125, 173, 174). Our findings in mice suggest that DER could be a more effective antiangiogenic therapy than bevacizumab for brain cancer management in humans. Moreover, the therapeutic efficacy of DER was not associated with diarrhea or other adverse effects, as occurs with the potent epidermal growth factor receptor (EGFR) inhibitor, gefitinib (175). In contrast, DER enhances general health. Although the molecular mechanisms by which DER reduces invasion are not yet fully described, these

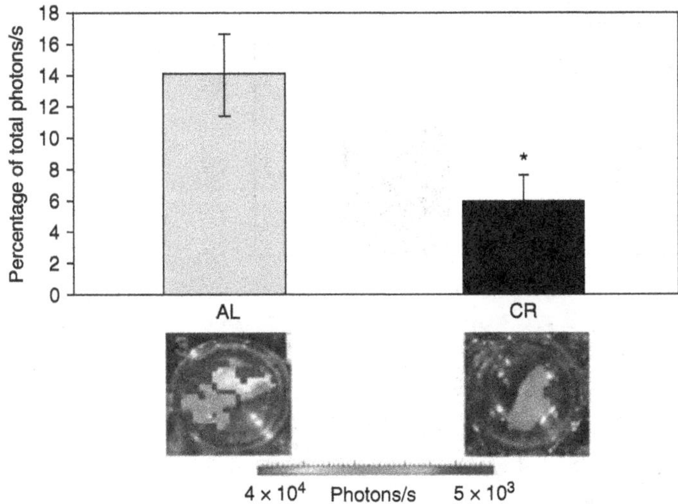

Figure 17.19 CR reduces invasion of VM-M3 GBM cells from one brain hemisphere to the other. VM-M3/Fluc (expressing firefly luciferase gene for bioluminescence detection) tumor fragments were implanted as described (172). Each brain hemisphere was imaged for bioluminescence ex vivo. The bioluminescence from each hemisphere was added together to obtain a total bioluminescence value (photons/s) for each brain. Data for the contralateral hemisphere was then expressed as the percentage of the total brain photons/s. Values represent the means ± SEM for 9–10 mice per group. Representative bioluminescence images are shown. The asterisks indicate that the CR values differ significantly from the AL control group at $P < 0.05$ using the two-tailed student's t-test. The results show that dietary energy restriction in the form of CR reduces interhemispheric invasion. *Source*: Reprinted with permission from Reference 172. See color insert.

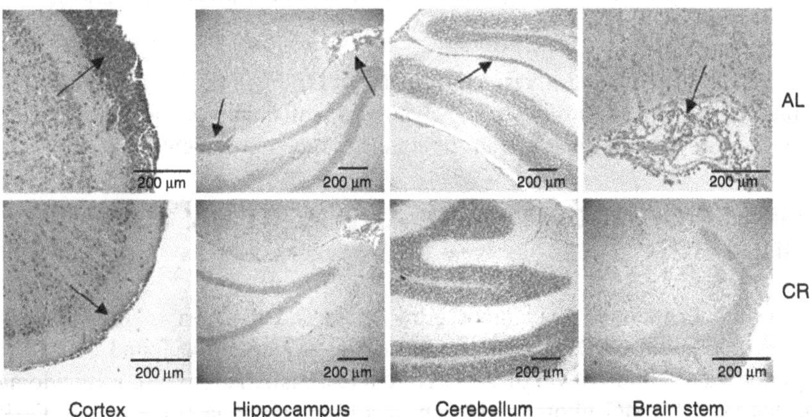

Figure 17.20 Influence of calorie restriction on VM-M3/Fluc tumor cell invasion to the contralateral hemisphere. VM-M3/Fluc tumor fragments were implanted as described (172). Histological analysis (H&E) was used to validate the presence of tumor cells under AL (top) and CR (bottom) in cerebral cortex (200×), hippocampus (100×), cerebellum (100×), and brain stem (200×). Results show that CR reduces VM-M3 cell invasion in the brain. Arrows indicate the presence of tumor cells. At least three samples were examined per group. *Source*: Reprinted with permission from Reference 172. See color insert.

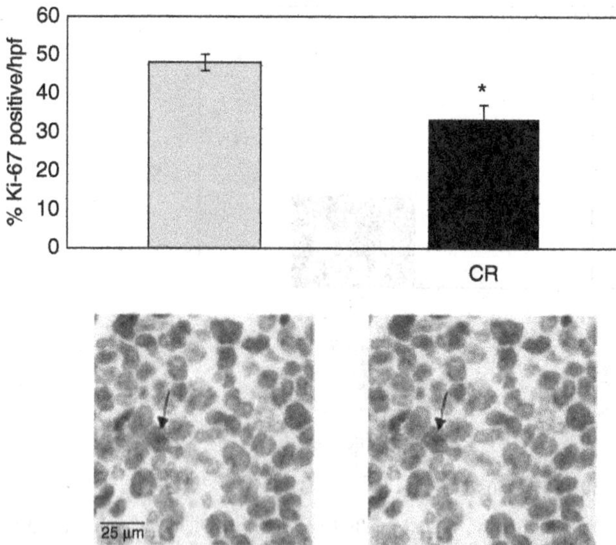

Figure 17.21 CR reduces VM-M3 glioblastoma cell proliferation. Ki-67 staining is a biomarker of cell proliferation. The qualitative and quantitative analysis of Ki-67-positive tumor cells in tissue sections was evaluated as we described (172). Ki-67-positive tumor cells were counted in three independent areas under high magnification and averaged for a single value per sample. Values represent the means ± SEM for three independent samples per group. The asterisks indicate that the values for the CR group differ from those of the AL control group at $P < 0.05$ using the two-tailed student's t-test. Representative immunohistological sections are shown. Images are shown at 400×. Ki-67-positive cells are indicated in brown and by the arrow. *Source*: Reprinted with permission from Reference 172. To see this figure in color please go to ftp://ftp.wiley.com/public/sci_tech_med/cancer_metabolic_disease.

findings indicate that the anti-invasive properties of DER can be due in large part to reductions in tumor cell proliferation, glycolysis, inflammation, and angiogenic factors in both the tumor cells and in the tumor microenvironment.

INFLUENCE OF GROWTH SITE AND HOST ON TUMOR PROGRESSION

It is important to mention that tumor growth site and the host might influence the therapeutic action of DER against cancer. For example, we found that DER significantly reduces the growth of a PTEN-deficient malignant mouse astrocytoma and the human U87-MG glioma, which have PI3K activation (Fig. 17.21). DER also reduces the growth of the mouse ependymal-cell brain tumor (EPEN). We have not yet found a brain tumor that is resistant to the growth inhibitory effects of DER when implanted in the orthotopic site.

However, our findings with these tumors differ from the findings of Kalaany and Sabatini, who showed that DR was ineffective in reducing the growth of the U87-MG and other human tumors when the tumors are grown in mice that

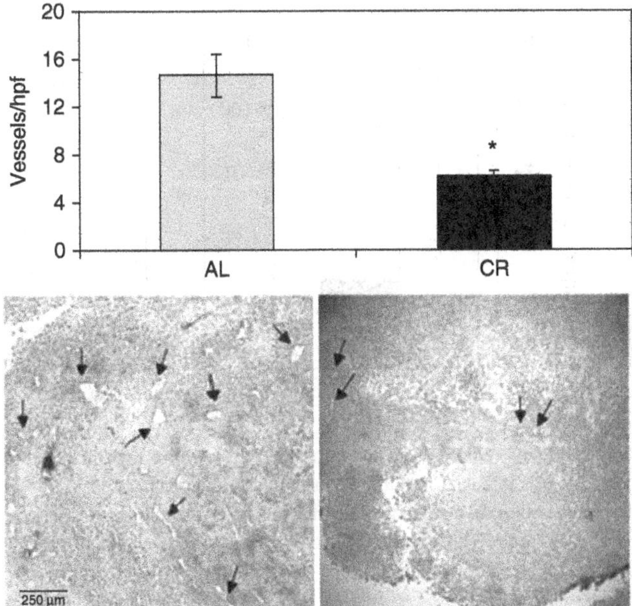

Figure 17.22 CR reduces vascularity in VM-M3 glioblastoma grown orthotopically in the syngeneic VM host. Blood vessels were stained with the factor VIII antibody as we described (172). Blood vessels were counted in three independent areas under high magnification and averaged for a single value per sample. Values represent the means ± SEM for three independent samples per group. The asterisks indicate that the values for the CR group differ from those of the AL control group at $P < 0.05$ using the two-tailed student's t-test. Representative immunohistological sections are shown. Images are shown at 100×. Arrows identify blood vessels. The results show that CR reduces angiogenesis in the VM-M3 glioblastoma. *Source*: Reprinted with permission from Reference 172. See color insert.

express characteristics of diabetes, that is, nonobese diabetic/SCID mice (176, 177) (Fig. 17.23). In contrast to the Kalaany and Sabatini study, we evaluated tumor growth in the orthotopic site (brain) and in mice that did not have characteristics of diabetes (31, 34, 77). As mentioned in Chapter 3, it is not clear why investigators would choose to evaluate human tumor growth in mice that have characteristics of diabetes. As I have mentioned in Chapter 3, human cells will express mouse-specific lipids when grown in mouse hosts. These findings indicate that the tumor implantation site and type of host could influence the effects of DER on tumor growth. Investigators will need to compare and contrast our findings with those of Kalaany and Sabatini to better judge the effects of DER on the growth of experimental tumors (31, 32, 34, 77, 177).

The most accurate information on tumor biology is obtained from those tumor models that are grown orthotopically in the syngeneic host, where the genetic background of the tumor and host is the same. While tumors certainly grow well in mice that express characteristics of diabetes, I think it is better to use mouse

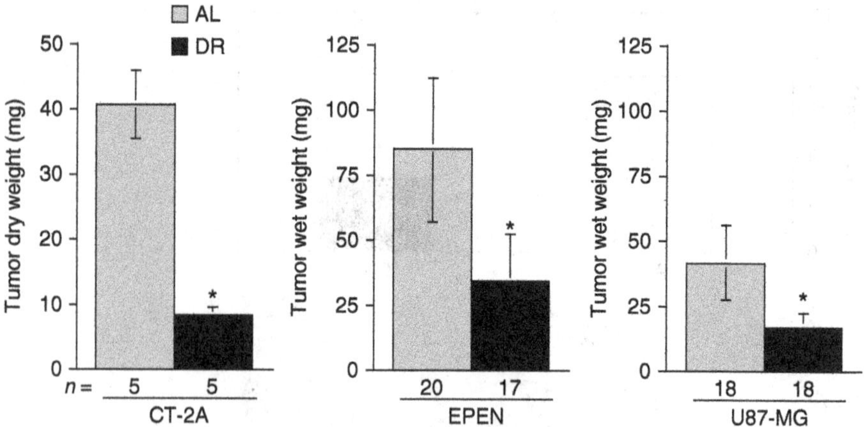

Figure 17.23 Dietary energy restriction (DR) reduces the intracerebral growth of experimental mouse and human brain tumors. DR was initiated 24 h after tumor implantation and continued for 15 days. Values are expressed as means with 95% confidence intervals, and n = the number of tumor-bearing mice examined in each group. The asterisks indicate that the tumor weights of the DR groups were significantly lower than those of the control groups at $P < 0.01$. *Source*: Reprinted with permission from Reference 31.

hosts that do not express these characteristics, especially if traits related to energy metabolism are examined.

IMPLICATIONS OF DIETARY ENERGY REDUCTION FOR ANTICANCER THERAPEUTICS

Our findings with mouse brain tumors are relevant to those in vivo studies where food intake and body weight are reduced in conjunction with antiangiogenic or anticancer therapies. For example, if a new antiangiogenic drug reduces both body weight and tumor growth in experimental test subjects, it is necessary for the investigators to demonstrate the extent to which the angiogenic effect is due specifically to the drug and not to DER. The previous studies of Tannenbaum and Mukherjee showed that tumor therapies, which secondarily restrict food intake or assimilation, might produce changes in tumor growth that could be mistaken for a primary effect (4, 11, 20).

The importance of including the appropriate controls for evaluation of new cancer drugs cannot be overstated. This point was missed in Dr. Ervin Epstein's presentation at the 2011 annual AACR meeting held on April 2–6 in Orlando, FL. Dr. Epstein mentioned that loss or distorted taste (dysgeusia), muscle cramps, and weight loss were common side effects in patients treated with the new hedgehog inhibitor, GDC-449, for nonmelanoma skin cancer. Twenty percent of treated patients discontinued GDC-0449 because of the side effects. Termination of treatment was associated with weight gain and tumor recurrence. Since no controls for dysgeusia, muscle cramps, and weight loss were included in the study design, it is not clear if the therapeutic effects observed were due to GDC-449 or to the side effects of the drug. *How many cancer drugs would the FDA approve if all appropriate control groups for adverse effects were included in experimental designs?*

Kerbel and coworkers (178) also stressed similar issues when they mentioned that many antitumor drugs could also have "accidental" antiangiogenic effects. We suggested that any cancer therapeutic that reduces tumor growth, while also reducing food intake and body weight, might operate in part through the antiangiogenic effects of DER (33). It remains to be determined how much of the therapy from certain toxic cancer drugs is due to a primary effect or to a secondary effect of DER and weight loss.

The inclusion of both pair-fed control groups and active body weight controls in the analysis of new experimental drugs could help distinguish specific effects of the drug from the nonspecific effects of DER. We found that complete starvation of mice for 2 days was necessary in order for an active body weight control group to match the weight loss in mice injected with temozolomide (100 mg/kg) (Mukherjee and Seyfried, unpublished observation). As some drugs may reduce food assimilation, active body weight controls must be evaluated together with pair-fed controls. Unfortunately, many scientific reports of new cancer drugs or therapies fail to include all the necessary control groups needed to distinguish specific from nonspecific effects. This is troubling.

TARGETING GLUCOSE

Dietary energy restriction specifically targets the IGF-1/PI3K/Akt/HIF-1α signaling pathway, which underlies several cancer hallmarks to include cell proliferation, evasion of apoptosis, and angiogenesis (4, 5, 9, 31, 32, 77, 92–97) (Fig. 17.11 and Table 17.2). DER also causes a simultaneous downregulation of multiple genes and metabolic pathways regulating glycolysis (100–102, 179). This is important, as enhanced glycolysis is required for the rapid growth and survival of many tumor cells (180–182). In addition, subsets of tumors have inherited or acquired mutations in the TCA cycle genes (147, 183). Such mutations are expected to limit the function of the TCA cycle, thus increasing the glycolytic dependence of these tumors. Tumors with these types of mutations could be especially vulnerable to management through DER. Hence, DER can be considered a broad-spectrum, nontoxic metabolic therapy that inhibits multiple signaling pathways required for progression of malignant tumors regardless of tissue origin. It is not clear to me so why many oncologists have difficulty appreciating this concept.

In addition to DER, several small molecules that target aerobic glycolysis or tumor cell energy metabolism are under consideration as novel tumor therapeutics (84, 87, 179, 184–187). Some of these molecules include the following:

1. 2-deoxyglucose (2-DG) (137, 187–201),
2. lonidamine (202–204),

3. 3-bromopyruvate (3-BP) (182, 205–207),
4. imatinib (208, 209),
5. oxythiamine (210–213),
6. 6-aminonicotinamide (186, 187),
7. dichloroacetate (214–219), and
8. resveratrol (50).

The review of Sonveaux and colleagues provides an overview of the many drugs that can be used to target tumor glycolysis (179). Toxicity can become an issue, however, as some of these compounds target pathways other than glycolysis or nucleotide synthesis and high dosages are sometimes required to achieve therapeutic efficacy in vivo (220). For example, a reformulation of the "red wine" drug resveratrol was discontinued as a therapy for multiple myeloma because some patients developed kidney failure (221). This is unfortunate, as resveratrol itself has many health benefits that can improve longevity, reduce inflammation, and help lower blood glucose (222).

Although CR mimetics can be effective in lowering blood glucose levels, further studies will be needed to determine their influence on blood ketones, which would protect normal cell metabolism under low glucose conditions. This is especially important for the brain, which depends heavily on glucose for normal function (8, 15, 84). Brain damage could arise if glucose is targeted without using ketones as a compensatory fuel. It should be possible to effectively target and kill tumor cells using a combination of CR mimetics and the restricted KD as we showed with 2-DG (30). Ketones will protect the brain during hypoglycemia. Ciraolo et al. (223) used a similar metabolic approach in dogs to starve cancer cells that were unable to utilize fat-derived fuels. I have discussed these issues more in Chapter 18.

METFORMIN

Metformin (Glucophage) is a widely used drug for reducing blood glucose levels in patients with type-2 diabetes. The actual mechanism of action is not known, but recent findings from Maria Mihaylova and Ruben Shaw show that metformin lowers blood glucose levels by targeting class IIa histone deacetylases in liver (224). Because of its ability to lower blood glucose levels, metformin has also be considered for treating glucose availability to tumor cells (185, 225–227). In addition to inhibiting gluconeogenesis, metformin also acts like insulin in facilitating glucose uptake into cells. Glucose transporters are upregulated in tumor cells. One side effect of metformin is lactic acidosis. Lactate is produced from glycolysis. Glycolysis drives tumor growth. It is therefore possible that metformin could enhance tumor cell glycolysis in some cancer patients.

Claffey and coworkers showed that metformin had only a marginal effect on the growth of primary breast cancer (triple-negative 66cl4 tumor cells from Balb/c mice), but had no significant effect on the metastatic spread of the tumor cells from fat pads to lung (227). In an attempt to lower blood glucose levels, we administered metformin to tumor-bearing mice that were also treated with CR (40%) (unpublished results). We found that the mice became lethargic and unhealthy in appearance when metformin was administered together with CR. The symptoms were consistent with lactic acidosis, but this was not confirmed.

We also do not think it is good to completely shutdown gluconeogenesis, as glucose is vital for CNS function. We stopped metformin treatment in the tumor-bearing mice under DER because of unacceptable drug toxicity. When used alone in AL-fed mice, however, metformin had no significant influence on brain tumor growth. In other words, our findings in brain tumor-bearing mice treated with metformin were similar to the recent findings of Phoenix and coworkers (227) involving breast cancer. While metformin can be effective in reducing very high blood glucose levels in diabetic and obese patients, I am unsure if metformin will be effective against brain tumor growth in patients. Personally, I would avoid using metformin as a cancer therapy until more extensive preclinical studies are conducted on mice with metastatic and highly invasive cancers.

In contrast to metformin, somatostatin might be more effective in lowering blood glucose levels for cancer patients without producing toxicity (223). Somatostatin targets glucagon naturally to lower blood glucose levels (228). Further studies are needed to assess the therapeutic effects of somatostatin, especially when combined with the KD-R.

SYNERGISTIC INTERACTION OF THE RESTRICTED KETOGENIC DIET (KD-R) AND 2-DEOXYGLUCOSE (2-DG)

Although DER is effective in reducing tumor growth and invasion, this therapeutic approach alone is unlikely to completely eradicate all types of malignant cancers (51, 229). I believe that metabolic diet therapies will be enhanced when combined with drugs that also target glucose energy metabolism. Support for my hypothesis comes from our study showing that the nonmetabolizable glycolysis inhibitor, 2-DG, works synergistically with the KD-R to reduce CT-2A astrocytoma growth (30). 2-DG is readily transported into cells, is phosphorylated by hexokinase, but cannot be metabolized further and thus accumulates in the cell [Fig. 17.24, figure from Aft et al. (188)]. This leads to ATP depletion and the induction of cell death. In this regard, 2-DG has been described as a CR mimetic, that is, a drug that mimics some aspects of CR (194, 200). However, treatment of animal models and cancer patients with relatively high doses of 2-DG (>200 mg/kg) was largely ineffective in managing tumor growth (189, 190, 230). Adverse effects of 2-DG included elevated blood glucose levels, progressive weight loss with lethargy, behavioral symptoms of hypoglycemia, and cardiac vacuolization (187, 189, 190, 220, 230, 231). These findings indicate that 2-DG used alone is not likely to be an effective therapy for most cancers.

Few studies have evaluated the therapeutic efficacy of antiglycolytic or anticancer drugs in combination with DER (51). Recent studies from Safdie

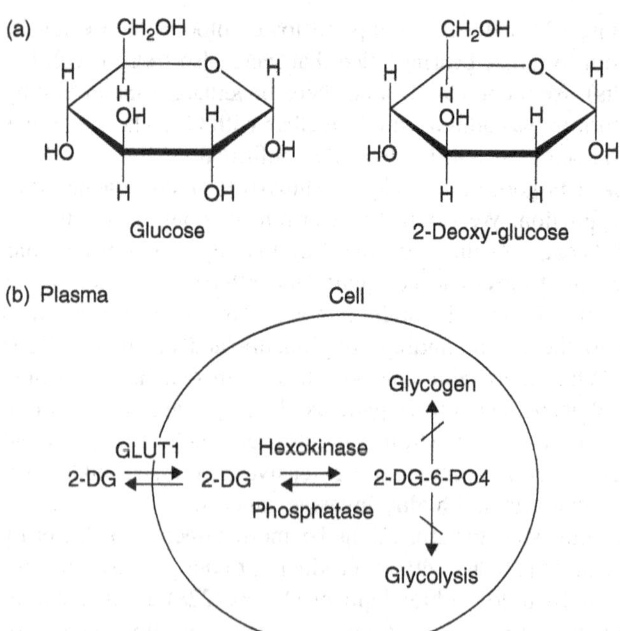

Figure 17.24 Structural comparison of glucose and 2-deoxy-D-glucose. 2-DG and glucose differ at the second carbon (a). Schematic diagram of 2-DG action (b). 2-DG enters the cell through the glucose transporter and is phosphorylated by hexokinase. Owing to low levels of intracellular phosphatase, 2-DG-PO4 is trapped in the cell. 2-DG-PO4 is unable to undergo further metabolism. High intracellular levels of 2-DG-6-PO4 cause allosteric and competitive inhibition of hexokinase. This results in inhibition of glucose metabolism. *Source*: Reprinted with permission from Reference 188.

and the Longo group suggest that CR and fasting can enhance patient health during chemotherapy (26, 232, 233). We were the first to show that the KD-R supplemented with 25 mg/dl of 2-DG was effective in reducing intracerebral tumor growth to a greater extent than was either 2-DG or KD-R when administered alone. These findings showed a powerful synergistic interaction between 2-DG and the diet (Fig. 17.25).

I should mention that some toxicity was seen in our mice treated with the drug–diet combination, as several mice died when given the combination. This was surprising to us, since no toxicity was observed in mice that received either therapy alone. It is unclear whether a similar phenomenon would occur in humans using this drug/diet combination. However, the approximate LD_{50} for 2-DG in humans is about 350 mg/kg (191, 234, 235). The toxicity seen in mice might relate to their high BMR, which is about sevenfold greater than that in humans (30, 37, 70). Yao et al. (236) also recently showed that 2-DG was nontoxic when used alone to treat mice with Alzheimer's disease. Also, I know that Dr. Bomar Herrin used 2-DG (40 mg/dl) with the KD to manage his multiple myeloma. Dr. Herrin has discussed his experiences using this therapeutic strategy in Chapter 20. The process

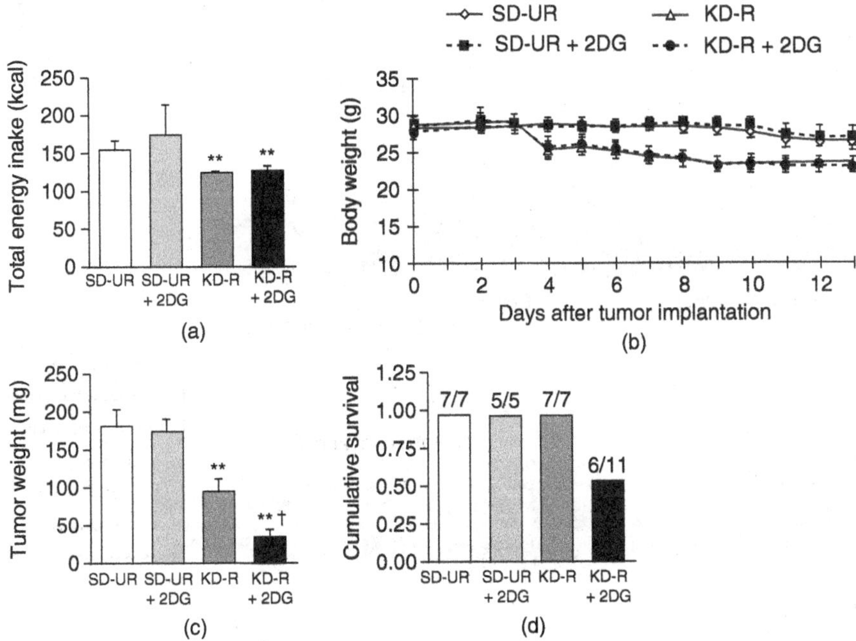

Figure 17.25 Influence of the restricted ketogenic diet with or without 2-DG on total energy intake (a), body weight (b), tumor growth (c), and cumulative survival (d) in mice bearing the orthotopically implanted CT-2A malignant astrocytoma. All mice were fed the standard high carbohydrate rodent diet in unrestricted (UR) amounts for the first 3 days after tumor implantation before their separation into one of four diet groups ($n = 5-11$ mice per group) fed either a standard high carbohydrate diet unrestricted (SD-UR) or a KD-R with or without 2-DG (25 mg/kg) for 10 days. The four groups were matched for body weight. 2-DG was initiated 6 days after tumor implantation and was continued for 7 days (b and c). As shown in (b), the feeding paradigm for the KD-R and KD-R+2-DG groups was designed to reduce body weights by ~20% relative to values recorded before the diet was initiated (3 days after tumor implantation). The average total energy intakes in (a) represent the number of kilocalories consumed by the indicated group over the dietary treatment period (day 3 to day 13). All values are expressed as the mean ± SEM. In (a and c), average values for the indicated group are significantly less than the average value for the SD-UR group at $^{**}P < 0.01$. The mean value for the KD-R+2-DG group is significantly lower than the mean value for the KD-R group at $^{\dagger}P < 0.01$. No significant differences were observed between the SD-UR and SD-UR+2-DG groups throughout the study. For (d), the number of tumor-bearing mice that were alive in each group at the conclusion of the study is listed as a ratio above each solid vertical bar (e.g., the "6/11" indicates that 6 of the 11 original mice were alive at the end of the study in the associated group). *Source*: Reprinted with permission from Reference 30. To see this figure in color please go to ftp://ftp.wiley.com/public/sci_tech_med/cancer_metabolic_disease.

by which the KD-R acts synergistically with 2-DG to reduce tumor is shown in Figure 17.26.

It is my opinion that few of the current CR-mimetic drugs will have major therapeutic effect against most advanced metastatic cancers, especially if used in the absence of some degree of DER. DER will enhance the therapeutic efficacy

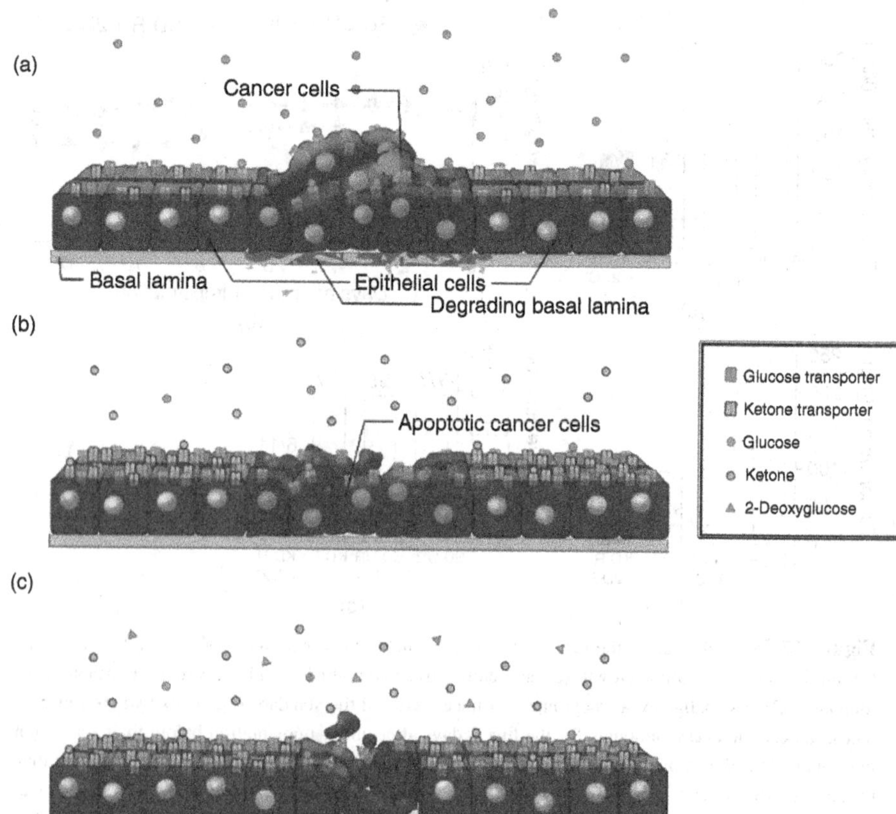

Figure 17.26 Proposed mechanism of tumor management using the KD-R and 2-DG. Tumors will grow fast under normal physiological conditions when glucose is high and ketones are low. Under these conditions, glucose transporter expression is greater in the tumor cells than in the normal cells. Basal lamina degradation is seen under expanding tumor. Tumors will grow slowly under the KD-R when glucose levels become low and ketone levels increase. Glucose transporter expression becomes reduced in tumor cells but becomes elevated in normal cells (Fig. 17.16). The increased glucose demand of normal cells under the KD-R will create increased competition between normal cells and tumor cells for available glucose, which is now in short supply. In addition to enhanced demand for glucose, normal cells increase their ability to metabolize ketone bodies for energy. Ketones gradually replace glucose for energy metabolism in normal cells. Tumor cells cannot switch to ketones for energy due to multiple defects in ketone body metabolism. This situation places much greater metabolic stress on tumor cells than on normal cells. The introduction of low dosage 2-DG will place even greater metabolic stress on tumor cells, as 2-DG will further block glucose utilization in tumor cells. Ketone metabolism will protect normal cells from the adverse effects of hypoglycemia and 2-DG. (a) Fed state, (b) restricted ketogenic diet, and (c) restricted ketogenic diet+2-deoxyglucose. *Source*: Reprinted with permission from Thomas N. Seyfried and Jeffery Ling. See color insert.

of CR-mimetic drugs. CR mimetics will also be more effective against advanced cancer if administered with drugs that also target glutamine, a major fuel for metastatic cancer (229).

I was quite impressed with the apparent antitumor effects of 3-BP presented by Dr. Ko from Peter Pedersen's laboratory (205). How many oncologists have looked at these effects? It is unclear why 3-BP is not in clinical trials at major medical centers. Some have suggested that this drug cannot be used because it is not patentable. Similar arguments were given for the use of dichloroacetate. I think 3-BP and possibly dichloroacetate could be even more effective cancer therapies if combined with if combined with the KD-R. Sometimes novel therapies can be effective without being toxic or expensive. It is time to move these drugs from the culture dish to the cancer patients who might benefit from them!

On the basis of our findings and those from the Longo group, it is clear that therapeutic fasting and DER can enhance the antitumor effects of chemotherapy and help patients tolerate the adverse effects of chemotherapy (26, 232, 233). The administration of antiglycolytic drugs together with energy-restricted diets could act as a powerful double "metabolic punch" for the rapid killing of glycolysis-dependent tumor cells. Combinations of CR mimetics with the restricted KD could open new avenues in cancer drug development, as many drugs that might have minimal therapeutic efficacy or high toxicity when administered alone could become therapeutically relevant and less toxic when combined with energy-restricted diets. Patient advocate groups should request clinical trials using these drug–diet combinations for treatment of advanced cancers.

CAN SYNERGY OCCUR WITH THE KD-R AND HYPERBARIC OXYGEN THERAPY?

We know that the KD-R is an effective therapeutic strategy for managing a broad range of cancers. We also know that synergy occurs between the KD-R and 2-DG for brain cancer management. These observations beg the question as to how the KD-R might work if combined with hyperbaric oxygen therapy (HBO). HBO involves the treatment of subjects at 1–2 atmospheres in 100% oxygen. This question arose during my conversations with Dr. Dominic D'Augostino from the University of South Florida. Dominic is an avid diver and knowledgeable in using HBO. HBO increases oxygen content of tissues and has been used to facilitate wound healing as well as treating several tumor types, including lung, breast, and glioma (237–240). It appears that many types of tumor cells are susceptible to the hyperoxia produced by HBO. Dominic also mentioned to me that HBO "explodes" mitochondria in cultured tumor cells. This would kill any tumor cell with marginal respiratory activity or is dependent on glutamine for mitochondrial fermentation.

I was also surprised to find that the influence of HBO on the tumor growth and vascularity is remarkably similar to the influence of DER on tumor growth. Like DER, hyperoxia targets tumor angiogenesis, while increasing tumor cell apoptosis (238). Stuhr and colleagues showed that HBO reduces microvessel density in

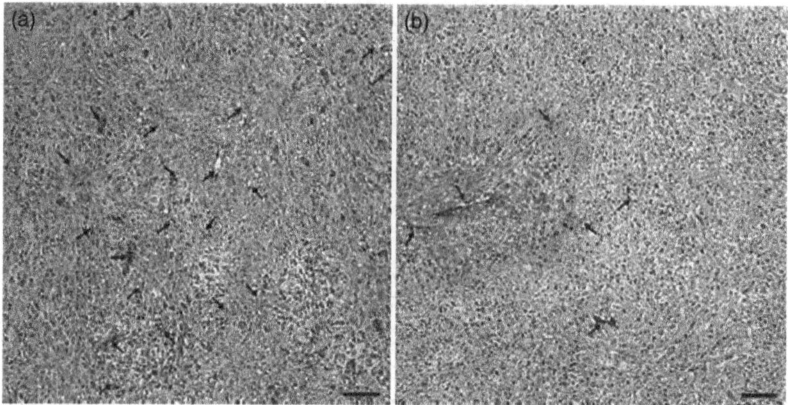

Figure 17.27 Hyperbaric hyperoxia targets glioma angiogenesis. Glioma tissue stained with von Willebrand factor (a) before and (b) after hyperbaric hyperoxia (2 bar, 100% O_2) treatment. Arrows indicate blood vessels. Fewer blood vessels are seen following hyperbaric hyperoxia treatment. The figures are scaled to the same magnification 10×(bar = 100 μm). *Source*: Reprinted with permission from Reference 239. To see this figure in color please go to ftp://ftp.wiley.com/public/sci_tech_med/cancer_metabolic_disease.

gliomas to a similar extent as we showed for DER (Fig. 17.27). This image can be compared with those in Figures 17.9 and 17.10. The antiangiogenic effects of HBO were also seen in other cancers. Besides reducing angiogenesis and increasing apoptosis, HBO also appears to target inflammation (241). Viewed together, these findings suggest that HBO and DER target tumor energy metabolism in similar ways.

How might synergy occur if HBO and the KD-R were to be used as a combined treatment for cancer? Both the KD-R and HBO target tumor glycolysis. The KD-R reduces glucose availability, while HBO targets hexokinase II (237). Hexokinase II is attached to the mitochondria and plays a significant role in stimulating glycolysis. Pedersen has done considerable work in describing the role of hexokinase II in the Warburg effect (206, 242). The restriction of glucose availability will also downregulate the pentose phosphate pathway, which is glucose dependent. The pentose phosphate pathway includes transketolase 1, an enzyme involved in driving glycolysis and the Warburg effect in malignant tumors (212, 243–245).

Besides supporting glycolysis, the pentose phosphate pathway also provides metabolites for DNA synthesis and NADPH for lipid synthesis. NADPH is essential for maintaining catalase activity (246). Catalase is needed to metabolize H_2O_2 to water and O_2. Tumor cells have excessive ROS due to respiratory dysfunction. NADPH depletion would therefore increase the vulnerability of tumor cells to ROS through linkage to catalase reduction. The KD-R would reduce levels of NADPH through the pentose phosphate pathway, thus reducing catalase activity, while HBO would elevate ROS, thus increasing risk of ROS-induced death. Ketones protect against ROS damage in cells with normal respiration because ketone metabolism in mitochondria oxidizes the coenzyme Q couple, thus decreasing the Q semiquinone, a major source of radical production in cells (54) (Fig. 4.4). On the basis of this information, I predict that treatment of cancer patients with a combination of the KD-R and HBO could be a new and effective therapeutic strategy for destroying tumor cells without harming normal cells. Administration of the KD-R with HBO is a practical and simple strategy with potentially powerful therapeutic action against all glycolytic cancers.

TARGETING GLUTAMINE

Although dietary energy restriction and antiglycolytic cancer drugs will have therapeutic efficacy against many tumors that depend on glycolysis and glucose for growth, these therapeutic approaches could be less effective against those tumor cells that also depend heavily on glutamine for energy (84, 144, 247–250). Glutamine is a major energy metabolite for many tumor cells and especially for cells of hematopoietic or myeloid lineage (144, 250–254). This is important as cells of myeloid lineage are considered the origin of many metastatic cancers (Chapter 13). Moreover, glutamine is necessary for the synthesis of those cytokines involved in cancer cachexia, including TNF-α, IL-1, and IL-6 (249, 253, 255, 256). This further indicates a metabolic linkage between metastatic cancer and myeloid cells, for example, macrophages. It therefore becomes important to also consider glutamine targeting for the metabolic management of metastatic cancer.

Although glutamine is widely recognized as a major metabolic fuel for cultured cancer cells (141, 144, 248, 257), there are reports suggesting that glutamine supplementation might be beneficial to some cancer patients (252, 258, 259). On the basis of these reports, it was not clear at the beginning of our experiments whether targeting glutamine would enhance or reduce progression of metastatic cancer. No prior studies to our knowledge had attempted to target glutamine in a natural in vivo mouse model of metastatic cancer (229).

Phenylbutyrate is a relatively nontoxic drug that can lower systemic glutamine levels in humans, but mice are unable to metabolize phenylbutyrate to phenylacetate (260). To lower blood glutamine levels in humans, phenylbutyrate must be first metabolized to phenylacetate. Phenylacetate then binds glutamine and is excreted in the urine as phenylacetylglutamine (261, 262). While phenylbutyrate would be effective in reducing circulating glutamine levels in patients, we were unable to test this treatment in our metastatic VM mouse model. Consequently, we tested the glutamine hypothesis using the glutamine analog drug, 6-diazo-5-oxo-L-norleucine (DON).

As DON had been used previously in mice, we decided to evaluate the influence of DON as a potential therapy for systemic metastatic cancer in our mouse model (229). DON is a glutamine antagonist that inhibits glutamine metabolism (229, 263). DON was effective in reducing colon and lung tumor growth in patients when administered with the glutamine-depleting enzyme PEG-PGA (263). Hence, we considered DON as a drug that could potentially target glutamine availability to reduce systemic metastatic cancer.

GLUTAMINE TARGETING INHIBITS SYSTEMIC METASTASIS

We evaluated the influence of DON and DER (CR in this study) on systemic metastasis using the VM-M3 tumor model. This metastatic VM model has been discussed in Chapters 3 and 13. Leanne Huysentruyt introduced the firefly luciferase gene into the metastatic VM tumor cells allowing for noninvasive detection of tumor growth and metastasis via bioluminescent imaging (159, 264). Figure 17.28 and Table 17.3 shows typical systemic metastatic spread and organ involvement of the VM-M3 tumor. Our experimental design for the DON and DER studies is shown in Figure 17.29a. DON was administered 5 days after flank implantation at an initial dose of 1 mg/kg/day. The body weights of DON-treated mice were similar during drug treatment compared to the control mice (Fig. 17.29b). Blood glucose levels were similar between

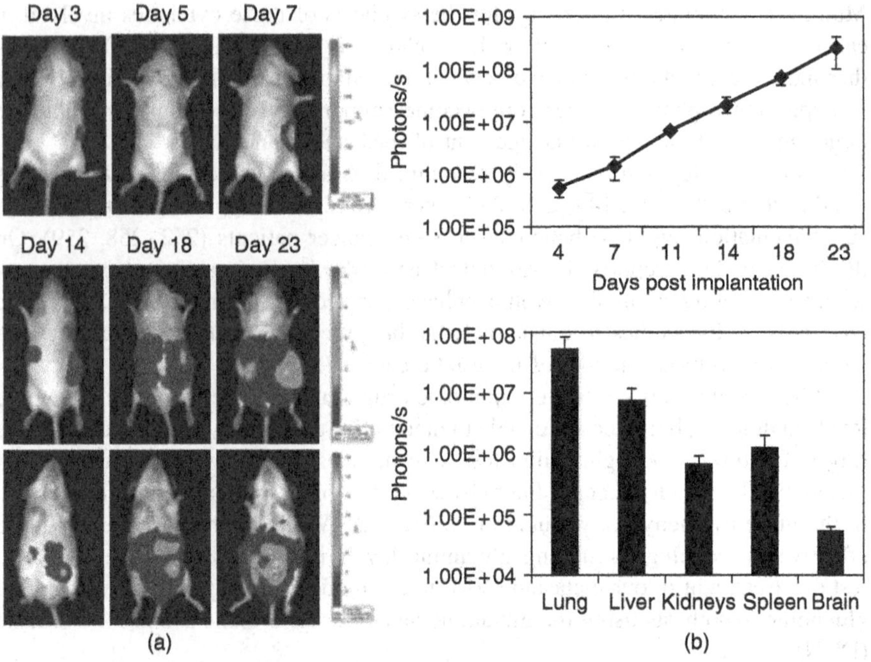

Figure 17.28 Growth and metastatic spread of VM-M3/Fluc tumor with bioluminescence imaging. Tumors were implanted subcutaneously (s.c.) and intracutaneously (i.c.) as we described (159). Multiple metastases were detected *in vivo* following implantation (a). Dorsal (upper panels) images were taken over 23 days. Ventral (lower panels) images were taken once metastasis was detectable (representative mice shown). Bioluminescence from the whole mouse (dorsal and ventral images added together) was quantified and plotted on a log scale (b). Organs were imaged *ex vivo* at the end of the study and metastasis was quantified as we described (c) (159). All values are expressed as the means of six independent samples +95% confidence interval. *Source*: Reprinted with permission from Reference 159. See color insert.

Table 17.3 Percentage of Animals with Metastasis to Organs[a]

Group (n)	Liver	Lung	Kidney	Spleen
control (19)	100	100	47	68
DON (12)	0*	0*	0*	50
DON + CR (11)	0*	0*	0*	27*

[a]The presence of metastasis was detected with bioluminescence imaging.

*The asterisk indicates that the DON or DON + CR group is significantly less than the control group as calculated by chi square analysis at the $p < 0.01$ level.

the DON group and control group, though blood glucose levels were significantly lower in the DER group compared to the control or DON groups (Fig. 17.29c).

Although DON and DER were effective in reducing the size of the primary tumor when grown in the flank of syngeneic VM mice (Fig. 17.30a), the antitumor effect of DON was greater than that of DER. Importantly, DER by itself was unable to prevent metastatic spread to distant organs (Fig. 17.30b). Our findings with the VM-M3 cells differed from the previous findings of Tannenbaum and Claffey and coworkers, who showed that CR or dietary energy restriction was effective in reducing metastatic spread of breast tumor cells to lung (29, 227). In contrast to the studies in breast cancer, we found that DER by itself was unable to reduce metastatic spread to lung or other organs.

It was this observation more than anything else that directed our attention to the possible role of glutamine in driving systemic metastasis. Although DER reduces blood glucose levels, it does not reduce blood glutamine levels. Indeed, blood glutamine levels might increase under DER in mice, as moderate physical activity can increase blood glutamine (72). Mice increase physical activity food foraging under DER. We knew that the VM-M3 tumor cells shared several characteristics with macrophages and that glutamine is a major fuel of immune cells, including macrophages. We also knew that transformed macrophages or their fusion hybrids are the origin of metastatic cancer cells (Chapter 13 and Reference 167). Hence, it would be important to determine if glutamine restriction might reduce systemic metastasis.

We found that the DON prevented metastatic spread to the liver, lung, and kidney (Fig. 17.30b). In addition, we examined liver histology because liver becomes heavily infiltrated with VM-M3 cells. Indeed, liver metastasis was found in 100% of the control mice. Liver is also a common site for many metastatic human cancers. Histological analysis confirmed the lack of tumor cells in the liver of the DON-treated mice in comparison to the control AL nontreated mouse and control and CR-treated groups (Fig. 17.31).

Interestingly, the DON-treated mice showed metastasis to the spleen. The spleen is recognized as a reservoir for monocytes and may represent a sanctuary for the myeloid-like metastatic cells (265). Previous studies showed increases in glutaminase activity in the spleens of tumor-bearing mice (266). Glutaminase

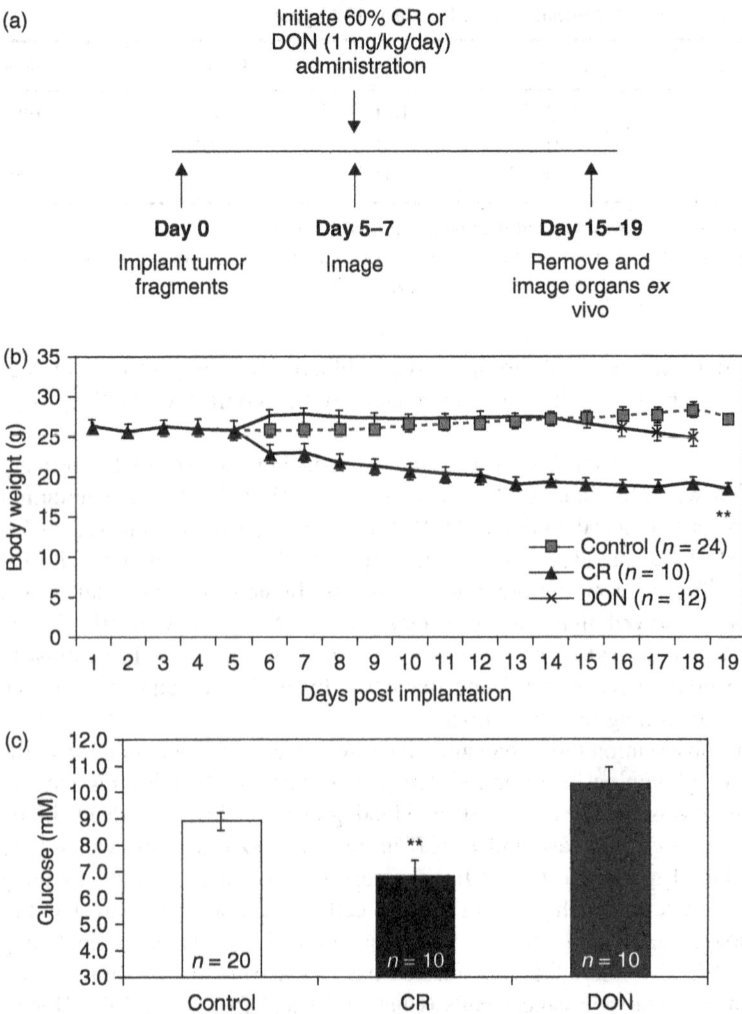

Figure 17.29 Experimental design and effect of calorie restriction or DON on body weights and blood glucose. (a) The injection and treatment protocol were as we described (229). (b) Body weights were monitored daily. Before treatment, the body weights of all mice were averaged for a single value. (c) Mice were sacrificed 15–19 days post implantation and blood was collected for the analysis of glucose levels. Values represent the mean ± SEM of 10–20 mice per group. The asterisks indicate that the CR values differ significantly from the AL control group at $P < 0.01$. *Source:* Reprinted with permission from Reference 229.

is the first enzyme involved in glutamine metabolism. We suggest that the spleen might be able to support metastatic VM-M3 cells due to an influx of glutamine, originally intended to support immune function (229, 252). Further studies will be necessary to determine if spleen sanctuary is unique to VM-M3 metastatic cells or is a general characteristic of other metastatic cancers.

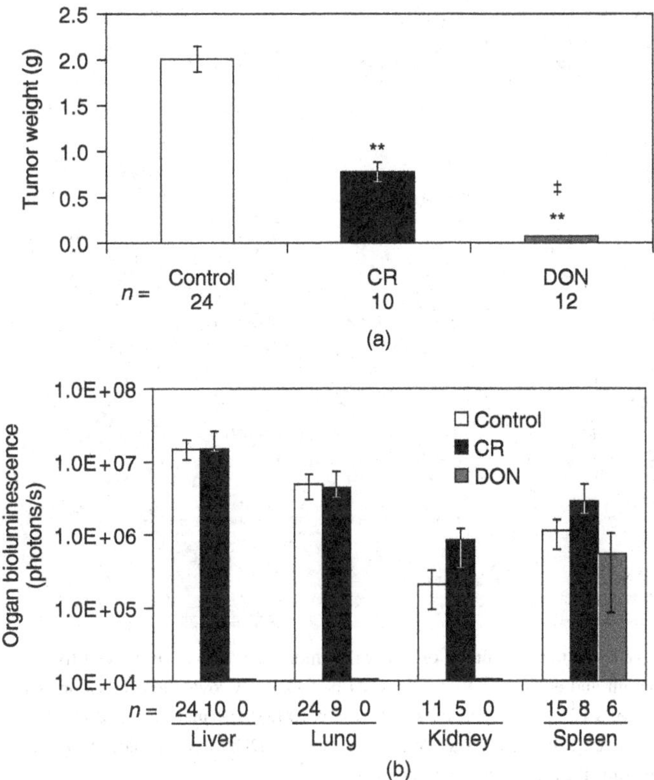

Figure 17.30 Effect of calorie restriction (CR) or DON on VM-M3 tumor growth and metastasis. (a) Mice were implanted subcutaneously (s.c.) with the VM-M3/Fluc tumor as we described (229). Mice were sacrificed 15–19 days post implantation and the tumors removed and weighed. The asterisks indicate that the CR or DON values differ significantly from the AL control group at $P < 0.01$. The symbol "‡" indicates that the DON values differ significantly from the CR values at $P < 0.01$. (b) At the time of sacrifice, the organs were removed and imaged *ex vivo*. Bioluminescence values were plotted on a log scale. All values represent the mean six SEM of 6–24 mice per group. No detectable bioluminescence was found in the lung, liver, or kidney of DON-treated mice. *Source:* Reprinted with permission from Reference 229.

Tolerance is generally good for human patients treated with DON and glutaminase inhibitor PEG-PGA (263). In contrast, we found that DON was toxic to the VM mice (229). Although DON was effective in reducing metastasis to liver, lung, and kidney, survival was less in the DON-treated mice than in the CR-treated mice (Fig. 17.32). Because glucose and glutamine act synergistically to drive VM-M3 growth in vitro, Laura Shelton and I developed a diet/drug combination to determine if CR and DON treatment could act synergistically to reduce systemic metastasis in vivo (72). We found that DON treatment, either alone or in combination with CR, significantly reduced tumor growth and metastasis. Moreover, less DON was used to achieve therapeutic effect in the DON+CR mice than in the

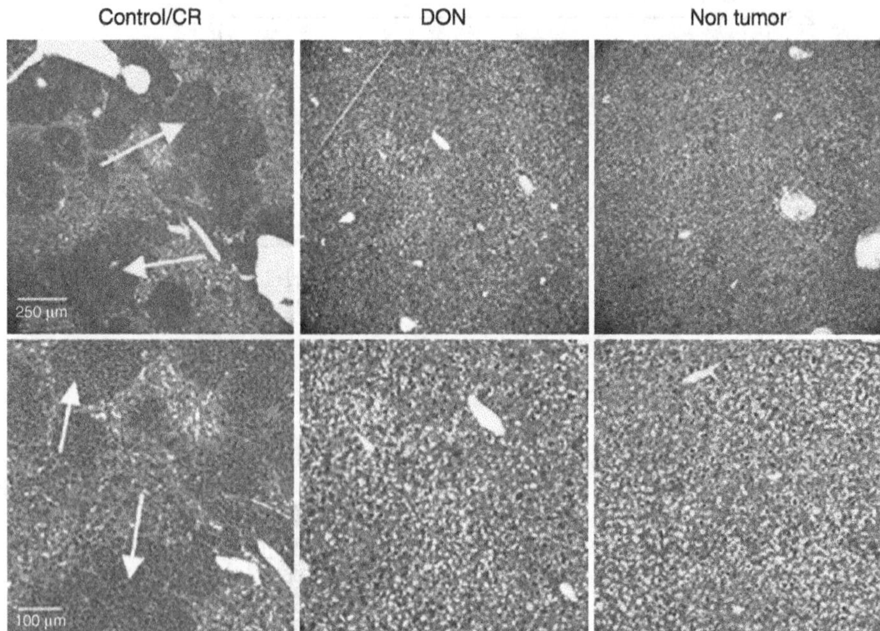

Figure 17.31 DON prevents metastatic spread of VM-M3 tumor cells to liver. Removed livers were stained with hematoxylin and eosin (H&E) as we described (229). Arrows indicate secondary tumor lesions in the control and CR group. Images are shown at 100×(top panel) and 200×(bottom panel). No evidence of metastatic tumor cells was detected in liver of DON-treated mice. *Source*: Reprinted with permission from Reference 229. See color insert.

mice treated with DON alone. We reduced the DON doses for the mice in order to reduce potential toxicities or extreme energy stress.

The mice treated with DON+CR were active throughout the study and maintained a healthy body weight. Interestingly, the mice on DON treatment alone showed a more adverse response to drug treatment than did the mice on DON+CR. The mice treated with DON alone dropped body weight and were lethargic over the last 3 days of the study (72). Toxicity in the mice treated with DON alone became more evident as the study progressed. Toxicity was reduced and survival was enhanced when DON was administered together with CR, as a lower drug dosage was needed to achieve therapeutic effect. Moreover, the incidence of metastasis to spleen was significantly lower in the DON+CR mice than in the mice treated with DON alone (Table 17.3). This implies that both glucose and glutamine are major energy metabolites for the VM-M3 tumor cells. These findings should stimulate renewed interest in DON and other glutamine targeting drugs for the treatment for human metastatic cancer, especially when combined with limited caloric intake and with other drugs that target glucose metabolism.

We also suspect that DON might be more effective in targeting metastasis in humans than in mice, especially if DON is administered together with

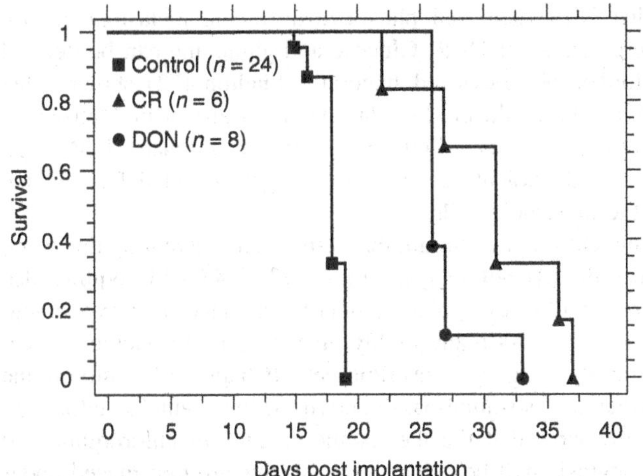

Figure 17.32 Influence of DON or CR on mouse survival. VM mice were implanted subcutaneously (s.c.) with the VM-M3/Fluc tumor as described (229). All control mice reached morbidity 15–19 days post implantation. Both the CR and DON-treated mice survived significantly longer than the untreated controls. Survival in the CR mice was not significantly longer than the DON-treated mice. *Source*: Reprinted with permission from Reference 229.

calorie-restricted diets, such as the KD-R. Glutamine restriction can increase glucose metabolism (28). This should not be a problem, as the KD-R targets glucose metabolism. Viewed collectively, these studies prove principle that targeting glutamine can be a powerful therapy for managing systemic metastatic cancer.

TARGETING PHAGOCYTOSIS

As mentioned in Chapter 13, phagocytosis is a characteristic of many types of metastatic cancers. Several investigators suggested that tumor cell phagocytosis could be targeted as a potential therapy for metastatic cancers (167). For example, Ghoneum and Gollapudi (267, 268) showed that MCF-7 breast cancer cells undergo apoptosis after engulfing yeast cells either in vitro or in vivo. Phagocytosis of yeast cells also effectively induces apoptosis in human cancers of the gastrointestinal tract, including tongue, squamous cell carcinoma, and colon adenocarcinoma (269). These reports suggest that the phagocytic behavior of metastatic tumor cells can be targeted for the development of new antimetastasis therapies.

Additionally, the phagocytic activity of metastatic melanoma cells is significantly increased when the cells are grown under low glucose conditions, suggesting that metastatic cells use phagocytosis as a way to "feed" when nutrient supplies are low (270, 271) (Chapter 13). These observations suggest that targeting phagocytosis could be effective in reducing metastasis of some cancers.

Although DER reduces circulating glucose levels, it did not reduce metastasis in our VM metastatic model. DER increases macrophage phagocytosis (272). It is therefore possible that DER increased phagocytosis in the metastatic mouse cells, thus reducing energy stress of DER. Glucose and glutamine can be derived from lysosomal digestion of phagocytosed materials. Shelton (72) showed that the metastatic VM tumor cells could produce lactate when grown in Matrigel in minimal media containing no glucose. Cells grown in the absence of Matrigel produced no lactate and died, indicating that the Matrigel provided fermentable energy metabolites for the metastatic cells.

The antimalarial drug, chloroquine, might be useful in circumventing this problem. Chloroquine reduces the pH within lysosomes (273, 274). Chloroquine also has demonstrated therapeutic efficacy against human brain cancer and experimental pancreatic cancer (275, 276). As many highly invasive and metastatic cancers can be derived from naturally phagocytic myeloid cells (Chapter 13), chloroquine could be effective in reducing lysosomal-based activities, for example, autophagy and phagocytosis. It is my view that the therapeutic efficacy of chloroquine and other potential antiphagocytosis will be enhanced if patients are first placed under DER.

TARGETING THE MICROENVIRONMENT

Some tumors behave as wounds that do not heal (277). Bissell and Hines (146) recently provided a compelling and provocative discussion on the role of the microenvironment in the initiation and progression of tumors. They showed that the microenvironment maintains tissue architecture, thus inhibiting cell growth and suppressing the malignant phenotype of cancer cells. Growth factors and cytokines released by fibroblasts and macrophages, cells programmed to heal wounds, can actually provoke chronic inflammation and tumor progression (146, 278, 279). Part of the wound healing process also involves degradation of the extracellular matrix and enhancement of angiogenesis, which further contribute to tumor progression (278, 280). Bissell and Hines (146) state, "*it has become clear that targeting the cells and components of the microenvironment is likely to provide profound clinical benefits.*" Our findings using DER provide direct support for their prediction.

As I have discussed in this chapter, DER targets inflammation in the tumor microenvironment and the signaling pathways involved with driving tumor angiogenesis and inflammation (21, 77). Kari and coworkers previously showed that CR significantly reduced the inflammatory properties of alveolar lung macrophages in response to *Streptococcus* infection (272). The transition in energy metabolism from glucose to ketones is powerfully anti-inflammatory to the tumor microenvironment. This is one reason why the KD is under consideration for use in numerous neurological and neurodegenerative diseases where inflammation is part of the pathology (164, 281, 282).

As DER is a systemic therapy that simultaneously targets both the tumor cells as well as the tumor microenvironment, this approach can be effective in retarding progression of many cancers. There is no drug therapy that I am aware of that can target as many proinflammatory mechanisms in the microenvironment as can DER.

I think real progress in tumor management will be achieved once patients and the oncology community come to recognize this fact.

DIETARY ENERGY REDUCTION AS A MITOCHONDRIAL ENHANCEMENT THERAPY (MET)

I have used the term *dietary energy reduction* to describe those therapies that use reduced dietary energy intake to treat cancer. This term is more appealing than the terms *calorie restriction* or *dietary restriction*. While these terms might be acceptable when discussing treatments for animal cancers, they are less appropriate in discussions for treating human cancer patients. Humans, and especially cancer patients, do want restrictions. These people are already suffering enough form their disease. The word "restriction" connotes a negative approach to their treatment. Even the term *dietary energy reduction* has a negative connotation. What would be a better term to convey a more positive approach to cancer management than terms involving food reduction or restriction? I suggest we consider the term *mitochondrial enhancement therapy* (MET) to denote this cancer therapeutic strategy in humans.

Indeed, the transition from glucose metabolism to ketone metabolism will reduce tissue inflammation and ROS, while enhancing the metabolic efficiency of the mitochondria. MET is a more appealing term for cancer management in humans because it eliminates terms involving dietary restrictions. MET is also more accurate, as this term addresses the mechanism by which the therapy actually works. It is my opinion that MET will eventually be recognized as the most effective therapeutic strategy for managing and preventing cancer in humans.

SUMMARY

This chapter summarizes the vast information showing that targeting energy metabolism can manage tumor growth and cancer. Because all tumor cells suffer common defects in their ability to process energy, they become highly vulnerable to therapies that target these processes. OxPhos insufficiency leads to a greater dependence on fermentation for survival. Therapies that target cancer cell fermentation will go far in managing the disease. Besides using drugs that target tumor cell fermentation, normal body cells can also be used to indirectly target the metabolically challenged tumor cells. Tumor cells are less metabolically flexible than normal cells. Hence, nontoxic metabolic therapies that exploit the energy flexibility of normal cells will target the entire microenvironment, thus driving tumor cells to extinction.

REFERENCES

1. Hursting SD, Kari FW. The anti-carcinogenic effects of dietary restriction: mechanisms and future directions. Mutat Res. 1999;443:235–49.

2. Jose DG, Good RA. Quantitative effects of nutritional protein and calorie deficiency upon immune responses to tumors in mice. Cancer Res. 1973;33:807–12.
3. Wheatley KE, Williams EA, Smith NC, Dillard A, Park EY, Nunez NP, et al. Low-carbohydrate diet versus caloric restriction: effects on weight loss, hormones, and colon tumor growth in obese mice. Nutr Cancer. 2008;60:61–8.
4. Mukherjee P, Sotnikov AV, Mangian HJ, Zhou JR, Visek WJ, Clinton SK. Energy intake and prostate tumor growth, angiogenesis, and vascular endothelial growth factor expression. J Natl Cancer Inst. 1999;91:512–23.
5. Kari FW, Dunn SE, French JE, Barrett JC. Roles for insulin-like growth factor-1 in mediating the anti-carcinogenic effects of caloric restriction. J Nutr Health Aging. 1999;3:92–101.
6. Mavropoulos JC, Buschemeyer WC 3rd, Tewari AK, Rokhfeld D, Pollak M, Zhao Y, et al. The effects of varying dietary carbohydrate and fat content on survival in a murine LNCaP prostate cancer xenograft model. Cancer Prev Res (Phila). 2009;2:557–65.
7. Bonorden MJ, Rogozina OP, Kluczny CM, Grossmann ME, Grambsch PL, Grande JP, et al. Intermittent calorie restriction delays prostate tumor detection and increases survival time in TRAMP mice. Nutr Cancer. 2009;61:265–75.
8. Seyfried TN, Mukherjee P. Targeting energy metabolism in brain cancer: review and hypothesis. Nutr Metab. 2005;2:30.
9. Thompson HJ, McGinley JN, Spoelstra NS, Jiang W, Zhu Z, Wolfe P. Effect of dietary energy restriction on vascular density during mammary carcinogenesis. Cancer Res. 2004;64:5643–50.
10. Kritchevsky D. Caloric restriction and experimental carcinogenesis. Toxicol Sci. 1999;52:13–6.
11. Tannenbaum A. Nutrition and Cancer. In: Homburger F, editor. Physiopathology of Cancer. New York: Paul B. Hober; 1959. p. 517–62.
12. Hursting SD, Lavigne JA, Berrigan D, Perkins SN, Barrett JC. Calorie restriction, aging, and cancer prevention: mechanisms of action and applicability to humans. Annu Rev Med. 2003;54:131–52.
13. Hursting SD, Smith SM, Lashinger LM, Harvey AE, Perkins SN. Calories and carcinogenesis: lessons learned from 30 years of calorie restriction research. Carcinogenesis. 2010;31:83–9.
14. Kritchevsky D. Caloric restriction and cancer. J Nutr Sci Vitaminol. 2001;47:13–9.
15. Seyfried TN, Sanderson TM, El-Abbadi MM, McGowan R, Mukherjee P. Role of glucose and ketone bodies in the metabolic control of experimental brain cancer. Br J Cancer. 2003; 89:1375–82.
16. Shelton HM. Fasting for Renewal of Life. Tampa (FL): American National Hygiene Society, Inc; 1974.
17. Greene AE, Todorova MT, Seyfried TN. Perspectives on the metabolic management of epilepsy through dietary reduction of glucose and elevation of ketone bodies. J Neurochem. 2003;86: 529–37.
18. Kritchevsky D. Fundamentals of Nutrition: Applications to Cancer Research. In: Heber D, Blackburn GL, Go VLW, editors. Nutritional Oncology. Boston: Academic Press; 1999. p. 5–10.
19. Weindruch R, Walford RL. The retardation of aging and disease by dietary restriction. Springfield (IL): Thomas; 1988.
20. Tannenbaum A. The genesis and growth of tumors: II. Effects of caloric restriction per se. Cancer Res. 1942;2:460–7.
21. Mulrooney TJ, Marsh J, Urits I, Seyfried TN, Mukherjee P. Influence of caloric restriction on constitutive expression of NF-kappaB in an experimental mouse astrocytoma. PloS One. 2011;6:e18085.
22. Duan W, Lee J, Guo Z, Mattson MP. Dietary restriction stimulates BDNF production in the brain and thereby protects neurons against excitotoxic injury. J Mol Neurosci. 2001;16:1–12.
23. Spindler SR. Rapid and reversible induction of the longevity, anticancer and genomic effects of caloric restriction. Mech Ageing Dev. 2005;126:960–6.
24. Chung HY, Kim HJ, Kim KW, Choi JS, Yu BP. Molecular inflammation hypothesis of aging based on the anti-aging mechanism of calorie restriction. Microsc Res Tech. 2002;59:264–72.

25. Birt DF, Yaktine A, Duysen E. Glucocorticoid mediation of dietary energy restriction inhibition of mouse skin carcinogenesis. J Nutr. 1999;129:571S–4S.
26. Raffaghello L, Safdie F, Bianchi G, Dorff T, Fontana L, Longo VD. Fasting and differential chemotherapy protection in patients. Cell Cycle. 2010;9:4474–6.
27. Meadows GG, Fu Y-M. Dietary Restriction of Specific Amino Acids Modulates Tumor and Host Interactions. In: Meadows GG, editor. Integration/Interaction of Oncologic Growth. 2nd ed. New York: Kluwer Academic; 2005; p. 271–83.
28. Liu X, Fu YM, Meadows GG. Differential effects of specific amino acid restriction on glucose metabolism, reduction/oxidation status and mitochondrial damage in DU145 and PC3 prostate cancer cells. Oncol Lett. 2011;2:349–55.
29. Tannenbaum A, Silverstone H. Effect of limited food intake on survival of mice bearing spontaneous mammary carcinoma and on the incidence of lung metastases. Cancer Res. 1953;13:532–36.
30. Marsh J, Mukherjee P, Seyfried TN. Drug/diet synergy for managing malignant astrocytoma in mice: 2-deoxy-D-glucose and the restricted ketogenic diet. Nutr Metab. 2008;5:33.
31. Mukherjee P, Abate LE, Seyfried TN. Antiangiogenic and proapoptotic effects of dietary restriction on experimental mouse and human brain tumors. Clin Cancer Res. 2004;10:5622–9.
32. Mukherjee P, El-Abbadi MM, Kasperzyk JL, Ranes MK, Seyfried TN. Dietary restriction reduces angiogenesis and growth in an orthotopic mouse brain tumour model. Br J Cancer. 2002;86:1615–21.
33. Seyfried TN, Mukherjee P. Anti-Angiogenic and Pro-Apoptotic Effects of Dietary Restriction in Experimental Brain Cancer: Role of Glucose and Ketone Bodies. In: Meadows GG, editor. Integration/Interaction of Oncologic Growth. 2nd ed. New York: Kluwer Academic; 2005. p. 259–70.
34. Zhou W, Mukherjee P, Kiebish MA, Markis WT, Mantis JG, Seyfried TN. The calorically restricted ketogenic diet, an effective alternative therapy for malignant brain cancer. Nutr Metab. 2007;4:5.
35. Ho VW, Leung K, Hsu A, Luk B, Lai J, Shen SY, et al. A low carbohydrate, high protein diet slows tumor growth and prevents cancer initiation. Cancer Res. 2011;71:4484–93.
36. Mantis JG, Centeno NA, Todorova MT, McGowan R, Seyfried TN. Management of multifactorial idiopathic epilepsy in EL mice with caloric restriction and the ketogenic diet: role of glucose and ketone bodies. Nutr Metab. 2004;1:11.
37. Mahoney LB, Denny CA, Seyfried TN. Caloric restriction in C57BL/6J mice mimics therapeutic fasting in humans. Lipids Health Dis. 2006;5:13.
38. Cahill GF Jr, Veech RL. Ketoacids? Good medicine? Trans Am Clin Climatol Assoc. 2003;114: 149–61. discussion 62–3.
39. Veech RL, Chance B, Kashiwaya Y, Lardy HA, Cahill GF Jr. Ketone bodies, potential therapeutic uses. IUBMB Life. 2001;51:241–7.
40. Williamson DH, Mellanby J, Krebs HA. Enzymic determination of D(−)-beta-hydroxybutyric acid and acetoacetic acid in blood. Biochem J. 1962;82:90–6.
41. Krebs HA, Williamson DH, Bates MW, Page MA, Hawkins RA. The role of ketone bodies in caloric homeostasis. Adv Enzyme Reg. 1971;9:387–409.
42. Cahill GF Jr. Fuel metabolism in starvation. Annu Rev Nutr. 2006;26:1–22.
43. Stafstrom CE, Rho JM. Epilepsy and the Ketogenic Diet. Totowa (NJ): Humana Press; 2004.
44. Hartman AL, Vining EP. Clinical aspects of the ketogenic diet. Epilepsia. 2007;48:31–42.
45. Zuccoli G, Marcello N, Pisanello A, Servadei F, Vaccaro S, Mukherjee P, et al. Metabolic management of glioblastoma multiforme using standard therapy together with a restricted ketogenic diet: case report. Nutr Metab. 2010;7:33.
46. Nebeling LC, Miraldi F, Shurin SB, Lerner E. Effects of a ketogenic diet on tumor metabolism and nutritional status in pediatric oncology patients: two case reports. J Am Coll Nutr. 1995;14:202–8.
47. Bear MF,Connors BW, Paradiso MA. Neuroscience: Exploring the Brain. 2nd ed. Baltimore, MD: Lippincot Williams & Wilkins; 2001.
48. Greene AE, Todorova MT, McGowan R, Seyfried TN. Caloric restriction inhibits seizure susceptibility in epileptic EL mice by reducing blood glucose. Epilepsia. 2001;42:1371–8.

49. Kossoff EH. International consensus statement on clinical implementation of the ketogenic diet: agreement, flexibility, and controversy. Epilepsia. 2008;49(Suppl 8):11–3.
50. Wittig R, Coy JF. The role of glucose metabolism and glucose-associated signaling in cancer. Persp Med Chem. 2007;1:64–82.
51. Seyfried TN, Kiebish MA, Marsh J, Shelton LM, Huysentruyt LC, Mukherjee P. Metabolic management of brain cancer. Biochim Biophys Acta. 2010;1807:577–94.
52. Stafford P, Abdelwahab MG, Kim do Y, Preul MC, Rho JM, Scheck AC. The ketogenic diet reverses gene expression patterns and reduces reactive oxygen species levels when used as an adjuvant therapy for glioma. Nutr Metab. 2010;7:74.
53. Wheless JW. The ketogenic diet: an effective medical therapy with side effects. J Child Neurol. 2001;16:633–5.
54. Veech RL. The therapeutic implications of ketone bodies: the effects of ketone bodies in pathological conditions: ketosis, ketogenic diet, redox states, insulin resistance, and mitochondrial metabolism. Prostaglandins Leukot Essent Fatty Acids. 2004;70:309–19.
55. Rebrin I, Kamzalov S, Sohal RS. Effects of age and caloric restriction on glutathione redox state in mice. Free Radical Biol Med. 2003;35:626–35.
56. Weindruch R, Walford RL, Fligiel S, Guthrie D. The retardation of aging in mice by dietary restriction: longevity, cancer, immunity and lifetime energy intake. J Nutr. 1986;116:641–54.
57. Jarrett SG, Milder JB, Liang LP, Patel M. The ketogenic diet increases mitochondrial glutathione levels. J Neurochem. 2008;106:1044–51.
58. Kim DY, Vallejo J, Rho JM. Ketones prevent synaptic dysfunction induced by mitochondrial respiratory complex inhibitors. J Neurochem. 2010;114:130–41.
59. Haces ML, Hernandez-Fonseca K, Medina-Campos ON, Montiel T, Pedraza-Chaverri J, Massieu L. Antioxidant capacity contributes to protection of ketone bodies against oxidative damage induced during hypoglycemic conditions. Exp Neurol. 2008;211:85–96.
60. Kim DY, Davis LM, Sullivan PG, Maalouf M, Simeone TA, Brederode JV, et al. Ketone bodies are protective against oxidative stress in neocortical neurons. J Neurochem. 2007;101:1316–26.
61. Yamada KA, Rensing N, Thio LL. Ketogenic diet reduces hypoglycemia-induced neuronal death in young rats. Neurosci Lett. 2005;385:210–4.
62. Masuda R, Monahan JW, Kashiwaya Y. D-beta-Hydroxybutyrate is neuroprotective against hypoxia in serum-free hippocampal primary cultures. J Neurosci Res. 2005;80:501–9.
63. Imamura K, Takeshima T, Kashiwaya Y, Nakaso K, Nakashima K. D-beta-Hydroxybutyrate protects dopaminergic SH-SY5Y cells in a rotenone model of Parkinson's disease. J Neurosci Res. 2006;84:1376–84.
64. Kashiwaya Y, Takeshima T, Mori N, Nakashima K, Clarke K, Veech RL. D-beta-Hydroxybutyrate protects neurons in models of Alzheimer's and Parkinson's disease. Proc Natl Acad Sci USA. 2000;97:5440–4.
65. Guzman M, Blazquez C. Ketone body synthesis in the brain: possible neuroprotective effects. Prostaglandins Leukot Essent Fatty Acids. 2004;70:287–92.
66. Wallace DC. A mitochondrial paradigm of metabolic and degenerative diseases, aging, and cancer: a dawn for evolutionary medicine. Annu Rev Genet. 2005;39:359–407.
67. Owen OE, Morgan AP, Kemp HG, Sullivan JM, Herrera MG, Cahill GF Jr. Brain metabolism during fasting. J Clin Invest. 1967;46:1589–95.
68. Bhagavan NV. Medical Biochemistry. 4th ed. New York: Harcourt; 2002.
69. Kashiwaya Y, Pawlosky R, Markis W, King MT, Bergman C, Srivastava S, et al. A ketone ester diet increased brain malonyl CoA and uncoupling protein 4 and 5 while decreasing food intake in the normal Wistar rat. J Biol Chem. 2010;285:25950–6.
70. Terpstra AH. Differences between humans and mice in efficacy of the body fat lowering effect of conjugated linoleic acid: role of metabolic rate. J Nutr. 2001;131:2067–8.
71. Sato K, Kashiwaya Y, Keon CA, Tsuchiya N, King MT, Radda GK, et al. Insulin, ketone bodies, and mitochondrial energy transduction. FASEB J. 1995;9:651–8.
72. Shelton LM. Targeting Energy Metabolism in Brain Cancer. Chestnut Hill (MA): Boston College; 2010.

73. MAURER GD, BRUCKER DP, BAEHR O, HARTER PN, HATTINGEN E, WALENTA S, et al. Differential utilization of ketone bodies by neurons and glioma cell lines: a rationale for ketogenic diet as experimental glioma therapy. BMC Cancer. 2011;11:315.
74. SKINNER R, TRUJILLO A, MA X, BEIERLE EA. Ketone bodies inhibit the viability of human neuroblastoma cells. J Pediatr Surg. 2009;44:212–6. discussion 6.
75. TISDALE MJ, BRENNAN RA. Loss of acetoacetate coenzyme A transferase activity in tumours of peripheral tissues. Br J Cancer. 1983;47:293–7.
76. FREDERICKS M, RAMSEY RB. 3-Oxo acid coenzyme a transferase activity in brain and tumors of the nervous system. J Neurochem. 1978;31:1529–31.
77. MARSH J, MUKHERJEE P, SEYFRIED TN. Akt-dependent proapoptotic effects of dietary restriction on late-stage management of a phosphatase and tensin homologue/tuberous sclerosis complex 2-deficient mouse astrocytoma. Clin Cancer Res. 2008;14:7751–62.
78. TISDALE MJ. Role of acetoacetyl-CoA synthetase in acetoacetate utilization by tumor cells. Cancer Biochem Biophys. 1984;7:101–7.
79. SEYFRIED TN. Mitochondrial glutamine fermentation enhances ATP synthesisin murine glioblastoma cells. Proceedings of the 102nd Annual Meeting of the American Association of Cancer Research, Orlando (FL); 2011.
80. WARBURG O. Revidsed Lindau Lectures: The prime cause of cancer and prevention - Parts 1 & 2. In: Burk D, editor. Meeting of the Nobel-Laureates Lindau, Lake Constance, Germany: K.Triltsch; 1969. p. http://www.hopeforcancer.com/OxyPlus.htm.
81. PENDERGRASS TW, MILSTEIN JM, GEYER JR, MULNE AF, KOSNIK EJ, MORRIS JD, et al. Eight drugs in one day chemotherapy for brain tumors: experience in 107 children and rationale for preradiation chemotherapy. J Clin Oncol (Official J Am Soc Clin Oncol). 1987;5:1221–31.
82. NEBELING LC, LERNER E. Implementing a ketogenic diet based on medium-chain triglyceride oil in pediatric patients with cancer. J Am Diet Assoc. 1995;95:693–7.
83. SCHMIDT M, PFETZER N, SCHWAB M, STRAUSS I, KAMMERER U. Effects of a ketogenic diet on the quality of life in 16 patients with advanced cancer: a pilot trial. Nutr Metab. 2011;8:54.
84. MATHEWS EH, LIEBENBERG L, PELZER R. High-glycolytic cancers and their interplay with the body's glucose demand and supply cycle. Med Hypotheses. 2011;76:157–65.
85. WARBURG O. On the origin of cancer cells. Science. 1956;123:309–14.
86. MCKENNA MC, GRUETTER R, SONNEWALD U, WAAGEPETERSEN HS, SCHOUSBOE A. Energy Metabolism of the Brain. In: SIEGEL GJ, ALBERS RW, BRADEY ST, PRICE DP, editors. Basic Neurochemistry: Molecular, Cellular, and Medical Aspects. New York: Elsevier Academic Press; 2006. p. 531–57.
87. LOPEZ-LAZARO M. A new view of carcinogenesis and an alternative approach to cancer therapy. Mol Med. 2010;16:144–53.
88. SHAW RJ. Glucose metabolism and cancer. Curr Opin Cell Biol. 2006;18:598–608.
89. DERR RL, YE X, ISLAS MU, DESIDERI S, SAUDEK CD, GROSSMAN SA. Association between hyperglycemia and survival in patients with newly diagnosed glioblastoma. J Clin Oncol. 2009;27: 1082–6.
90. MCGIRT MJ, CHAICHANA KL, GATHINJI M, ATTENELLO F, THAN K, RUIZ AJ, et al. Persistent outpatient hyperglycemia is independently associated with decreased survival after primary resection of malignant brain astrocytomas. Neurosurgery. 2008;63:286–91. discussion 91.
91. WOODS SC, STRICKER EM. Food Intake and Metabolism. In: ZIGMOND MJ, editor. Fundemental Neuroscience. New York: Academic Press; 1999. p. 1091–109.
92. PELICANO H, XU RH, DU M, FENG L, SASAKI R, CAREW JS, et al. Mitochondrial respiration defects in cancer cells cause activation of Akt survival pathway through a redox-mediated mechanism. J Cell Biol. 2006;175:913–23.
93. YOUNG CD, ANDERSON SM. Sugar and fat—that's where it's at: metabolic changes in tumors. Breast Cancer Res. 2008;10:202.
94. THOMPSON HJ, JIANG W, ZHU Z. Mechanisms by which energy restriction inhibits carcinogenesis. Adv Exp Med Biol. 1999;470:77–84.
95. THOMPSON HJ, ZHU Z, JIANG W. Dietary energy restriction in breast cancer prevention. J Mammary Gland Biol Neoplasia. 2003;8:133–42.

96. THOMPSON HJ, ZHU Z, JIANG W. Identification of the apoptosis activation cascade induced in mammary carcinomas by energy restriction. Cancer Res. 2004;64:1541–5.
97. ZHU Z, JIANG W, MCGINLEY J, WOLFE P, THOMPSON HJ. Effects of dietary energy repletion and IGF-1 infusion on the inhibition of mammary carcinogenesis by dietary energy restriction. Mol Carcinog. 2005;42:170–6.
98. DUNN SE, KARI FW, FRENCH J, LEININGER JR, TRAVLOS G, WILSON R, et al. Dietary restriction reduces insulin-like growth factor I levels, which modulates apoptosis, cell proliferation, and tumor progression in p53-deficient mice. Cancer Res. 1997;57:4667–72.
99. JAMES SJ, MUSKHELISHVILI L, GAYLOR DW, TURTURRO A, HART R. Upregulation of apoptosis with dietary restriction: implications for carcinogenesis and aging. Environ Health Perspect. 1998;106:307–12.
100. HAGOPIAN K, RAMSEY JJ, WEINDRUCH R. Influence of age and caloric restriction on liver glycolytic enzyme activities and metabolite concentrations in mice. Exp Gerontol. 2003;38:253–66.
101. LEE CK, KLOPP RG, WEINDRUCH R, PROLLA TA. Gene expression profile of aging and its retardation by caloric restriction. Science. 1999;285:1390–3.
102. LEE CK, WEINDRUCH R, PROLLA TA. Gene-expression profile of the ageing brain in mice. Nat Genet. 2000;25:294–7.
103. LAKKA SS, RAO JS. Antiangiogenic therapy in brain tumors. Expert Rev Neurother. 2008;8: 1457–73.
104. FOLKMAN J. The role of angiogenesis in tumor growth. Semin Cancer Biol. 1992;3:65–71.
105. LEON SP, FOLKERTH RD, BLACK PM. Microvessel density is a prognostic indicator for patients with astroglial brain tumors. Cancer. 1996;77:362–72.
106. WESSELING P, RUITER DJ, BURGER PC. Angiogenesis in brain tumors; pathobiological and clinical aspects. J Neurooncol. 1997;32:253–65.
107. ASSIMAKOPOULOU M, SOTIROPOULOU BONIKOU G, MARAZIOTIS T, PAPADAKIS N, VARAKIS I. Microvessel density in brain tumors. Anticancer Res. 1997;17:4747–53.
108. NISHIE A, ONO M, SHONO T, FUKUSHI J, OTSUBO M, ONOUE H, et al. Macrophage infiltration and heme oxygenase-1 expression correlate with angiogenesis in human gliomas. Clin Cancer Res. 1999;5:1107–13.
109. IZYCKA-SWIESZEWSKA E, RZEPKO R, BOROWSKA-LEHMAN J, STEMPNIEWICZ M, SIDOROWICZ M. Angiogenesis in glioblastoma—analysis of intensity and relations to chosen clinical data. Folia Neuropathol. 2003;41:15–21.
110. TAKANO S, YOSHII Y, KONDO S, SUZUKI H, MARUNO T, SHIRAI S, et al. Concentration of vascular endothelial growth factor in the serum and tumor tissue of brain tumor patients. Cancer Res. 1996;56:2185–90.
111. SEYFRIED TN, MUKHERJEE P. Ganglioside GM3 is antiangiogenic in malignant brain cancer. J Oncol. 2010;2010:961243.
112. VREDENBURGH JJ, DESJARDINS A, HERNDON JE 2nd, DOWELL JM, REARDON DA, QUINN JA, et al. Phase II trial of bevacizumab and irinotecan in recurrent malignant glioma. Clin Cancer Res. 2007;13:1253–9.
113. HSU SC, VOLPERT OV, STECK PA, MIKKELSEN T, POLVERINI PJ, RAO S, et al. Inhibition of angiogenesis in human glioblastomas by chromosome 10 induction of thrombospondin-1. Cancer Res. 1996;56:5684–91.
114. CHENG SY, HUANG HJ, NAGANE M, JI XD, WANG D, SHIH CC, et al. Suppression of glioblastoma angiogenicity and tumorigenicity by inhibition of endogenous expression of vascular endothelial growth factor. Proc Natl Acad Sci USA. 1996;93:8502–7.
115. KIRSCH M, SCHACKERT G, BLACK PM. Anti-angiogenic treatment strategies for malignant brain tumors. J Neuro Oncol. 2000;50:149–63.
116. ROUS P. The influence of diet on transplanted and spontaneous mouse tumors. J Exp Med. 1914;20:433–51.
117. JENDRASCHAK E, SAGE EH. Regulation of angiogenesis by SPARC and angiostatin: implications for tumor cell biology. Semin Cancer Biol. 1996;7:139–46.
118. SUNDERKOTTER C, STEINBRINK K, GOEBELER M, BHARDWAJ R, SORG C. Macrophages and angiogenesis. J Leukoc Biol. 1994;55:410–22.

119. Powolny AA, Wang S, Carlton PS, Hoot DR, Clinton SK. Interrelationships between dietary restriction, the IGF-I axis, and expression of vascular endothelial growth factor by prostate adenocarcinoma in rats. Mol Carcinog. 2008;47:458–65.
120. Samuel DP, Wen PY, Kieran MW. Antiangiogenic (metronomic) chemotherapy for brain tumors: current and future perspectives. Expert Opin Investig Drugs. 2009;18:973–83.
121. di Tomaso E, Snuderl M, Kamoun WS, Duda DG, Auluck PK, Fazlollahi L, et al. Glioblastoma recurrence after cediranib therapy in patients: lack of "rebound" revascularization as mode of escape. Cancer Res. 2011;71:19–28.
122. Iwamoto FM, Abrey LE, Beal K, Gutin PH, Rosenblum MK, Reuter VE, et al. Patterns of relapse and prognosis after bevacizumab failure in recurrent glioblastoma. Neurology. 2009;73:1200–6.
123. Reardon DA, Turner S, Peters KB, Desjardins A, Gururangan S, Sampson JH, et al. A review of VEGF/VEGFR-targeted therapeutics for recurrent glioblastoma. J Natl Compr Canc Netw. 2011;9:414–27.
124. Parekh C, Jubran R, Erdreich-Epstein A, Panigrahy A, Bluml S, Finlay J, et al. Treatment of children with recurrent high grade gliomas with a bevacizumab containing regimen. J Neuro Oncol. 2011;103:673–80.
125. Zhang W, Lin Y, Chen B, Song SW, Jiang T. Recurrent glioblastoma of childhood treated with bevacizumab: case report and molecular features. Childs Nerv Syst. 2010;26:137–43.
126. Verhoeff JJ, Lavini C, van Linde ME, Stalpers LJ, Majoie CB, Reijneveld JC, et al. Bevacizumab and dose-intense temozolomide in recurrent high-grade glioma. Ann Oncol. 2010;21:1723–7.
127. Puchowicz MA, Xu K, Sun X, Ivy A, Emancipator D, LaManna JC. Diet-induced ketosis increases capillary density without altered blood flow in rat brain. Am J Physiol Endocrinol Metab. 2007;292:E1607–15.
128. De Bock K, Cauwenberghs S, Carmeliet P. Vessel abnormalization: another hallmark of cancer? Molecular mechanisms and therapeutic implications. Curr Opin Genet Dev. 2010;21: 73–9.
129. Jain RK. Normalization of tumor vasculature: an emerging concept in antiangiogenic therapy. Science. 2005;307:58–62.
130. Urits I, Mukherjee P, Meidenbauer J, Seyfried TN. Dietary restriction promotes vessel maturation in a mouse astrocytoma. J Oncol. 2012;201210. Article ID 264039.
131. Verbeek MM, Otte-Holler I, Wesseling P, Ruiter DJ, de Waal RM. Induction of alpha-smooth muscle actin expression in cultured human brain pericytes by transforming growth factor-beta 1. Am J Pathol. 1994;144:372–82.
132. Purow B. The elephant in the room: do microRNA-based therapies have a realistic chance of succeeding for brain tumors such as glioblastoma? J Neuro Oncol. 2011;103:429–36.
133. Denny CA, Heinecke KA, Kim YP, Baek RC, Loh KS, Butters TD, et al. Restricted ketogenic diet enhances the therapeutic action of N-butyldeoxynojirimycin towards brain GM2 accumulation in adult Sandhoff disease mice. J Neurochem. 2010;113:1525–35.
134. Lawrence T, Gilroy DW. Chronic inflammation: a failure of resolution? Int J Exp Pathol. 2007;88:85–94.
135. She QB, Solit DB, Ye Q, O'Reilly KE, Lobo J, Rosen N. The BAD protein integrates survival signaling by EGFR/MAPK and PI3K/Akt kinase pathways in PTEN-deficient tumor cells. Cancer Cell. 2005;8:287–97.
136. Hammerman PS, Fox CJ, Thompson CB. Beginnings of a signal-transduction pathway for bioenergetic control of cell survival. Trends Biochem Sci. 2004;29:586–92.
137. Mukherjee P, Mulrooney TJ, Marsh J, Blair D, Chiles TC, Seyfried TN. Differential effects of energy stress on AMPK phosphorylation and apoptosis in experimental brain tumor and normal brain. Mol Cancer. 2008;7:37.
138. Ruggeri BA, Klurfeld DM, Kritchevsky D. Biochemical alterations in 7,12-dimethylbenz[a]-anthracene-induced mammary tumors from rats subjected to caloric restriction. Biochim Biophys Acta. 1987;929:239–46.

139. Mies G, Paschen W, Ebhardt G, Hossmann KA. Relationship between of blood flow, glucose metabolism, protein synthesis, glucose and ATP content in experimentally-induced glioma (RG1 2.2) of rat brain. J Neuro Oncol. 1990;9:17–28.
140. Spitz DR, Sim JE, Ridnour LA, Galoforo SS, Lee YJ. Glucose deprivation-induced oxidative stress in human tumor cells. A fundamental defect in metabolism? Ann N Y Acad Sci. 2000;899:349–62.
141. Dang CV. Glutaminolysis: supplying carbon or nitrogen or both for cancer cells? Cell Cycle. 2010;9:3884–86.
142. Seyfried TN, Shelton LM. Cancer as a metabolic disease. Nutr Metab. 2010;7:7.
143. Semenza GL, Artemov D, Bedi A, Bhujwalla Z, Chiles K, Feldser D, et al. 'The metabolism of tumours:' 70 years later. Novartis Found Symp. 2001;240:251–60. discussion 60–4.
144. Yuneva M. Finding an "Achilles' heel" of cancer: the role of glucose and glutamine metabolism in the survival of transformed cells. Cell Cycle. 2008;7:2083–9.
145. Yudkoff M, Daikhin Y, Melo TM, Nissim I, Sonnewald U, Nissim I. The ketogenic diet and brain metabolism of amino acids: relationship to the anticonvulsant effect. Annu Rev Nutr. 2007;27:415–30.
146. Bissell MJ, Hines WC. Why don't we get more cancer? A proposed role of the microenvironment in restraining cancer progression. Nat Med. 2011;17:320–9.
147. Fosslien E. Cancer morphogenesis: role of mitochondrial failure. Ann Clin Lab Sci. 2008;38:307–29.
148. Mantovani A, Sica A. Macrophages, innate immunity and cancer: balance, tolerance, and diversity. Curr Opin Immunol. 2010;22:231–7.
149. Hanahan D, Weinberg RA. Hallmarks of cancer: the next generation. Cell. 2011;144:646–74.
150. Karin M. Nuclear factor-kappaB in cancer development and progression. Nature. 2006;441:431–6.
151. Atkinson GP, Nozell SE, Harrison DK, Stonecypher MS, Chen D, Benveniste EN. The prolyl isomerase Pin1 regulates the NF-kappaB signaling pathway and interleukin-8 expression in glioblastoma. Oncogene. 2009;28:3735–45.
152. Portnow J, Suleman S, Grossman SA, Eller S, Carson K. A cyclooxygenase-2 (COX-2) inhibitor compared with dexamethasone in a survival study of rats with intracerebral 9L gliosarcomas. Neuro Oncol. 2002;4:22–5.
153. Badie B, Schartner JM, Hagar AR, Prabakaran S, Peebles TR, Bartley B, et al. Microglia cyclooxygenase-2 activity in experimental gliomas: possible role in cerebral edema formation. Clin Cancer Res. 2003;9:872–7.
154. Wang JB, Erickson JW, Fuji R, Ramachandran S, Gao P, Dinavahi R, et al. Targeting mitochondrial glutaminase activity inhibits oncogenic transformation. Cancer cell. 2010;18:207–19.
155. Takano T, Lin JH, Arcuino G, Gao Q, Yang J, Nedergaard M. Glutamate release promotes growth of malignant gliomas. Nat Med. 2001;7:1010–5.
156. Seyfried TN, Shelton LM, Mukherjee P. Does the existing standard of care increase glioblastoma energy metabolism? Lancet Oncol. 2010;11:811–3.
157. Mauro C, Leow SC, Anso E, Rocha S, Thotakura AK, Tornatore L, et al. NF-kappaB controls energy homeostasis and metabolic adaptation by upregulating mitochondrial respiration. Nat Cell Biol. 2011;13:1272–9.
158. Drage MG, Holmes GL, Seyfried TN. Hippocampal neurons and glia in epileptic EL mice. J Neurocytol. 2002;31:681–92.
159. Huysentruyt LC, Mukherjee P, Banerjee D, Shelton LM, Seyfried TN. Metastatic cancer cells with macrophage properties: evidence from a new murine tumor model. Int J Cancer. 2008;123:73–84.
160. Mantovani A, Allavena P, Sica A, Balkwill F. Cancer-related inflammation. Nature. 2008;454:436–44.
161. Higami Y, Barger JL, Page GP, Allison DB, Smith SR, Prolla TA, et al. Energy restriction lowers the expression of genes linked to inflammation, the cytoskeleton, the extracellular matrix, and angiogenesis in mouse adipose tissue. J Nutr. 2006;136:343–52.

162. Christofk HR, Vander Heiden MG, Harris MH, Ramanathan A, Gerszten RE, Wei R, et al. The M2 splice isoform of pyruvate kinase is important for cancer metabolism and tumour growth. Nature. 2008;452:230–3.
163. Bluemlein K, Gruning NM, Feichtinger RG, Lehrach H, Kofler B, Ralser M. No evidence for a shift in pyruvate kinase PKM1 to PKM2 expression during tumorigenesis. Oncotarget. 2011;2:393–400.
164. Maalouf M, Sullivan PG, Davis L, Kim DY, Rho JM. Ketones inhibit mitochondrial production of reactive oxygen species production following glutamate excitotoxicity by increasing NADH oxidation. Neuroscience. 2007;145:256–64.
165. Puchowicz MA, Zechel JL, Valerio J, Emancipator DS, Xu K, Pundik S, et al. Neuroprotection in diet-induced ketotic rat brain after focal ischemia. J Cereb Blood Flow Metab. 2008;28:1907–16.
166. Lee C, Safdie FM, Raffaghello L, Wei M, Madia F, Parrella E, et al. Reduced levels of IGF-I mediate differential protection of normal and cancer cells in response to fasting and improve chemotherapeutic index. Cancer Res. 2010;70:1564–72.
167. Huysentruyt LC, Seyfried TN. Perspectives on the mesenchymal origin of metastatic cancer. Cancer Metastasis Rev. 2010;29:695–707.
168. Huysentruyt LC, Akgoc Z, Seyfried TN. Hypothesis: are neoplastic macrophages/microglia present in glioblastoma multiforme? ASN Neuro. 2011;3.
169. Shelton LM, Mukherjee P, Huysentruyt LC, Urits I, Rosenberg JA, Seyfried TN. A novel pre-clinical in vivo mouse model for malignant brain tumor growth and invasion. J Neuro Oncol. 2010;99:165–76.
170. Scherer HJ. A critical review: the pathology of cerebral gliomas. J Neurol Neuropsychiat. 1940;3:147–77.
171. Zagzag D, Esencay M, Mendez O, Yee H, Smirnova I, Huang Y, et al. Hypoxia- and vascular endothelial growth factor-induced stromal cell-derived factor-1alpha/CXCR4 expression in glioblastomas: one plausible explanation of Scherer's structures. Am J Pathol. 2008;173: 545–60.
172. Shelton LM, Huysentruyt LC, Mukherjee P, Seyfried TN. Calorie restriction as an anti-invasive therapy for malignant brain cancer in the VM mouse. ASN Neuro. 2010;2:e00038.
173. de Groot JF, Fuller G, Kumar AJ, Piao Y, Eterovic K, Ji Y, et al. Tumor invasion after treatment of glioblastoma with bevacizumab: radiographic and pathologic correlation in humans and mice. Neuro Oncol. 2010;12:233–42.
174. Verhoeff JJ, van Tellingen O, Claes A, Stalpers LJ, van Linde ME, Richel DJ, et al. Concerns about anti-angiogenic treatment in patients with glioblastoma multiforme. BMC Cancer. 2009;9:444.
175. Uhm JH, Ballman KV, Wu W, Giannini C, Krauss JC, Buckner JC, et al. Phase II evaluation of gefitinib in patients with newly diagnosed grade 4 astrocytoma: Mayo/North Central Cancer Treatment Group study N0074. Int J Radiat Oncol Biol Phys. 2010;80:347–53.
176. Chaparro RJ, Konigshofer Y, Beilhack GF, Shizuru JA, McDevitt HO, Chien YH. Nonobese diabetic mice express aspects of both type 1 and type 2 diabetes. Proc Natl Acad Sci USA. 2006;103:12475–80.
177. Kalaany NY, Sabatini DM. Tumours with PI3K activation are resistant to dietary restriction. Nature. 2009;458:725–31.
178. Kerbel RS, Viloria-Petit A, Klement G, Rak J. 'Accidental' anti-angiogenic drugs. Anti-oncogene directed signal transduction inhibitors and conventional chemotherapeutic agents as examples. Eur J Cancer. 2000;36:1248–57.
179. Porporato PE, Dhup S, Dadhich RK, Copetti T, Sonveaux P. Anticancer targets in the glycolytic metabolism of tumors: a comprehensive review. Front Pharmacol. 2011;2:49.
180. Ortega AD, Sanchez-Arago M, Giner-Sanchez D, Sanchez-Cenizo L, Willers I, Cuezva JM. Glucose avidity of carcinomas. Cancer Lett. 2009;276:125–35.
181. Altenberg B, Greulich KO. Genes of glycolysis are ubiquitously overexpressed in 24 cancer classes. Genomics. 2004;84:1014–20.

182. Xu RH, Pelicano H, Zhou Y, Carew JS, Feng L, Bhalla KN, et al. Inhibition of glycolysis in cancer cells: a novel strategy to overcome drug resistance associated with mitochondrial respiratory defect and hypoxia. Cancer Res. 2005;65:613–21.
183. Dang L, White DW, Gross S, Bennett BD, Bittinger MA, Driggers EM, et al. Cancer-associated IDH1 mutations produce 2-hydroxyglutarate. Nature. 2009;462:739–44.
184. Lane MA, Roth GS, Ingram DK. Caloric restriction mimetics: a novel approach for biogerontology. Methods Mol Biol. 2007;371:143–9.
185. Ingram DK, Zhu M, Mamczarz J, Zou S, Lane MA, Roth GS, et al. Calorie restriction mimetics: an emerging research field. Aging Cell. 2006;5:97–108.
186. Yeluri S, Madhok B, Prasad KR, Quirke P, Jayne DG. Cancer's craving for sugar: an opportunity for clinical exploitation. J Cancer Res Clin Oncol. 2009;135:867–77.
187. Pelicano H, Martin DS, Xu RH, Huang P. Glycolysis inhibition for anticancer treatment. Oncogene. 2006;25:4633–46.
188. Aft RL, Zhang FW, Gius D. Evaluation of 2-deoxy-D-glucose as a chemotherapeutic agent: mechanism of cell death. Br J Cancer. 2002;87:805–12.
189. Cay O, Radnell M, Jeppsson B, Ahren B, Bengmark S. Inhibitory effect of 2-deoxy-D-glucose on liver tumor growth in rats. Cancer Res. 1992;52:5794–6.
190. Dills WL Jr, Kwong E, Covey TR, Nesheim MC. Effects of diets deficient in glucose and glucose precursors on the growth of the Walker carcinosarcoma 256 in rats. J Nutr. 1984;114:2097–106.
191. Dwarakanath BS. Cytotoxicity, radiosensitization, and chemosensitization of tumor cells by 2-deoxy-D-glucose in vitro. J Cancer Res Ther. 2009;5(Suppl 1): S27–31.
192. Dwarkanath BS, Zolzer F, Chandana S, Bauch T, Adhikari JS, Muller WU, et al. Heterogeneity in 2-deoxy-D-glucose-induced modifications in energetics and radiation responses of human tumor cell lines. Int J Radiat Oncol Biol Phys. 2001;50:1051–61.
193. Jelluma N, Yang X, Stokoe D, Evan GI, Dansen TB, Haas-Kogan DA. Glucose withdrawal induces oxidative stress followed by apoptosis in glioblastoma cells but not in normal human astrocytes. Mol Cancer Res. 2006;4:319–30.
194. Kang HT, Hwang ES. 2-Deoxyglucose: an anticancer and antiviral therapeutic, but not any more a low glucose mimetic. Life Sci. 2006;78:1392–9.
195. Lopez-Rios F, Sanchez-Arago M, Garcia-Garcia E, Ortega AD, Berrendero JR, Pozo-Rodriguez F, et al. Loss of the mitochondrial bioenergetic capacity underlies the glucose avidity of carcinomas. Cancer Res. 2007;67:9013–7.
196. Lyamzaev KG, Izyumov DS, Avetisyan AV, Yang F, Pletjushkina OY, Chernyak BV. Inhibition of mitochondrial bioenergetics: the effects on structure of mitochondria in the cell and on apoptosis. Acta Biochim Pol. 2004;51:553–62.
197. Mohanti BK, Rath GK, Anantha N, Kannan V, Das BS, Chandramouli BA, et al. Improving cancer radiotherapy with 2-deoxy-D-glucose: phase I/II clinical trials on human cerebral gliomas. Int J Radiat Oncol Biol Phys. 1996;35:103–11.
198. Rhodes CG, Wise RJ, Gibbs JM, Frackowiak RS, Hatazawa J, Palmer AJ, et al. In vivo disturbance of the oxidative metabolism of glucose in human cerebral gliomas. Ann Neurol. 1983;14:614–26.
199. Sandulache VC, Ow TJ, Pickering CR, Frederick MJ, Zhou G, Fokt I, et al. Glucose, not glutamine, is the dominant energy source required for proliferation and survival of head and neck squamous carcinoma cells. Cancer. 2011;117:2926–38.
200. Zhu Z, Jiang W, McGinley JN, Thompson HJ. 2-Deoxyglucose as an energy restriction mimetic agent: effects on mammary carcinogenesis and on mammary tumor cell growth in vitro. Cancer Res. 2005;65:7023–30.
201. Loar P, Wahl H, Kshirsagar M, Gossner G, Griffith K, Liu JR. Inhibition of glycolysis enhances cisplatin-induced apoptosis in ovarian cancer cells. Am J Obstet Gynecol. 2010;202: 371e1–8.
202. Paggi MG, Carapella CM, Fanciulli M, Del Carlo C, Giorno S, Zupi G, et al. Effect of lonidamine on human malignant gliomas: biochemical studies. J Neuro Oncol. 1988;6:203–9.
203. Floridi A, Paggi MG, Fanciulli M. Modulation of glycolysis in neuroepithelial tumors. J Neurosurg Sci. 1989;33:55–64.

204. Oudard S, Poirson F, Miccoli L, Bourgeois Y, Vassault A, Poisson M, et al. Mitochondria-bound hexokinase as target for therapy of malignant gliomas. Int J Cancer. 1995;62:216–22.
205. Ko YH, Smith BL, Wang Y, Pomper MG, Rini DA, Torbenson MS, et al. Advanced cancers: eradication in all cases using 3-bromopyruvate therapy to deplete ATP. Biochem Biophys Res Commun. 2004;324:269–75.
206. Pedersen PL. Warburg, me and hexokinase 2: Multiple discoveries of key molecular events underlying one of cancers' most common phenotypes, the "Warburg Effect", i.e., elevated glycolysis in the presence of oxygen. J Bioenerg Biomembr. 2007;39:211–22.
207. Pedersen PL. The cancer cell's "power plants" as promising therapeutic targets: an overview. J Bioenerg Biomembr. 2007;39:1–12.
208. Hambley TW, Hait WN. Is anticancer drug development heading in the right direction? Cancer Res. 2009;69:1259–62.
209. Kominsky DJ, Klawitter J, Brown JL, Boros LG, Melo JV, Eckhardt SG, et al. Abnormalities in glucose uptake and metabolism in imatinib-resistant human BCR-ABL-positive cells. Clin Cancer Res. 2009;15:3442–50.
210. Frohlich E, Fink I, Wahl R. Is transketolase like 1 a target for the treatment of differentiated thyroid carcinoma? A study on thyroid cancer cell lines. Invest New Drugs. 2009;27: 297–303.
211. Kroemer G. Mitochondria in cancer. Oncogene. 2006;25:4630–2.
212. Langbein S, Zerilli M, Zur Hausen A, Staiger W, Rensch-Boschert K, Lukan N, et al. Expression of transketolase TKTL1 predicts colon and urothelial cancer patient survival: Warburg effect reinterpreted. Br J Cancer. 2006;94:578–85.
213. Lopez-Lazaro M. The warburg effect: why and how do cancer cells activate glycolysis in the presence of oxygen? Anticancer Agents Med Chem. 2008;8:305–12.
214. Bonnet S, Archer SL, Allalunis-Turner J, Haromy A, Beaulieu C, Thompson R, et al. A mitochondria-K+ channel axis is suppressed in cancer and its normalization promotes apoptosis and inhibits cancer growth. Cancer Cell. 2007;11:37–51.
215. Chen Y, Cairns R, Papandreou I, Koong A, Denko NC. Oxygen consumption can regulate the growth of tumors, a new perspective on the warburg effect. PloS One. 2009;4:e7033.
216. Michelakis ED, Sutendra G, Dromparis P, Webster L, Haromy A, Niven E, et al. Metabolic modulation of glioblastoma with dichloroacetate. Sci Trans Med. 2010;2:31ra4.
217. Pan JG, Mak TW. Metabolic targeting as an anticancer strategy: dawn of a new era? Sci STKE. 2007;2007:pe14.
218. Papandreou I, Goliasova T, Denko NC. Anti-cancer drugs that target metabolism, is dichloroacetate the new paradigm? Int J Cancer. 2010;128:1001–8.
219. Stockwin LH, Yu SX, Borgel S, Hancock C, Wolfe TL, Phillips LR, et al. Sodium dichloroacetate selectively targets cells with defects in the mitochondrial ETC. Int J Cancer. 2010;127:2510–9.
220. Minor RK, Smith DL Jr, Sossong AM, Kaushik S, Poosala S, Spangler EL, et al. Chronic ingestion of 2-deoxy-D-glucose induces cardiac vacuolization and increases mortality in rats. Toxicol Appl Pharmacol. 2010;243:332–9.
221. Whalen J, Loftus P. 'Red wine' drug trial halted by Glaxo. Wall Street Journal. 2010 May 5;Sect. B1–B2.
222. Maroon J. The Longevity Factor: How Resveratrol and Red Wine Activate Genes for a Longer and Healthier Life. New York: ATRIA; 2009.
223. Ciraolo ST, Previs SF, Fernandez CA, Agarwal KC, David F, Koshy J, et al. Model of extreme hypoglycemia in dogs made ketotic with (R,S)-1,3-butanediol acetoacetate esters. Am J Physiol. 1995;269:E67–75.
224. Mihaylova MM, Vasquez DS, Ravnskjaer K, Denechaud PD, Yu RT, Alvarez JG, et al. Class IIa histone deacetylases are hormone-activated regulators of FOXO and mammalian glucose homeostasis. Cell. 2011;145:607–21.
225. Beckner ME, Gobbel GT, Abounader R, Burovic F, Agostino NR, Laterra J, et al. Glycolytic glioma cells with active glycogen synthase are sensitive to PTEN and inhibitors of PI3K and gluconeogenesis. Lab Invest. 2005;85:1457–70.

226. Oleksyszyn J. The complete control of glucose level utilizing the composition of ketogenic diet with the gluconeogenesis inhibitor, the anti-diabetic drug metformin, as a potential anti-cancer therapy. Med Hypotheses. 2011;77:171–3.
227. Phoenix KN, Vumbaca F, Fox MM, Evans R, Claffey KP. Dietary energy availability affects primary and metastatic breast cancer and metformin efficacy. Breast Cancer Res Treat. 2010;123:333–44.
228. Del Guercio MJ, di Natale B, Gargantini L, Garlaschi C, Chiumello G. Effect of somatostatin on blood sugar, plasma growth hormone, and glucagon levels in diabetic children. Diabetes. 1976;25:550–3.
229. Shelton LM, Huysentruyt LC, Seyfried TN. Glutamine targeting inhibits systemic metastasis in the VM-M3 murine tumor model. Int J Cancer. 2010;127:2478–85.
230. Landau BR, Laszlo J, Stengle J, Burk D. Certain metabolic and pharmacologic effects in cancer patients given infusions of 2-deoxy-D-glucose. J Natl Cancer Inst. 1958;21:485–94.
231. Singh D, Banerji AK, Dwarakanath BS, Tripathi RP, Gupta JP, Mathew TL, et al. Optimizing cancer radiotherapy with 2-deoxy-d-glucose dose escalation studies in patients with glioblastoma multiforme. Strahlenther Onkol. 2005;181:507–14.
232. Safdie FM, Dorff T, Quinn D, Fontana L, Wei M, Lee C, et al. Fasting and cancer treatment in humans: a case series report. Aging (Albany, NY). 2009;1:988–1007.
233. Raffaghello L, Lee C, Safdie FM, Wei M, Madia F, Bianchi G, et al. Starvation-dependent differential stress resistance protects normal but not cancer cells against high-dose chemotherapy. Proc Natl Acad Sci USA. 2008;105:8215–20.
234. Dwarakanath BS, Singh D, Banerji AK, Sarin R, Venkataramana NK, Jalali R, et al. Clinical studies for improving radiotherapy with 2-deoxy-D-glucose: present status and future prospects. J Cancer Res Ther. 2009;5(Suppl 1): S21–6.
235. Singh D. Dose esclation studies in patients of glioblastoma. Applications of 2-deoxy-D-glucose in management of cancer. Delhi, India Inst. Nuc. Med & Allied Sci. 2006; 37.
236. Yao J, Chen S, Mao Z, Cadenas E, Brinton RD. 2-Deoxy-d-glucose treatment induces ketogenesis, sustains mitochondrial function, and reduces pathology in female mouse model of Alzheimer's disease. PloS One. 2011;6:e21788.
237. Moen I, Oyan AM, Kalland KH, Tronstad KJ, Akslen LA, Chekenya M, et al. Hyperoxic treatment induces mesenchymal-to-epithelial transition in a rat adenocarcinoma model. PloS One. 2009;4:e6381.
238. Raa A, Stansberg C, Steen VM, Bjerkvig R, Reed RK, Stuhr LE. Hyperoxia retards growth and induces apoptosis and loss of glands and blood vessels in DMBA-induced rat mammary tumors. BMC Cancer. 2007;7:23.
239. Stuhr LE, Raa A, Oyan AM, Kalland KH, Sakariassen PO, Petersen K, et al. Hyperoxia retards growth and induces apoptosis, changes in vascular density and gene expression in transplanted gliomas in nude rats. J Neuro Oncol. 2007;85:191–202.
240. Margaretten NC, Witschi H. Effects of hyperoxia on growth characteristics of metastatic murine tumors in the lung. Cancer Res. 1988;48:2779–83.
241. Wilson HD, Wilson JR, Fuchs PN. Hyperbaric oxygen treatment decreases inflammation and mechanical hypersensitivity in an animal model of inflammatory pain. Brain Res. 2006;1098:126–8.
242. Arora KK, Pedersen PL. Functional significance of mitochondrial bound hexokinase in tumor cell metabolism. Evidence for preferential phosphorylation of glucose by intramitochondrially generated ATP. J Biol Chem. 1988;263:17422–8.
243. Langbein S, Frederiks WM, zur Hausen A, Popa J, Lehmann J, Weiss C, et al. Metastasis is promoted by a bioenergetic switch: new targets for progressive renal cell cancer. Int J Cancer. 2008;122:2422–8.
244. Otto C, Kaemmerer U, Illert B, Muehling B, Pfetzer N, Wittig R, et al. Growth of human gastric cancer cells in nude mice is delayed by a ketogenic diet supplemented with omega-3 fatty acids and medium-chain triglycerides. BMC Cancer. 2008;8:122.
245. Sun W, Liu Y, Glazer CA, Shao C, Bhan S, Demokan S, et al. TKTL1 is activated by promoter hypomethylation and contributes to head and neck squamous cell carcinoma carcinogenesis through

increased aerobic glycolysis and HIF1alpha stabilization. Clin Cancer Res (Official J Am Assoc Cancer Res). 2010;16:857–66.

246. Scott MD, Zuo L, Lubin BH, Chiu DT. NADPH, not glutathione, status modulates oxidant sensitivity in normal and glucose-6-phosphate dehydrogenase-deficient erythrocytes. Blood. 1991;77:2059–64.
247. Kaadige MR, Elgort MG, Ayer DE. Coordination of glucose and glutamine utilization by an expanded Myc network. Transcription. 2010;1:36–40.
248. DeBerardinis RJ, Cheng T. Q's next: the diverse functions of glutamine in metabolism, cell biology and cancer. Oncogene. 2010;29:313–24.
249. Yang C, Sudderth J, Dang T, Bachoo RG, McDonald JG, Deberardinis RJ. Glioblastoma cells require glutamate dehydrogenase to survive impairments of glucose metabolism or Akt signaling. Cancer Res. 2009;69:7986–93.
250. Reitzer LJ, Wice BM, Kennell D. Evidence that glutamine, not sugar, is the major energy source for cultured HeLa cells. J Biol Chem. 1979;254:2669–76.
251. Yuneva M, Zamboni N, Oefner P, Sachidanandam R, Lazebnik Y. Deficiency in glutamine but not glucose induces MYC-dependent apoptosis in human cells. J Cell Biol. 2007;178:93–105.
252. Medina MA. Glutamine and cancer. J Nutr. 2001;131:2539S–42S. Discussion 50S–1S.
253. Newsholme P. Why is L-glutamine metabolism important to cells of the immune system in health, postinjury, surgery or infection? J Nutr. 2001;131:2515S–22S. Discussion 23S–4S.
254. DeBerardinis RJ. Is cancer a disease of abnormal cellular metabolism? New angles on an old idea. Genet Med. 2008;10:767–77.
255. Argiles JM, Moore-Carrasco R, Fuster G, Busquets S, Lopez-Soriano FJ. Cancer cachexia: the molecular mechanisms. Int J Biochem Cell Biol. 2003;35:405–9.
256. Tijerina AJ. The biochemical basis of metabolism in cancer cachexia. Dimens Crit Care Nurs. 2004;23:237–43.
257. Shanware NP, Mullen AR, DeBerardinis RJ, Abraham RT. Glutamine: pleiotropic roles in tumor growth and stress resistance. J Mol Med. 2011;89:229–36.
258. Souba WW. Glutamine and cancer. Ann Surg. 1993;218:715–28.
259. Kuhn KS, Muscaritoli M, Wischmeyer P, Stehle P. Glutamine as indispensable nutrient in oncology: experimental and clinical evidence. Eur J Nutr. 2010;49:197–210.
260. James MO, Smith RL, Williams RT, Reidenberg M. The conjugation of phenylacetic acid in man, sub-human primates and some non-primate species. Proc R Soc Lond B Biol Sci. 1972;182:25–35.
261. Darmaun D, Welch S, Rini A, Sager BK, Altomare A, Haymond MW. Phenylbutyrate-induced glutamine depletion in humans: effect on leucine metabolism. Am J Physiol. 1998;274: E801–7.
262. Phuphanich S, Baker SD, Grossman SA, Carson KA, Gilbert MR, Fisher JD, et al. Oral sodium phenylbutyrate in patients with recurrent malignant gliomas: a dose escalation and pharmacologic study. Neuro Oncol. 2005;7:177–82.
263. Mueller C, Al-Batran S, Jaeger E, Schmidt B, Bausch M, Unger C, et al. A phase IIa study of PEGylated glutaminase (PEG-PGA) plus 6-diazo-5-oxo-L-norleucine (DON) in patients with advanced refractory solid tumors. ASCO: J Clin Oncol. 2008;26.
264. Huysentruyt LC, Shelton LM, Seyfried TN. Influence of methotrexate and cisplatin on tumor progression and survival in the VM mouse model of systemic metastatic cancer. Int J Cancer. 2010;126:65–72.
265. Swirski FK, Nahrendorf M, Etzrodt M, Wildgruber M, Cortez-Retamozo V, Panizzi P, et al. Identification of splenic reservoir monocytes and their deployment to inflammatory sites. Science. 2009;325:612–6.
266. Aledo JC, Segura JA, Barbero LG, Marquez J. Early differential expression of two glutaminase mRNAs in mouse spleen after tumor implantation. Cancer Lett. 1998;133:95–9.
267. Ghoneum M, Gollapudi S. Phagocytosis of *Candida albicans* by metastatic and non metastatic human breast cancer cell lines in vitro. Cancer Detect Prev. 2004;28:17–26.
268. Ghoneum M, Wang L, Agrawal S, Gollapudi S. Yeast therapy for the treatment of breast cancer: a nude mice model study. In Vivo. 2007;21:251–8.

269. Ghoneum M, Hamilton J, Brown J, Gollapudi S. Human squamous cell carcinoma of the tongue and colon undergoes apoptosis upon phagocytosis of *Saccharomyces cerevisiae*, the baker's yeast, in vitro. Anticancer Res. 2005;25:981–9.
270. Lugini L, Matarrese P, Tinari A, Lozupone F, Federici C, Iessi E, et al. Cannibalism of live lymphocytes by human metastatic but not primary melanoma cells. Cancer Res. 2006;66:3629–38.
271. Fais S. A role for ezrin in a neglected metastatic tumor function. Trends Mol Med. 2004;10:249–50.
272. Dong W, Selgrade MK, Gilmour IM, Lange RW, Park P, Luster MI, et al. Altered alveolar macrophage function in calorie-restricted rats. Am J Respir Cell Mol Biol. 1998;19:462–9.
273. Amaravadi RK, Yu D, Lum JJ, Bui T, Christophorou MA, Evan GI, et al. Autophagy inhibition enhances therapy-induced apoptosis in a Myc-induced model of lymphoma. J Clin Invest. 2007;117:326–36.
274. Giulian D, Chen J, Ingeman JE, George JK, Noponen M. The role of mononuclear phagocytes in wound healing after traumatic injury to adult mammalian brain. J Neurosci. 1989;9:4416–29.
275. Yang S, Wang X, Contino G, Liesa M, Sahin E, Ying H, et al. Pancreatic cancers require autophagy for tumor growth. Genes Dev. 2011;25:717–29.
276. Briceno E, Reyes S, Sotelo J. Therapy of glioblastoma multiforme improved by the antimutagenic chloroquine. Neurosurg Focus. 2003;14:e3.
277. Dvorak HF. Tumors: wounds that do not heal. Similarities between tumor stroma generation and wound healing. N Engl J Med. 1986;315:1650–9.
278. Joyce JA, Pollard JW. Microenvironmental regulation of metastasis. Nat Rev. 2009;9:239–52.
279. Seyfried TN. Perspectives on brain tumor formation involving macrophages, glia, and neural stem cells. Perspect Biol Med. 2001;44:263–82.
280. Greenberg JI, Cheresh DA. VEGF as an inhibitor of tumor vessel maturation: implications for cancer therapy. Expert Opin Biol Ther. 2009;9:1347–56.
281. Maalouf M, Rho JM, Mattson MP. The neuroprotective properties of calorie restriction, the ketogenic diet, and ketone bodies. Brain Res Rev. 2009;59:293–315.
282. Ruskin DN, Kawamura M, Masino SA. Reduced pain and inflammation in juvenile and adult rats fed a ketogenic diet. PloS One. 2009;4:e8349.

Chapter 18

Patient Implementation of Metabolic Therapies for Cancer Management

INTRODUCTION

The optimal treatment strategy for metastatic cancer is one that will kill the tumor cells without harming normal cells. While this has been the stated goal of the cancer industry, there are few known therapies that can effectively eliminate all metastatic cells without causing some toxicity to normal cells and to the cancer patient. As I have described in this treatise, most cancers, regardless of tissue origin, depend on fermentation energy for growth and survival. Glucose and glutamine are the major fermentable fuels for most cancer cells. Restriction of these fuels becomes a viable therapeutic strategy for management of most, if not all, cancers.

Dietary energy reduction (DER) creates the metabolic environment that targets tumor cell energy metabolism. DER creates a physiological environment where competition for available nutrients increases among all cells (1). While drugs mimicking the global therapeutic effects of DER would certainly be optimal, no drugs are presently known that can produce these effects. Most calorie-restriction mimetic drugs fail to elevate ketones, which will protect normal cells from hypoglycemia. My colleagues and I have recently developed a series of guidelines for treating glioblastoma multiforme with the restricted ketogenic diet (KD-R) (2). As all cancers suffer from a common biochemical malady, that is, respiratory insufficiency with compensatory fermentation, these guidelines are also applicable for treating most advanced or metastatic cancers. These guidelines will be effective, however, only for those individuals who can actively participate in managing their disease. If the medical establishment had an effective cure for malignant cancer, then 560,000

Cancer as a Metabolic Disease: On the Origin, Management and Prevention of Cancer, First Edition.
Thomas Seyfried.

people would not die each year from the disease. In light of these numbers, and the inadequacy of current treatments, patients with advanced cancer might want to reconsider the role they play in treating their disease. Ultimately, each person is in control of his or her own destiny.

GUIDELINES FOR IMPLEMENTING THE RESTRICTED KETOGENIC DIET AS A TREATMENT STRATEGY FOR CANCER

We believe that implementation of a calorie-restricted ketogenic diet would be an effective initial treatment strategy for targeting energy metabolism in most malignant and metastatic cancers (3, 4). The KD-R diet therapy would be even more effective when administered together with drugs that could also restrict the availability of glucose and glutamine. The protocol for using the KD-R could differ from one patient to the next, depending on the age and health status of the patient. Consequently, the information presented in the following protocol can be modified for individual cases.

Phase 1: Initiation

Phase 1 of the treatment strategy would require cancer patients to gradually lower their circulating glucose levels while concurrently elevating their circulating ketone (β-hydroxybutyrate, β-OHB) levels. The procedures used for measuring blood glucose and ketone levels in cancer patients are essentially the same as those that would be used by individuals with diabetes. The Medisense Precision Xtra blood glucose and ketone monitor (Abbott Laboratories) is suggested for measuring blood ketones and glucose, but any meter that can measure glucose and ketones in the blood would be adequate. Patients can measure their blood glucose three times per day preferably before breakfast and about 2 h after lunch and dinner. It is essential for cancer patients to keep accurate records in order to identify any foods that might spike blood glucose levels.

Although it is better to measure ketone levels in blood than in urine (5–7), it may aid compliance to track urine ketones at frequent intervals during the early stages of implementation or until patients become familiar with the procedures for using the blood ketone meter. Thereafter, urine testing may be used in addition to blood testing as an additional measure of dietary compliance. As finger blood withdrawal is more easily tolerated in adults than in children, the protocol can be modified for children (less frequent or modified testing). There might also be blood glucose monitors available that do not require blood extraction for analysis.

Blood glucose ranges between 3.0 and 3.5 mM (55–65 mg/dl) and β-OHB ranges between 4 and 7 mM should be effective for reducing tumor growth in most patients. These values are within normal physiological ranges of glucose and ketones in humans and, based on our findings in mice, should have antiangiogenic, anti-inflammatory, and proapoptotic effects (Chapter 17). This treatment will induce

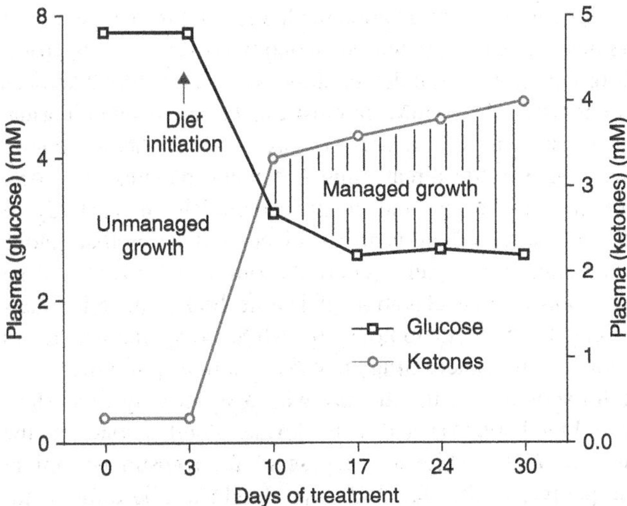

Figure 18.1 Relationship of circulating levels of glucose and ketones (β-hydroxybutyrate, β-OHB) to brain tumor management. The glucose and ketone values are within normal physiological ranges under fasting conditions in humans and will produce antiangiogenic, anti-inflammatory, and proapoptotic effects. We refer to this state as the zone of metabolic management. Metabolic stress will be greater in tumor cells than in normal cells when humans with brain cancer are in the metabolic zone. The values for blood glucose in mg/dl can be estimated by multiplying the milimolar values by 18. The glucose and ketone levels predicted for brain tumor management in human brain cancer patients are 3.1–3.8 mM (55–65 mg/dl) and 2.5–7.0 mM, respectively. *Source*: Modified from a previous version (8). See color insert.

metabolic isolation and significant growth arrest of tumor cells. We refer to these glucose and ketone levels as the zone of metabolic management (Fig. 18.1).

Blood ketone levels are higher when the ketogenic diet (KD) is administered in restricted amounts than when it is administered in unrestricted amounts (9, 10). Ketones that are not used for energy are excreted in the urine. This happens when the KD is consumed in unrestricted or elevated amounts. When the KD is consumed in small amounts, the ketones generated are retained because they are needed to supply normal cells with energy as glucose availability decreases (Fig. 17.3). The inclusion of ketone esters to the KD might also be helpful (11). However, ketone esters have yet to be evaluated as a form of chemotherapy in animals or patients. While ketone esters could have some therapeutic benefit, it is unlikely that they will have any major benefit unless glucose levels are also reduced. This comes from our findings that show that blood glucose levels remain high when the KD is consumed in unrestricted amounts. However, ketone esters could be taken in pill form during fasting. This could elevate circulating levels of ketones while maintaining low glucose levels.

It is important for patients and physicians to recognize that circulating ketone levels will rarely exceed 7–9 mM in most nondiabetic states. Although elevated ketone bodies are often associated with diabetic states, ketone body elevation in people with normal physiology is considered "good medicine" and is therapeutic for a broad range of cardiac, neurological, and neurodegenerative diseases (12–16). It is unlikely that the KD-R or ketone ester supplementation will elevate blood ketones to pathological levels (greater than 15–20 mM) in most cancer patients with normal physiology (15). A concern in using elevated ketone as a cancer therapy comes from fear of ketoacidosis, which is life threatening in diabetic patients (15). Any excess ketones are excreted in the urine, and do no accumulate in the body of most people with normal physiology. There is also evidence indicating that ketone bodies inhibit viability of human tumor cells (neuroblastoma), but not of normal cells (17). These findings indicate that elevation of ketone bodies in individuals with normal physiology can be toxic to tumor cells while being therapeutic to normal cells. I do not know of any cancer drugs that have similar properties.

Patients in good health should start the therapy with a water-only fast. Therapeutic fasting will lower blood glucose and will elevate blood ketones to the therapeutic ranges within 48–72 h. While this degree of food abstinence might sound draconian to some people, fasting for 2–3 days should not be difficult for those individuals in good physical health and who are already familiar with the health benefits of fasting. Fasting is often used to initiate the KD as a therapy for managing refractory seizures in children with epilepsy (18). Mathews et al. (1) recently described how blood glucose levels could be reduced in order to place metabolic stress on high glycolytic tumor cells without harming normal body tissues.

Diet initiation can be done in a hospital, in a cancer care clinic, or under the guidance of a physician in the home environment. I suggest that all cancer patients read the book "Fasting for Renewal of Life," by Shelton (19). The information in this book will allay concerns regarding adverse effects of fasting and will highlight the multiple health benefits associated with reduced food intake. The book by John Freeman and colleagues, "The Ketogenic Diet: A Treatment for Children and Others with Epilepsy," also provides excellent information on how the KD can be implemented and discusses the role of fasting in jump-starting the necessary metabolic adjustments for the diet therapy.

A gradual introduction of the KD without fasting might be necessary for those patients who are fragile or in poor health. For those patients not conducting a water-only fast, a restriction of carbohydrates to <12 g/day and a limitation of protein to about 0.8–1.2 g/kg of body weight/day is one way to potentially enter the therapeutic ranges for blood glucose and ketones. All cancer patients will need to carefully measure the amounts of food consumed each day according to the guidelines in Freeman's book.

Consumption of fats and oils can be used to make up the balance of energy needs. This approach will, however, require longer periods of time (perhaps several weeks) to reach the therapeutic ranges. Attending physicians should determine which course of action would be best suited for each patient regarding initiation of the KD, that is, a full therapeutic fast or a more gradual introduction of the RKD. However, it is imperative that blood glucose levels be reduced to the therapeutic range as quickly as possible in order to limit tumor progression.

Once patients get their blood glucose levels to 55–65 mg/dl range and get their blood ketones to the 3–5 mM range, they can then maintain this metabolic state using various ketogenic diets and caloric adjustments. As the KD can have a diuretic effect, it is best to avoid diuretic drugs such as Lasix (the prescribing physician should be consulted here). Indeed, it is best to monitor the effects of all medications and to keep dosages at minimum levels while on the RKD. Electrolyte levels should also be monitored or replenished if needed while on the diet for extended periods.

Many recipes for ketogenic meals are available on the Charlie Foundation website (http://www.charliefoundation.org). The Charlie Foundation was established to provide information on how the KD is used to manage refractory seizures in children. Beth Zupec-Kania is the chief dietitian and nutritionist of the Charlie Foundation; she also has practical experience in administering the KD to brain cancer patients. Beth provides her perspective on using the KD to treat brain cancer patients in Chapter 20.

It is our opinion that any ketogenic diet, consumed in restricted amounts, will be effective in maintaining reduced glucose and elevated ketones. The composition of fats in the diet can be flexible as long as blood glucose and ketones are maintained in the therapeutic ranges. Like the KD, low glycemic diets have also been used to manage seizures in children (20). Low glycemic diets might also be effective in helping to maintain low blood glucose levels in some cancer patients, as glucose is released slowly from low glycemic foods. It remains to be determined, however, if low glycemic diets will be as effective as the KD-R for maintaining glucose levels low enough to kill tumor cells.

Most cancer patients, who have not read the Shelton or Freeman books, will require some degree of professional nutritional guidance, particularly in the first few weeks of diet implementation. The key is to maintain a KD that is nutritious but is consumed in limited amounts. The definition of "ketogenic diet" allows for considerable leeway in food choices as long as the individual has reduced blood glucose and is producing ketones. Ghee, a clarified butter, can be combined with egg yolk as a ketogenic option. I am aware of a physician who used an egg yolk–butter mixture to maintain viability of a 66-year-old patient with a malignant brain tumor (glioblastoma multiforme). The patient remained alive for 37 months after diagnosis. The total daily caloric intake for this patient was about 1200–1300 kcal/day. As blood glucose levels were not measured, it was not clear if this patient remained in the zone over the entire 36 months. According to the statistics on survival, 36 months is considered long-term survival for this type of cancer (21). The key is to maintain a KD that is nutritious, but is consumed in limited amounts (restricted intake).

Coconut oil, safflower oil, and sunflower oil can also be included as part of the KD. Medium chain triglyceride (MCT) oil is another choice, as MCTs are transported directly from the small intestine to the liver, where they are metabolized to ketones (22). However, some gastrointestinal problems might arise from very rapid introduction of MCT to the diet or from prolonged use. If this happens, patients can switch to other KD food options. Patients can also supplement natural ketogenic foods with commercial ketogenic preparations such as KetoVolve from Solace Nutrition.

Consumption of the KD in unrestricted amounts will prevent blood glucose from reaching the reduced levels needed to target tumor progression and can have adverse effects for some patients. All cancer patients and their physicians should know that "less is better" when it comes to using the KD for managing cancer growth. The "less-is-better" concept cannot be overemphasized, especially when patients might be hungry or craving for a particular food item. Some patients might consider that if the KD is effective in killing tumor cells, eating more should be better. This is clearly not the case and should be avoided. Excessive or unrestricted KD consumption can cause insulin resistance and hyperglycemia (9). This would accelerate tumor progression and the demise of the patient.

It is helpful for cancer patients to keep accurate food records during diet implementation. This information should be shared with health-care professionals who are experienced in implementing very low carbohydrate therapies. In order to maintain compliance, patients can use the "KetoCalculator" (see Charlie Foundation web site for information on the KetoCalculator, http://www.ketocalculator.com/ketocalc/diet.asp) to facilitate menu planning and to help identify foods that are noncompliant with a therapeutic KD.

After the initial acclimation period, heavier individuals can safely lose up to 2 lbs/week until they are in the lower end of the normal range. Slight or severely compromised individuals should be monitored carefully to avoid a very rapid weight loss. As ketogenic diets are deficient in select vitamins and minerals (23), vitamin/minerals supplementation will be needed during a sustained KD-R. Sugar-free multivitamins and calcium are the standard supplements when on the KD (23). These supplements could include centrum (one tablet daily), calcium with vitamin D, that is, caltrate+ D, one tablet twice daily, omega-3 fatty acid, that is, nordic naturals omega-3, one capsule twice daily, and vitamin D 2000 IU daily. According to my graduate associate, Roberto Flores, supplementation with B vitamins will enhance metabolism in cells capable of respiration while possibly stressing metabolism in the tumor cells. While the KD-R can be supplemented, it is important to make sure that the supplements do not inadvertently elevate blood glucose levels. Patients can use their glucose monitors to determine if various supplements elevate glucose levels.

Variability in Caloric Adjustments and Weight Loss

As calories can be metabolized differently among individuals (metabolic heterogeneity), the calorie adjustments needed to achieve the glucose/ketone metabolic state will differ among patients. In mice, we used body weight as the independent variable for adjusting the degree of calorie restriction for brain tumor management (9, 10). To do this, we set a specific body weight reduction, for example, 20%, as a target. Owing to differences in the basal metabolic rate between humans and mice (seven times lower in man), this practice might not be effective in humans. Some patients might achieve the therapeutic glucose/ketone range without significant weight reduction, while other patients might require significant weight reduction to achieve the metabolic state.

Importantly, the weight loss associated with the KD-R is part of a metabolically appropriate response to calorie restriction. In contrast, the weight loss often seen in cancer patients following radiation or chemotherapy is due to toxicity and to the effects of the therapy on appetite. I find it interesting that some oncologists would criticize the KD-R for causing weight loss, but readily accept weight loss as a normal part of conventional cancer therapies. Some cancer patients are given high caloric drinks to prevent the weight loss from toxic cancer therapies. Worse yet, some cancer patients are given steroids to reduce nausea and vomiting from toxic chemotherapy. It is my view that high energy drinks and steroids will help rescue some cancer cells from the therapies used to kill them. It is not surprising to me that metastatic cancer returns in many patients who consume high energy drinks while they are treated with toxic therapies. Some chemotherapy could be more effective if patients fast during the treatment. Indeed, the work of Fernando Safdie and the Longo group has shown that fasting was therapeutic for patients undergoing chemotherapy (24).

I know from the results of Dr. Nebeling's case studies with two children, and from adults with malignant brain cancer, that tumor progression can be reduced if blood glucose is lowered and ketones are elevated (5, 25). I also know from our extensive work in mice that brain tumor growth is not slowed if blood glucose remains high despite elevations in ketones and the persistence of normal body weight (9, 26). In other words, the therapeutic efficacy of the metabolic KD therapy will be enhanced if blood glucose can be maintained in the low ranges (55–65 mg/dl). Given the wide variations in age, body type, weight, and metabolic status that we are likely to encounter in humans as compared to mouse models, I anticipate the need to individualize the degree of caloric restriction to lower glucose and to elevate ketones to ranges that will retard cancer progression. Frequent blood glucose measurements, as described previously, will help refine this process. Considering that "personalized therapy" is the new mantra for cancer management (27), I predict that personalized adjustment in food content and composition will help maximize the therapeutic benefit of the KD-R.

Symptoms of Glucose Withdrawal

Some patients might experience light-headedness, nausea, headache, and so on in the first few days of the KD-R, especially if they initiate the therapy with a multiday fast. These symptoms are generally transient and are associated more with glucose withdrawal than with adverse effects of the diet. Evidence suggests that the human brain can become addicted to glucose from a lifelong consumption of energy-dense foods of low nutritional value (28). Consequently, the abrupt cessation of food intake may produce temporary withdrawal symptoms similar to those experienced from cessation of any addictive substance. This is one reason why considerable personal discipline and motivation is needed to follow the KD regimen.

Glucose withdrawal symptoms can be greater in those individuals who have never fasted than in individuals who have experience with fasting. As most people in modern industrial societies do not practice therapeutic fasting as a lifestyle, glucose withdrawal symptoms will likely be encountered in most patients who attempt the KD-R as a cancer therapy. These symptoms could also be greater in older individuals than in younger ones. Indeed, fasting might not be possible in some older people who have lived a food-rich life of excess.

When compared to the debilitating effects of conventional chemotherapies and radiation, however, the symptoms associated with the KD-R are relatively mild and will pass after 2–3 days for most people. Nevertheless, glucose withdrawal symptoms and the feeling of hunger are simply too uncomfortable for some people regardless of the potential therapeutic benefits. It is therefore important for physicians to recognize that some cancer patients might be unable or unwilling to implement the KD-R for various reasons. Some individuals are simply incapable of fasting. Hence, the standard of care becomes the only therapeutic option for these patients.

It is not good to force the KD-R on any patient who does not want this therapy. The KD-R should be used only for those patients who are motivated, disciplined, and healthy enough to make the necessary changes to diet and lifestyle. Unfortunately, many cancer patients are either incapable or are unwilling to meet these requirements. The media has indoctrinated some people into thinking that fat consumption is unhealthy. For a variety of reasons, there will be some people who simply cannot maintain an KD-R. Patient education, engagement in the process, and family support are therefore keys to the success of this kind of cancer therapy.

The KD-R can reduce the feeling of hunger while maintaining reduced glucose and elevated ketone body levels. A recent study in rats suggests that diets supplemented with ketone esters might produce physiological effects similar to those for the KD-R, but without significant food restriction (11). However, administration of ketone esters has not yet been tested in cancer patients. It would be important to evaluate the influence of ketone esters on blood glucose and ketones during therapeutic fasting.

Dosages of any medications will need to be monitored carefully under the KD-R. We have shown that therapeutic action for 2-deoxyglucose (2-DG) was greater when administered with the KD-R than when administered with an unrestricted KD (Chapter 17). It is not clear, however, if the toxic effects seen in some mice treated with the drug/diet combination will also be seen in humans who might use this therapy (29). I know of one person who experienced no adverse effects in using the KD-R with 2-DG (40 mg/kg) to manage multiple myeloma. I believe that 2-DG administered with the KD-R will be effective in reducing the growth of many tumors that depend heavily on glycolysis for growth and survival.

The KD-R will not be effective in the presence of dexamethasone (decadron) or other steroid medications. Patients using steroids with the KD-R are unable to lower glucose to the therapeutic ranges. Steroid medications prevent glucose levels from falling into the therapeutic zone and therefore antagonize the therapeutic effects of KD-R. While steroids can rapidly mitigate some aspects of the tumor-related symptoms over the short term (paralysis, edema, appetite, etc.), chronic steroidal use will ultimately accelerate the growth of surviving tumor cells and thus the demise of cancer patients. Administering steroids to cancer patients would be like adding gasoline to a fire. The KD-R therapy with its neuroprotective and neurotherapeutic effects will not harm patients as can high dose dexamethasone.

Role of Exercise

Exercise during the fast should be fine, as long as the exercise is not too vigorous. Vigorous exercise will increase blood glucose due to muscle release of lactate and amino acids including glutamine. Moreover, excessive exercise will activate circulating monocytes that will leave the blood and enter the tumor. Monocyte activation can be part of the problem and not the solution for managing advanced cancer. I would recommend walking, not running. Moderate exercise will not stress the body's immune system and should have therapeutic benefit.

I am not sure why some cancer patients feel the need to engage in excessive exercise. They have a life-threatening disease. The cancer patient should maintain a relaxed state. Patient education is the key to the success of this metabolic strategy.

Phase 2: Surgery

Phase two of the treatment strategy would involve surgical resection of tumor tissue. We have suggested surgical resection as an option for cancer patients after first implementing the KD-R therapy (30). This option will be possible only if there is an opportunity for a "watchful waiting" period prior to scheduled surgery. This option will not be possible for those patients in a critical condition at the time of presentation. Dietary energy reduction and the KD-R will reduce tumor vascularization and inflammation and will more clearly delineate tumor tissue from the surrounding normal tissue, as we have shown in mice with brain cancer (31, 32) (Fig. 17.18). It remains to be determined whether the KD-R will produce the same effects in all human cancers, but I believe it will. This could be assessed in patients through histological evaluation of their tumor cells and through periodic magnetic resonance (MR) or position emission tomography (PET) imaging analyses of their tumor (5) (Fig. 2.4).

Surgical teams should recognize that less invasive, smaller tumors with reduced vascularity and clearly circumscribed boundaries should be easier to resect than larger tumors with poorly circumscribed boundaries and extensive vascularization and invasion into surrounding tissues. Tumors subjected to KD-R therapy for several weeks should have reduced inflammation and angiogenesis. This will ensure greater tumor removal, thereby increasing the likelihood of longer-term survival or even cure.

The urge to resect malignant tumors as soon as possible after diagnosis may not be in the best interests of all patients and could actually exacerbate disease progression by inducing inflammation in the microenvironment (19, 33). The KD-R could confer an additional advantage for some cancer patients, as surgical resection alone can alter the microenvironment, thus enhancing the invasive behavior of tumor cells.

The practice of surgical resection as soon as possible after tumor diagnosis could also be counterproductive for a subset of patients, especially those with lower grade tumors. The metabolic diet therapy will slow tumor progression naturally. This will provide more time before moving to the surgical option. It is therefore possible that progression-free survival could be extended in some cancer patients with advanced or metastatic cancers if an aggressive metabolic therapy were implemented prior to surgery. Surgery will cure cancer especially if the entire tumor can be resected.

Phase 3: Maintenance

Finally, phase three of the treatment strategy is designed to maintain metabolic pressure on surviving tumor cells. The glioblastoma patient that we had treated initiated a fasting regime and the KD-R within days following debulking surgery (5). Metabolic pressure could also involve carefully executed diet-cycling strategies (5, 34, 36). Diet cycling for cancer patients could involve weekly transitions from calorically KD-R to nutritious, low calorie, and low glycemic diets. Patients should continue monitoring their blood glucose and ketone levels for as long as possible or until disease resolution is achieved. The longer the patients can maintain metabolic stress on their tumor, the better will be their long-term prognosis. All this must be done in a nutritional environment that also promotes health, that is, no nutritional deficiencies. Periodic MR or PET imaging analysis including MR spectroscopy (once every 3–6 months) can help in assessing the therapeutic progress of some tumors (5, 37).

While the KD-R will target energy metabolism and improve progression-free survival in cancer patients, we do not believe that the KD-R, used as a singular therapy, will provide complete disease resolution for most patients. The goal of the maintenance strategy is to increase the probability of survival for at least 36 months in patients with advanced metastatic cancer. Recent studies suggest that many patients with advanced cancer should be given the details of their condition (35). Advanced cancer patients should know that they can be considered long-term survivors if they can live for at least 36 months beyond diagnosis. Patients with advanced cancers might be more motivated to comply with the KD-R requirements and adhere to the protocol if they recognize that 36 months would mark them as long-term survivors.

In order to significantly extend patient survival, we recommend combining the KD-R therapy with drugs that also target glucose and glutamine. The KD-R can be administered together with 2-DG (30–40 mg/kg) and with phenylbutyrate (15 g/day) as a diet drug cocktail for targeting both glucose and glutamine in cancer patients. As I have mentioned in Chapter 17, 2-DG will target glucose metabolism and glycolysis, while phenylbutyrate will help lower circulating glutamine levels. Glutamine works synergistically with glucose to drive rapid tumor growth (Chapters 8 and 17).

Phenylbutyrate is metabolized to phenylacetate, which binds to glutamine for elimination in the urine (38). According to Henri Brunengraber, glycerol phenylbutyrate could be more effective in reducing systemic glutamine than sodium phenylbutyrate (buphenyl), as nontoxic dosing can be higher for glycerol phenylbutyrate than for buphenyl (personal communication). The drug AN-113 might access the brain better than phenylbutyrate and could therefore be more effective than phenylbutyrate in reducing brain levels of glutamine (39). As we have not yet tested the therapeutic effects of AN-113 in our metastatic model of cancer, our recommendation for using this drug as a therapy for metastatic cancer remains speculative at this time.

I anticipate the development of new drugs that will more effectively target glutamine than those currently available. Interestingly, the glutamine analog drug, 6-diazo-5-*oxo*-L-norleucine (DON) appears to have less toxicity in humans than in mice. In contrast to the DON toxicity, we found when treating mice with metastatic cancer (38) that DON, used with a glutaminase inhibitor, was well tolerated in patients with advanced colon and lung cancer (40). Drugs that can simultaneously target glucose and glutamine should be effective in managing advanced metastatic cancers.

In light of the nontoxic therapeutic efficacy of the KD-R, as we have demonstrated in preclinical studies (Chapter 17), I believe that this metabolic therapy could be used together with a broad range of drugs that also target cancer energy metabolism. These drugs have been discussed in Chapter 17. Johannes Coy and coworkers have also shown therapeutic efficacy against gastric cancer in mice treated with omega-3 fatty acids and the KD (41). While some calorie-restriction mimetic drugs might have little therapeutic efficacy or express unacceptable toxicity when used alone, their therapeutic benefit might be significantly enhanced and their toxicity reduced when combined with the KD-R, especially since the KD-R would allow for the use of lower dosage levels. For example, clinical trials were halted due to unacceptable toxicity for SRT501, a reformulation of the "red wine" drug, resveratrol (42). This drug, along with others that are too toxic when used alone, could have renewed use if used together with the KD-R.

Despite the adverse effects of radiation and many current chemotherapies, it is unlikely that the oncology field will abandon these profitable therapies anytime soon. It is more probable that oncologists will opt to use these therapies in conjunction with the KD-R. Indeed, findings from Longo and colleagues have shown that fasting could improve response to chemotherapy at lower drug dosages (24, 43). Moreover, radiation and toxic drug therapy will remain as a mainstay for those patients who are incapable or unwilling to use the KD-R as a treatment strategy. Radiation therapy can be delayed for 4–6 weeks following brain tumor surgery without affecting tumor growth (44). This could give patients an opportunity to consider whether radiation therapy or the KD-R might be best for their situation. In light of the vast data showing that cancer is primarily a metabolic disease and that current treatment strategies for advanced cancer result in consistently poor outcomes, it is only a matter of time before the standard therapeutic practices are revised.

COMPLICATING ISSUES FOR IMPLEMENTING THE KD-R AS A TREATMENT STRATEGY FOR CANCER

Several issues can complicate attempts to implement KD-R as a treatment strategy for advanced or metastatic cancers. One issue is the nonconventional and nonpharmacological nature of the metabolic therapy. Modern medicine has not looked favorably toward metabolic diet therapies for managing complex diseases especially when well-established procedures for acceptable clinical practice are available, regardless of how toxic or ineffective these procedures might be in managing the disease (30). In the case of cancer management, these approved practices generally involve maximal surgical resection followed a few weeks later by either radiation therapy alone or a combination of radiation and chemotherapy. Many cancer patients receive corticosteroids, which significantly elevate blood glucose levels. The type of therapy given will usually depend on the size and location of the tumor as well as on the age and health status of the patient.

Some cancer patients are considered hopeless cases before initiation of treatment. For example, the number of older patients with glioblastoma that are either offered no therapy or who choose no therapy appears to be increasing (45). Significant neurological damage can occur in children who survive malignant brain cancer while the risk of developing long-term morbidity and mortality is greatly enhanced (25, 46–49). Worse yet, some conventional therapeutic protocols involving combinatorial radiotherapy with chemotherapy or antiangiogenic therapy may actually exacerbate the disease (50–53). These situations are unacceptable and highlight the inadequacies of conventional approaches for treating invasive metastatic cancers in either adults or children.

Availability of a drug that would mimic the global therapeutic effects of the KD-R would certainly be the easiest way to implement a metabolic therapy. However, as I have mentioned above, no drugs are known that can lower glucose levels, while simultaneously elevating ketones in the absence of some form of restricted food intake. Difficulty with implementation and a paucity of experienced practitioners remain as complicating issues in adapting the KD-R as a standard therapy for malignant cancers. It is troubling that many oncologists are not familiar with Otto Warburg or his ideas on the origin of cancer. It is my opinion that successful management of cancer will be better than current practices if all oncologists were to read Warburg's papers on the origin of cancer before starting their practice.

RADIATION AND CHEMOTHERAPY IS A STANDARD TREATMENT FOR MANY MALIGNANT CANCERS

Although radiation therapy is commonly used to treat cancer, I believe that radiation therapy places some patients at risk of developing more aggressive cancers in the future. Radiation damages mitochondria in normal cells while creating an inflamed microenvironment. Inflammation enhances glucose and glutamine energy metabolism that further damages mitochondria. Respiratory insufficiency with compensatory fermentation is the origin of most cancers as I have described in Chapters 9 and 14. It makes little sense to treat delicate tissue with a therapy that is toxic to normal cells, enhances growth of the surviving tumor cells, and increases risk for new cancers. Although some chemotherapy will increase patient survival for advanced cancers, the benefits have been marginal at best (Chapter 16). Like radiation therapy, chemotherapy can also cause tissue necrosis and inflammation. Considering that healthy, long-term survivors of the conventional treatment strategies for advanced cancers are more the exception than the rule, the KD-R administered with energy-targeting drugs could represent a novel alternative treatment strategy to the conventional therapeutic approaches.

COMPLIANCE

Strict compliance with the requirements of the KD-R poses the most significant challenge for implementation. Consequences of noncompliance are not as obvious to cancer patients as they are to patients with epilepsy who also use the KD for seizure control. Breakthrough seizures, which are immediate and disturbing for both the patient and the family, are the consequence of noncompliance for the epilepsy patient in using ketogenic diets. More specifically, the epilepsy patient experiences an immediate and unambiguous consequence of noncompliance. The consequence of noncompliance for cancer patients would be a subtle increase in tumor progression that would not be immediately obvious to either the patient or the family. In contrast to breakthrough seizures for the epilepsy patient, shortened overall survival would be the expected consequence of noncompliance for the cancer patient.

I also recognize that maintaining dietary compliance in the home environment might be more difficult than maintaining compliance in a controlled medical clinic. Some patients can feel alone and isolated when attempting to use the KD in the home environment. Distractions in the home environment could also interfere with the strict guidelines necessary to maintain compliance. Cancer is a disease of the entire body and is best treated in a relaxing and stress-free environment. Patient education in using the metabolic therapy would also be better in a clinic setting than in the home environment. Consequently, metabolic therapies for cancer treatments would be best administered in specialized clinics that deal specifically with cancer as a metabolic disease. This does not mean that the KD-R cannot be administered in the home environment, but only that administration and success of outcome could be better for some patients in the specialized clinic environment than in the home environment. Certainly, consultation with registered dieticians, nutritionists, and physicians familiar with the concepts are in the best interest of all patients. This is discussed more in Chapter 20.

CANCER AS A GENETIC DISEASE

Another complicating issue for implementing the KD-R as a cancer therapy is the persistent view that all cancers are a genetic disease (Chapter 9). Why should the oncology field switch to metabolic therapies for a disease considered to be genetic in origin? The view of cancer as a genetic disease is the driving force for investment in targeted molecular therapies and for the idea that cancer treatments should be personalized in order to target defective signaling pathways within tumors. More importantly, the persistent failure of the targeted gene therapies to resolve the cancer problem justifies the continuation of radiotherapy and toxic drug therapies.

The evidence I present in this treatise shows that cancer is primarily a metabolic disease and that the vast numbers of mutations found in tumor cells arise as downstream epiphenomenon of mitochondrial damage. The recent studies of Stratton (54, 55) indicate that over one million mutations can be found in the cells of most tumors. How will it be possible to target all these mutations to achieve a cure? The suggestion that some of these mutations are drivers and others are passengers is nonsense. The nuclear transfer experiments described in Chapter 11 clearly show that nuclear expressed genes do not drive the disease. The idea that cancer can be identified and managed using genetic strategies has been an enormous failure (56). Once cancer becomes recognized as a metabolic disease rather than as a genetic disease, more effective and less toxic therapeutic strategies will emerge. Only then will we be able to abandon radiation and the poisonous drugs currently being used to treat advanced cancers.

MECHANISM OF ACTION?

Another concern in implementing the KD-R for cancer management involves the mechanism of action. How can the process of targeting glucose and glutamine availability, while elevating ketone bodies, be an effective management for the majority of malignant and metastatic cancers? The mechanism of action is rooted in the well-established scientific principle that tumor cells largely use fermentation energy for their growth and survival, as I have described in previous chapters. Glucose and glutamine drive cancer cell fermentation through substrate-level phosphorylation. Because tumor cells are less flexible than normal cells in using alternative energy substrates (ketones), tumor cells will experience more energy stress when their access to fermentable fuels becomes restricted. Despite the recognized effectiveness and mechanisms of action, the general use of this nontoxicity therapeutic strategy for cancer management could go underutilized because of its simplicity and cost effectiveness.

CACHEXIA

Another concern is how a metabolic therapy that reduces food intake and body weight can be recommended to patients who might be losing body weight because of cancer cachexia (57). Cancer cachexia generally involves anorexia, anemia, weight loss, and muscle atrophy (33, 58, 59). Although some cancer patients could be obese, a very rapid weight loss from cachexia involving both proteins and fat is a health concern (58). Cachexia is not common in patients with glioblastoma, but prognosis is worse in glioblastoma patients that express higher levels of IL-6, a biomarker of cachexia (60). Other procachexia molecules such as proteolysis-inducing factor are released from the tumor cells into the circulation and contribute to the cachexia phenotype (61–63). The KD-R will reduce inflammation and expression of IL-6 (9, 64–66). IL-6 also increases expression of hepcidin, which contributes to the anemia seen in many cancer patients (67). By killing the fermenting tumor cells that produce procachexia molecules, the KD-R can potentially reduce tumor cachexia (26, 57, 63, 68). Once tumor growth becomes arrested, patients can increase caloric consumption, which will accelerate weight gain and improve health. Nebeling and Tisdale used the KD to improve the nutritional status of cancer patients (25, 69). Hence, restricted consumption of ketogenic diets could be effective, in principle, for managing tumor growth in those cancer patients that express biomarkers of cachexia (25, 61, 62).

In contrast to most conventional cancer therapies that expose both normal cells and tumor cells to toxic assaults, dietary restriction and particularly the KD-R, are the only known therapies that can target tumor cells while enhancing the health and vitality of normal cells (5, 9, 25, 34). In this regard, the KD-R as a cancer therapy is conceptually superior to many current conventional cancer therapies.

Patient Information

How can effective nontoxic metabolic therapies be introduced as part of the standard clinical practice in oncology? It is incumbent upon oncologists to notify patients that effective alternatives to the current standards of care exist for treating highly invasive and metastatic cancers. Cancer patients should be aware of all potential therapeutic options for treating their disease, and not just the conventional treatment strategies (69). Patients should also know that the KD-R would retard tumor growth without producing toxic adverse effects. The therapy could be especially powerful when combined with drugs that also target glucose and glutamine.

It should be up to the patient and their family to decide whether or not the KD-R is a viable therapeutic option for their situation. Patients with invasive and metastatic tumors should have the opportunity to compare and contrast the results from recent drug studies (70, 71), with those of metabolic therapy using restricted diets (5, 25). Why are most cancer patients not offered this information? While standard practices within the field and a paucity of education regarding dietary metabolic therapies might make it difficult for some oncologists to suggest the KD-R as a therapeutic option for cancer management, I remain hopeful that all oncologists will eventually come to recognize the potential value of the KD-R as an effective treatment strategy for malignant cancers.

SUMMARY

In this chapter, I have provided information on a new, alternative treatment strategy for highly malignant cancers that targets tumor energy metabolism. We have recently published this protocol for glioblastoma in Treatment Strategies Oncology (4). The objective of this new therapeutic strategy is to change the metabolic environment of the tumor and the host. Access to glucose and glutamine within the tumor microenvironment provides neoplastic tumor cells with fermentable fuels necessary for their survival and growth. The low carbohydrate, high fat ketogenic diet KD will reduce circulating glucose levels and will elevate circulating levels of ketone bodies especially when consumed in restricted amounts. A transition from glucose to ketone bodies will restrict glucose availability to the malignant tumor cells. Ketone elevation protects and enhances the health and vitality of normal cells. The therapeutic efficacy of the KD-R against malignant cancers can be enhanced when combined with drugs that also target or reduce access to glucose and glutamine. A use protocol is presented to help oncologists and cancer patients implement the KD-R as a treatment strategy. Although the KD-R is less toxic and potentially more effective in managing advanced cancers than the conventional standard of care, considerable patient education, motivation, and discipline will be necessary for implementing this therapy. Considering the poor prognosis of most patients with metastatic cancers, metabolic treatment strategies could be an attractive alternative or complimentary option for many patients with malignant cancers. It is unnecessary in my opinion that cancer patients be charged excessive fees to suffer physical and mental pain in order to have their tumors managed.

REFERENCES

1. Mathews EH, Liebenberg L, Pelzer R. High-glycolytic cancers and their interplay with the body's glucose demand and supply cycle. Med Hypotheses. 2011;76:157–65.
2. Huysentruyt LC, Akgoc Z, Seyfried TN. Hypothesis: are neoplastic macrophages/microglia present in glioblastoma multiforme? ASN Neuro. 2011;3. In press.
3. Seyfried TN, Marsh J, Shelton LM, Huysentruyt LC, Mukherjee P. Is the restricted ketogenic diet a viable alternative to the standard of care for managing malignant brain cancer? Epilepsy Res. 2011.
4. Seyfried TN, Mukherjee P, Kalamian M, Zuccoli G. The Restricted Ketogenic Diet: An Alternative Treatment Strategy for Glioblastoma Multiforme. In: Holcroft R, editor. Treatment Stratagies Oncology. London: Cambridge Research Center; 2011. p.24–35. Available at http://viewer.zmags.com/publication/e119d6eb.
5. Zuccoli G, Marcello N, Pisanello A, Servadei F, Vaccaro S, Mukherjee P, et al. Metabolic management of glioblastoma multiforme using standard therapy together with a restricted ketogenic diet: case report. Nutr Metab. 2010;7:33.
6. Taboulet P, Deconinck N, Thurel A, Haas L, Manamani J, Porcher R, et al. Correlation between urine ketones (acetoacetate) and capillary blood ketones (3-beta-hydroxybutyrate) in hyperglycaemic patients. Diabetes Metab. 2007;33:135–9.
7. Turan S, Omar A, Bereket A. Comparison of capillary blood ketone measurement by electrochemical method and urinary ketone in treatment of diabetic ketosis and ketoacidosis in children. Acta Diabetol. 2008;45:83–5.
8. Seyfried TN, Kiebish M, Mukherjee P, Marsh J. Targeting energy metabolism in brain cancer with calorically restricted ketogenic diets. Epilepsia. 2008;49 (Suppl 8):114–6.
9. Zhou W, Mukherjee P, Kiebish MA, Markis WT, Mantis JG, Seyfried TN. The calorically restricted ketogenic diet, an effective alternative therapy for malignant brain cancer. Nutr Metab. 2007;4:5.

10. Mantis JG, Centeno NA, Todorova MT, McGowan R, Seyfried TN. Management of multifactorial idiopathic epilepsy in EL mice with caloric restriction and the ketogenic diet: role of glucose and ketone bodies. Nutr Metab. 2004;1:11.
11. Kashiwaya Y, Pawlosky R, Markis W, King MT, Bergman C, Srivastava S, et al. A ketone ester diet increased brain malonyl CoA and uncoupling protein 4 and 5 while decreasing food intake in the normal Wistar rat. J Biol Chem. 2010;285:25950–6.
12. Balietti M, Giorgetti B, Di Stefano G, Casoli T, Platano D, Solazzi M, et al. A ketogenic diet increases succinic dehydrogenase (SDH) activity and recovers age-related decrease in numeric density of SDH-positive mitochondria in cerebellar Purkinje cells of late-adult rats. Micron. 2010;41:143–8.
13. Mahoney LB, Denny CA, Seyfried TN. Caloric restriction in C57BL/6J mice mimics therapeutic fasting in humans. Lipids Health Dis. 2006;5:13.
14. VanItallie TB, Nufert TH. Ketones: metabolism's ugly duckling. Nutr Rev. 2003;61:327–41.
15. Veech RL, Chance B, Kashiwaya Y, Lardy HA, Cahill GF Jr. Ketone bodies, potential therapeutic uses. IUBMB Life. 2001;51:241–7.
16. Cahill GF Jr, Veech RL. Ketoacids? Good medicine? Trans Am Clin Climatol Assoc. 2003;114:149–61. Discussion 62–3.
17. Skinner R, Trujillo A, Ma X, Beierle EA. Ketone bodies inhibit the viability of human neuroblastoma cells. J Pediatr Surg. 2009;44:212–6. Discussion 6.
18. Freeman JM, Freeman JB, Kelly MT. The Ketogenic Diet: A Treatment for Epilepsy. 3rd ed. New York: Demos; 2000.
19. Shelton HM. Fasting for Renewal of Life. Tampa (FL): American National Hygiene Society, Inc; 1974.
20. Kossoff EH, Zupec-Kania BA, Rho JM. Ketogenic diets: an update for child neurologists. J Child Neurol. 2009;24:979–88.
21. Krex D, Klink B, Hartmann C, von Deimling A, Pietsch T, Simon M, et al. Long-term survival with glioblastoma multiforme. Brain. 2007;130:2596–606.
22. Nebeling LC, Lerner E. Implementing a ketogenic diet based on medium-chain triglyceride oil in pediatric patients with cancer. J Am Diet Assoc. 1995;95:693–7.
23. Freeman JM, Kossoff EH, Freeman JB, Kelly MT. The Ketogenic Diet: A Treatment for Children and Others with Epilepsy. 4th ed. New York: Demos; 2007.
24. Safdie FM, Dorff T, Quinn D, Fontana L, Wei M, Lee C, et al. Fasting and cancer treatment in humans: a case series report. Aging (Albany NY). 2009;1:988–1007.
25. Nebeling LC, Miraldi F, Shurin SB, Lerner E. Effects of a ketogenic diet on tumor metabolism and nutritional status in pediatric oncology patients: two case reports. J Am Coll Nutr. 1995;14:202–8.
26. Seyfried TN, Sanderson TM, El-Abbadi MM, McGowan R, Mukherjee P. Role of glucose and ketone bodies in the metabolic control of experimental brain cancer. Br J Cancer. 2003;89:1375–82.
27. Hayden EC. Personalized cancer therapy gets closer. Nature. 2009;458:131–2.
28. Morgan D, Sizemore GM. Animal models of addiction: fat and sugar. Curr Pharm Des. 2011;17:1168–72.
29. Marsh J, Mukherjee P, Seyfried TN. Drug/diet synergy for managing malignant astrocytoma in mice: 2-deoxy-D-glucose and the restricted ketogenic diet. Nutr Metab. 2008;5:33.
30. Seyfried TN, Kiebish MA, Marsh J, Shelton LM, Huysentruyt LC, Mukherjee P. Metabolic management of brain cancer. Biochim Biophys Acta. 2010;1807:577–94.
31. Mulrooney TJ, Marsh J, Urits I, Seyfried TN, Mukherjee P. Influence of caloric restriction on constitutive expression of NF-kappaB in an experimental mouse astrocytoma. PloS One. 2011;6:e18085.
32. Shelton LM, Huysentruyt LC, Mukherjee P, Seyfried TN. Calorie restriction as an anti-invasive therapy for malignant brain cancer in the VM mouse. ASN Neuro. 2010;2:e00038.
33. Kern KA, Norton JA. Cancer cachexia. JPEN. 1988;12:286–98.
34. Seyfried TN, Mukherjee P. Targeting energy metabolism in brain cancer: review and hypothesis. Nutr Metab. 2005;2:30.

35. Russell BJ, Ward AM. Deciding what information is necessary: do patients with advanced cancer want to know all the details? Cancer Management Res. 2011;3:191–9.
36. Cleary MP, Jacobson MK, Phillips FC, Getzin SC, Grande JP, Maihle NJ. Weight-cycling decreases incidence and increases latency of mammary tumors to a greater extent than does chronic caloric restriction in mouse mammary tumor virus-transforming growth factor-alpha female mice. Cancer Epidemiol Biomarkers Prev. 2002;11:836–43.
37. Yang I, Aghi MK. New advances that enable identification of glioblastoma recurrence. Nat Rev Clin Oncol. 2009;6:648–57.
38. Shelton LM, Huysentruyt LC, Seyfried TN. Glutamine targeting inhibits systemic metastasis in the VM-M3 murine tumor model. Int J Cancer. 2010;127:2478–85.
39. Entin-Meer M, Rephaeli A, Yang X, Nudelman A, Nudelman A, Haas-Kogan DA. AN-113, a novel prodrug of 4-phenylbutyrate with increased anti-neoplastic activity in glioma cell lines. Cancer Lett. 2007;253:205–14.
40. Mueller C, Al-Batran S, Jaeger E, Schmidt B, Bausch M, Unger C, et al. A phase IIa study of PEGylated glutaminase (PEG-PGA) plus 6-diazo-5-oxo-L-norleucine (DON) in patients with advanced refractory solid tumors. ASCO: J Clin Oncol. 2008;26.
41. Otto C, Kaemmerer U, Illert B, Muehling B, Pfetzer N, Wittig R, et al. Growth of human gastric cancer cells in nude mice is delayed by a ketogenic diet supplemented with omega-3 fatty acids and medium-chain triglycerides. BMC Cancer. 2008;8:122.
42. Whalen J, Loftus P. 'Red wine' drug trial halted by Glaxo. Wall Street J. 2010:B1–B2.
43. Raffaghello L, Lee C, Safdie FM, Wei M, Madia F, Bianchi G, et al. Starvation-dependent differential stress resistance protects normal but not cancer cells against high-dose chemotherapy. Proc Natl Acad Sci USA. 2008;105:8215–20.
44. Lawrence YR, Blumenthal DT, Matceyevsky D, Kanner AA, Bokstein F, Corn BW. Delayed initiation of radiotherapy for glioblastoma: how important is it to push to the front (or the back) of the line? J Neuro Oncol. 2011;105:1–7.
45. Lowry JK, Snyder JJ, Lowry PW. Brain tumors in the elderly: recent trends in a Minnesota cohort study. Arch Neurol. 1998;55:922–8.
46. Morris EB, Gajjar A, Okuma JO, Yasui Y, Wallace D, Kun LE, et al. Survival and late mortality in long-term survivors of pediatric CNS tumors. J Clin Oncol. 2007;25:1532–8.
47. Bowers DC, Liu Y, Leisenring W, McNeil E, Stovall M, Gurney JG, et al. Late-occurring stroke among long-term survivors of childhood leukemia and brain tumors: a report from the childhood cancer survivor study. J Clin Oncol. 2006;24:5277–82.
48. Clarson CL, Del Maestro RF. Growth failure after treatment of pediatric brain tumors. Pediatrics. 1999;103:E37.
49. Birkholz D, Korpal-Szczyrska M, Kaminska H, Bien E, Polczynska K, Stachowicz-Stencel T, et al. Influence of surgery and radiotherapy on growth and pubertal development in children treated for brain tumour. Med Wieku Rozwoj. 2005;9:463–9.
50. de Groot JF, Fuller G, Kumar AJ, Piao Y, Eterovic K, Ji Y, et al. Tumor invasion after treatment of glioblastoma with bevacizumab: radiographic and pathologic correlation in humans and mice. Neuro Oncol. 2010;12:233–42.
51. Verhoeff JJ, van Tellingen O, Claes A, Stalpers LJ, van Linde ME, Richel DJ, et al. Concerns about anti-angiogenic treatment in patients with glioblastoma multiforme. BMC Cancer. 2009;9:444.
52. Claes A, Wesseling P, Jeuken J, Maass C, Heerschap A, Leenders WP. Antiangiogenic compounds interfere with chemotherapy of brain tumors due to vessel normalization. Mol Cancer Ther. 2008;7:71–8.
53. Cahill DP, Levine KK, Betensky RA, Codd PJ, Romany CA, Reavie LB, et al. Loss of the mismatch repair protein MSH6 in human glioblastomas is associated with tumor progression during temozolomide treatment. Clin Cancer Res. 2007;13:2038–45.
54. Stratton MR. Exploring the genomes of cancer cells: progress and promise. Science. 2011;331: 1553–8.
55. Stratton MR, Campbell PJ, Futreal PA. The cancer genome. Nature. 2009;458:719–24.
56. Kolata G. How bright promise in cancer testing fell apart. New York Times. 2011 July 7.

57. Seyfried TN, Shelton LM. Cancer as a metabolic disease. Nutr Metab. 2010;7:7.
58. Tisdale MJ. Cancer anorexia and cachexia. Nutrition. 2001;17:438–42.
59. Argiles JM, Moore-Carrasco R, Fuster G, Busquets S, Lopez-Soriano FJ. Cancer cachexia: the molecular mechanisms. Int J Biochem Cell Biol. 2003;35:405–9.
60. Chaffer CL, Weinberg RA. A perspective on cancer cell metastasis. Science. 2011;331:1559–64.
61. Pouliquen DL. Hepatic mitochondrial function and brain tumours. Curr Opin Clin Nutr Metab Care. 2007;10:475–9.
62. Todorov PT, Wyke SM, Tisdale MJ. Identification and characterization of a membrane receptor for proteolysis-inducing factor on skeletal muscle. Cancer Res. 2007;67:11419–27.
63. Tisdale MJ. Biology of cachexia. J Natl Cancer Inst. 1997;89:1763–73.
64. Mavropoulos JC, Buschemeyer WC 3rd, Tewari AK, Rokhfeld D, Pollak M, Zhao Y, et al. The effects of varying dietary carbohydrate and fat content on survival in a murine LNCaP prostate cancer xenograft model. Cancer Prev Res (Phila). 2009;2:557–65.
65. Ruskin DN, Kawamura M, Masino SA. Reduced pain and inflammation in juvenile and adult rats fed a ketogenic diet. PloS One. 2009;4:e8349.
66. Spaulding CC, Walford RL, Effros RB. Calorie restriction inhibits the age-related dysregulation of the cytokines TNF-alpha and IL-6 in C3B10RF1 mice. Mech Ageing Dev. 1997;93:87–94.
67. Ward DG, Roberts K, Brookes MJ, Joy H, Martin A, Ismail T, et al. Increased hepcidin expression in colorectal carcinogenesis. World J Gastroenterol. 2008;14:1339–45.
68. Stehle G, Sinn H, Wunder A, Schrenk HH, Stewart JC, Hartung G, et al. Plasma protein (albumin) catabolism by the tumor itself—implications for tumor metabolism and the genesis of cachexia. Crit Rev Oncol Hematol. 1997;26:77–100.
69. Fearon KC, Borland W, Preston T, Tisdale MJ, Shenkin A, Calman KC. Cancer cachexia: influence of systemic ketosis on substrate levels and nitrogen metabolism. Am J Clin Nutr. 1988;47:42–8.
70. Vredenburgh JJ, Desjardins A, Herndon JE 2nd, Dowell JM, Reardon DA, Quinn JA, et al. Phase II trial of bevacizumab and irinotecan in recurrent malignant glioma. Clin Cancer Res. 2007;13:1253–9.
71. Uhm JH, Ballman KV, Wu W, Giannini C, Krauss JC, Buckner JC, et al. Phase II evaluation of gefitinib in patients with newly diagnosed grade 4 astrocytoma: Mayo/North central cancer treatment group study N0074. Int J Radiat Oncol Biol Phys. 2010;80:347–53.

Chapter 19

Cancer Prevention

I have presented substantial evidence showing that respiratory damage underlies the origin of cancer. Cancer is a disease of energy metabolism. If respiratory injury is the prime cause of cancer then protecting mitochondria and respiration from damage becomes the prime means of preventing cancer (1). It is well documented that the incidence of cancer can be significantly reduced by avoiding exposure to those agents or conditions that provoke tissue inflammation, such as smoking, excessive alcohol consumption, carcinogenic chemicals, ionizing radiation, and obesity (2–5).

Elevated levels of inflammation biomarkers (IL-6, IL-8, C-reactive protein, etc.) predict increased risk of cancer (6). Chronic inflammation, regardless of its origin, damages tissue morphogenetic fields and the epithelial and mesenchymal cells within the field (7–15). Most importantly, inflammation damages cellular mitochondria, thus reducing the efficiency of OxPhos. Reduced OxPhos efficiency initiates a mitochondrial stress response (RTG signaling) within cells (Chapter 10). RTG signaling is needed to upregulate either glycolysis in the cytoplasm or amino acid fermentation in the mitochondria. Only those cells that can enhance their fermentation in response to respiratory damage will survive. Cells incapable of enhancing fermentation will die from energy failure. As mitochondrial function maintains the differentiated state, cells that upregulate fermentation for survival are at increased risk of becoming less differentiated and ultimately transformed. Prolonged reliance on fermentation destabilizes the nuclear genome, thus initiating the path to carcinogenesis and frank neoplasia. Inflammation damages cellular respiration; damaged respiration is the origin of cancer.

Prevention of inflammation and damage to the tissue microenvironment will go far in reducing the incidence of most cancers. Vaccines against oncogenic viruses can also reduce the incidence of some cancers, as viruses can damage mitochondria in infected tissues (Chapter 9). It is known that avoidance of cancer risk factors, which produce chronic inflammation and mitochondrial damage, will reduce the incidence of at least 80% of all cancers (3, 4). In principle, there are few chronic diseases that are more easily preventable than cancer (1).

Cancer as a Metabolic Disease: On the Origin, Management and Prevention of Cancer, First Edition.
Thomas Seyfried.

If cancer is so easy to prevent in principle, why does the number of new cancer cases increase each year (Table 1.1)? There are several reasons that can contribute to the rising incidence of cancer in industrial societies. First, the emphasis on prevention has always been less important than the emphasis on management. Government health agencies and the media often focus more attention on potential cures, remedies, and new causes than on ways to prevent cancer. While screening programs for various cancers have helped in preventing some cancers, these procedures have had little impact on the yearly cancer death rates. Second, more research funds are given to finding cures than to exploring preventions. The American Institute of Cancer Research (AICR) is one of the few foundations devoted to funding research on cancer prevention. More emphasis is needed on strategies for prevention. Finally, several cancer risk factors are associated with activities that make life pleasurable such as smoking, eating, drinking alcohol, and sex. Most people I know do not think about cancer when engaging in these activities. Life is hard for many people. The pleasure of the moment will often override mental calculations of future cancer risk. While antismoking and safe-sex campaigns have succeeded in reducing cancer risk associated with these activities, less success has been achieved in reducing cancer risk from behaviors associated with overeating and excessive drinking.

Inflammation damages cellular respiration. Many cancers arise from protracted respiratory damage. I was surprised to read Dr. Harold Varmus's statement, "we don't really understand what obesity contributes to cancer causation" (Science, 333: 397, 2011). This issue is also the NCI provocative question #1: How does obesity contribute to cancer risk? (provocativequestions.nci.nih.gov). It is well documented that obesity enhances body inflammation (4, 16). Chronic inflammation can cause cancer through the mechanisms described in this treatise. It is my opinion that the incidence of cancer will continue to increase each year until people become knowledgeable of how cancer risk factors influence cellular respiration. Whether this information will help people avoid the risks is another question (17). However, obesity becomes a nonissue if people lose weight. It is not necessary to invest millions of federal tax dollars into the cancer–obesity issue if the solution is simply to have people move more and eat less. It remains to be determined, however, if members of our species are willing and motivated enough to make the lifestyle changes or sacrifices necessary to prevent cancer. This issue also addresses provocative question #4.

CELL PHONES AND CANCER

Discussions linking cell phone use to brain cancer have generated considerable controversy among researchers and anxiety among users. According to Jane Brody of the *New York Times*, widespread fear that excessive cell phone use could cause brain cancer began in 1993 when David Raynard sued the cell phone industry on the grounds that excessive cell phone use killed his wife who died of brain cancer (18). The recent study by the World Health Organization (WHO) has raised new concerns regarding cancer risk from cell phone use (19). The cell phone is now considered to be a carcinogen in the same category as chloroform, formaldehyde, and lead. The risk of developing brain cancer from cell phone use will depend on gene-environmental interactions similar to the risk factors for developing any cancer.

The editors of the Wall Street Journal have described the WHO report on cancer risk and cell phones as "a needless cancer scare" (Saturday/Sunday, June 4–5, 2011). Their main criticism was the lack of a credible explanation on how mobile signals might lead to the cellular mutations that cause cancer. This is a reasonable criticism considering that most people think mutations cause cancer. As was made clear in my treatise, however, it is not mutations that cause most cancers, but damaged respiration. The more relevant question is how prolonged cell phone use might damage cellular respiration.

Knowing what I know about the origin of cancer, it is clear how excessive cell phone use could cause cancer in some people. Cell phones produce what is called extremely low frequency electromagnetic fields (ELF-EMF). These frequencies are in the range of those found in microwave ovens and television transmitters (20). Persistent tissue exposure to ELF-EMF produces thermogenesis (heat) in affected areas. While the increased temperature is slight, frequent and prolonged temperature shifts can influence CNS energy metabolism (21). Tissue thermogenesis will activate macrophages that then release inflammatory cytokines (22, 23). These cytokines will induce inflammation in the tissue microenvironment, thus disturbing the integrity of the tissue morphogenetic field.

Inflammation will damage respiration in the cells of the field (Chapter 10). The path to carcinogenesis often begins with damaged or insufficient respiration. Mutations arise as an epiphenomenon of persistent fermentation, which ultimately arises from insufficient cellular respiration. The cell phone risk for cancer should be viewed in terms of respiratory insufficiency and disturbed energy metabolism in exposed cells rather than in terms of DNA damage and mutations. Hence, cell phone use can be linked to cancer risk through inflammation and injury to respiration in those individuals that are prone to focal CNS inflammation from increased temperature.

ALZHEIMER'S DISEASE AND CANCER RISK

It appears that the risk of cancer in persons with Alzheimer's disease (AD) is significantly less than that in persons without the disease (24). How do we explain such a phenomenon? The NCI provocative question #6 suggests that if we understood in molecular terms why patients with AD have altered risk for cancer development, we might find leads for cancer prevention or treatment. While this question is provocative if cancer were a genetic disease, the question is less provocative when recognizing cancer as a metabolic disease. It is known that AD is a type of hypometabolic disorder (25). Loss of appetite with accompanying reduced body weight and blood glucose levels are seen in many patients with AD (25). As glucose drives tumor cell growth (Chapter 17), hypometabolism and reduced glucose would create a type of calorie-restricted environment. Such an environment would naturally inhibit tumor initiation and growth. In contrast to calorie restriction, however, the hypometabolism in AD is not associated with elevated ketone bodies. In her book, "*Alzheimer's Disease: What If There Was a Cure? The Story of Ketones*", Dr. Mary Newport addresses the importance of elevating ketone bodies as a therapy for AD. Reduced cancer risk for AD patients is likely due to loss of appetite, which reduces body weight and blood glucose levels. Low blood glucose would reduce the risk of inflammation and cancer. A more provocative question than the NCI PQ6 is how hypometabolism becomes a common phenotype in those with AD. It is easier to address the NCI provocative questions when considering cancer as a metabolic disease than as a genetic disease.

KETONE METABOLISM REDUCES CANCER RISK

In addition to avoiding established cancer risk factors, the metabolism of ketone bodies protects the mitochondria from inflammation and damaging reactive oxygen species (ROS). ROS production increases naturally with age and damages cellular proteins, lipids, and nucleic acids. Accumulation of ROS decreases the efficiency of mitochondrial energy production, thus requiring compensatory fermentation. Cancer risk increases with age and accumulation of ROS. Ketone metabolism enhances mitochondrial function, thus preventing fermentation. Ketone body metabolism, especially when glucose levels are reduced, will go far in preventing genomic instability and reducing cancer risk (26).

The origin of mitochondrial ROS comes largely from the spontaneous reaction of molecular oxygen (O_2) with the semiquinone radical of coenzyme Q, that is, $^\bullet$QH (see Fig. 4.4). This interaction will generate the superoxide radical $O_2^{-\bullet}$ (27–29). Coenzyme Q is a hydrophobic molecule that resides in the inner mitochondrial membrane and is essential for electron transfer. Ketone body metabolism increases the ratio of the oxidized form to the fully reduced form of coenzyme Q ($CoQ/CoQH_2$) (28). Oxidation of the coenzyme Q couple reduces the amount of the semiquinone radical, thus decreasing the probability of superoxide production (27). Ketone body metabolism reduces ROS and enhances mitochondrial energy efficiency, thus reducing cancer risk.

In addition to reducing ROS, ketone body metabolism also increases the reduced form of glutathione since the cytosolic-free $NADP^+/NADPH$ concentration couple is in near equilibrium with the glutathione couple (27, 30, 31). More specifically, ketone body metabolism facilitates destruction of hydrogen peroxide. The reduction of free radicals through ketone body metabolism helps maintain the inner mitochondrial membrane integrity. This enhances the energy efficiency of mitochondria. As ROS also induce tissue inflammation, reduced ROS will reduce tissue inflammation. Ketone bodies are not only a more efficient metabolic fuel than glucose but also possess anti-inflammatory potential (Chapters 17 and 18). Metabolism of ketone bodies for energy will maintain mitochondrial health and efficiency, thus reducing the incidence of cancer. How simple is this?

MITOCHONDRIAL ENHANCEMENT THERAPY

The simplest means of initiating the metabolism of ketone bodies is through dietary energy reduction with *adequate nutrition*. It is important to emphasize adequate nutrition, as reduced caloric intake associated with malnutrition can potentially increase cancer incidence (32–34). DER should not produce malnutrition! Consequently, consumption of foods containing the active groups of respiratory enzymes (iron salts, riboflavin, nicotinamide, and pantothenic acid) will be effective in maintaining health when combined with dietary energy restriction (3). Vitamin D is also known to enhance mitochondrial efficiency. Indeed, any food item that can enhance mitochondrial respiratory energy efficiency will be effective in reducing the risk of cancer.

Reducing blood glucose levels through DER facilitates ketone body uptake and metabolism for use as an alternative respiratory fuel (27, 35, 36). It is important to remember that tumor cells cannot effectively use ketone bodies for energy because of their injury to respiration. The metabolism of ketone bodies increases succinate dehydrogenase activity while enhancing the overall efficiency of energy production through respiration (37, 38). The supplementation of DER with ketone esters could be even more effective as a respiratory enhancement therapy (39). The drug 1,3-butanediol could also help elevate ketone bodies to reduce inflammation and cancer.

Specifically, DER and ketone body metabolism delays entropy (40). Entropy is the bioenergetic signature of cancer (1). Entropy refers to the degree of disorder in systems and is the foundation of the second law of thermodynamics (41). Szent-Gyorgyi has described cancer as an increased state of entropy, where randomness and disorder predominate (42). As cancer is a disease of accelerated entropy, DER targets the very essence of the disease (1).

THERAPEUTIC FASTING AND CANCER PREVENTION

It is well documented that DER reduces the incidence of both inherited and acquired cancers in experimental animals (33, 43–48). Evidence also indicates that DER can reduce the incidence of several human cancers (49, 50). However, a 40% DER in rodents is comparable to water-only therapeutic fasting or to very low caloric diets (500–600 kcal/day) in humans (40). This is due to differences in the basal metabolic rate, which is about seven times less in humans than in mice (Chapter 17). Consequently, DER is tolerated better and is more effective in preventing cancer in humans than in mice. The implementation of periodic DER, which targets multiple cancer-provoking factors, can be a simple and cost-effective lifestyle change that is capable of reducing the incidence of cancer.

Humans have evolved to function for prolonged periods in the absence of food. Herbert Shelton described how most adults in good general health can function normally after fasting (water only) for as long as 30–40 days (51). While total food abstinence for this long might seem impossible to many people, the evidence presented showing that this is possible is quite compelling. George Cahill and

Oliver Owen have also shown that many overweight people could be fasted for prolonged periods (months) without adverse effects (52). Owen and Cahill were also the first to show that ketone bodies become the major fuel for the brain during periods of starvation (53). "Danjiki" is the Japanese term for therapeutic fasting and is known to produce numerous health benefits including prevention of cancer. Humans are capable of conducting prolonged fasts without harm.

It is important to mention that therapeutic fasting is not the same as starvation. Although the terms fasting and starvation are often used interchangeably, they represent different physiological states. Starvation is a pathological state where the body suffers from energy imbalance and is deprived of key minerals and vitamins necessary for maintaining metabolic homeostasis. Fasting, on the other hand, is therapeutic and maintains metabolic homeostasis. Vitamins A, D, E, and K are stored in liver and body fat, and are released slowly during fasting. Minerals are stored in the bones and are also released slowly during fasting. Only the water-soluble vitamins C and B-complex vitamins would require supplementation after a 10–14-day fast. Periodic therapeutic fasting is extremely healthy for the body. Although weight loss will occur following therapeutic fasting, the weight loss associated with fasting is natural and nontoxic. Fasting-associated weight loss contrasts markedly with chemotherapy-associated weight loss, which is unnatural and often linked to toxic poisoning.

Blood Glucose and Ketone Levels During Fasting

I have recorded the blood glucose and ketone levels in several of my students who have voluntarily fasted for up to 6 days. The students were all healthy young adults (males and females) between 21 and 28 years of age. The students consumed only water or decaffeinated green tea during the fast. All students, both males and females, were able to bring their blood glucose and ketone levels into the therapeutic ranges within 3 days (Chapter 18). Most cancer patients should have a similar experience as long as they are not taking any interfering medications.

Glucose withdrawal symptoms were experienced by most of the students over the first couple of days, but these symptoms were transient and gradually subsided after 2 days. It is interesting that glucose withdrawal symptoms (anxiety, headache, nausea, etc.) are also seen in many persons following withdrawal from other addictive substances such as alcohol, tobacco, and drugs. Some of the students felt energetic after 5 days of fasting. They all learned that fasting is therapeutic and not harmful.

One of my graduate students, Julian Arthur, lowered his blood glucose to 39 mg/dl by the third day of the fast. I asked Julian how he felt walking around with such low blood glucose levels. He said, "I feel fine, no problems." Julian's blood ketones were also at 1.1 mmol, which would compensate for low glucose and prevent adverse effects of hypoglycemia. Hypoglycemia is a concern only for those individuals who lower glucose levels without also elevating their blood ketone levels. The gradual transition from glucose to ketone metabolism protects tissues from the effects of hypoglycemia. George Cahill and colleagues have documented these observations (52, 54, 55).

Another student, Ivan Urits, was unable to lower his glucose to the metabolic range despite 6 days of fasting and elevated ketone levels (2–3 mmol). His glucose was reduced only to 68 mg/dl during the fast. It turned out that Ivan was drinking caffeinated black coffee, rather than drinking only water during the fast. Caffeine can prevent glucose levels from entering the therapeutic zone necessary to target the energy metabolism of tumor cells. Herbert Shelton argues against coffee consumption during fasting (51). It would be better to consume calorie-free decaffeinated beverages than caffeinated beverages. I suggest that persons avoid caffeinated beverages if they plan to use the restricted ketogenic diet (KD-R) as an approach to prevent cancer. It will be up to each person to know what they can or cannot do to maintain their blood glucose within the therapeutic ranges as described in Chapter 18.

Mr. Jimmy Moore also described his experience with a 7-day, mostly water-only fast in a podcast video (http://livinlavidalowcarb.com/blog/jimmy-moores-at-least-one-week-fasting-experiment-begins/10484). Mr. Moore is a well-known blogger for the health benefits of low carbohydrate diets. He was able to document the physiological changes he experienced during the fast in nontechnical language. Although Mr. Moore followed most of what Herbert Shelton would consider standard practices (51), Mr. Moore included bullion cubes in the fast. Chicken or beef bullion contains some calories and salts, which might prevent glucose from reaching the lowest levels needed to put maximum metabolic pressure on tumor cells. However, Mr. Moore's blood glucose levels did fall within the required therapeutic ranges for tumor management during his fast. Further research is needed to document the influence of bullion and other low calorie and low carbohydrate food items on blood glucose and ketone levels during food fasting. Nevertheless, it is important for cancer patients to recognize from Mr. Moore's podcast that fasting is not harmful.

AUTOPHAGY AND AUTOLYTIC CANNIBALISM: A THERMODYNAMIC APPROACH TO CANCER PREVENTION

Autophagy is the process by which cells break down and recycle energy-rich molecules from inefficient organelles (56, 57). The deficient organelles fuse with endosomes or lysosomes for the digestive process. Autolytic cannibalism is the process by which the body digests whole cells and tissues that are metabolically inefficient relative to normal healthy cells and tissues. Both processes can occur under DER. DER creates global metabolic stress on all cells and organ systems in the body. The nutrients contained in the metabolically deficient cells and tissues are then redistributed through the circulation to normal cells in order to sustain the vitality of the body under energy stress. The weak cells are revealed only after metabolic energy stress is placed on the whole society of cells.

I predict that normal cells of the body will use dysplastic tissue (precancerous) as a source of energy in order to maintain body heat and metabolic homeostasis under energy stress. The body temperature is lower during therapeutic fasting or DER than during unrestricted feeding (40, 58, 59). In order to maintain temperature, the body will metabolize the stored energy (fat) or dysplastic tissue.

Unlike normal cells that transition to ketone bodies for energy under DER, dysplastic tumor cells have inefficient respiration and are in the early stages of fermentation dependence. The transition from glucose to ketones as a major energy source is an evolutionary conserved adaptation to food deprivation that permits the survival of normal cells during extreme shifts in a nutritional environment. The metabolism of ketones spares protein and protects the brain. Only those cells with flexible genomes, honed through millions of years of environmental forcing and variability selection, can readily transition from one energy state to another (Chapter 15).

I propose a thermodynamic mechanism of cancer prevention under DER that exploits the metabolic flexibility of normal cells at the expense of the genetically defective and metabolically challenged dysplastic cells. Tumor cells will be less able to survive energy stress than normal cells, thus allowing the normal cells to use the energy metabolites of the dysplastic tissue for maintaining body heat and organ homeostasis. In other words, the body will cannibalize dysplastic tissue through autolytic processes in order to supply the normal cells with energy. *The strong devour the weak for the good of the whole*. This process would occur, however, only under conditions of energy stress. Under conditions of energy excess, cancer cells would persist in the body and possibly thrive. The thermodynamic mechanism I propose for DER-induced cancer prevention is similar in concept to the role of CR-induced hormesis, and "vitagenes" for enhancing longevity (60).

CANCER PREVENTION BY FOLLOWING RESTRICTED KETOGENIC DIET

I know of a situation where a person used the KD-R metabolic therapy to treat cervical dysplasia. Cervical dysplasia refers to abnormal changes in the cells on the surface of the cervix and can range from mild to severe. Severe dysplasia can sometimes be considered precancerous. An abnormal Pap smear, colposcopy, and biopsy had indicated that this person had high grade squamous intraepithelial lesions. The person then conducted an KD-R for 4 weeks prior to further scheduled biopsy. As the KD-R is difficult to self-administer in the home setting, the person and her male friend did the metabolic therapy together.

Interestingly, the follow-up biopsy after the KD-R revealed only a few regions of dysplasia, none of which were considered to be high grade. The attending physician was miffed at the dramatic change in diagnosis. While this is simply an anecdotal report with no hard proof that the KD-R was responsible for the effect, the findings suggest that the KD-R could have been responsible. No treatment other than the KD-R was administered between the first and the second biopsy.

Further studies in more patients will be needed to confirm whether the KD-R is an effective treatment for cervical dysplasia and possibly cancer prevention. Each patient could serve as his/her own control for these studies. This therapy could be done in combination with various other cancer-screening procedures, for example, breast or lung biopsies and colonoscopies. Why use surgery or toxic drugs to treat preneoplasia if a nontoxic metabolic therapy such as fasting or the KD-R is effective in removing the neoplasia or nodules? Those interested in cancer prevention should know about this.

How Long Should People Fast or Remain on the Restricted Ketogenic Diet to Prevent Cancer?

The length of therapeutic fasting or administration of the KD-R for cancer prevention could vary from one person to the next. In general, a 7-day, water-only fast done once per year would be sufficient for the body to consume dysplastic or precancerous tissue. It usually requires 2–3 days for the blood glucose to reach the therapeutic levels of 55–65 mg/dl and for ketones to reach the 3–5 mmol therapeutic levels. Once the body reaches this metabolic state, autophagy and autolytic cannibalism will begin purging the body of neoplastic tissue.

For those individuals incapable of conducting longer fasts, several shorter fasts (2–3 days) done two to three times per year should also be effective in preventing cancer. A ketogenic diet consumed for 1 week should also be an effective cancer-prevention strategy as long as the blood glucose and ketones are maintained within the therapeutic ranges. It is clear that there are many variations to the metabolic approach to cancer prevention. It is my opinion that these procedures, while simple in principle, are difficult to conduct in practice. Most people simply lack the desire or motivation to undergo these practices. Consequently, health clinics that focus on energy metabolism and metabolic therapies could help people prevent cancer or manage their disease.

Experimental support for my ideas would have far-reaching significance as an alternative means of preventing and treating cancer. This approach exploits the metabolic flexibility of normal cells at the expense of the genetically defective and metabolically challenged tumor cells in order to maintain energy homeostasis. Whereas most tumor therapies rely on external agents, for example, radiation, chemicals, and stem cells, our approach relies on the energy transformations occurring within the tumor and the host tissues under caloric stress. These energy transformations will cause greater autolysis in tumor tissue than in normal tissue. Reduced growth of tumor tissue and improved energy homeostasis will be the outcome. Support for my ideas can lead to new cancer therapies that are less toxic and more effective than those presently available. This therapeutic approach to cancer management would be empowering to those individuals who want to control their destiny. The information in this chapter addresses the NCI provocative questions #1, #4, and #6 (provocativequestions.nci.nih.gov).

REFERENCES

1. SEYFRIED TN, SHELTON LM. Cancer as a metabolic disease. Nutr Metab. 2010; 7: 7.
2. ARMSTRONG GT, LIU Q, YASUI Y, NEGLIA JP, LEISENRING W, ROBISON LL, et al. Late mortality among 5-year survivors of childhood cancer: a summary from the Childhood Cancer Survivor Study. J Clin Oncol. 2009; 27: 2328–38.
3. WARBURG O. Revidsed Lindau Lectures: The prime cause of cancer and prevention - Parts 1 & 2. In: Burk D, editor. Meeting of the Nobel-Laureates Lindau, Lake Constance, Germany: K.Triltsch; 1969. http://www.hopeforcancer.com/OxyPlus.htm.
4. ANAND P, KUNNUMAKKARA AB, SUNDARAM C, HARIKUMAR KB, THARAKAN ST, LAI OS, et al. Cancer is a preventable disease that requires major lifestyle changes. Pharm Res. 2008; 25: 2097–116.
5. DOLL R, PETO R. The causes of cancer: quantitative estimates of avoidable risks of cancer in the United States today. J Natl Cancer Inst. 1981; 66: 1191–308.
6. PINE SR, MECHANIC LE, ENEWOLD L, CHATURVEDI AK, KATKI HA, ZHENG YL, et al. Increased levels of circulating interleukin 6, Interleukin 8, C-Reactive protein, and risk of lung cancer. J Natl Cancer Inst. 2011; 103: 1112–22.
7. BISSELL MJ, HINES WC. Why don't we get more cancer? A proposed role of the microenvironment in restraining cancer progression. Nat Med. 2011; 17: 320–9.
8. BISSELL MJ, RADISKY DC, RIZKI A, WEAVER VM, PETERSEN OW. The organizing principle: microenvironmental influences in the normal and malignant breast. Differentiation. 2002; 70: 537–46.
9. DOLBERG DS, HOLLINGSWORTH R, HERTLE M, BISSELL MJ. Wounding and its role in RSV-mediated tumor formation. Science. 1985; 230: 676–8.
10. SIEWEKE MH, BISSELL MJ. The tumor-promoting effect of wounding: a possible role for TGF-beta-induced stromal alterations. Crit Rev Oncog. 1994; 5: 297–311.
11. STERNLICHT MD, BISSELL MJ, WERB Z. The matrix metalloproteinase stromelysin-1 acts as a natural mammary tumor promoter. Oncogene. 2000; 19: 1102–13.
12. STERNLICHT MD, LOCHTER A, SYMPSON CJ, HUEY B, ROUGIER JP, GRAY JW, et al. The stromal proteinase MMP3/stromelysin-1 promotes mammary carcinogenesis. Cell. 1999; 98: 137–46.
13. COUSSENS LM, WERB Z. Inflammation and cancer. Nature. 2002; 420: 860–7.
14. COLOTTA F, ALLAVENA P, SICA A, GARLANDA C, MANTOVANI A. Cancer-related inflammation, the seventh hallmark of cancer: links to genetic instability. Carcinogenesis. 2009; 30: 1073–81.
15. SONNENSCHEIN C, SOTO AM. The Society of Cells: Cancer and the Control of Cell Proliferation. New York: Springer; 1999.
16. CLEMENT K, VIGUERIE N, POITOU C, CARETTE C, PELLOUX V, CURAT CA, et al. Weight loss regulates inflammation-related genes in white adipose tissue of obese subjects. FASEB J. 2004; 18: 1657–69.
17. MATHEWS EH, LIEBENBERG L, PELZER R. High-glycolytic cancers and their interplay with the body's glucose demand and supply cycle. Med Hypotheses. 2011; 76: 157–65.
18. BRODY JE. Cellphone: A convenience, a hazard, or both? New York Times. 2002.
19. DELLORTO D. WHO: Cell Phone Use Can Increase Possible Cancer Risk. CNN; 2011. http://www.cnn.com/2011/HEALTH/05/31/who.cell.phones/index.html.
20. MOULDER JE, FOSTER KR, ERDREICH LS, MCNAMEE JP. Mobile phones, mobile phone base stations and cancer: a review. Int J Radiat Biol. 2005; 81: 189–203.
21. KWON MS, VOROBYEV V, KANNALA S, LAINE M, RINNE JO, TOIVONEN T, et al. GSM mobile phone radiation suppresses brain glucose metabolism. J Cereb Blood Flow Metab: Official J Int Soc Cereb Blood Flow Metab. 2011; 31: 2293–301.
22. KLOSTERGAARD J, BARTA M, TOMASOVIC SP. Hyperthermic modulation of respiratory inhibition factor- and iron releasing factor-dependent macrophage murine tumor cytotoxicity. Cancer Res. 1989; 49: 6252–7.
23. LYONS BE, OBANA WG, BORCICH JK, KLEINMAN R, SINGH D, BRITT RH. Chronic histological effects of ultrasonic hyperthermia on normal feline brain tissue. Radiat Res. 1986; 106: 234–51.

24. Behrens MI, Lendon C, Roe CM. A common biological mechanism in cancer and Alzheimer's disease? Curr Alzheimer Res. 2009; 6: 196–204.
25. Landin K, Blennow K, Wallin A, Gottfries CG. Low blood pressure and blood glucose levels in Alzheimer's disease. Evidence for a hypometabolic disorder? J Intern Med. 1993; 233: 357–63.
26. Raffoul JJ, Guo Z, Soofi A, Heydari AR. Caloric restriction and genomic stability. J Nutr Health Aging. 1999; 3: 102–10.
27. Veech RL. The therapeutic implications of ketone bodies: the effects of ketone bodies in pathological conditions: ketosis, ketogenic diet, redox states, insulin resistance, and mitochondrial metabolism. Prostaglandins Leukot Essent Fatty Acids. 2004; 70: 309–19.
28. Veech RL, Chance B, Kashiwaya Y, Lardy HA, Cahill GF Jr. Ketone bodies, potential therapeutic uses. IUBMB Life. 2001; 51: 241–7.
29. Chance B, Sies H, Boveris A. Hydroperoxide metabolism in mammalian organs. Physiol Rev. 1979; 59: 527–605.
30. Ziegler DR, Ribeiro LC, Hagenn M, Siqueira IR, Araujo E, Torres IL, et al. Ketogenic diet increases glutathione peroxidase activity in rat hippocampus. Neurochem Res. 2003; 28: 1793–7.
31. Seyfried TN, Mukherjee P. Targeting energy metabolism in brain cancer: review and hypothesis. Nutr Metab. 2005; 2: 30.
32. Elias SG, Peeters PH, Grobbee DE, van Noord PA. Breast cancer risk after caloric restriction during the 1944–1945 Dutch famine. J Natl Cancer Inst. 2004; 96: 539–46.
33. Hursting SD, Forman MR. Cancer risk from extreme stressors: lessons from European Jewish survivors of World War II. J Natl Cancer Inst. 2009; 101: 1436–7.
34. Qiao YL, Dawsey SM, Kamangar F, Fan JH, Abnet CC, Sun XD, et al. Total and cancer mortality after supplementation with vitamins and minerals: follow-up of the Linxian general population nutrition intervention trial. J Natl Cancer Inst. 2009; 101: 507–18.
35. Seyfried TN, Sanderson TM, El-Abbadi MM, McGowan R, Mukherjee P. Role of glucose and ketone bodies in the metabolic control of experimental brain cancer. Br J Cancer. 2003; 89: 1375–82.
36. Luo Y, Zhou H, Krueger J, Kaplan C, Lee SH, Dolman C, et al. Targeting tumor-associated macrophages as a novel strategy against breast cancer. J Clin Invest. 2006; 116: 2132–41.
37. Balietti M, Fattoretti P, Giorgetti B, Casoli T, Di Stefano G, Solazzi M, et al. A ketogenic diet increases succinic dehydrogenase activity in aging cardiomyocytes. Ann N Y Acad Sci. 2009; 1171: 377–84.
38. Sato K, Kashiwaya Y, Keon CA, Tsuchiya N, King MT, Radda GK, et al. Insulin, ketone bodies, and mitochondrial energy transduction. FASEB J. 1995; 9: 651–8.
39. Kashiwaya Y, Pawlosky R, Markis W, King MT, Bergman C, Srivastava S, et al. A ketone ester diet increased brain malonyl CoA and uncoupling protein 4 and 5 while decreasing food intake in the normal Wistar rat. J Biol Chem. 2010; 285: 25950–6.
40. Mahoney LB, Denny CA, Seyfried TN. Caloric restriction in C57BL/6J mice mimics therapeutic fasting in humans. Lipids Health Dis. 2006; 5: 13.
41. Schneider ED, Sagan D. Into the Cool: Energy Flow, Thermodynamics, and Life. Chicago: University of Chicago Press; 2005.
42. Szent-Gyorgyi A. The living state and cancer. Proc Natl Acad Sci USA. 1977; 74: 2844–7.
43. Cleary MP, Jacobson MK, Phillips FC, Getzin SC, Grande JP, Maihle NJ. Weight-cycling decreases incidence and increases latency of mammary tumors to a greater extent than does chronic caloric restriction in mouse mammary tumor virus-transforming growth factor-alpha female mice. Cancer Epidemiol Biomarkers Prev. 2002; 11: 836–43.
44. Kritchevsky D. Fundamentals of nutrition: applications to cancer research. In: Heber D, Blackburn GL, Go VLW, editors. Nutritional Oncology. Boston: Academic Press; 1999. p. 5–10.
45. Kritchevsky D. Caloric restriction and experimental mammary carcinogenesis. Breast Cancer Res Treat. 1997; 46: 161–7.
46. Hopper BD, Przybyszewski J, Chen HW, Hammer KD, Birt DF. Effect of ultraviolet B radiation on activator protein 1 constituent proteins and modulation by dietary energy restriction in SKH-1 mouse skin. Mol Carcinog. 2009; 48: 843–52.

47. Tannenbaum A. Nutrition and cancer. In: Homburger F, editor. Physiopathology of Cancer. New York: Paul B. Hober; 1959. p 517–62.
48. Cleary MP, Grossmann ME. The manner in which calories are restricted impacts mammary tumor cancer prevention. J Carcinog. 2011;10:21.
49. Steinbach G, Heymsfield S, Olansen NE, Tighe A, Holt PR. Effect of caloric restriction on colonic proliferation in obese persons: implications for colon cancer prevention. Cancer Res. 1994;54:1194–7.
50. Albanes D. Caloric intake, body weight, and cancer: a review. Nutr Cancer. 1987;9:199–217.
51. Shelton HM. Fasting for Renewal of Life. Tampa (FL): American Natural Hygene Society, Inc.; 1974.
52. Cahill GF Jr. Starvation in man. N Engl J Med. 1970;282:668–75.
53. Owen OE, Morgan AP, Kemp HG, Sullivan JM, Herrera MG, Cahill GF Jr. Brain metabolism during fasting. J Clin Invest. 1967;46:1589–95.
54. Cahill GF Jr. Fuel metabolism in starvation. Annu Rev Nutr. 2006;26:1–22.
55. Cahill GF, Veech RL Jr. Ketoacids? Good medicine?. Trans Am Clin Climatol Assoc. 2003;114:149–61. discussion 62-3.
56. Klionsky DJ. Cell biology: regulated self-cannibalism. Nature. 2004;431:31–2.
57. Singh R, Cuervo AM. Autophagy in the cellular energetic balance. Cell Metab. 2011;13:495–504.
58. Duffy PH, Leakey JEA, Pipkin JL, Turturro A, Hart RW. The physiologic, neurologic, and behavioral effects of caloric restriction related to aging, disease, and environmental factors. Environ Res. 1997;73:242–8.
59. Ingram DK, Zhu M, Mamczarz J, Zou S, Lane MA, Roth GS, et al. Calorie restriction mimetics: an emerging research field. Aging Cell. 2006;5:97–108.
60. Calabrese V, Cornelius C, Cuzzocrea S, Iavicoli I, Rizzarelli E, Calabrese EJ. Hormesis, cellular stress response and vitagenes as critical determinants in aging and longevity. Mol Aspects Med. 2011;32:279–304.

Chapter 20

Case Studies and Personal Experiences in Using the Ketogenic Diet for Cancer Management

The purpose of this chapter is to present evidence for the translational application of the principles outlined in the treatise for cancer management in humans. If cancer is indeed a metabolic disease involving damaged cellular respiration, then therapies that exploit this damage should be effective in managing the disease. The key to management would center on targeting glucose and glutamine availability to the tumors. These fuels drive cancer cell fermentation when respiration is insufficient to maintain cellular energy homeostasis. This chapter includes information gathered from the experiences of individuals who have attempted this metabolic approach to cancer management.

EFFECTS OF A KETOGENIC DIET ON TUMOR METABOLISM AND NUTRITIONAL STATUS IN PEDIATRIC ONCOLOGY PATIENTS: COMMENTS FROM DR. LINDA NEBELING

Background

Our pilot study was the first to explore the effects of diet composition on tumor glucose metabolism in pediatric oncology patients. The ketogenic diet (KD) protocol was based on extensive literature on KD and epileptic seizure control. This pilot

Cancer as a Metabolic Disease: On the Origin, Management and Prevention of Cancer, First Edition. Thomas Seyfried.

project was designed to test the feasibility and the effects of KD on tumor glucose metabolism in children with certain types of brain tumors. Could using a high fat KD impact the rate of tumor glucose use at the tumor site? The protocol, as tested, demonstrated some sustained reduction in glucose utilization rates at the tumor site when monitored by a PET scanner using FDG-18-labeled glucose. Theoretically, the effect on the rate of glucose use at the tumor site may impact the rate of tumor growth. That said, the protocol was not designed to reverse tumor growth or treat specific types of cancer.

Lessons Learned When Introducing the Diet

The protocol required that the children were stable and able to tolerate an oral diet and had a stable home environment with committed parents or caretakers who were willing to support the diet protocol. Implementation of the KD required care and attention of an experienced dietitian to monitor the establishment of the state of ketosis in each child (1). Gradual introduction to the diet over a 4- to 6-day period minimized transient gastrointestinal disturbances, such as nausea, vomiting, diarrhea, or constipation, that may occur if the diet is introduced to quickly. Following the diet at home, for any period of time, requires considerable enthusiasm, oversight, and training of the primary caretaker.

Patient Selection and Response to Diet

The KD has been implemented successfully for decades to control children with epilepsy who are resistant to seizure control with medications. The diet and the children who participated in the project were selected because their specific tumors were resistant to chemotherapy, there were limited surgical options, and they had experienced seizures. None of the subjects were receiving active radiation treatment at the time they were following the diet (2). The protocol was developed for short-term use. Serum lipids, glucose, ketones, insulin, and protein levels were monitored weekly. As expected, blood ketone levels were extremely sensitive indicators to the degree of ketosis and dietary adherence. In spite of the considerable support and instruction provided, and routine follow-up visits, adherence to the KD protocol was not perfect. A wrong snack or a can of soda would impact ketosis, requiring the patient and caretaker to focus on dietary adherence the next day. Consumption of the diet was not a major limitation for the patients, but if we could have developed a ketosis-compliant Oreo cookie, it would have been a big hit. Overall lipid levels in the subjects enrolled in the pilot study were not adversely affected. No toxicities were documented.

Patient 1 had experienced repeated seizures before starting the KD pilot study. She did not experience further seizures during the duration of the protocol. There was an overall improvement in her quality of life during this period as assessed by her medical team and her primary caretaker.

What We Learned

The data from this pilot project suggested that the KD was able to meet the overall energy and nutrient needs of the cancer patients. Tolerance to the diet was established within 4–5 days without much difficulty. The fact that dietary-induced ketosis could be established in children with advanced-stage metastatic disease was a major accomplishment. Modifications to the dietary composition were required during periods when a child was ill with a cold or sinus infection in order to sustain the state of ketosis. The use of low carbohydrate forms of vitamin and mineral supplements was essential to maintain nutrient adequacy (1). Caloric intake and body weights remained fairly stable during the trial period. Blood glucose values declined to below normal levels. Blood ketone levels were elevated 20- to 30-fold (2).

Use of PET technology helped to assess changes in glucose utilization rates at the site of the tumor during the pilot study. PET scan data indicated a decrease by approximately 22% in FDG update in both patients, reflecting glucose metabolism at the site of tumor (2).

There were many limitations to this pilot study. The sample size did not allow for statistically measurable results. The specific type of patient population, that is, children with advanced-stage cancer, was another limitation. Because patients were required to have measurable evidence of disease and a stable environment at home that would support adherence to the dietary protocol, the patient recruitment period took over 2 years. The use of PET technology was a major asset to this project to enable the investigators to better assess the effects of the diet at the tumor site.

Since the publication of this pilot study, new scientific advances have expanded our understanding of cancer metabolism (3, 4). Interest in this protocol remains, and I have continued to receive communications from interested patients and oncologists around the world. The KD protocol is not a treatment itself but may be considered a possible compliment to the treatment directed by the oncology team. Current clinical trials will expand our understanding of the relationship between KD and cancer metabolism (http://www.clinicaltrials.gov).

Linda Nebeling, Ph. D., MPH, RD, FADA, Chief, Health Behaviors Research Branch, Behavioral Research Program, Division of Cancer Control and Population Sciences, National Cancer Institute.

RAFFI'S STORY: COMMENTS FROM MIRIAM KALAMIAN

In the December of 2004, our 4-year-old son Raffi was diagnosed with a brain tumor. Although the biopsy concluded that it was a low grade glioma, its huge size, delicate location, and atypically aggressive nature suggested a relatively poor prognosis. Like most parents, we placed our child's fate in the hands of specialists who immediately implemented the "gold standard" of care and promised to keep us abreast of any cutting-edge breakthroughs. Unfortunately for Raffi, the "gold standard" protocol failed and no new technologies came galloping to the rescue.

Instead, we were forced to watch as progressively more aggressive interventions dismantled our child, piece by precious piece. Vision, language, cognition, motor skills, and endocrine function deteriorated. This took a huge toll on his quality of life. Our little fighter had done everything that was asked of him, but his tumor was clearly winning.

In the March of 2007, our world changed abruptly. Researchers at Boston College published a study that described the effect of carbohydrate- and calorie-restricted diets on implanted brain tumors in mice. As expected, dietary restriction resulted in a shift to ketosis, a metabolic state generally associated with starvation. However, this KD also reduced the amount of serum glucose available for tumor tissue metabolism, thus slowing the rate of disease progression.

Although it is generally accepted that tumor tissue thrives on glucose, this study was one of the first to suggest that there might be a connection between dietary intake of carbohydrates and tumor progression. An earlier case study involving two children with late-stage astrocytomas had shown that an 8-week trial of a KD met the children's nutritional needs while simultaneously reducing glucose uptake at the tumor site. We learned that the KD protocol had been safely and successfully implemented for over 80 years as a treatment for intractable pediatric epilepsy. Could this simple yet novel dietary approach succeed where more aggressive therapies had failed?

Despite Raffi's deteriorating condition, his cancer specialist refused to engage in any discussion of diet. After several weeks of researching the topic, we decided to initiate the KD on our own. We cobbled together information provided by physicians, parents, and organizations dedicated to promoting the classic KD as a treatment for epilepsy. An online support group for parents introduced us to the real-world challenges and strategies inherent in implementing this therapy.

By this point, we had spent over 6 months away from home and it was time to return to our rural community in Montana. Raffi's pediatrician and local oncologist listened to our reasoned argument in favor of a trial of this dietary therapy. Although they were skeptical, they both viewed the KD as a "do no harm" intervention and agreed to support us if we implemented it concurrent with the treatment. Within days, Raffi was placed on a low dose, low toxicity chemotherapy drug (note that this same drug had previously failed to stop Raffi's tumor growth).

With this thin framework of support, we managed to clear the many hurdles inherent in the first few months of implementation. Miraculously, an MRI at 3 months showed that the tumor had shrunk by 15%. This led me to boldly ask for help from Beth Zupec-Kania, the Charlie Foundation's dedicated KD specialist. Beth answered our many questions, corrected our calculation errors, and granted us access to a web-based meal-planning tool. Now, we are a part of an emerging community of parents and professionals who have had firsthand experience with this kinder, gentler management strategy.

Eventually, Raffi's oncologist discontinued chemotherapy and the KD was continued as a stand-alone therapy. Over the next several years, Raffi's general health and neurological status continued to improve, but ultimately, damage done by the tumor and/or surgeries proved to be progressive and irreversible. This begs the question, "What would have been the outcome if the KD had been offered as part of that initial gold standard protocol?"

It has been more than 6 years since our son's diagnosis, yet it is impossible to recall those first few months without reliving the pain that accompanies such devastating news. KD allowed us to step away from our role as passive bystanders and take action that significantly improved our son's quality of life. Despite our success with this therapy, most medical professionals still view KD as "too difficult" or "too restrictive" for mainstream use. Granted, KD is not an option in every case, but we strongly believe that it deserves to be included in every initial discussion of treatment options.

KD as a Dietary Therapy for Cancer

Raffi's success with the KD has changed my life. First and foremost, it saved my son from further harm and gave him a better quality of life. It also fueled my passion to fine-tune this dietary approach to cancer management so that I could assist families that were facing similar struggles. To that end, I enrolled in a rigorous program of study that led to an MS degree in Human Nutrition.

Through my studies, I quickly learned that the internal regulation of our multiple metabolic pathways is comparable to that of an exquisite hybrid engine, shifting seamlessly between fuel sources to keep the system in an optimal energy state. As my knowledge and insight into the intricacies of both normal and tumor metabolism expanded, so did my understanding of the advantages inherent in a metabolic approach to cancer management. Raffi, among others, was a living proof that even the brain and CNS can readily utilize ketones while intrinsically producing just enough glucose to meet highly specific system requirements.

Unfortunately, I also learned that few health professionals would consider a change in practice that contradicts their commonly held beliefs, even when those beliefs are based on faulty or outdated information. Worse yet, oncologists and other specialists often actively discourage their patients from engaging in any adjunct therapy, citing a lack of evidence or voicing unfounded concerns over safety.

Understanding Ketosis

While still a student, I took every opportunity to examine fed-state ketosis as a remarkable adaptation instead of an aberrant metabolic state. I learned that health care professionals needed help in understanding the critical differences between "benign ketosis" and "diabetic ketoacidosis." The former describes the shift to ketosis during periods of reduced carbohydrate intake; the latter refers to a life-threatening condition associated with poorly controlled diabetes mellitus.

Cells derive energy from a variety of sources, and under normal physiological conditions, glucose is the primary fuel. When glucose-yielding carbohydrates are in short supply, the liver easily converts both dietary and stored fats to three types of ketone bodies, one of which is especially adept at meeting the energy needs of most cells. Another process, known as *beta-oxidation*, also utilizes fats for fuel, primarily in response to the energy needs of muscle tissue. Even the amino acids found in proteins, which are generally reserved for tissue repair and maintenance, can be diverted to glucose production if needed to meet these specialized requirements.

During carbohydrate restriction, the liver assumes primary control of glucose homeostasis by gauging how much glucose is needed and then by manufacturing it in a process known as *gluconeogenesis* (literally, "making new glucose"). Gluconeogenesis also occurs in the adrenal cortex of the kidney, which is particularly sensitive to steroid hormones, such as cortisol. In addition, the Cori cycle converts lactic acid, a cellular waste product, into glucose. These processes combine to provide the glucose needed to nourish certain glucose-dependent cells and tissue.

Implementation and Compliance

Currently, implementation of the KD is often chaotic. Almost all individuals who contact me for help are also investigating other therapies, both conventional and alternative. With no formal guidance on how to identify a "best-practice" protocol, it is sometimes left to the client or caregivers to piece together a plan from a variety of disjointed and possibly conflicting sources. In contrast, individuals receiving conventional cancer care rely on clinics for education, coordination of services, delivery of care, and assessment of treatment outcomes. Furthermore, most costs are covered by health insurance or other medical aid sources, thus easing both the emotional and financial burden for those receiving care. Obviously, conventional care clinics need to offer the KD as part of their integrated cancer treatment plans, thus making this option accessible to more people.

To meet the current need for patient education and support, I make available an extensive list of resources. Some of this material was developed for purposes other than cancer management (e.g., epilepsy, diabetes); others are my own "works in progress" that I refine and rewrite based on emerging evidence and personal experience. If requested, I put together a "Starter Kit" to help my clients make the needed changes. Firsthand experience with my own son provided me with these insights:

- Despite my hard work and best intentions, 100% compliance to the KD is neither reasonable nor feasible.
- I need tools to respond to real-world problems, such as calculation errors, malfunctioning scales, and the misguided actions of others.
- Blood glucose levels may rise due to factors beyond my control, such as illness, injury, comorbidities, or prescribed medications.
- Most individuals benefit from a support network that keeps them connected to a community of people facing similar struggles.

Although I adapt the KD to meet each client's specific needs, I start with a certain set of goals and objectives in the areas of implementation, oversight, and control of the physical and psychosocial environments. I discuss the impact that stress has on glucose-regulating hormones and the ways in which attitude affects compliance. Rather than attempting to control for every contingency, I focus instead on providing clients with strategies that help them to cope with their personal set of challenges.

Some clients begin the KD with a complete fast lasting one or more days. This quick shift to ketosis can be quite validating, but it is not feasible in every situation. Alternatively, one can ease into ketosis over a period of several weeks. How a client initiates the KD is not as important as whether or not they have a grasp on the details and are engaged in making the needed changes in their diet.

As clients work through these early stages, questions inevitably arise. Follow-up support is especially critical in the first few weeks. I keep the communication lines open by asking questions such as How are you feeling? What is your weight status? Are you using a kitchen scale to weigh your foods? Are you utilizing the meal-planning tools? What are your blood glucose and ketone levels? This interaction not only provides me with information needed to refine the diet plan but also cues me as to how engaged the individual is in the process.

Most of my clients have some initial resistance to consuming the recommended amount of fats and oils. I suspect this is the case if a client states that they are hungry or if weight loss has been too rapid. This prejudice against fat is not surprising, given the amount of press devoted to low fat campaigns equating fat consumption with heart disease and certain cancers. This reasoning assumes "cause and effect." Instead, these diseases may arise from a disruption in metabolism commonly found in individuals whose diets mainly consist of easily digestible carbohydrates.

"Numbers" are important indicators of compliance. I ask clients to test blood glucose and ketone levels on arising, in midafternoon, and in the early evening. I expect to see a quicker shift to ketosis in individuals who fast or in young children (who are more metabolically flexible). I methodically review food choices, exercise habits, stress levels, and changes in drugs if these numbers are not within the therapeutic range. Often, a pattern emerges that suggests the needed changes.

Long-term adherence to KD involves a variety of strategies that I refer to as *boxing with the tumor*. For example, I may suggest alternating between a classic KD and a more calorie-restricted one. Occasional short fasts may also be utilized to maintain metabolic pressure on the tumor or to simply strengthen ketosis.

All my clients undergo routine assessments that are part of their cancer treatment protocol. Generally, these assessments include MR or PET imaging as well as lab tests that measure biomarkers (if available) for their specific cancer type. Patients undergoing chemotherapy also have routine complete blood counts and comprehensive metabolic panels to monitor the negative impact of therapy on the body's organs and systems.

Chronic stress undermines one's ability to maintain low blood glucose levels. Notably, physical and emotional stress stimulates the production of cortisol, which results in excessive gluconeogenesis. Stressors also keep the sympathetic nervous system on high alert. One effect is that fat and glucose are liberated in anticipation of extraordinary muscular demands. Few individuals with cancer can claim to be stress free, so the effect stress may exert must be considered when evaluating an individual's response to treatment.

I am very grateful when caregivers are supportive of the KD, as this support increases compliance. However, I speak to many caregivers whose loved ones ultimately decide not to implement the KD. For some, poor general health is the driving factor, and there may be little to no gain in either quality or quantity of life by slowing the progression of the disease. Others are skeptical of the potential efficacy of such a simple therapy. Still others are simply not willing to make the needed changes in their lifestyle. I sympathize with caregivers, but in my experience, individuals who are pressured into adopting the KD rarely comply and therefore have little chance of success. Their conclusion: "I tried it and it didn't work." This has the potential to be damaging since their experience is likely to color the perception of other individuals with cancer who are considering the KD.

Effects of Treatment

Most clients note some type of initial discomfort as they move into ketosis. The most common complaint is headache, followed by fatigue and mood changes. Children, who generally make the shift faster, may experience lethargy, nausea, and vomiting. These symptoms are transient and easily remedied by following the initiation algorithm. In the short term, the KD may cause an increase in lipid levels, and over the long term, it may raise the risk of developing kidney stones or osteoporosis. Even these risks may be lowered through modifications to the general protocol.

There is one adverse effect of KD that is specific to the pediatric population: children who adhere to a KD long term generally experience a decrease in linear growth (as shown by data gathered from children on KD as a treatment for epilepsy). Keep in mind that conventional therapies also impact growth and have other far more serious adverse effects than those associated with the KD.

Proper implementation of the KD does not guarantee slowed tumor progression. On the other hand, I have had clients who strayed from the classic, for example, consuming more than the recommended amount of protein, yet still enjoyed good outcomes. Perhaps collection and analysis of data, combined with research that examines metabolic response to treatment, will begin to reveal some of the variables that undoubtedly influence individual outcomes.

One advantage of the KD as an adjunct therapy (particularly for brain tumors) is that ketones are neuroprotective, possibly mitigating some of the damage done by conventional therapies. Concurrent use of the KD may also have a synergistic effect on tumor metabolism, allowing for the luxury of starting with less aggressive therapies. Keep in mind that treatment plans are considered successful if they extend life by only a few weeks or months. Obviously, these therapies value *quantity* over *quality* of life.

Challenges

There are significant challenges to gaining the support of conventional cancer specialists. Most daunting is the requirement that the KD undergoes the type of clinical trials that lead to evidence of improved outcomes. Double-blind randomized controlled trials (RCTs), such as those funded by pharmaceutical companies, are the gold standard here. Unfortunately, RCTs are a logistical nightmare for the KD. Consider how little funding there is to conduct trials of any nondrug therapy, then factor in the burden of blinding the study by using prepackaged foods, such as liquid meal replacements, which are associated with a high dropout rate in studies involving free-living adults. Also, as our understanding of tumor cell metabolism evolves, we will need to test the variables that may impact outcomes specific to certain subsets of tumors.

As a dietary therapist, I need more and better tools to help my clients gain the support of cancer specialists. As head of the treatment team, the cancer specialist has the authority to order needed tests and the skills to monitor general health. If the specialist is not willing or able to engage in this process, I suggest that my clients seek out health care professionals who will agree to provide oversight.

Family support for KD is a critical component for success. In my experience, families are either lifelines or saboteurs. An additional challenge for children lies in the dietary habits of siblings and peers, and in the school's commitment to following the plan. I urge my clients to join monitored online support groups, as it is vitally important for new clients to know that others have walked the same (or similar) path.

To the best of my knowledge, few people in the United States employ KD as a stand-alone therapy. Therefore, it is difficult to objectively assess its therapeutic efficacy. However, the same could be said about conventional therapies where the KD is used as an adjunct. Clients often state that their specialist has noted an unusually good response to treatment without granting a passing nod to the role that the KD may have played.

In concluding, I believe that KD is poised to gain greater acceptance as an adjunct therapy in the treatment of cancer. What is needed now is a surge in research that will clarify a best-practice protocol. Such a protocol may include low toxicity drugs that will work synergistically to take advantage of the tumor tissue's compromised metabolic state and inferior adaptive mechanisms. In the meantime, those of us who are passionate about the KD should collaborate to form a more unified and accessible "keto" community charged with providing education and support.

Miriam Kalamian, Dietary Therapies LLC, Hamilton, MT (http://dietarytherapies.com).

BIOLOGICAL PLAUSIBILITY THAT CANCER IS A METABOLIC DISEASE DEPENDENT FOR GROWTH ON GLUCOSE AND GLUTAMINE: COMMENTS FROM DR. BOMAR HERRIN

Two years ago while lifting weights my right arm snapped. This was a "before cancer (BC)/after diagnosis (AD)" event for me. Life changed and what was to follow included surgical repair of the pathological fracture, radiation therapy to this and other cancerous lesions, and a footrace to avoid the natural course of this illness, described in the medical literature as *incurable*. I was diagnosed with multiple myeloma.

The most obvious difference to me between my cancer cells and normal cells was the glowing PET scan in the areas of malignancy. My cancer cells were picking up the radiolabeled glucose molecules to a much greater extent than my healthy cells. This observation initiated my internet search for a solution. Websites with pastel backgrounds and claims of "miracle cures" were abundant, but I was repeatedly drawn to the black-and-white articles with small print and real data. There was good science there! There was also a trail of research from the early observations of Otto Warburg to the work of Dr. Thomas Seyfried. A metabolic vulnerability of cancer cells made sense. It had biological plausibility and a risk/reward profile, which was compelling.

After multiple e-mail exchanges with Dr. Seyfried, I decided on a substrate-utilization assault on the malignant tumor cells using a combination of a KD, 2-deoxyglucose (2-DG), and phenylbutyrate (PB). There were protocols for 2-DG and PB being used separately in clinical trials for patients with advanced cancer, but none involved combining these agents with a KD. Not wanting to reach the advanced disease stage described in these studies, I asked several institutions to host a "Cancer Cell Shindig." I would bring the malignant cells, the guinea pig, and the money if they would organize the party. There were no takers.

Using an aggressive Atkins Diet and physical activity, I was able to push my glucose down into the 48–70 mg/dl range while achieving elevated urinary ketosis. I followed published protocols for 2-DG and PB. I selected 2-DG (40 mg/kg) each morning and PB (5 g three times/day). This triad therapy was well tolerated, but I had difficulty maintaining the degree of hypoglycemia and ketosis needed to metabolically stress the tumor cells. This may be both a metabolic and social issue. Not eating creates a type of isolation from friends and family, and my body seemed to be on a mission to return glucose levels to 90 mg/dl at every opportunity. I am convinced that the combination of diet and/or drugs, which disrupt glucose metabolism, will alter both cancer recurrence and cancer survival rates. As my internet search continues, the marked decrease in cancer rates in patients using metformin, a diabetic medication altering glucose metabolism, and its activity against multiple cancer cell lines, offers more evidence that Dr. Seyfried is barking up the right tree!

This week (July 11, 2011) will be the 2-year anniversary of my cancer-induced arm break. Although my disease remains in the "incurable" column, there appears to have been no progression. The most recent laboratory work showed improved results over those of a year ago. Is this dumb luck or biological plausibility? More studies need to be done.

Dr. Bomar Herin, Physician and cancer patient.

USING THE RESTRICTED KETOGENIC DIET FOR BRAIN CANCER MANAGEMENT: COMMENTS FROM NEURO-ONCOLOGIST, DR. KRAIG MOORE

I can describe three patients who I have treated with the restricted ketogenic diet (KD-R) according to the recommendations of the Seyfried group (4). All three patients have been diagnosed with glioblastoma multiforme (GBM). All three patients underwent the standard therapy of surgery for tumor debulking followed by standard external beam radiation therapy (XRT) given concurrently with low dose temozolomide (Temodar) followed by monthly adjuvant Temodar. All three patients started the KD-R following completion of the standard therapy. Only one of the three patients (patient 1) underwent post-KD-R PET scan evaluation. Unfortunately, only the official interpretation is available. Thus, patient 1 is presented as a small case report. At present, patients 2 and 3 are still on the KD-R.

Patient 1 is a 40-year-old male who in 2008 presented with word-finding difficulties, confusion, and blurred vision. Imaging revealed a left parietal heterogeneously enhancing mass. He underwent a gross total resection with placement of Gliadel Wafers in early 2009. Final pathology was reported as GBM. He was treated postoperatively with standard XRT with Temodar followed by 12 cycles of adjuvant monthly Temodar. The patient tolerated without difficulty the 12 cycles of monthly adjuvant Temodar, which was discontinued after 12 cycles due to stable disease. The patient did well for several months after Temodar treatment. MRI performed in July 2010 as part of a routine follow up revealed a new area of enhancement. He was subsequently started on Avastin. The patient started the KD-R in July 2010. He remained compliant until November 2010. For the most part, he tolerated the KD-R without difficulty. No significant fatigue or difficulties with mental capacity were reported. He continued to work in addition to maintaining his exercise regimen. The major difficulty encountered was in maintaining his blood glucose in the target range of 55–65 mg/dl. Despite significant reductions in daily caloric intake, his blood glucose ranged from 50 to 91 mg/dl.

The morning glucose was consistently the best ranging from 55 to 70 mg/dl on average; however, glucose readings measured at midday or in the evening were often above the therapeutic range (55–65 mg/dl). The maintenance of ketosis was less of a problem than the maintenance of low glucose. The patient maintained a blood ketone level in the 4 mM range without signs or symptoms of pathological ketosis. The patient's major complaint was becoming accustomed to eating smaller portions, high in fat and low in carbohydrate. He found it very difficult to find foods that either had very low carbohydrates or no carbohydrates at all. He experienced some success in lowering the blood sugar slightly when products such as toothpaste, mouthwash, soaps were changed to products that had little or no carbohydrates or preservatives such as Arm & Hammer, and Ivory soap. Although not large in magnitude, he did manage to lower his blood glucose by a point or two.

The patient was adherent to the KD-R for 4 months. In September 2010, follow-up MRI of the brain was performed and it revealed a new area of enhancement. No PET scan or MR spectroscopy was performed at that time to determine if the new area of enhancement was disease progression or tumor necrosis. In my experience, patients with high grade glioma that have Gliadel Wafers placed at surgery should be investigated with either a PET scan or an MR spectroscopy. Patients who undergo the standard therapy of postoperative radiation/chemotherapy followed by adjuvant chemotherapy are slightly more prone to tumor necrosis and/or new areas of tumor enhancement. Avastin was discontinued in September 2010, and the patient enrolled in a treatment at another medical center, but remained on the KD-R until the new therapy was started in November 2010. As part of the pretreatment imaging, the patient underwent a CT/PET scan at this institution. The official interpretation from the PET scan is as follows:

The study does not show any abnormal hyper metabolic activity to suggest the presence of any metabolically active tumor. However, there is decreased activity present in the left temporo-parietal region, which corresponds with the abnormal enhancement noted on MRI of the brain.

Despite the CT/PET results, which are most consistent with tumor necrosis and not disease recurrence or progression, the new treatment was instituted. Unfortunately, the new therapy required 5% dextrose during the infusion, and the patient was placed on Decadron. This of course required cessation of the KD-R, as blood glucose cannot be lowered in the presence of Decadron. The patient was lost to follow up for several months. In the spring of 2011, I was informed that the patient had progressed (just 6 weeks following institution of the therapy). He was unable to restart the KD-R because he required Decadron to control tumor-related swelling. Despite attempts to contact the patient to inquire about his present condition, no response has been obtained at the time of this transcript.

Patients 2 and 3 are presently on the KD-R and for the most part doing well without side effects. Both have been on the KD-R for at least 2 months. Patient 3 has been able to keep an exquisite and detailed diary of his experience with the RKD thus far. Like patients 1 and 2, he has had difficulty finding meals low in carbohydrate, although he has been able to maintain ketosis with a blood ketone level of 4.4 mM. In fact, when his blood ketones dropped to 3.1 mM and below, he noted increase in seizure activity. Once his ketone level increased back to above 4.0 mM, his seizure activity returned to baseline. Neither patient 2 nor 3 has had post-KD-R imaging at this time.

Thus far, the following has been the consensus.

1. Creating a calorie-restricted diet that is very low in carbohydrate is difficult especially in finding appropriate foods.
2. Maintaining the present target glucose range of 55–65 mg/dl and having this target range the same for everyone presents a problem. All three patients have had difficulty hitting the 55–65 mg/dl blood glucose target range. Everyone is different. In the case of the three patients, one is an elderly female weighing about 115 lbs. The others are men over 6-ft height and

averaging 180–200 lbs in weight. The age, height, and weight are different, as are the physiology of each patient. Trying to fit all patients into the 55–65 mg/dl blood glucose range caused difficulty. We are at present using the basal metabolic rate (BMR) to determine the amount of calories per day the person can consume to start the KD-R. The BMR takes into account age, height, and weight. We calculate the BMR minus 25–35%. This is used as the daily number of calories the patient is allowed when starting the KD-R. The ratio of fats/carbohydrate+protein diet is maintained at 4:1. This system is not perfect and variations, suggestions, and improvements are always welcome, but for now, I feel this is the best way to start the patient on the KD-R because it does allow for physiologic variations between patients. Once the number of calories has been calculated and 4:1 fats to carbohydrate/protein diet initiated, the patient then measures his/her blood glucose two to three times per day to determine the lowest glucose range for them and the diet is adjusted according to these glucose measurements.

3. Maintenance of the ketone level and ketosis must be stressed. I feel it is critical that the patient obtain a blood ketone level of greater than 4 mM. Obtaining this level of ketosis was not difficult for any of the three patients. Patient 3, in particular, was able to control his seizures better with a blood ketone level of greater than 4.0 mM. None of the patients reported signs and symptoms of pathological ketosis.
4. It is difficult to find neuro-oncologists or medical oncologists who will attempt implementing the KD-R.
5. Exercise? Exercise in general is good. The exercising muscle does lower the blood glucose, which is our goal. Also, since the patient is consuming a diet high in fat, exercise will certainly control the triglyceride and cholesterol levels, which can be a concern for anyone on a diet high in fat. The caveat here is the Cori cycle. The Cori cycle takes the lactic acid generated by the muscle during exercise and to the liver. The liver converts the lactic acid into a *new* glucose molecule in a process referred to as *gluconeogenesis* (new glucose). The new glucose is then transported via blood from the liver back to the exercising muscle. In general, this should not be of concern; however, patient 1 exercised every evening, resulting in slightly higher night glucose and morning glucose readings. We had him change his regimen to continue his exercise, but in the morning. We obtained a slightly lower glucose reading in the evenings, and his morning glucose (which was consistently near or within the target range) went back down to a more acceptable level for a person on the KD-R. Exercise may or may not be a factor. Everyone is different. In the case of patient 1, it did make a difference of one to three points on the glucose readings, again, a caveat that should be considered.

I think the KD-R has thus far proven to be well tolerated with regard to side effects. Patients were able to obtain the target blood ketone level of 4.0 mM without difficulty and at least thus far have yet to report any signs or symptoms of pathological ketosis. The most common problem encountered by all three patients is creating a diet that is calorically restricted and low in carbohydrate. However, the biggest obstacle was in maintaining the blood glucose in the target range. Despite the fact that all three patients were aware of their life-threatening condition and that the present standard of care has yet to make a substantial increase in the overall survival, it is still necessary to improve the KD-R treatment in order to make it more acceptable to patients.

At present, I personally believe that the KD-R can be used as a concurrent therapy along with the standard treatment for patients diagnosed with a high grade glioma (grade III or IV) as well as for those patients diagnosed with low grade gliomas (grade II). I am very encouraged by the post-KD-R imaging performed on patient 1, which showed *no abnormal hypermetabolic activity to suggest the presence of metabolically active tumor*. These are the results expected in high grade glioma patients treated successfully with the KD-R.

The KD-R may have an even greater therapeutic benefit for those diagnosed with the slower-growing low grade gliomas (grade II astrocytoma, oligodendroglioma, etc.). Since the RKD directly attacks tumor cell metabolism, it may prevent the progression of a grade II glioma to higher grades, that is, anaplastic or GBM, which is the course of the vast majority of low grade gliomas. I encourage *all* patients diagnosed with low grade gliomas to strongly consider starting the KD-R. Although much work needs to be done with emphasis on the individualization and the tolerability of the KD-R, I feel the KD-R should be instituted in *all* patients as part of the standard of care for patients diagnosed with glioma of all grades and types. In closing, this point must be stressed, the KD-R should not be done without medical supervision. Yes, the KD-R is a diet basically under patient control, but there are potential side effects. Please resist the temptation just to start the KD-R on your own without medical supervision. The patient's blood glucose and ketone readings must be followed closely. Laboratory values such as electrolytes, triglyceride, cholesterol must be closely monitored. It is my opinion that the probability of success in using this therapy for brain cancer management will be greater if done under medical supervision than done without supervision. In this way, it can become a component of standard treatment.

Dr. Kraig Moore, Brain Cancer Oncologist (Kraig Moore, braintumorphysicians@gmail.com).

THE KETOGENIC DIET FOR BRAIN CANCER MANAGEMENT: COMMENTS FROM BETH ZUPEC-KANIA

My experience has been largely with KD therapy for children with epilepsy through the Charlie Foundation. It is amazing to see the improvements in seizure control in the majority of children who use this therapy. I have also worked with many children who have both epilepsy and autism, and the behavior and quality of sleep improvement in these children is especially impressive. The metabolic regulation that results from the KD, although beyond our complete comprehension, deserves attention and respect. It is no surprise to me that the diet is effective for many brain disorders including brain cancer.

I have been contacted by 10 people who have been diagnosed with various stages of glioma(s), have failed traditional treatment, and are interested in KD therapy. The willingness among these individuals to follow a restrictive diet is remarkable. Although these accounts are not scientifically recorded, a summary of the protocol and results in this small population follows.

The protocol that I have used to initiate KD therapy in people who I have neither met in person nor have a complete medical history on is designed to prevent adverse events. A written consent is first obtained absolving me from the responsibility of the effects of an "experimental diet," and an agreement to discuss the use of this therapy with the primary oncologist was also obtained. An "Intake" is obtained, which includes information on diet, weight, height, history of weight loss, medications, and supplements. Once these have been completed, a modified KD is provided. This is a simple, two-page guide with specific instructions of the quantity and types of foods recommended. This diet not only includes whole foods and eliminates most processed foods but also restricts carbohydrates to approximately 50 g daily (note that the typical carbohydrate intake is above 300 g daily). Healthy fats and a serving of protein are recommended with each meal. If after 2 weeks, the individual is further interested in advancing to a more restrictive diet, a calculated KD is created.

The KD requires weighing foods on a gram scale and is in a ratio of 2:1, 3:1, or 4:1 depending on protein needs (higher protein requires lower ratio). Calories are controlled and divided evenly into three meals. The diet is initiated beginning with one ketogenic meal daily with two regular meals (of the user's choice). The second week is advanced to two ketogenic meals with one regular, and the third week is the complete KD. Among the 10 people described above, 4 have chosen to transition to the calculated KD. Glucose is measured twice daily, and adjustments in calories or carbohydrates are made to control glucose between 55 and 65 mg/dl. Medium chain fats are included to assist with maintaining stable glucose and ketosis and for their laxative effect.

During the KD treatment, I advise good quality, low carbohydrate supplements: 2000 IU of vitamin D, Dietary Reference Intake levels for calcium and micronutrients, and a separate phosphorus supplement. These specific recommendations have been made based on my analysis of optimally selected KDs of different ratios and the consistent lack of nutrients among them. In addition to nutrition supplementation, avoidance of caffeine and diuretics is also advised due to the strong diuretic effect of this low carbohydrate diet. Consumption of adequate carbohydrate-free fluids is also advised. Fiber supplementation is also suggested to prevent constipation, which is the most common adverse effect of the diet. Sick day guidelines including low carbohydrate electrolyte replacement beverages are provided.

Access to the online KetoCalculator program (www.ketocalculator.com) is provided to each person so that they can be independent in creating meal plans. This program also allows me to supervise their work and to edit it if necessary. Communication with these participants has been initially through an hour-long phone call then subsequently via e-mail. Some have been very consistent in apprising me of their progress; others have contacted me only few times. Those who have opted to transition to the calculated diet have been most communicative. Of the four people who have opted to follow the calculated diet, three have consistently contacted me with questions and progress reports. All three have had success with "stable tumor" or "atrophy of tumor" determined by an MRI. Two of the individuals have managed the diet for several years and are still alive despite being told initially that they had "only a few months" to live. One individual has since expired; he had advanced-stage cancer with metastases to the lungs and other organs (before starting KD) and had outlived his prognosis by a year. His wife managed his diet and was able to enjoy an "active and alert man" until the last 2 months of his life.

Although I know that the KD has been instrumental in helping these individuals, I cannot ignore the notion that early intervention may have been more beneficial. If cancer is a metabolic syndrome then why are we not promoting diet therapy as a strong preventative measure against cancer growth, especially against glioblastoma? Additionally, some of the chemotherapy agents impair appetite in many of these people leading to difficulty in compliance to diet therapy despite strong motivation on the part of the patient. If the chemotherapeutic agent is a "last-resort" treatment, perhaps the diet could be given a chance ahead of this agent.

Charlie Foundation Consultant, Beth Zupec-Kania, RD, CD (www.charlie foundation.org).

SUMMARY

It is clear from the experiences of these caregivers, physicians, and patients that the KD has potential as either an alternative or complimentary therapy for cancer. One key point was the difficulty encountered in having cancer patients maintain their blood glucose levels in the therapeutic ranges. Although the KD was capable of elevating circulating ketone levels, it was less effective in maintaining reduced blood glucose levels. As Beth Zupec-Kania mentioned in her comments, it is likely that certain medications might prevent glucose from reaching therapeutic levels necessary to kill tumor cells. This is especially the case for those patients taking steroids, which prevent glucose levels from entering the therapeutic zone. It is unclear if seizure medications or certain chemotherapies also prevent glucose levels from reaching the therapeutic zone. Most of my healthy students who have fasted or used the KD had no trouble reaching the therapeutic zones of glucose and ketones.

I agree with Dr. Moore that the KD-R should have its greatest potential in managing lower grade tumors when used alone or when combined with nontoxic drugs that also target tumor cell energy metabolism. It is tragic that many patients with low grade tumors are not aware or notified of the potential therapeutic efficacy of the KD-R. Hopefully, this situation will change in the future.

REFERENCES

1. Nebeling LC, Lerner E. Implementing a ketogenic diet based on medium-chain triglyceride oil in pediatric patients with cancer. J Am Diet Assoc. 1995;95:693–7.
2. Nebeling LC, Miraldi F, Shurin SB, Lerner E. Effects of a ketogenic diet on tumor metabolism and nutritional status in pediatric oncology patients: two case reports. J Am Coll Nutr. 1995;14:202–8.
3. Seyfried TN, Shelton LM. Cancer as a metabolic disease. Nutr Metab. 2010;7:7.
4. Seyfried TN, Kiebish MA, Marsh J, Shelton LM, Huysentruyt LC, Mukherjee P. Metabolic management of brain cancer. Biochim Biophys Acta. 2010;1807:577–94.

Chapter 21

Conclusions

The journal *Science* recently commemorated the fortieth anniversary of the US National Cancer Act with a series of articles that were thought to encapsulate the state of the field (1). Initiation of this act during the Nixon administration provided a massive stimulus for cancer research. Despite the massive research effort, major questions related to the origin and management of cancer posed in 1971 remain unanswered in 2012. Some of these questions include How do abnormalities in chromosome number arise in tumor cells? Can tissue-specific markers be used to determine the epithelial versus mesenchymal origin of a solid tumor? Can the immune system be manipulated so that it recognizes tumor cells as foreign invaders that must be eliminated from the body? Do viruses play a role in human cancer? The answer is a definite yes to the question of whether viruses play a role in human cancer (Chapter 9). Answers to the other questions have not been forthcoming. On the basis of the information in this treatise, I can provide credible answers to the other outstanding questions. Moreover, my treatise addresses many of the NCI "provocative questions" from the NCI web site (provocativequestions.nci.nih.gov).

I addressed in Chapter 10 the answer to the question of how abnormalities in chromosome number arise in tumor cells. Basically, the stability of chromosome number and the integrity of the genome are dependent on the integrity OxPhos. Spindle assembly and the fidelity of chromosomal segregation during mitosis are dependent on the energy of OxPhos. Injury to cellular respiration with compensatory fermentation will cause genomic instability including aneuploidy and mutations. It is the efficiency of mitochondrial respiration that maintains cellular differentiation and prevents tumorigenesis and dedifferentiation.

The information presented in Chapter 13 addresses the question of how tissue-specific markers can be used to determine the epithelial versus mesenchymal origin of solid tumors. Metastatic tumor cells arise from respiratory damage to myeloid cells, which are already mesenchymal. Many of the biomarkers expressed in metastatic cancer cells are also expressed in macrophages. While epithelial tumor

Cancer as a Metabolic Disease: On the Origin, Management and Prevention of Cancer, First Edition.
Thomas Seyfried.

cells proliferate rapidly, they do not generally metastasize unless they fuse with a cell of myeloid origin. Tissue biomarkers of myeloid cells are expressed in many metastatic cancer cells.

I present information in Chapters 13 and 17 that addresses the question of whether the immune system can be manipulated so that it recognizes tumor cells as foreign invaders that must be eliminated from the body. According to my view, metastatic cancer cells arise from cells of the immune system (macrophages). While it might be difficult to induce nonneoplastic macrophages to recognize neoplastic macrophages as foreign invaders, it might be easier to eliminate the metastatic cells of immune origin by targeting their energy metabolism and capacity for phagocytosis.

A significant emphasis of the anniversary issue of *Science* was devoted to how targeted drugs could be combined to stop resistant tumors. "Even the most successful targeted therapies lose potency with time. Researchers hope to figure out how tumors escape; they aim to turn months of survival into years" (2). I have real difficulty with these statements. Any successful therapy for advanced metastatic cancer *should* provide long-term management for the disease. That this seldom happens indicates that few successful targeted drug therapies are currently available. It is, therefore, misleading to imply that successful therapies for *advanced* cancers are available.

It is clear to me how tumors escape from the so-called "successful therapies". Cancer cells will escape as long as they can maintain their ability to ferment. Fermentation energy (glycolysis) underlies drug resistance (3). If tumor cells cannot ferment, they will die. How many of the targeted therapies actually shut down glucose and glutamine fermentation? A statement was made indicating that "uncontrolled cell growth is often driven by an aberrant protein in the cell membrane that transmits a spurious signal to the nucleus instructing it to divide" (2). This is nonsense. Proliferation is the default state of cells. Respiration maintains growth regulation and the differentiated state. Fermentation drives unbridled proliferation. Uncontrolled cell growth is not driven by an aberrant protein but by *insufficient respiration with compensatory fermentation*. Rational drug therapies will be realized once this concept becomes more widely recognized.

It is important to recognize that my view of cancer as a metabolic disease is not part of the mainstream view of cancer, which is viewed as an incomprehensively complex genetic disease. Support for my position comes from a perusal of the articles in the *Science* issue commemorating the anniversary of the US National Cancer Act. No aspect of cancer metabolism was mentioned in this issue. As I mentioned in Chapter 10, *the failure to discuss the role of energy metabolism in the origin of cancer would be like failing to discuss the role of the sun in the origin of the solar system.* Should we be surprised that the same questions remain unresolved after 40 years? Should we be surprised that most targeted therapies developed from the cancer genome projects have been a costly waste of time? Should we be surprised that so little progress has been made in managing advanced cancers?

The following is a summary of major conclusions from my treatise. While some of these conclusions are subject to debate and further verification, I believe that they are supported by facts and will be confirmed in time.

MAJOR CONCLUSIONS

1. No real progress has been made in the management of advanced or metastatic cancer for more than 40 years. The number of people dying each year and each day has changed little in more than 10 years.
2. Most of the conceptual advances made in understanding the mechanisms of cancer have more to do with nonmetastatic tumors than with metastatic tumors.
3. Most cancer, regardless of cell or tissue origin, is a singular disease of respiratory insufficiency coupled with compensatory fermentation.
4. Some factors that can cause respiratory insufficiency and cancer include age, viral infections, hypoxia, inflammation, rare inherited mutations, radiation, and carcinogens.
5. The genomic instability seen in tumor cells is a downstream epiphenomenon of respiratory insufficiency and enhanced fermentation.
6. Genomic instability makes cancer cells vulnerable to metabolic stress.
7. Cancer cells do not have a growth advantage over normal cells.
8. Cancer progression is not Darwinian but Lamarckian.
9. The view that most cancer is a genetic disease is no longer credible.
10. Respiratory injury can explain Szent-Gyorgyi's oncogenic paradox.
11. Most metastatic cancers arise from respiratory injury in cells of myeloid origin, possibly involving hybridization events between macrophages and neoplastic epithelial cells.
12. Cancer cells depend largely on glucose and glutamine metabolism for survival, growth, and proliferation.
13. Restricted access to glucose and glutamine will compromise cancer cell growth and survival.
14. Enhanced fermentation is largely responsible for tumor cell drug resistance.
15. Protection of mitochondria from oxidative damage will prevent or reduce risk of cancer.
16. Life style changes will be needed to manage and prevent cancer.
17. Mitochondrial enhancement therapies administered together with drugs that target glucose and glutamine metabolism will go far as a nontoxic, cost-effective solution to the cancer problem.
18. A new era will emerge for cancer management and prevention, once cancer becomes recognized as a metabolic disease.

REFERENCES

1. Kiberstis P, Marshall E. Cancer crusade at 40. Celebrating an anniversary. Introduction. Science. 2011;331:1539.
2. Kaiser J. Combining targeted drugs to stop resistant tumors. Science. 2011;331:1542–5.
3. Xu RH, Pelicano H, Zhou Y, Carew JS, Feng L, Bhalla KN, et al. Inhibition of glycolysis in cancer cells: a novel strategy to overcome drug resistance associated with mitochondrial respiratory defect and hypoxia. Cancer Res. 2005;65:613–21.

Index

Page numbers with *t* and *f* indicates table and figure

Cancer as a Metabolic Disease: On the Origin, Management and Prevention of Cancer, First Edition. Thomas Seyfried.